Frontiers of Virology 3

Y. Becker · G. Darai (Eds.)

Pathogenicity of Human Herpesviruses due to Specific Pathogenicity Genes

With 86 Figures and 29 Tables

Springer-Verlag
Berlin Heidelberg New York London Paris
Tokyo Hong Kong Barcelona Budapest

Editors:
Professor Yechiel Becker
Department of Molecular Virology
Institute of Microbiology
Faculty of Medicine
The Hebrew University of Jerusalem
POB 12272
Jerusalem 91120, Israel

Professor Gholamreza Darai
Institut für Medizinische Virologie
Universität Heidelberg
Im Neuenheimer Feld 324
D-69120 Heidelberg, Germany

ISBN-13:978-3-642-85006-6 e-ISBN-13:978-3-642-85004-2
DOI: 10.1007/978-3-642-85004-2

Library of Congress Cataloging-in-Publication Data. Pathogenicity of human herpesviruses due to specific pathogenicity genes / Y. Becker, G. Darai (eds.). p. cm. — (Frontiers of virology:3) Includes bibliographical references and index. ISBN-13:978-3-642-85006-6

1. Herpesvirus infections—Genetic aspects. 2. Viral genetics. 3. Herpesvirus infections—Pathogenesis. I. Becker, Yechiel. II. Darai, Gholamreza. III. Series. QR201. H48P38 1994 616.9' 25042—dc 20 93-38678

Softcover reprint of the hardcover 1st edition 1994

Typesetting: Thomson Press (India) Ltd., New Delhi
27/3130/SPS – 5 4 3 2 1 0 – Printed on acid free paper

Frontiers of Virology: Aims and Plans

Virology at the end of the twentieth century is still a frontier of the biological sciences with new viruses emerging as disease-causing agents of man, animal, and plant. Ever since 1796 when Edward Jenner named the causative agent of smallpox "virus" (toxin), knowledge on virus diseases has slowly accumulated. The vast influenza epidemic (the "Spanish flu") at the end of World War I revealed that virologists lacked the scientific tools to identify the epidemic and protect against it. During the middle of the twentieth century, the methods for the diagnosis of viruses and for research were markedly improved. New technologies were developed to identify the structure and organization of viruses, ranging from bacterial to plant viruses and from human to animal viruses. In the past 50 years, virology has developed as a part of immunology and molecular biology, providing these biological sciences with tools and at the same time utilizing their biochemical, molecular, and immunological advancements to further knowledge on the mechanisms by which viruses cause a variety of discases. Thus, virology remains in the forefront of science.

With this idea that virology is constantly developing, the present series, *Frontiers of Virology*, was conceived. We intend to select topics on which knowledge from a number of biological fields of research, including conceptual and technical breakthroughs, can merge in the field of virology and move it further ahead. We plan to put the emphasis on discoveries which will help to curb virus diseases. We hope that *Frontiers of Virology* will be of interest not only to virologists, molecular biologists, and immunologists, but also to physicians with expertise in infectious disccases.

We wish to express our thanks to Dr. Jürgen Wieczoreck, Springer-Verlag, for his general support and encouragement to develop the series *Frontiers of Virology*.

Y. Becker, Jerusalem
G. Darai, Heidelberg

Preface

Six members of the Herpesviridae family are human pathogens, including herpes simplex virus 1 and 2 (HSV-1 and 2), Epstein-Barr virus (EBV), varicella zoster virus (VZV), human cytomegalovirus (HCMV), and human herpesvirus 6 (HHV 6). Each of these viruses is capable of causing distinct diseases of varying severity in children, young adults, and the aged. The diseases range from infection of epithelial tissue to the infection of internal organs and white blood cells. A common feature of the six pathogenic human herpesviruses is their ability to latently infect different cell types in which the viral DNA is not integrated and is unable to express its pathogenicity. Reactivation of the herpesviruses is a result of cellular processes which reactivate viral genes, leading to virus progeny and to signs of infection. Due to their ability to become latent after initial infection, once the pathogenic herpesviruses infect children they are maintained throughout life, having the potential of cause various diseases upon reactivation.

The existence of herpesvirus genomes, which can be reactivated, in nervous tissues and other cell types for a lifetime in most members of the human population indicates that herpesviruses have adapted themselves to the human host. A combination of the activity of the virus genes and the molecular control mechanisms in the infected cells determines the balance between health and disease. The extent of the severity of herpesvirus-related disease is controlled by the cellular and humoral immune responses of the host which govern recovery from infection and disease. The mechanisms of infection control in the herpesvirus-infected immunocompetent host are shuttered in the immunocompromised host due to infection with human immunodeficiency virus (HIV) and resulting aquired immunodeficiency syndrome (AIDS). In this situation, latent herpesviruses acquired long before HIV infection have the ability to reactivate and endanger the life of the immunocompromised individual. The questions of how the pathogenic human herpesviruses cause diseases and what should be done to prevent infection and disease have prompted scientific interest in the molecular basis of the functions of all human herpesvirus genes in various tissues and organs of the infected host. It is hoped that the understanding of the role of herpesvirus genes and their involvement in the disease processes will lead to new approaches to prevent and cure human herpesvirus diseases.

Volume 3 of *Frontiers of Virology* has been devoted to scientifically defining the involvement of herpesvirus genes in human diseases caused by infection with herpesviruses. The studies on virus pathogenicity genes must be carried out in experimental animals, which serve as a model system of the human body, since these particular viral genes are functionally active in specific cell types of differentiated tissues of the susceptible organs in humans. In many chapters knowledge is compiled on the pathogenicity genes of herpes simplex virus types 1 and 2, the viruses which have been most actively investigated at the molecular level. One or two chapters are devoted to each of the other human herpesviruses to obtain a more comprehensive understanding of the relationship between virus genes and the specific disease caused by a herpesvirus. The involvement of structural and nonstructural herpesvirus genes in disease is dealt with at different levels of the infection, from attachment of virions to cell membrane-bound heparan sulfate to the entry of viral nucleic acid and the expression in tissue cells of virus genes coding for virus enzymes or the expression of the HSV-1 latency gene in ganglia. Special attention is given to HSV-1 genes coding for proteins involved in intraperitoneal or intracerebral pathogenicity of herpes simplex virus. Genes in EBV, HCMV, and HHV6 are presented and their role in disease is discussed.

The understanding of the involvement of herpesvirus genes in the development of human diseases is of particular importance, since it focuses on the pathomechanisms of herpesvirus infections and provides new tools for the development of live herpesvirus vaccines. The identified viral pathogenicity genes could be deleted or inactivated in the herpesvirus genomes in order to generate apathogenic, but immunogenic, virus strains.

In view of the lack of effective human herpesvirus vaccines, the advantages of the knowledge of virus pathogenicity genes for vaccine development are discussed in comparison to the use of naturally occurring apathogenic virus strains. In *Frontiers of Virology*, Vol. 1, *Diagnosis of Human Viruses by Polymerase Chain Reaction Technology*, diagnostic techniques to identify herpesvirus in infected tissues are presented. In *Frontiers of Virology*, Vol. 2, *Molecular Aspects of Human Cytomegalovirus Disease*, the involvement of HCMV in human disease is discussed in detail. The present Volume of *Frontiers of Virology*, Vol. 3, *Pathogenicity of Human Herpesviruses and Specific Pathogenicity Genes* adds to the understanding of herpesvirus gene involvement in diseases and shows the potential of this knowledge for planning new strategies for the development of vaccines in the future.

The authors of Vol. 3 join the contributors to the previous volumes of *Frontiers of Virology* in presenting an additional comprehensive view of virus genes and their involvement in diseases, a subject of interest to physicians dealing with infectious diseases, to virologists and molecular biologists, and to immunologists and vaccine developers. The editors thank all the contributors of the chapters and the staff of Springer-Verlag, especially Dr. R. Lange, for their interest and help in making the publication of this volume possible. The editorial assistance of Waltraud Janssen is much appreciated.

January 1994

Y. Becker, Jerusalem
G. Darai, Heidelberg

Contents

B. Epstein Barr Virus

C. Human Cytomegalovirus

D. Human Herpesvirus 6

E. Trends

B. Peptide Hormones

List of Authors

Abrams, W.R., School of Dental Medicine, University of Pennsylvania, Philadelphia, PA 19104 and Center for Oral Health Research, University of Pennsylvania, Philadelphia, PA 19104, USA

Ambrosino, C., Dipartimento di Medicina Sperimentale e Clinica, Università di Reggio Calabria, 88100 Catanzaro, Italy

Asher, Y., Department of Molecular Virology, Faculty of Medicine, The Hebrew University of Jerusalem, P.O. Box 12272, 91120 Jerusalem, Israel

Aurelian, L., Virology/Immunology Laboratories, Departments of Pharmacology and Microbiology, The University of Maryland School of Medicine, Baltimore, MD 21201 and Departments of Comparative Medicine and Biochemistry, The Johns Hopkins Medical Institutions, Baltimore, MD 21205, USA

Baldassarre, F., Dipartimento di Medicina Sperimentale e Clinica, Università di Reggio Calabria, 88100 Catanzaro, Italy

Becker, Y., Department of Molecular Virology, Faculty of Medicine, The Hebrew University of Jerusalem, P.O. Box 12272, 91120 Jerusalem, Israel

Ben-Hur, T., Department of Neurology, Hadassah University Hospital, P.O. Box 12000, 91120 Jerusalem, Israel

Boname, J.M., Department of Microbiology and Immunology, Stanford University School of Medicine, Stanford, CA 94305-5402, USA

Brandt, C.R., Departments of Ophthalmology and Medical Microbiology/Immunology, University of Wisconsin-Madison, 6630 Medical Sciences Center, 1300 University Avenue, Madison, WI 53706-1532, USA

Carrigan, D.R., Department of Pathology and the Bone Marrow Transplant Program, Medical College of Wisconsin, Milwaukee, WI 53226, USA

Chiang, H.-Y., School of Dental Medicine, University of Pennsylvania, Philadelphia, PA 19104, and Center for Oral Health Research, University of Pennsylvania, Philadelphia, PA 19104, USA

Cohen, G.H., School of Dental Medicine, University of Pennsylvania, Philadelphia, PA 19104 and Center for Oral Health Research, University of Pennsylvania, Philadelphia, PA 19104, USA

Crumpacker, C.S., Beth Israel Hospital and Harvard Medical School, Boston, MA 02115, USA

Darai, G., Institut für Medizinische Virologie der Universität Heidelberg, Im Neuenheimer Feld 324, 69120 Heidelberg, FRG

Desai, P., Department of Human Genetics, University of Michigan, Ann Arbor, MI 48109, and Department of Molecular Genetics and Biochemistry, University of Pittsburgh School of Medicine, Pittsburgh, PA 15261, USA

Dolter, K.E., Department of Immunology and Microbiology, Wayne State University School of Medicine, Detroit, MI 48201, USA
Eisenberg, R., Laboratory of Microbiology, Department of Pathobiology, School of Veterinary Medicine, University of Pennsylvania, 3800 Spruce Street, Philadelphia, PA 19104 and Center for Oral Health Research, University of Pennsylvania, Philadelphia, PA 19104, USA
Farrell, P.J., Ludwig Institute for Cancer Research, St. Mary's Hospital Medical School, Norfolk Place, London W2 1PG, UK
Fink, D.J., Department of Neurology and VA Medical Center, University of Michigan, Ann Arbor, MI 48109, USA
Giordano, V., Dipartimento di Medicina Sperimentale e Clinica, Università di Reggio Calabria, 88100 Catanzaro, Italy
Glorioso J.C., Department of Molecular Genetics and Biochemistry, University of Pittsburgh School of Medicine, Pittsburgh, PA 15261, USA
Goins, W.F., Department of Molecular Genetics and Biochemistry, University of Pittsburgh School of Medicine, Pittsburgh, PA 15261, USA
Grifman, M., Department of Molecular Virology, Faculty of Medicine, The Hebrew University of Jerusalem, P.O. Box 12272, 91120 Jerusalem, Israel
Haarr, L., National Centre for Research in Virology and The Gade Institute, University of Bergen, 5020 Bergen, Norway
Hidaka, Y., Department of Virology, Faculty of Medicine, Kyushu University, 3-1-1, Maidashi, Higashi-ku, Fukuoka 812, Japan
Jacquemin, M., Ludwig Institute for Cancer Research, St. Mary's Hospital Medical School, Norfolk Place, London W2 1PG, UK
Kehm R., Institut für Medizinische Virologie der Universität Heidelberg, Im Neuenheimer Feld 324, 69120 Heidelberg, FRG
Kintner, R., Department of Medical Microbiology/Immunology, University of Wisconsin-Madison 6630 Medical Sciences Center, 1300 University Avenue, Madison WI 53706-1532, USA
Kleinman, Y., Department of Internal Medicine, Bikur Holim Hospital, Jerusalem, Israel
Knopf, K.W., Deutsches Krebsforschungszentrum, Angewandte Tumorimmunologie, Im Neuenheimer Feld 280, 69120 Heidelberg, FRG
Kumano, Y., Department of Virology, Faculty of Medicine, Kyushu University, 3-1-1, Maidashi, Higashi-ku, Fukuoka 812, Japan
Lagenaur, A., Department of Microbiology and Immunology, Stanford University School of Medicine, Stanford, CA 94305-5402, USA
Lamadé, W., Institut für Medizinische Virologie der Universität Heidelberg, Im Neuenheimer Feld 324, 69120 Heidelberg, FRG
Langeland, N., Department of Medicine, Haukeland Hospital, 5021 Bergen, Norway
Latchman, D.S., Division of Molecular Pathology, University College and Middlesex School of Medicine, The Windeyer Building, Cleveland Street, London W1P 6DB, UK
Levine, M., Department of Human Genetics, University of Michigan, Ann Arbor, MI 48109, USA

Long, D., School of Dental Medicine, University of Pennsylvania, Philadelphia, PA 19104, and Center for Oral Health Research, University of Pennsylvania, Philadelphia, PA 19104, USA

Lorentzen, E., Institut für Medizinische Virologie der Universität Heidelberg, Im Neuenheimer Feld 324, 69120 Heidelberg, FRG

Mallardo, M., Dipartimento di Medicina Sperimentale e Clinica, Università di Reggio Calabria, 88100 Catanzaro, Italy

Marsden, H.S., Medical Research Council, Institute of Virology, Glasgow G11 5JR, UK

Muggeridge, M.I., School of Dental Medicine, University of Pennsylvania, Philadelphia, PA 19104 and Center for Oral Health Research, University of Pennsylvania, Philadelphia, PA 19104, USA

Minagawa, H., Department of Virology, Faculty of Medicine, Kyushu University, 3-1-1, Maidashi, Higashi-ku, Fukuoka 812, Japan

Mocarski, E.S., Department of Microbiology and Immunology, Stanford University School of Medicine, Stanford, CA 94305-5402, USA

Mori, R., Department of Virology, Faculty of Medicine, Kyushu University, 3-1-1, Maidashi Higashi-ku, Fukuoka 812, Japan

Moyal, M., Department of Molecular Virology, Faculty of Medicine, The Hebrew University of Jerusalem, P.O. Box 12272, 91120 Jerusalem, Israel

Øyan, A.M., National Centre for Research in Virology and The Gade Institute, University of Bergen, 5020 Bergen, Norway

Pumfery, A.M., Department of Medical Microbiology/Immunology, University of Wisconsin-Madison, 6630 Medical Sciences Center, 1300 University Avenue, Madison, WI 53706-1532, USA

Quinto, I., Dipartimento di Medicina Sperimentale e Clinica, Università di Reggio Calabria, 88100 Catanzaro, Italy

Ramakrishnan, R., Department of Human Genetics, University of Michigan, Ann Arbor, MI 48109, USA

Razzaque, A., Division of Viral Products, CBER, FDA, Bethesda, MD 20892, USA

Rösen-Wolff, A., Institut für Medizinische Virologie der Universität Heidelberg, Im Neuenheimer Feld 324, 69120 Heidelberg, FRG

Ruocco, M.R., Dipartimento di Medicina Sperimentale e Clinica, Università di Reggio Calabria, 88100 Catanzaro, Italy

Scala, G., Dipartimento di Biochimica e Biotecnologie Mediche, Università Federico II, Via S. Pansini, 5, 80131 Naples, Italy

Sinclair, A.J., Ludwig Institute for Cancer Research, St. Mary's Hospital Medical School, Norfolk Place, London W2 1PG, UK

Sodora, D.L., School of Dental Medicine, University of Pennsylvania, Philadelphia, PA 19104, and Center for Oral Health Research, University of Pennsylvania, Philadelphia, PA 19104, USA

Sprecher, E., Department of Molecular Virology, Faculty of Medicine, The Hebrew University of Jerusalem, P.O. Box 12272, 91120 Jerusalem, Israel

Strick, R., Deutsches Krebsforschungszentrum, Angewandte Tumorimmunologie, Im Neuenheimer Feld 280, 69120 Heidelberg, FRG

Tabor, E., Department of Molecular Virology, Faculty of Medicine, The Hebrew University of Jerusalem, P.O. Box 12272, 91120 Jerusalem, Israel

Tenser, R.B., Departments of Medicine (Neurology) and Microbiology and Immunology, Pennsylvania State University College of Medicine, Hershey, PA 17033, USA

Toh, Y., Department of Virology, Faculty of Medicine, Kyushu University, 3-1-1, Maidashi, Higashi-ku, Fukuoka 812, Japan

Venuta, S., Dipartimento di Medicina Sperimentale e Clinica, Università di Reggio Calabria, 88100 Catanzaro, Italy

Visalli, R.J., Department of Medical Microbiology/Immunology, University of Wisconsin-Madison, 6630 Medical Sciences Center, 1300 University Avenue, Madison, WI 53706-1532, USA

Wagner, E.K., Department of Molecular Biology and Biochemistry and the Program in Animal Virology, University of California, Irvine, CA 92717, USA

Wilcox, W.C., Laboratory of Microbiology, Department of Pathobiology, School of Veterinary Medicine, University of Pennsylvania, 3800 Spruce Street, Philadelphia, PA 19104 and Center for Oral Health Research, University of Pennsylvania, Philadelphia, PA 19104, USA

Yamanishi, K., Department of Virology, Research Institute for Microbial Diseases, Osaka University, Osaka, Japan

Yayon, A., Department of Chemical Immunology, Weizmann Institute of Science, Rehovot, Israel

A. Herpes Simplex Virus

I. Genes Involved in Entry

Chapter 1 Entry of Herpes Simplex Virus Type 1 into Cells – Early Steps in Virus Pathogenicity

Yechiel Becker[1], Eynat Tabor[1], Yael Asher[1], Mirta Grifman[1], Yosef Kleinman[2], and Avner Yayon[3]

Summary

The entry of herpes simplex virus type 1 (HSV-1) virions into cells represents the initial stage in the process by which HSV-1 introduces its genetic information into a living cell. The molecular basis of virus entry into cells is the ability of the virion's structural proteins to recognize and interact with protein components of the cell membrane serving as virus receptors. Electron microscopic studies of the cell-associated virions revealed that many virions bind to the cell membrane, but some virions bind to pits in the cell membrane which are either uncoated or clathrin-coated pits (Rosenthal et al. 1984). The virions that bind to the cell membrane enter the cell cytoplasm by a fusion mechanism, while the virions attached to the membrane pits enter the cell cytoplasm by the endocytosis mechanism.

In the present chapter, several aspects of HSV-1 entry into cells is discussed, based on reported studies and on studies performed by us. A comprehensive analysis is presented of the molecular mechanisms involved in HSV-1 entry by fusion of virion envelopes with the cell membrane and by entry through coated pits by an endocytosis process. The interaction of HSV-1 virions with the heparan sulfate moiety on the proteoglycans present in the cell membrane was reported (WuDunn and Spear 1989). In the present study, results showing a markedly reduced adsorption of HSV-1 virions to a Chinese hamster ovary (CHO) mutant in the synthesis of heparan sulfate are presented, together with evidence that a specific amino acid sequence in HSV-1 glycoprotein B (gB) – and similar sequences in HSV-1 gC and gH – are responsible for the virion's attachment to heparan sulfate. A synthetic peptide which contains ten amino acids of the above domain of HSV-1 gB interferes with HSV-1 infection of cells. It is also suggested that the N terminus of the virion's gD polypeptide contains a fusion sequence which enters the cell membrane and causes its destabilization (Rosenthal et al. 1984). The fusion of the virion envelope with the cell membrane is then possible, leading to uncoating of the virion and transport of the viral capsid through the cytoplasm, thus releasing the viral DNA into the nucleus to allow viral gene expression. The effects of inhibitors of enzymatic activities in the cell membrane (quercetin, amiloride, neomycin) on virus infection are discussed.

---

[1] Department of Molecular Virology, Faculty of Medicine, The Hebrew University of Jerusalem, Jerusalem, Israel
[2] Department of Internal Medicine, Bikur Holim Hospital, Jerusalem, Israel
[3] Department of Chemical Immunology, Weizmann Institute of Science, Rehovot, Israel

It is suggested that the interaction of HSV-1 virions with coated pits occurs either after binding of virions to heparan sulfate on cellular proteoglycans or directly with the proteins present in coated pits on the cell membrane. The amino acid sequence in HSV-1 gB, which binds to heparan sulfate, was shown to resemble the amino acid sequence in apolipoprotein E (apoE), which itself has been shown to interact with the low-density lipid receptor (LDLR) in coated pits (Dyer and Curtiss 1991). We suggest that HSV-1 may have the ability to interact with LDLR in coated pits due to the interaction of domains in gB, gC, or gD that resemble the apoE attachment sequence to LDLR. Like LDL, HSV-1 virions are incorporated into the cell cytoplasm by an endocytosis process. The fate of virions entering cells by endocytosis seems to differ from the fate of virions entering the cells by fusion. The significance of endocytosis in the penetration of HSV-1 virions into cells through coated pits in the cell membrane is not yet known and requires special attention.

Introduction

The morphology of herpes simplex virus type 1 (HSV-1) virions interacting with the cell membrane of cultured cells has been the subject of numerous electron microscopic studies. The interaction of virions with the cell membrane of target cells was noted to occur by two mechanisms (Rosenthal et al. 1989): (1) entry of virions after fusion with the outer membrane of the infected cell (Epstein et al. 1964; Morgan et al. 1968; Smith and de Harven 1974; Rosenthal et al. 1989) and (2) entry of virions attached to clathrin-coated and uncoated pits in the cell membrane by an endocytotic mechanism (Hummeler et al. 1969; Dales and Silverberg 1969; Smith and de Harven 1974; Campadelli-Fiume et al. 1988; Rosenthal et al. 1989).

Studies on the biochemical events occurring during the entry of HSV-1 virions through the outer membrane of infected cells (reviewed by Rosenthal et al. 1989) revealed that: (a) transfer of fluorescent energy from fluorescein-labeled virions to rhodamine-labeled cell membrane occurred during the entry process (Keller et al. 1977), (b) entry of HSV-1 modulated surface protein mobility in the membrane of the infected cell (Rosenthal et al. 1989), and (c) virions' glycoproteins were incorporated into the infected cell membrane during the entry process in the absence of cellular protein synthesis (Para et al. 1980). It was noted that the entry of HSV-1 virions into the cell occurs after a 5-min lag, and by 10 min after addition of HSV-1 virions to cells, the structural components of the virus are associated with the cells. The entry process of the virions was found to be affected by the pH of the culture medium. At pH 6.3, 10 min after infection at least half of the virions in the process of penetration were found to be associated with cell surface pits or as enveloped virions in vesicles. At 30 min, approximately two-thirds of the virions were in cell surface pits and in intracytoplasmic vesicles, suggesting an endocytosis process. In contrast, at neutral pH of the medium, entry of virions was by fusion with the cell membrane (Rosenthal et al. 1989). Entry through pits in the cell membrane by an endocytosis process was demonstrated by Campadelli-Fiume et al. (1988) using BJ cells which constitutively expressed HSV-1 glycoprotein D (gD) gene and had the viral gD inserted in the cell membrane. Virions interacting with clathrin-coated pits can be seen in the electron micrographs, and intracytoplasmic, endocytic vacuoles containing virions are evident.

The entry of HSV-1virions into cells is preceded by the adsorption and binding of the virions to a cell surface component. This step in virion–cell membrane interaction is a very rapid step which can be circumvented by heparin. The ability of heparin to interfere with HSV-1 infection of cells has already been reported (Valeri and Cantell 1963; Nahmias and Kibrik 1963, 1964; Takameto and Fabisch 1964; Horvath and Hadhazy 1965, Benda 1966; Lehel and Hadhazy 1966) and led to the demonstration of the fact that exposure of herpes virus-infected cultures to heparin resulted in a complete inhibition of virus replication. It was reported by Hochberg and Becker (1968) that heparin affected virus adsorption to cells, but had no effect on virus replication after virus entry into infected cells. Heparin distinguished between virions which became attached to the surface of the cell membranes by a heparin-sensitive mechanism and virions which were firmly attached to the cells in a heparin-resistant mechanism or were already incorporated into the cell cytoplasm. The virions on the surface of the infected cell membrane which were displaced by heparin were regarded as electrostatically-bound virions (Hochberg and Becker 1968). More recently, WuDunn and Spear (1989) reported that cell surface heparan sulfate serves as the initial receptor for HSV-1 and HSV-2. Heparin, a related glycosaminoglycan, blocks HSV adsorption to cells. WuDunn and Spear (1989) concluded that both HSV-1 and HSV-2 initiate infection by binding first to cell surface heparan sulfate prior to interaction with receptors.

Kaner et al. (1990) studied the role of basic fibroblast growth factor (FGF) receptor as the entry receptor for HSV-1 using bovine arterial smooth muscle cells. An inhibitor of basic FGF binding to its receptors (a polypeptide fragment of basic FGF consisting of amino acids 103–120) which binds to the basic FGF receptor and inhibits the binding of ^{125}I-basic FGF to high-affinity receptors interfered with the uptake of HSV-1 virions by the cells. However, the authors based their conclusion on the attachment of virions to cells at 2 h postinfection (p.i.). This time period is far beyond the 5- to 10-min entry period described by Rosenthal et al. (1989). To test their conclusion regarding the importance of basic FGF receptors as HSV-1 receptors, these authors used a full-length cDNA clone of a high-affinity murine fibroblast basic FGF receptor (Baird et al. 1990), which was transfected and expressed in a Chinese hamster ovary (CHO) cell line. The murine FGF receptor protein is a tyrosine kinase type receptor with about 90% similarity to the chicken FGF receptor and the human fms-like gene (FLG) tyrosine kinase. The receptor is phosphorylated upon stimulation with both basic FGF and Kaposi FGF Mansukhani et al. 1990). Kaner et al. (1990) described the wild-type (wt) CHO cells as cells which (a) contain fewer than 1000 FGF receptors per cell as a background level, (b) do not respond mitogenically to FGF, and (c) are unable to take up HSV-1 at the end of a 2 h incubation period with labeled HSV-1. The wt CHO cells were transfected with the basic FGF receptor cDNA clone, two stable transfectant lines were isolated and found to express 31000 and 100000 high-affinity receptors per cell. The control cells which were transfected with the antisense cDNA construct of the receptor did not express functional FGF receptors. Kaner et al. (1990) reported a greater than ten-fold increase of labeled HSV-1 uptake in the transfectants which expressed the high-affinity FGF receptors compared to the controls. The synthetic FGF peptide (amino acids 103–120) that binds to the high-affinity receptor blocked the increase in the amount of adsorbed virus. Kaner et al. (1990) indicated that the basis FGF receptor might not be the only port of entry for HSV-1 and the virus might use unidentified FGF receptors. Kaner et al. (1990) suggested that HSV-1 binds to the heparan sulfate on the FGF (WuDunn and Spear 1989). To substantiate the role of basic FGF receptors in HSV-1 adsorption, Baird et al. (1990) reported that the herpes virions contained cell-derived basic FGF (18-kDa immunoreactive protein), which ensures the ability of the virions to interact with the high-affinity FGF receptor.

Two studies, by Muggeridge et al. (1992) and by Mirda et al. (1992), and experiments in the present study provided evidence that rules out the need for basic FGF receptor for the binding and entry of HSV-1 virions into cells. Lycke et al. (1991) corroborated the finding of WuDunn and Spear that cellular heparan sulfate is the initial binding molecule on cell membranes involved in HSV-1 adsorption to cells.

The attachment of virions to heparan sulfate, which is covalently bound to proteoglycans situated in the cell membrane, is the rapid initial step of the virus–cell interaction. The entry of the virions into the cells either by fusion of the virion envelope with the outer cell membrane or by interaction with receptors in the coated pits on the cell membrane is controlled by the extracellular pH, the composition of the cell membrane, and receptors in coated pits. The binding of the virions to heparan sulfate on cellular proteoglycans in the cell membrane is mediated by the viral structural glycoproteins which are located in the virion envelope (Stannard et al. 1987). The structural glycoproteins projecting from the virion envelope were visualized by electron microscopy using gold-labeled monoclonal antibodies to gB, gC, and gD (Stannard et al. 1987). These three viral glycoproteins as well as gH were implicated in the attachment to heparan sulfate and penetration. However, HSV-1 gC and gE are thought not to be essential to virus adsorption, since virions lacking these glycoproteins are infectious (see Stannard et al. 1987).

The aim of the present study is to analyze the properties of HSV-1 gB, gC, gD, and gH with respect to the ability of the virions to interact with heparan sulfate and with the outer membrane components of the cell in the processes leading to virion–cell membrane fusion, or interaction of virions with coated pits leading to virus entry by endocytosis. It will be noted that an amino acid sequence in gB polypeptide is involved in the attachment of the virions to heparan sulfate on proteoglycans in the cell membrane. A similar amino acid sequence in HSV-1 gC and gH may also contribute to the attachment of the virions to heparan sulfate. The amino acid sequence in gB is most probably the critical heparan sulfate-binding domain. A synthetic peptide was synthesized based on the gB-binding domain (Becker 1992). This synthetic peptide successfully interfered with the attachment of the heparin to heparan sulfate (A. Yayon and Y. Becker, in preparation). The HSV-1 gB heparan sulfate-binding domain resembles an amino acid sequence in apolipoprotein E(apoE), which was reported to be able to bind to low-density lipoprotein (LDL) receptors (LDLR) (Dyer and Curtiss 1991) in coated pits. Computer analysis (Becker 1992) of HSV-1 mature gD (Eisenberg et al. 1984) suggested that the N terminus of the mature gD polypeptide may act as a fusion peptide which facilitates the fusion of the virion envelope with the cell membrane. The fates of the HSV-1 virions which enter the cell cytoplasm by either fusion or endocytosis differ, and the molecular events leading to the uncoating of the virions' DNA remains to be studied.

Experimental Approaches

Herpes Simplex Virus Type 1 Attachment to Chinese Hamster Ovary Cell Lines

The following CHO cell lines were used:

1. CHO KI – wt containing low-affinity FGF receptors
2. CHO KIFR2 – CHO cells transfected with the high-affinity FGF receptor gene which express both low- and high-affinity FGF receptors
3. CHO 803 – mutation in heparan sulfate synthesis-only 10% of heparan sulfate is present on the cell surface
4. Vero cells – Green monkey cell line

Radiolabeled Herpes Simplex Virus Type 1 Virions

Vero cells were infected with HSV-1 (KOS) at 10 pfu/cell. At 6 h p.i., 10 μCi [^{3}H]thymidine/ml (specific activity 5000 m Ci/m*M*, Nuclear Research Center, Negev) was added and the cultures were incubated at 37°C. The infected cell cultures were harvested at 24 h p.i. and sonicated, and the [^{3}H]dT-labeled virions (HSV-1 (KOS)) were isolated as a distinct band by centrifugation in 12% –52% w/w sucrose gradients.

Time Course of Virion Adsorption to Chinese Hamster Ovary Cells

Fifteen minutes prior to the addition of the labeled virus, 50 μl heparin (Sigma) was added to half the cultures, with the other half being mock treated. The labeled purified virus was added to the cultures, and samples of infected cells were taken at different time intervals (2, 5, 10, 20, 30 and 60 min). Each sample was treated with 1% sodium dodecyl sulfate (SDS) and the trichloroacetic acid (TCA)-precipitated radioactivity was determined in a Kontron scintillation counter.

Effect of Glycoprotein B Peptide on Viral Penetration

A synthetic oligopeptide homologous to a domain of HSV-1 gB (Y. Becker, to be published) was used. Cells were incubated in the presence of 0.5 mg synthetic peptide/culture plate (2 cm diameter) for 30 min and then infected with 100 plaque-forming units (PFU) HSV-1 (F) for a further 30 min. Cells were washed with Dulbecco's modified Eagle's medium (DMEM) and overlayed with 1% Nobel agar-containing DMEM. Viral plaques were screened 3 days p.i.

Computer Analysis of Herpes Simplex Virus Type 1 Glycoproteins

The compilation of seven algorithms in one program by Wolf et al. provides information for analyzing the properties of amino acids in a polypeptide chain by using the primary amino acid sequence of the peptide. The following properties of amino acids were studied: (a) hydrophilicity, according to Hopp and Woods or Kyte and Doolittle; (b) surface probability, according to Emini et al.; (c) chain flexibility, according to Karplus and Schultz; (d) secondary structure, according to Chou and Fasman, and Garnier-Osguthorpe-Robson; and (e) antigenicity index, according to Wolf et al. All of these provide quantitative estimates for hydrophilicity, surface probability, chain flexibility, and antigenicity index (for references, see Becker 1992).

Results

Interaction of Herpes Simplex Virus Type 1 with the Cell Membrane Leads to Entry by the Fusion Mechanism

Heparan Sulfate on Cell Membrane Proteoglycans in Chinese Hamster Ovary Cells Serves for Herpes Simplex Virus Type 1 Virion Adsorption

The results presented in Fig. 1 reveal the time course of HSV-1 virion adsorption to CHO wt cells. At 30 min p.i., 90% of the radioactive virions were absorbed to the cells. Seventy percent of the total amount of the radioactive virions which were adsorbed at 30 min were already attached to the cell surface of the CHO KI wt cells at 10 min p.i. These results indicated that virion attachment to the cells is a rapid process, in agreement with Rosenthal et al. (1984, 1989). Addition of heparin to the CHO KI cells resulted in the displacement of 59% of the attached virus. Thirty percent of the radioactive virions were bound to cellular receptors in a way that resisted heparin treatment. This result indicated that two subpopulations of virions are present on the CHO cells, one quickly becoming heparin resistant and the other remaining heparin sensitive.

Decreased Herpes Simplex Virus Type 1 Adsorption to Chinese Hamster Ovary Mutant

CHO 803 cells have a mutation which reduces the synthesis of heparan sulfate, resulting in 10% of the heparan sulfate being bound to the proteoglycan protein

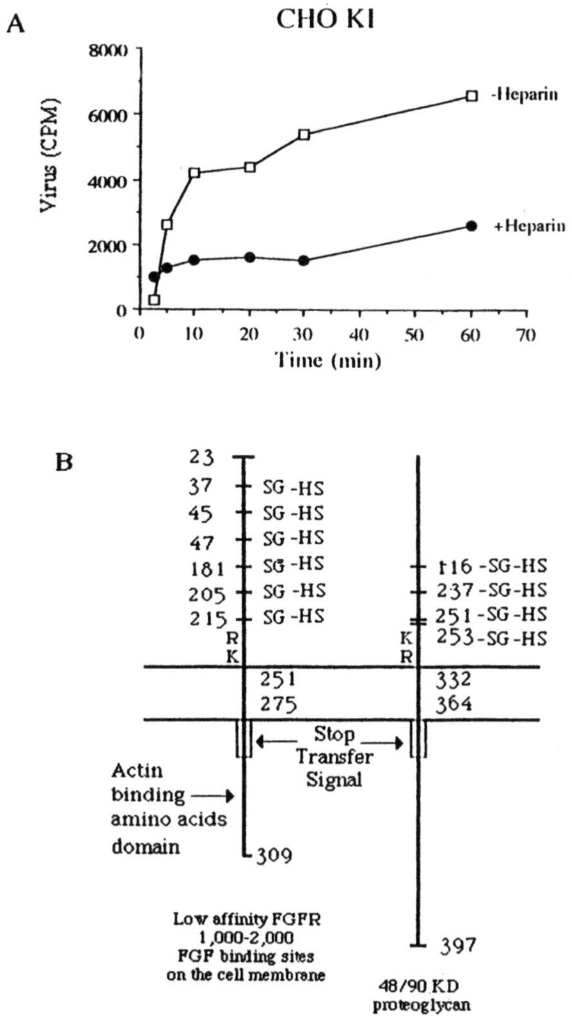

Fig. 1. **A** Time course of herpes simplex virus adsorption to wild-type Chinese hamster ovary (*CHO*) cells. **B** Structure of heparan sulfate on proteoglycans – low-affinity fibroblast growth factor (*FGF*), receptor *FGFR* – in the cell membrane. *SG* are the serine–glycine amino acids to which heparan sulfate (*HS*) is enzymatically covalently attached. *R*, arginine; *K*, lysine

cores as compared to the wt CHO. As can be seen in Fig. 2, the adsorption of HSV-1 virions to CHO 803 cells was markedly reduced. Most of the adsorbed virions were resistant to treatment with heparin. It is possible that the heparin-resistant virions which are bound to the cell membrane lacking heparan sulfate are the virions which are associated with the coated pits in the cell membrane. Further studies are needed to confirm this.

A

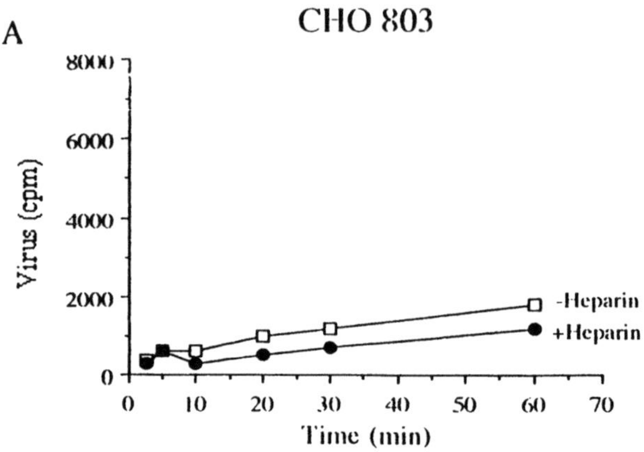

B

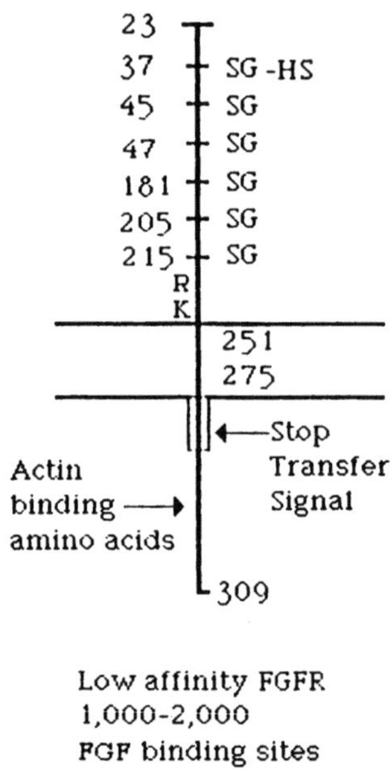

Fig. 2. **A** Time course of herpes simplex virus adsorption to heparan sulfate (*HS*)-deficient Chinese hamster ovary (*CHO*) cells. **B** HS-deficient proteoglycans present in the cell membrane of the mutant CHO; low-affinity fibroblast growth factor (*FGF*) receptor, *FGFR* lacking HS is suggested. *SG*, serine–glycine; *R*, arginine; *K*, lysine

Time Course of Herpes Simplex Virus Type 1 Virion Adsorption to Chinese Hamster Ovary Cells Transfected with the High-Affinity fibroblast Growth Factor Receptor Gene CHO KIFR2

The time course of HSV-1 virion adsorption to CHO cells which express the high-affinity FGF receptors (Fig. 3) did not differ from that of CHO KI wt (Fig. 1). Most (55%) of the adsorbed virions were displaced from the cell surface by

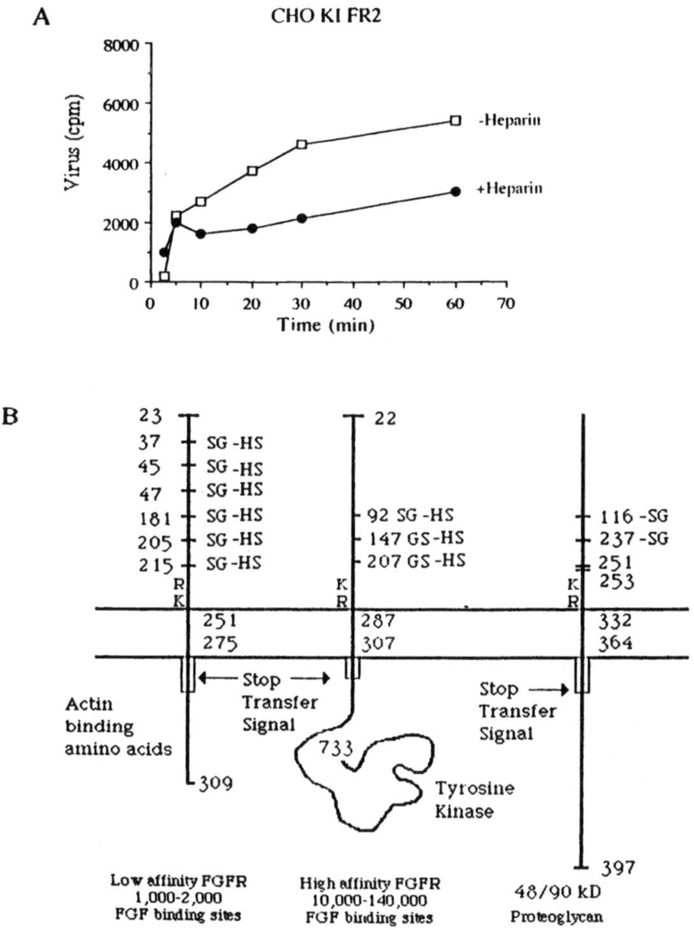

Fig. 3. A Time course of herpes simplex virus adsorption to Chinese hamster ovary (*CHO*) cells which were engineered to express the high-affinity fibroblast growth factor (*FGF*) receptor (*FGFR*) gene. **B** Possible heparan sulfate (*HS*)-carrying cellular proteoglycans (low-affinity FGFR, high affinity FGFR, and 48/90 kDa proteoglycan as possible HS carriers in the cell membrane of this engineered cell line). *R*, arginine; *K*, lysine

heparin. The virions were already bound to the cells by a heparin-resistant mechanism at 5 min. p.i.

The Amino Acid Sequence in Herpes Simplex Virus 1 Glycoprotein B Involved in Binding to Heparan Sulfate

Computer Analysis of Herpes Simplex Virus Type 1 Glycoprotein B Polypeptide

To assess the properties of the topogenic and antigenic domains in HSV-1 gB, computer analysis was performed (Becker 1992). The computer analysis predicted

Table 1. Amino acid sequences in HSV-1 glycoproteins B, C, F, and H, which may be involved in heparan sulfate binding

Virus	Amino acid sequence
HSV-1 (KOS) gB	(66) P K P K K N K K P K N P T P P R P (82)
HSV-1 gC	(88) P K P T S T P K S P P T S T P D P K P K (107)
HSV-1 gF	(67) K P K K N K K P N N P P P P R P (82)
HSV-1 gH	(53) P N T P D P Q K P P R (63)

HSV-1, herpes simplex virus type 1; g, glycoprotein.

antigenic domains which matched well with the experimentally derived antigenic domains. Special attention was given to domain 1, which was highly antigenic and rich in prolines and lysines (Table 1). According to the computer analysis, the amino acid domain has a β-turn conformation and a high surface probability value, as compared to other domains in the extracellular portion of gB that anchor the virions to heparan sulfate on the cell membrane.

Antiviral Activity of a Synthetic Peptide Made According to the Domain in Glycoprotein B

A synthetic peptide designated "gB peptide" (designed according to amino acids in domain 1) was added to HSV-1-infected BSC-1 cells in 2-cm-diameter plastic culture plates. The effect of gB peptide on the number of HSV-1 plaques was determined (Table 2). It was noted that gB peptide at a concentration of 0.5 mg/plate markedly inhibited plaque formation, and the same effect was obtained after removing the peptide by washing the infected cultures 30 min after virus addition. Two synthetic peptides (RGDD and RGEK) had no effect on the number of the plaques. These results suggest that the amino acids in antigenic domain 1 of HSV-1 gB (Becker 1992) contain the domain for attachment to heparan sulfate. To corroborate this finding, the effect of gB peptide on heparin binding to heparan sulfate was investigated. It was found that the gB peptide interfered with heparin binding (A. Yayon and Y. Becker, in preparation).

Table 2. Effect of gB peptide on HSV-1 infection of cells

Peptide	Concentration (mg/ml)	Experiment No. 1 HSV-1F	2 HSV-1F	3 HSV-1F	4 HSV-1HFEM
	0	64	214	77	39
gB Peptide	0.5	11	19	8	29
gB Peptide	0.5[b]	–	–	10	7
RGDD	0.5	70	–	–	–
RGEK	0.5	–	212	–	–

HSV-1, herpes simplex virus type 1; g, glycoprotein; RGDD and RGEK, synthetic peptides.
[a] Each number is an average of the plaques in four plates.
[b] After 30 min with glycoprotein B peptides, the cell cultures were carefully washed and reincubated in fresh medium.

Additional Herpes Simplex Virus Type 1 Glycoproteins Contain an Amino Acid Domain for Heparan Sulfate Binding

Herpes Simplex Virus Type 1 Glycoprotein C Contains a Heparan Sulfate-Binding Domain

Table 1 describes an amino acid domain in HSV-1 gC that has a high surface probability value and a predicted β-turn conformation by computer analysis. This amino acid domain resembles the gB-binding domain, which may explain the possible role of gC in the binding of HSV-1 virions to heparan sulfate and which is in agreement with the findings of Campedelli-Fiume et al. (1990) on the role of virion gC in binding to cells. However, it was reported that gC-negative HSV-1 has the ability to enter the human nervous system to reactivate and cause an ocular infection despite the lack of gC (Chap. 10, this volume). It is possible that the heparan sulfate-binding domain in gB fulfills the requirement of the virions to bind to heparan sulfate on the cell membrane. HSV-1 gC may help in the anchoring of virions to heparan sulfate, but its presence in the virions may not be essential for the adsorption process.

Herpes Simplex Virus Type 1 Glycoproteins F and H Contain a Putative Heparan Sulfate-Binding Domain

Computer analysis of HSV-1 gF reveals an amino acid domain which markedly resembles the heparan sulfate-binding domain of gB (Table 2). In contrast, HSV-1 gH contains an amino acid domain that only partially resembles the heparan sulfate-binding domain present in gB (Table 1). The importance of these domains in the attachment of the HSV-1 virions to cells is not yet known.

Fusion of the Envelope of Herpes Simplex Virus Type 1 with Virions and the Role of Glycoprotein D

The studies by Ligas and Johnson (1988) and Campadelli-Fiume et al. (1988) revealed the importance of HSV-1 gD in the entry of HSV-1 virions by the mechanism of fusion of the virion envelope with the cell membrane. Ligas and Johnson (1988) constructed a virus with a deletion in the viral gD gene. Virions lacking gD adsorbed to the cell surface, but were unable to enter the cells; only treatment with polyethylene glycol enhanced the entry of virions into cells, and formation of plaques were evident only with this treatment. Campedelli-Fiume et al. (1988) used BJ cells which constitutively express the HSV-1 gD gene and contained viral gD on the cell membrane. HSV-1 infected BJ cells, but the virus entered the cell by endocytosis only. Virions were noted to be associated with coated pits on the cell membrane, and not with the cytoplasmic membrane as seen in infection of permissive cells. It was suggested that the presence of gD on the BJ cell membrane prevented the fusion mechanism, while the endocytosis mechanisms was the only way for virion entry into the cells (see below).

Computer Analysis of Herpes Simplex Virus Type 1 Glycoprotein D

Computer analysis of HSV-1 gD polypeptide revealed the hydrophilicity and hydrophobicity of the amino acids in the polypeptide chain. This analysis (Becker 1991) identified the hydrophobic N-terminal amino acid sequence which may serve as a signal sequence for the cellular system that recognizes the signal sequence of a glycoprotein and transports the polypeptide through a cell membrane. The N-terminal signal sequence is cleaved by the signal peptidase on the outer side of the membrane. Indeed, Eisenberg et al. (1984) reported that the 22 N-terminal amino acids are absent from the mature HSV-1 gD present in virions. The new N terminus of the mature gD polypeptide (from amino acid 23) contains five amino acids with a putative β-turn conformation followed by nine hydrophobic amino acids with an α-helix conformation (Table 3). It is hypothesized that the N terminus of the mature gD in HSV-1 virions may have the ability to serve as a fusion sequence which is capable of being inserted into the cell membrane of the infected cells. The insertion of the mature gD N terminus into the cell membrane is thought to be the reason for the destabilization of the membrane fluidity reported by Rosenthal et al. (1984). For comparison, the fusion sequence of Newcastle disease virus (NDV) has an α-helix conformation, while the fusion sequence of respiratory syncytial virus (RSV) has a β-sheet conformation at the N terminus of the fusion sequence. Brasseur et al. (1988) reported the mode of insertion of the NDV fusion sequence into the cell membrane; using infrared spectroscopy and conformational analysis, it was shown that the viral fusion peptide adopted an oblique orientation with respect to the lipid acyl chains. These authors suggested that the unusual orientation of the inserted N terminus of the viral fusion sequence may induce disorganization of the surrounding lipids, which is one of the initial events in virus-cell fusion. Further studies are needed to elucidate the mode of the fusion which occurs between the virion envelope gD and the cell membrane.

Table 3. The amino acid sequence of the N terminus of the mature from of HSV-1 gD and known viral fusion sequences

HSV-1 gD	(23) V R S K Y	A L V D A S L K M	A D P N R F R G (44)
Hydrophilicity	(+)	(−)	(+)
Conformation	β-Turn	α-Helix	β-Turn
Surface probability		Low	High
Newcastle disease virus	F I G A I I G S V A L G V A		
Hydrophilicity	(−)		
Conformation	α-Helix		
Respiratory syncytial virus	F L G F L L G V G S A I A S G V A V S		
Hydrophilicity	(−)		
Conformation	β-Sheet		

HSV-1, herpes simplex virus type 1; g, glycoprotein.

Effect of Inhibitors of Cell Membrane-Associated Enzymes on Virion Entry

Two inhibitors were selected for the study: (1) amiloride, which directly inhibits growth factor receptor tyrosine kinase activity (Davis and Czech 1985) and serves as a Na^+/H^+ antiporter (L'Allemine et al. 1984) and (2) quercetin, which selectively inhibits cyclic nucleotide-independent protein kinase (Cochet et al. 1982) and the phosphorylation activity of the protein coded by Rous Sarcoma virus-transforming gene (Graziani et al. 1983). Quercetin also inhibits protein kinase C (Christensen et al. 1987).

Effect of Amiloride on Herpes Simplex Virus Type 1 Infection of Cells

Infection of cells with HSV-1 in the presence of amiloride resulted in a marked reduction of virus titer at 12 h p.i., while the inhibitory effect was much reduced at 24 h p.i. (Fig. 4). Since the diluent of amiloride had a similar depressing effect on virus titer at 12 h and 24 h p.i., it is concluded that amiloride had no specific

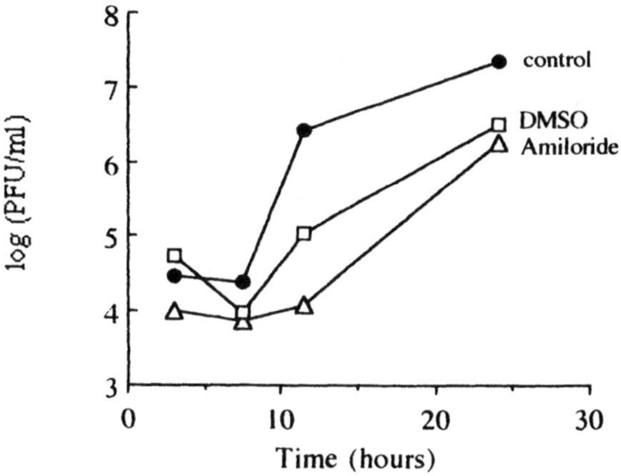

Fig. 4. Effect of amiloride on herpes simplex virus type 1 (HSV-1) replication. •, Untreated HSV-1-infected cells; ❑ cell treated with dimethylsulfoxide (*DMSO*)-containing medium during HSV-1 infection; Δ, amiloride-treated HSV-1 infected cells

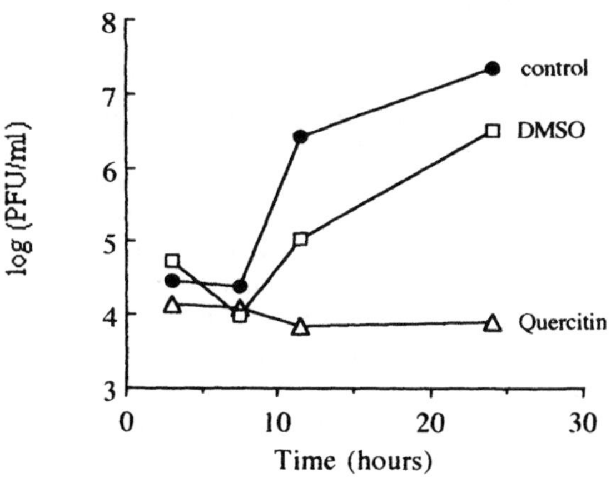

Fig. 5. Effect of quercetin on herpes simplex virus type 1 (HSV-1) replication. •, Untreated HSV-1- infected cells; ❑, cell cultures treated with dimethylsulfoxide (*DMSO*)-containing medium; Δ, quercitin-treated HSV-1 infected cells

effect on HSV-1 replication and that, therefore, the tyrosine kinase activity of the growth factor receptor, which is inhibited by the drug, was not involved in virus entry into cells.

Effect of Quercetin on Herpes Simplex Virus Type 1 Entry and Replication

The results represented in Fig. 5 revealed that quercetin completely inhibited the ability of HSV-1 to enter into and to replicate in the infected cells. The results of this experiment suggest that the activity of the cyclic nucleotide-independent protein kinase in the cell membrane is essential for virus entry into the cell. The exact mechanism and role of the kinase in virus entry is not yet known.

Interaction of Herpes Simplex Virus Type 1 Virions with Coated Pits and Virus Entry by Endocytosis

Interaction of Herpes Simplex Virus Type 1 with Coated Pits

The electron microscopic studies showing virions associated with clathrin-coated pits have already been discussed above. The nature of the proteins present in the clathrin-coated pits should be of interest for the evaluation and characterization of HSV-1 entry by the interaction with coated pits in the cell membrane. Russell et al. (1984) and Yamamoto et al. (1986) reported on the domains in LDLR glycoprotein, which is situated in the outer membrane of the clathrin-coated pits. ApoB of E, which carry cholesterol, interact with LDLR in the coated pits. This process is followed by endocytosis, which incorporates the receptor and the ligand into the cell cytoplasm (Goldstein and Brown 1977). It is therefore necessary to ask how HSV-1 virions interact with the coated pits.

The Herpes Simplex Virus Type 1 Glycoprotein Amino Acid Domain for Heparan Sulfate Binding Resembles the Low-Density Lipoprotein Receptor-Binding Domain in Apolipoprotein E

In a study on the properties of HSV-1 gB by computer analysis (Becker 1992), the amino acid sequence of gB (which was shown to interfere with the ability of HSV-1 to infect cells) was compared to the amino acid sequence of apoE reported by Dyer and Curtiss (1991; Table 1). It was noted that the apoE amino acid domain and HSV-1 gB resemble each other. It is hypothesized that HSV-1 can attach to the LDLR in coated pits with the same amino acid sequence present in gB, gC, or gF, which interact with heparan sulfate binding. It is proposed that most of the virions attach to heparan sulfate on the cell membrane, while a minority of the virions attach directly to LDLR and possibly to other proteins in the coated pits. The latter are the virions which are resistant to heparin treatment and can be incorporated into the cell cytoplasm by the endocytosis mechanism, in agreement with the finding of Campadelli-Fiume et al. (1988). Further studies are in progress to elucidate the fate of HSV-1 virions entering the cells by endocytosis through coated pits.

Conclusions

The present study presents an analysis of the mechanisms which are involved with the adsorption and entry of HSV-1 virions into cells. The virion-associated glycoproteins B and C have an amino acid domain that interacts with either of two receptors on the cell membrane: (1) heparan sulfate on proteoglycans present on the cell membrane, to which most of virions attach, and (2) LDLR and/or other proteins present in coated pits in the cell membrane, to which a minority of virions attach.

Virions which adsorb to the heparan sulfate moiety on proteoglycans adsorb very rapidly, within 5–10 min after addition of the virus to the cells. Following the adsorption step, fusion of the virion envelope with the cell membrane takes place. It is suggested, based on computer analysis, that the gB and gC in the virion envelope interact with heparan sulfate on the cell membrane, a process followed by gD N terminus fusing with the cell membrane, allowing the viral capsid to enter the cell cytoplasm to initiate a lytic infection.

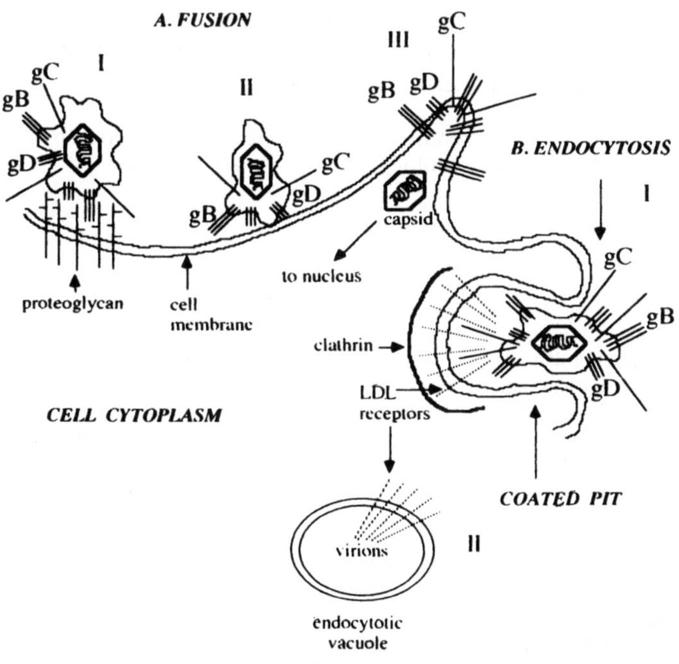

Fig. 6. Herpes simplex virus type 1 (HSV-1) virions interaction with and entry into cells by fusion (*A*) and by endocytosis (*B*). *A,* Three steps of virion interaction with the *cell membrane* are shown: the interaction of HSV-1 virions' glycoproteins *gB* and *gC* polypeptide with heparan sulfate on *proteoglycans* present in the cell membrane (*I*); the interaction of the virion *gD* polypeptide with the cell membrane (*II*); the complete fusion of the virion envelope with the cell membrane and entry of the *capsid* into the *cell cytoplasm. B,* Two steps in the endocytosis process are shown: the binding of a virion to a *coated pit*; most probably virion's *gB* binds to low-density lipoprotein (*LDL*) receptors in the cell membrane (*I*); The formation of an *endocytotic vacuole* that contains the endocytosed virion (*II*)

Virions that interact with the coated pits in the cells membrane most probably mimic the binding of apoE to LDLR present in the coated pits. The similarity of the amino acid sequence in the LDLR-binding domain in apoE and the amino acid domain in HSV-1 gB polypeptide suggests that HSV-1 virions may be incorporated into the cells by the endocytotic process that incorporates the apoE bound to the LDLR. The HSV-1 virions which interact with the coated pits were also found in endocytotic vesicles in the cell cytoplasm (Campadelli-Fiume et al. 1988). A model of HSV-1 virion interaction with cells is presented in Fig. 6.

Various aspects of virus entry into cells (e.g., the process of fusion of the virion envelopes with the cell membrane) are still unknown. The studies by Rosenthal et al. (1984) provided information on changes in the cell membrane after attachment of virions to the membrane. The demonstration that quercetin, which selectively inhibits cyclic nucleotide-independent protein kinase, can prevent HSV-1 infection of cells may suggest that cellular membrane-associated enzymes are involved in the penetration of HSV-1 virions by the fusion process. Information on the molecular details of the fusion process is needed. Less is known on the interaction of HSV-1 virions with the coated pits. The suggestions on the mode of recognition of the coated pits by HSV-1 virions presented in the study are putative, and additional experimental results are needed to understand the significance of virion entry into cells by endocytosis. The fate of the virions which enter the cells by the two different mechanisms requires further studies.

Acknowledgements. The experimental work reported in this chapter was supported by the Foundation for Molecular Virology and Cell Biology, Phoenix, Arizona, USA.

References

Baird A, Florkiewicz RZ, Maher PA, Kaner RJ, Hajjar DP (1990) Mediation of virion penetration into vascular cells by association of basic fibroblast growth factor with herpes simplex virus type 1. Nature 348: 344–346

Becker Y (1991) Computer prediction of antigenic domains in herpes simplex virus types 1 and 2 glycoprotein D as compared with experimentally proven domains. Virus Genes 5: 367–375.

Becker Y (1992) Computer prediction of antigenic and topogenic domains in HSV-1 and HSV-2 glycoprotein B (gB) Virus Genes 6: 131–141

Benda R (1966) Effect of heparin on B virus multiplication in vitro. Acta Virol Prague 10: 376

Brasseur R, Lorge P, Goormaghtigh E, Ruysschaert J-M, Espian D, Burney A (1988) The mode of insertion of the paramyxovirus F1 N-terminus into lipid matrix, an initial step in host cell/virus fusion. Virus Genes 1: 325–332

Campedelli-Fiume G, Arsenakis M, Farabegoli F, Roizman B. (1988) Entry of herpes simplex 1 in BJ cells that constitutively express viral glycoprotein D is by endocytosis and results in degradation of the virus. J Virol 62: 159–167

Campedelli-Fiume G, Stripe D, Boscaro A, Avitabile E, Foá-Tomasi L, Barker D, Roizman B (1990) Glycoprotein C-dependent attachment of herpes simplex virus to susceptible cell leading to productive infection. Virology 178: 213–222

Christensen RL, Shade DL, Graves CB, McDonald JM (1987) Evidence that protein kinase C is involved in regulating glucose transport in the adipocyte. Int J Biochem 19: 259–265

Cochet C, Feige JJ, Pirollet F, Keramidas M, Chambaz EM (1982) Selective inhibition of a cyclic nucleotide independent protein kinase (G type casein kinase) by quercitin and related compounds. Biochem Pharmacol 31: 1357–1361

Dales S, Silverberg H (1969) Viropoxis of herpes simplex virus by HeLa cells. Virology 37: 475–480

Davis RJ, Czech MP (1985) Amiloride directly inhibits growth factor tyrosine kinase activity. J Biol Chem 260: 2543–2551

Dyer CA, Curtiss A (1991) A synthetic peptide mimic of plasma apolipoprotein E that binds LDL receptor. J Biol Chem 266: 22803–22806

Eisenberg RJ, Long D, Hogue-Angeletti R, Cohen GH (1984) Amino terminal sequence of glycoprotein D of simplex virus types 1 and 2. J Virol 49: 265–268

Epstein MA, Hummeler K, Berkalof A (1964) The entry and distribution of herpes virus and colloidal gold in HeLa cells after contact in suspension. J Exp Med 119: 291–302

Goldstein JL, Brown MS (1977) The low-density lipoprotein pathway and its relation to atherosclerosis. Annu Rev Biochem 46: 897–930

Graziani Y, Erikson E, Erikson RL (1983) The effect of quercitin on the phosphorylation activity of the Rous sarcoma virus transforming gene product in vitro and in vivo. Eur J Biochem 135: 583–589

Hadhazy G, Horvath F, Gergely L (1966) Simultaneous inhibitory action on virus multiplication of interferon and some natural mucopolysaccharides (heparin hyaluronic acid). Acta Microbiol Hung 13: 193–196

Hochberg E, Becker Y (1968) Adsorption, penetration and uncoating of herpes simplex virus. J Gen Virol 2: 231–241

Horvath F, Hadhazy G (1965) Effect of heparin on the growth of herpes group of viruses. Acta Microbiol Hung 12: 145–149

Hummeler K, Tomassini M, Zajac B (1969) Early events in herpes simplex virus infection: a radioautographic study. J Virol 4: 67–74

Kaner RJ, Baird A, Mansukhan A, Basilico C, Summers, BD, Florkiewicz RZ, Maher PA, Hajjar DP (1990) Fibroblast growth factor receptor is a portal of cellular entry for herpes simplex virus type 1. Science 248: 1410–1413

Keller PM, Person S, Snipes W (1977) A fluorescence enhancement assay of cell fusion. J Cell Sci 28: 167–177

L'Allemain G, Paris S, Pouyssegur J (1984) Growth factor action and intracellular pH regulation in fibroblasts. Evidence for a major role of the Na^+/H^+ antiport. J Biol Chem 259: 5809–5815

Lehel F, Hadhazy G (1966) Effect of heparin on herpes simplex virus infection in the rabbit. Acta Microbiol Hung. 13: 197–203

Ligas MW, Johnson DC (1988) A herpes simplex virus mutant in which glycoprotein D sequences are replaced by beta galactosidase sequences binds to but is unable to penetrate into cells. J Virol 62: 1486–1494

Lycke E, Johansson M, Svennerholm B, Lindahl U (1991) Binding of herpes simplex virus to cellular heparan sulfate, an initial step in the adsorption process. J Gen Virol 72: 1131–1137

Mansukhani A, Moscatelli D, Talarico D, Levytska V, Basilico C (1990) A murine fibroblast growth factor (FGF) receptor expressed in CHO cells is activated by basic FGF and Kaposi FGF. Proc Natl Acad Sci USA 87: 4378–4382

Mirda DP, Navarro D, Paz P, Lee PL, Pereira L, Williams LT (1992) The fibroblast growth factor receptor is not required for Herpes simplex virus type 1 infection. J Virol 66: 448–457

Morgan C, Rose HM, Mednis B (1968) Electron microscopy of herpes simplex virus I entry. J Virol 2: 507–516

Muggeridge MI, Cohen GH, Eisenberg RJ (1982) Herpes simplex virus infection can occur without involvement of fibroblast growth factor receptor. J Virol 66: 824–830

Nahmias AJ, Kibrick S. (1963) Inhibitory effect of heparin on certain viruses. Bacteriol Proc 63: 145

Nahmias AJ, Kibrick S (1964) Inhibitory effect of heparin on herpes simplex virus. J Bacteriol 87: 1060–1066

Para MF, Bancke RB, Spear PG (1980) Immunoglobulin G (FC)-binding receptors on virions of herpes simplex virus type 1 and transfer of these receptors to the cell surface by infection. J Virol 34: 512–520

Rosenthal KS, Leuther MD, Barisas BG (1984) Herpes simplex virus binding and entry modulate cell surface protein mobility. J Virol 49: 480–493

Rosenthal KS, Killius J, Hodnichak CM, Venetta TM, Gyurgyik L, Janiga K (1989) Mild acidic pH inhibition of the major pathway of herpes simplex virus entry into HEP-2 cells. J Gen Virol 70: 857–867

Russel DW, Schneider WJ, Yamamoto T, Luskey KL, Brown MS, Goldstein JL (1984) Domain map of LDL receptor: sequence homology with the epidermal growth factor receptor precursor. Cell 37: 577–585

Smith JD, de Harven E (1974) Herpes simplex virus and cytomegalovirus replication in WI-38 cells. II. An ultrastructural study of viral penetration. J Virol 14: 945–956

Stannard IM, Fuller AO, Spear PG (1987) Herpes simplex virus glycoproteins associated with different morphological entities projecting from the virion envelope. J Gen Virol 68: 715–725

Takameto KK, Fabisch P. (1964) Inhibition of herpesvirus by natural and synthetic acid polysaccharides. Proc Soc Exp Biol Med 116, 140–144

Vaheri A, Cantell K (1963) The effect of heparin on herpes simplex virus. Virology 21: 661–662

WuDunn D, Spear PG (1989) Initial interaction of herpes simplex virus with cells is binding to heparan sulfate. J Virol 63: 52–58

Yamamoto T, Bishop RW, Brown MS, Goldstein JL, Russel DW, (1986) Deletion in cysteine-rich region of LDL receptor impedes transport to cell surface in WHHL rabbit. Science 232: 1230–1237

Chapter 2 Neomycin and Herpes Simplex Virus Receptor Binding: The Role of Glycoprotein C

Anne M. Øyan[1], Lars Haarr[1], Joseph C. Glorioso[2], Karen E. Dolter[3], Howard S. Marsden[4], Clyde S. Crumpacker[5], and Nina Langeland[1,6]

Summary

This chapter reviews studies on attachment of herpes simplex virus types 1 and 2 (HSV-1 and HSV-2) to their cellular receptors using the polycations neomycin and polylysine to differentiate between binding of the two viruses. These compounds have been found to inhibit adsorption of HSV-1, while HSV-2 is still able to attach and to infect the cells. Mapping of the genes conferring the neomycin effect localized initially a region on the HSV genome containing eight genes. Analysis of this region led to the conclusion that the *UL44* gene, encoding glycoprotein C (gC), renders HSV-2 resistant to neomycin and that the resistance is mediated by the N-terminal half of the gC-2 protein. The homolog protein gC-1 of HSV-1, however, seems not to be involved in the neomycin-sensitive attachment of that virus. Deletion of gC-2 made attachment of HSV-2 as sensitive as that of HSV-1 and led to the suggestion that HSV-2 has two adsorption pathways, one being neomycin resistant and gC-2 dependent and the other being neomycin sensitive and common for HSV-1 and HSV-2. This is consistent with the observations that preadsorption of HSV-2 inhibits subsequent infection by either HSV-1 or HSV-2, while HSV-1 as the first virus inhibits adsorption of HSV-1

Introduction

Entry of an animal virus into a host cell is a multistep process including attachment to the cellular receptor and penetration through the cell membrane. Specific receptors have been identified for several viruses. Examples are the CD4 glycoprotein on T lymphocytes, which is responsible for binding of human immunodefi-

[1] National Centre for Research in Virology and The Gade Institute, University of Bergen, 5020 Bergen, Norway
[2] Department of Molelular Genetics and Biochemistry, University of Pittsburgh School of Medicine, Pittsburgh, PA 15261, USA
[3] Department of Immunology and Microbiology, Wayne State University School of Medicine, Detroit, MI 48201, USA
[4] MRC Virology Unit, Institute of Virology, Glasgow G11 5JR, Scotland
[5] Beth Israel Hospital and Harvard Medical School, Boston, MA 02115, USA
[6] Department of Medicine, Haukeland Hospital, 5021 Bergen, Norway

ciency virus (HIV) (Dalgleish et al. 1984; Maddon et al. 1986), the receptor for the complement component C3b on B lymphocytes, which also acts as a receptor for Epstein-Barr virus (Fingeroth et al. 1984; Tanner et al. 1987), and the intercellular adhesion molecule 1 (ICAM-1), which serves as the major receptor for rhinovirus (Greve et al. 1989; Staunton et al. 1989). No protein has been identified as receptor for herpes simplex virus (HSV), but heparan sulphate is crucial for attachment (WuDunn and Spear 1989; Shieh et al. 1992).

Search for ligands on the surface of HSV has mainly focused on glycoproteins. Seven of these glycoproteins gB, gC, gD, gE, gG, gH, and gI, have been extensively studied (for review see Marsden 1987), while gJ, gK, and gL were identified more recently (Roizman and Sears 1990; Hutchinson et al. 1992 a, b). Glycoproteins B and C bind to heparin-sepharose (Herold et al. 1991) suggesting roles for them in attachment. Penetration rather than attachment is affected by deleting gB or gD (Cai et al. 1988; Ligas and Johnson 1988), or by the presence of antibodies specific for gB, gD or gH (Fuller and Spear 1987; Highlander et al. 1987, 1988; Fuller et al. 1989). Three glycoproteins, gB, gD and gH (and possibly also gL; Hutchinson et al. 1992 a), are essential for replication of virus in tissue culture (Sarmiento et al. 1979; Little et al. 1981; Cai et al. 1987; 1988; Desai et al. 1988; Ligas and Johnson 1988).

Some compounds inhibit the interaction between HSV and the host cell and are thus potential tools for studying the receptor or the ligand, particularly heparin, polylysine and neomycin. This chapter reviews the studies with neomycin which led to the conclusion that gC of HSV-2 is a major ligand for adsorption of virus to the host cell.

Properties of Neomycin

Neomycin is an aminoglycoside antibiotic which exerts its antimicrobial effect by binding to the 30S ribosomal subunit of prokaryotic cells, thus inhibiting protein synthesis (Pestka 1971). The compound has been used as an inhibitor of phosphoinositide-mediated signal transduction in eukaryotic cells; it inhibits thrombin-induced cell growth (Carney et al. 1985), Ca^{2+} -dependent secretion in mast cells (Cockcroft and Gomperts 1985), inositol phosphate release in erythrocyte ghosts (Downes and Michell 1981), and Ca^{2+} -dependent exocytosis in sea urchin eggs (Whitaker and Aitchison 1985). In kidney brush border membranes, the cellular binding site for neomycin was shown to be a phosphoinositide (Sastrasinh et al. 1982). Nuclear magnetic resonance studies have revealed very rapid complexing kinetics between neomycin and phosphatidylinositol 4,5-bisphosphate (PIP_2) (Reid and Gajjar 1987).

Our original intention was to test whether binding of HSV to the host cell started a signal transduction process mediated by phosphoinositides. Since an increase in PIP_2 was observed in cells after infection with HSV (Langeland et al. 1986b), we tested neomycin as an inhibitor of PIP_2 metabolism in an attempt to analyse further the PIP_2 variation.

Effects of Neomycin on Herpes Simplex Virus Infection

Inhibition of Infection by Herpes Simplex Virus Type 1, but Not Type 2

A surprising result of these studies was that neomycin inhibited infection by HSV-1, but not HSV-2 (Langeland et al. 1986a). The compound was present from 30 min before infection until 1h postinfection, and the effect was recorded as the number of plaques obtained in the absence or presence of drug (Fig. 1A; Langeland et al. 1986a). At 3 m*M* and higher concentrations, neomycin had a marked effect on HSV-1 infection, while HSV-2 was unaffected at all concentrations tested (up to 7.5 m*M*).

Site of Action

Neomycin had to be present at the time of infection in order to obtain maximal plaque reduction (Fig. 1B), suggesting its effect was on attachment and/or penetration. Further studies with the drug present only during adsorption at 4°C,

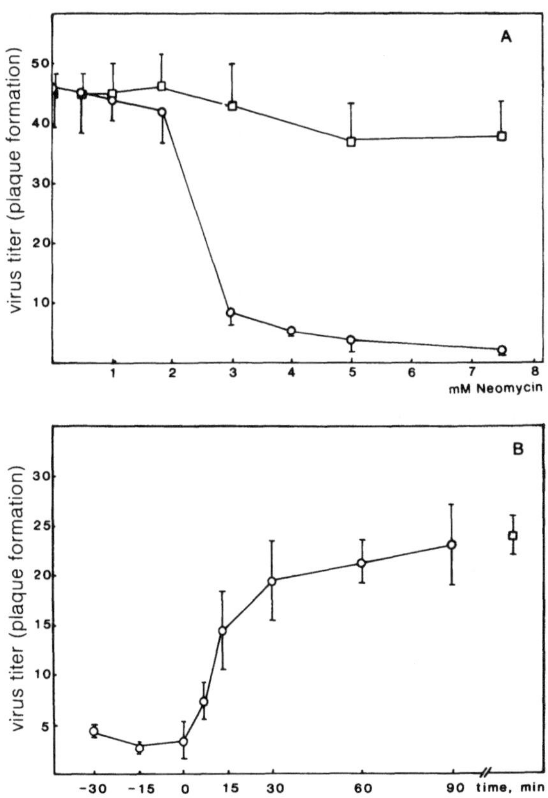

Fig. 1 A, B. Effect of neomycin on virus infection. **A** Plaques formed at different concentrations of neomycin. Baby hamster kidney cells were incubated with the indicated final concentration of neomycin 30 min prior to infection with HSV-1 (O) or HSV-2 (❑). Equal amounts of infective virus were added, and the dishes incubated for 48 h at 37°C, after which plaques were counted. Each point is the average of three individual determinations (± S.D.). **B** Time course of neomycin effect on HSV-1 plaque formation. Neomycin at the final concentration of 5m*M* was added at the indicated times before and after infection. The cells were infected with equal amounts of HSV-1 and further treated as indicated in A. The *square* at the *right* in the figure indicates parallel incubation without neomycin added. Data are means ± S.D. for three separate dishes

or both during adsorption and penetration at 37°C, gave similar reduction in plaque formation, indicating inhibition of attachment (Langeland et al. 1988). Binding of radiolabeled virus to cells at either temperature was blocked by the drug, thus providing additional support for the conclusion that neomycin acts on attachment. Addition of neomycin after adsorption at 4°C had no effect on the subsequent penetration (Langeland et al. 1988). The effect of the aminoglycoside was reversible in that pretreatment of cells or virus, and subsequent removal of the compound, resulted in the same number of plaques as were found in the absence of any treatment. Because of the reversible action of the drug, it was difficult to determine whether the site of action was on the virus or on the cell. One approach to distinguish between these possibilities was to use polylysine, which acts exactly like neomycin in inhibiting HSV-1 infection, but not HSV-2 infection of baby hamster kidney (BHK) cells; the same region on the HSV genome confers both the neomycin and the polylysine effects (Campadelli-Fiume et al. 1990; Langeland et al. 1990). Polylysine and neomycin appeared to have additive effects on HSV-1 infection, and the former compound was shown to act on the cell rather than on the virus (Langeland et al. 1988). Both neomycin and polylysine seem, therefore, to interact with the cellular receptor for HSV-1.

Other polyamino acids and aminoglycosides also inhibit HSV-1 infection, e.g. polyarginine, gentamycin and streptomycin, but these compounds have not been studied in detail (Langeland et al. 1987, 1988). All of them are polycations, but cationic charge alone is not sufficient to explain the inhibition on HSV-1 infection. Polyhistidine, glucosamine and spermine had only marginal effects (Langeland et al. 1988).

The polyanionic compound heparin blocks adsorption of HSV-1 as well as HSV-2 to cells (WuDunn and Spear 1989; Herold et al. 1991;). So far there is no experimental evidence for a possible link between the action of heparin and that of neomycin. Such studies are complicated by the fact that the inhibition exerted by both compounds is reversible. When present simultaneously their individual effects are abolished, most likely due to ionic interactions between the two compounds.

Although plaque reduction by neomycin (and polylysine) has been observed in several different cell lines, specifically BHK cells, rabbit skin cells, human glioma cells and normal human glial cells (Langeland et al. 1990), the phenomenon is apparently not common to all types of cells. Infection of Hep-2 cells by HSV-1 seems to be impaired also at a step later than adsorption (Garcin et al. 1990). WuDunn and Spear (1989) found that polylysine partially blocked adsorption of radiolabeled HSV-1, but not HSV-2, to Hep-2 cells, but infection by both serotypes were equally inhibited in plaque assay experiments. The authors concluded that polylysine inhibited a postadsorption step for both HSV-1 and HSV-2. Campadelli-Fiume et al. (1990) reported no difference in the sensitivity of infection of Vero cells by HSV-1 and HSV-2 to either neomycin or polylysine. The reason for this variation between cell lines is unknown; a possible explanation would be the presence of slightly different receptors for HSV.

Mapping of a Gene on the Herpes Simplex Virus Genome Conferring the Neomycin Effect

Identification of a Region on the Genome

HSV-1 and HSV-2 intertypic recombinants were used independently by two groups to map the same region on the HSV genome responsible for the effects of polylysine and neomycin (Campadelli-Fiume et al. 1990; Langeland et al. 1990). Both groups confirmed their results by transfecting the relevant fragment of HSV-2 DNA (BamHI A) into the HSV-1 genome such that the recombinant virus became resistant to neomycin. The mapped region contained eight genes, from *UL40* to *UL47*. Since the compounds affected the interaction between ligand and receptor, the gene or genes involved were expected to encode a ligand (or ligands). Only one glycoprotein gene, *UL44* (gC), was present in the mapped region. *UL40* encodes a subunit of ribonucleotide reductase (Preston et al. 1984), *UL42* encodes the polymerase-accessory and DNA-binding protein (Parris et al. 1988; Gottlieb et al. 1990), and *UL47* encodes a tegument protein (McLean et al. 1990). The predicted product of *UL43* is a membrane protein (McGeoch et al. 1988).

Fine Mapping and Localisation of Neomycin Resistance to Glycoprotein C-2

Resistance to neomycin was transferred from HSV-2 to HSV-1 by transfection with a HSV-2 DNA fragment containing the entire gC gene and portions of the flanking genes *UL43* and *UL45* (Campadelli-Fiume et al. 1990) Glycoprotein C was implicated in attachment, since gC-1 negative mutants adsorbed more slowly than the parental strain of HSV-1 to the host cells (Langeland et al. 1990; Herold et al. 1991). Moreover, our unpublished data showed that mutants in genes *UL41* or *UL43* were not different from wild-type virus in their sensitivity to neomycin (Table 1). These findings suggested that the gC gene was of particular interest. However, gC-1 did not appear to confer sensitivity to neomycin or polylysine upon HSV-1. Mutants not specifying gC-1 and various HSV-1 mutants resistant to monoclonal antibodies against gC, were as sensitive to the drug as their parental HSV-1 strain (Campadelli-Fiume et al. 1990; Langeland et al. 1990).

Table 1. Neomycin Resistance of recombinant virus

Recombinants	Plaque-forming units (%)
HSV-1 wild type	7
UL43 deletion mutant	10
17G41	24
17(41)-G41	18
HSV-2 wild type	87

One hundred percent plaque-forming units was defined as the number of plaques obtained without neomycin treatment. Results are means of three separate experiments.

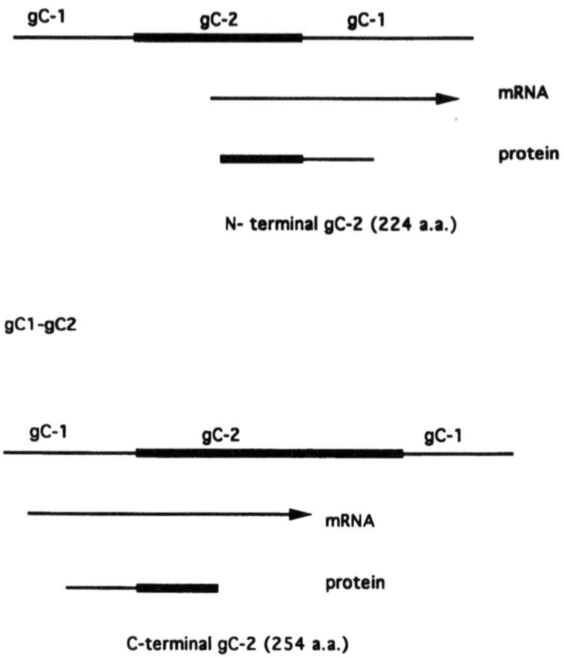

Fig. 2. Map of the chimeric glycoprotein C (gC) in the recombinants *gC1–gC2* and *gC2–gC1* (Dolter et al. 1992). These two recombinants are completely type 1 except for gC, which express the gC-2–gC-1 chimeric proteins with gC-2 in either N- or C-terminal end. The location of RNAs and translated proteins were determined by Northern blotting and mapping with various monoclonal antibodies against the different antigenic sites. gC2–gC1 contains the N-terminal 224 amino acids (*a.a.*) from gC-2, and gC1–gC2 the C-terminal 254 a.a. from gC-2

An explanation of these rather surprising results was suggested by Campadelli-Fiume et al. (1990): the mapping experiments using intertypic recombinant had identified a region on the HSV-2 genome conferring neomycin resistance rather than a region on the HSV-1 genome conferring sensitivity. These authors showed that several gC-negative mutants of HSV-2 were sensitive to the compound. They also observed that HSV-1 became resistant by incorporating into the genome a fragment from HSV-2 spanning the gC gene and a portion of the flanking gene *UL45*. Consistent with these results, we have observed that the reinsertion of the gC-2 gene into a gC-2 deletion mutant of HSV-2 restores the resistance of HSV-2 to neomycin (Øyan et al. 1993).

To map the resistance within the gC-2 gene we used a set of HSV-1-HSV-2 recombinants carrying chimeric gC genes in an otherwise HSV-1 genome (Dolter et al. 1992). As shown in Fig. 2 the gC proteins encoded by these genes were type 2 specific in the N-terminal or C-terminal half. The first recombinant was neomycin resistant, while the other one was neomycin sensitive (Øyan et al. 1991). This result further delimits neomycin resistance to the 224 N-terminal amino acids of HSV-2.

The Role of Glycoprotein C in Adsorption and Glycoprotein C-Independent Adsorption

Taken together, the results described above (Langeland et al. 1987, 1988, 1990; Campadelli-Fiume et al. 1990; Herold et al. 1991) indicate that both gC-1 and gC-2 are involved in attachment of virus to the cell, but their mechanisms of action are different. HSV-1 has one pathway for adsorption which is inhibited by neomycin and polylysine. The ligand involved in this pathway is not gC-1, since deletion of the protein has no effect on the neomycin inhibition (Campadelli-Fiume et al. 1990; Langeland et al. 1990). Adsorption of HSV-2 in the presence of neomycin or polylysine is apparently mediated by gC-2 (Campadelli-Fuime et al. 1990) and involves the N-terminal portion of the glycoprotein (Øyan et al. 1991). A role for other HSV-2 proteins in neomycin resistance was excluded in the experiments, because the chimeric virus contains only a portion of the gC-2 gene in an otherwise HSV-1 genome.

Additional evidence for a role of gC-2 in attachment was obtained from studies showing that a gC-2 deletion mutant adsorbed more slowly than the parental virus (Øyan et al. 1993). However, the fact the deletion mutant does still attach and that deletion of gC-2 (Campadelli-Fiume et al. 1990), or replacement of the N-terminal part of the protein with the homolog from gC-1 (Øyan et al. 1991), rendered the virus sensitive to neomycin clearly indicates that HSV-2 has an additional pathway of attachment.

Evidence exists for separate adsorption mechanisms for HSV-1 and HSV-2. Vahlne et al. (1979) showed that UV-irradiated HSV-1 blocked subsequent adsorption of HSV-1, but not of HSV-2. Addison and coworkers (1984) obtained similar results using a temperature-sensitive mutant of HSV-1, which at nonpermissive temperature blocked adsorption of HSV-1 but not HSV-2. Experiments with polylysine and neomycin, as reviewed above, support this conclusion. Other data, however, suggest a common adsorption mechanism for HSV-1 and HSV-2 (WuDunn and Spear 1989). The data presented in Fig. 3 show that both views may be correct. In agreement with earlier findings HSV-1 inhibits superinfection by HSV-1 but not HSV-2. However, HSV-2 reduces further binding of both HSV-1 and HSV-2. This result can be explained by the existence of a major,

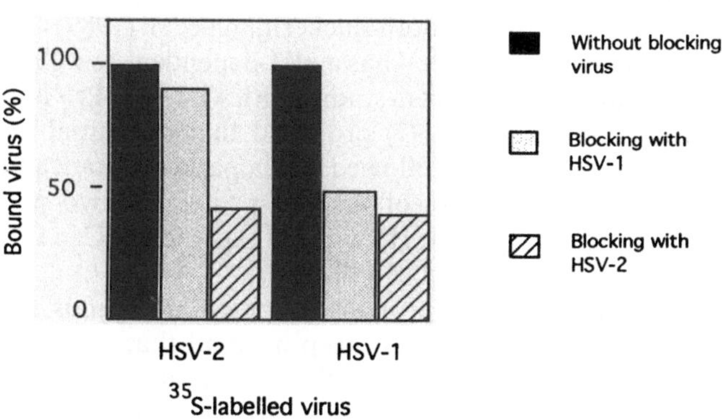

Fig. 3. Adsorption of ^{35}S-labeled HSV type 1 or 2 to baby hamster kidney (BHK) cells after preincubation with unlabeled virus. BHK cells were preincubated at 4°C with unlabeled HSV-1 17^{+} or HSV-2 HG52 at a multiplicity of more than 200 plaque-forming units per cell. After 2 h of incubation the cells were superinfected with partially purified radiolabeled HSV-1 or HSV-2. The bound radioactivity was compared to that obtained without preincubation of unlabeled virus (1500–2000 cpm per sample). The adsorption of virus is given as percentages of maximal binding

neomycin-sensitive receptor for HSV-1 to which also HSV-2 can bind, and by an additional receptor for HSV-2 which is neomycin resistant.

The ability of neomycin to block functional receptors is not unique for HSV. Vassbotn et al. (1992) have recently demonstrated that the receptor for the platelet-derived growth factor (PDGF) on murine embryonic 10T 1/2 cells and porcine aortic endothelial cells expressing PDGF α- and β-receptor can also be blocked reversibly by neomycin.

It is interesting to note the difference between gC-1 and gC-2 in their responce to neomycin and polylysine, and that the resistance of gC-2 is conferred by the N-terminal portion of the protein. There is a high degree of amino acid homology between the two glycoproteins, especially in the carboxyterminal region, but gC-2 is 31 amino acids shorter in the N-terminal end (Frink et al. 1983; Dowbenko and Lasky 1984; Swain et al. 1985; Homa et al. 1986; Seidel-Dugan et al. 1990). Both gC-1 and gC-2 are dispensible for infection in tissue culture (Draper et al. 1984; Zezulak and Spear 1984; Homa et al. 1986; Herold et al. 1991). Although most clinical isolates of HSV-1 express gC, mutants devoid of either gC-1 or gC-2 cause infections in mice (Johnson et al. 1986; Margolis et al. 1992), and gC-negative mutants have occasionally been isolated from humans (Hidaka et al. 1990, 1991). Spear and collaborators (Herold et al. 1991) have suggested that gB may substitute for gC in its absence.

Glycoproteins homologous to gC of HSV are encoded by pseudorabies virus (PrV) (Robbins et al. 1986a) as well as by bovine herpesvirus (BHV) (Liang et al. 1991). Both are designated gIII and neither is essential for virus growth in tissue culture (Robbins et al. 1986 b; Mettenleiter et al. 1990; Liang et al. 1991). They are, however, important for the attachment of virus to the cell, and the functions of gIII of PrV has been studied extensively. Virus mutants devoid of gIII attach more slowly than wild-type virus (Schreurs et al. 1988; Zuckermann et al. 1989; Mettenleiter et al. 1990). Heparin inhibits adsorption of wild-type virus but not of the mutants, and gIII-negative mutants do not compete with wild-type virus for the cellular receptor (Mettenleiter et al. 1990). Both Zuckermann et al. (1989) and Mettenleiter et al. (1990) have suggested that PrV has a gIII-dependent as well as a gIII-independent pathway for attachment. From studies with wild-type PrV and various gIII deletion mutants Zsak et al. (1991) proposed that the initial attachment is reversible and sensitive to heparin, followed by a heparin-resistant and irreversible binding involving the middle part of the gIII protein. A two-step attachment has recently been suggested also for HSV-1, but the role of gC in this process has not been investigated (Fuller and Lee 1992).

Much attention has been paid to glycoproteins as potential ligands for attaching HSV to a cellular receptor. One should keep in mind that the virus produced under conditions giving reduced glycosylation is still able to attach (Campadelli-Fiume et al. 1982; Spivack et al. 1982; Svennerholm et al. 1982; Kuhn et al. 1988;). Nonglycosylated virion proteins may therefore play a role in adsorption. In addition to the known virion glycoproteins there are several genes in the HSV-1 genome encoding potentially membrane-associated proteins (McGeoch et al. 1988). Also virion proteins anchored in ways other than by hydrophobic amino acid sequences must be considered. Thus, although gC-1 is not the neomycin-sensitive ligand, there are many alternative candidates.

References

Cai W, Person S, Warner SC, Zhou J, DeLuca NA (1987) Linker-insertion nonsense and restriction-site deletion mutations of the gB glycoprotein gene of herpes simplex virus type 1. J Virol 61: 714–721

Cai W, Gu B, Person S (1988) Role of glycoprotein B of herpes simplex virus type 1 in viral entry and cell fusion. J Virol 62: 2596–2604

Campadelli-Fiume G, Poletti L, Dall'Olio F, Serafini-Cessi F (1982) Infectivity and glycoprotein processing of herpes simplex virus type 1 grown in a ricin-resistant cell line deficient in N-acetylglucosaminyl transferase I. J Virol 43: 1061–1071

Campadelli-Fiume G, Stirpa D, Boscaro A, Avitabile E, Foa-Tomasi L, Barker D, Roizman B (1990) Glycoprotein C-dependent attachment of herpes simplex virus to susceptible cells leading to productive infection. Virology 178: 213–222

Carney DH, Scott DL, Gordon EA, LaBelle EF (1985) Phosphoinosotides in mitogenesis; neomycin inhibits thrombin-stimulated phosphinositide turnover and initiation of cell proliferation. Cell 42: 479–488

Cockcroft S, Gomperts BD (1985) Role of guanine nucleotide binding protein in the activation of polyphosphoinositide phosphodiesterase. Nature 314: 534–536

Dalgleish AG, Beverley PCL, Clapham RP, Crawford DH, Greaves MF, Weiss RA (1984) The CD4 (T4) antigen is an essential component of the receptor for the AIDS retrovirus. Nature 312: 763–767

Desai PJ, Schaffer PA, Minson AC (1988) Excretion of non-infectious virus particles lacking glycoprotein H by a temperature-sensitive mutant of herpes simplex virus type 1: evidence that gH is essential for virion infectivity. J Gen Virol 69: 1147–1156

Dolter KE, Goins WF, Levine M, Glorioso JC (1992) Genetic analysis of type-specific antigenic determinants of herpes simplex virus glycoprotein C. J Virol 66: 4864–4873

Dowbenko DJ, Lasky LA (1984) Extensive homology between the herpes simplex virus type 2 glycoprotein F gene and herpes simplex virus type 1 glycoprotein C gene. J Virol 52: 154–163

Downes CP, Michell RH (1981) The polyphosphoinositide phosphodiesterase of erythrocyte membranes. Biochem J 198: 133–140

Draper KG, Costa RH, Lee GT-Y, Spear PG, Wagner EK (1984) Molecular basis of the glycoprotein-C-negative phenotype of herpes simplex virus type 1 macroplaque strain. J Virol 51: 578–585

Fingeroth J, Weis J, Tedder T, Strominger J, Bird P, Fearon D (1984) Epstein-Barr virus receptor on human B lymphocytes is the C3d receptor CR2. Proc Natl Acad Sci USA 81: 4510–4516

Frink RJ, Eisenberg R, Cohen G, Wagner EK (1983) Detailed analysis of the portion of the herpes simplex virus type 1 genome encoding glycoprotein C. J Virol 45: 634–647

Fuller AO, Spear PG (1987) Anti-glycoprotein D antibodies that permit adsorption but block infection by herpes simplex virus 1 prevent virion-cell fusion at the cell surface. Proc Natl Acad Sci USA 84: 5454–5458

Fuller AO, Lee W-C (1992) Herpes simplex virus type 1 entry through a cascade of virus-cell interactions requires different roles of gD and gH in penetration. J Virol 66: 5002–5012

Fuller AO, Santos RE, Spear PG (1989) Neutralizing antibodies specific for glycoprotein H of herpes simplex virus permit viral attachment to cells but prevent penetration. J Virol 63: 3435–3443

Garcin D, Masse T, Madjar J-J, Jaquemont B (1990) Herpes simplex virus type-1 immediate-early gene expression and shut off of host protein synthesis are inhibited in neomycin-treated human epidermoid carcinoma 2 cells. Eur J Biochem 194: 279–286

Gottlieb J, Marcy AI, Coen DM, Challberg MD (1990) The herpes simplex virus type 1 UL42 gene product: a subunit of DNA polymerase that functions to increase processivity. J Virol 64: 5976–5987

Greve JM, Davis G, Meyer AM, Forte CP, Yost SC, Marlor CW, Kamarck ME, McClelland A (1989) The major human rhinovirus receptor is ICAM- I. Cell 56: 839–847

Herold BC, WuDunn D, Soltys N, Spear PG (1991) Glycoprotein C of herpes simplex virus type 1 plays a principal role in the adsorption of virus to the cells and in infectivity. J Virol 65: 1090–1098

Hidaka Y, Sakuma S, Kumano Y, Minagawa H, Mori R (1990) Characterization of glycoprotein C-negative mutants of herpes simplex virus type 1 isolated from a patient with keratitis. Arch Virol 113: 195–207

Hidaka Y, Sakai Y, Toh Y, Mori R (1991) Glycoprotein C of herpes simplex virus type 1 is essential for the virus to auade antibody-independent complement-mediated virus inactivation and lysis of virus-infected cells. J Gen Virol 72: 915–921

Highlander S, Sutherland SL, Gage PJ, Johnson DC, Levine M, Glorioso GC (1987) Neutralizing monoclonal antibodies specific for herpes simplex virus glycoprotein D inhibit virus penetration. J Virol 61: 3356–3364

Highlander SL, Cai W, Person S, Levine M, Glorioso JC (1988) Monoclonal antibodies define a domain on herpes simplex virus glycoprotein B involved in virus penetration. J Virol 62: 1881–1888

Homa FL, Purifoy DJM, Glorioso JC, Levine M (1986) Molecular basis of the glycoprotein C-negative phenotypes of herpes simplex virus type 1 mutants selected with a virus-neutralizing monoclonal antibody. J Virol 58: 281–289

Hutchinson L, Browne H, Wargent V, Davis-Poynter N, Primorac S, Goldsmith K, Minson AC, Johnson DC (1992a) A novel herpes simplex virus glycoprotein, gL, forms a complex with glycoprotein H (gH) and affects normal folding and surface expression of gH. J Virol 66: 2240–2250

Hutchinson L, Goldsmith K, Snoddy D, Ghosh H, Graham FL, Johnson DC (1992b) Identification and characterization of a novel herpes simplex virus glycoprotein, gK, involved in cell fusion. J Virol 66: 5603–5609

Johnson DC, McDermott MR, Chrisp C, Glorioso JC (1986) Pathogenicity in mice of herpes simplex virus type 2 mutants unable to express glycoprotein C. J Virol 58: 36–42

Kuhn JE, Eing BR, Brossmer R, Munk K, Braun RW (1988) Removal of N-linked carbohydrates decreases the infectivity of herpes simplex virus type 1. J Gen Virol 69: 2847–2858

Langeland N, Haarr L, Holmsen H (1986a) Evidence that neomycin inhibits HSV-1 infection of BHK cells. Biochem Biophys Res Commun 141: 198–203

Langeland N, Haarr L, Holmsen H (1986b) Polyphosphoinositide metabolism in baby-hamster kidney cells infected with herpes simplex virus type 1. Biochem J 237: 707–712

Langeland N, Holmsen H, Lillehaug JR, Haarr L (1987) Evidence that neomycin inhibits binding of herpes simplex virus type 1 to the cellular receptor. J Virol 61: 3388–3393

Langeland N, Moore LJ, Holmsen H, Haarr L (1988) Interaction of polylysine with the cellular receptor for herpes simplex virus type 1. J Gen Virol 69: 1137–1145

Langeland N, Øyan AM, Marsden HS, Cross A, Glorioso JC, Moore LJ, Haarr L (1990) Localization on the herpes simplex virus type 1 genome of a region encoding proteins involved in adsorption to the cellular receptor. J Virol 64: 1271–1277

Liang X, Babiuk LA, van Drunen Littel-van den Hurk S, Fitzpatrick DR, Zamb TJ (1991) Bovine herpesvirus 1 attachment to permissive cells is mediated by its major glycoproteins gI, gIII and gIV. J Virol 65: 1124–1132

Ligas MW, Johnson DC (1988) A herpes simplex virus mutant in which glycoprotein D sequences are replaced by β-galactosidase sequences binds to but is unable to penetrate into cells. J Virol 62: 1486–1494

Little SP, Jofre JT, Courtney RT, Schaffer PA (1981) A virion-associated glycoprotein essential for infectivity of herpes simplex virus type 1. Virology 115: 149–160

Maddon PJ, Dalgleish AG, McDougal JS, Clapham PR, Weiss RA, Axel R (1986) The T4 gene encodes the AIDS virus receptor and is expressed in the immune system and the brain. Cell 47: 333–348

Margolis TP, Sedarati F, Dobson AT, Feldman LT, Stevens JG (1992) Pathway of viral gene expression during acute neuronal infection with HSV-1. Virology 189: 150–160

Marsden HS (1987) Herpes simplex virus glycoproteins and pathogenesis. Molecular basis of virus disease. Fortieth Symposium of the Society of General Microbiology Cambridge University Press, Cambridge

McGeoch DJ, Dalrymple MA, Davison AJ, Dolan A, Frame MC, McNab D, Perry LJ, Scott JE, Taylor P (1988) The complete DNA sequence of the long unique region in the genome of herpes simplex virus type 1. J Gen Virol 69: 1531–1574
McLean G, Rixon F, Langeland N, Haarr L, Marsden H (1990) Identification and characterization of the virion protein products of herpes simplex virus type 1 gene UL 47. J Gen Virol 71: 2953–2960
Mettenleiter TC, Zsak L, Zuckermann F, Sugg N, Kern H, Ben-Porat T (1990) Interaction of glucoprotein gIII with a cellular heparinlike substance mediates adsorption of pseudorabies virus. J Virol 64: 278–286
Parris DS, Cross A, Haarr L, Orr A, Frame MC, Murphy M, McGeoch DJ, Marsden HS (1988) Identification of the gene encoding the 65-kilodalton DNA-binding protein of herpes simplex virus type 1. J Virol 62: 818–825
Pestka S (1971) Inhibitors of ribosome functions. Annu Rev Biochem 40: 697–710
Preston VG, Palfreyman JW, Dutia BM (1984) Identification of a herpes simplex virus type 1 polypeptide which is a component of the virus-induced ribonucleotide reductase. J Gen Virol 65: 1456–1466
Reid DG, Gajjar K (1987) A proton and carbon 13 nuclear magnetic resonance study of neomycin B and its interaction with phosphatidylinositol 4,5-bisphosphate. J Biol Chem 262: 7967–7972
Robbins AK, Watson RJ, Whealy ME, Hays WW, Enquist LW (1986 a) Characterization of a pseudorabies virus glycoprotein gene with homology to herpes simplex virus type 1 and type 2 glycoprotein C. J Virol 58: 339–347
Robbins AK, Whealy ME, Watson RJ, Enquist LW (1986 b) Pseudorabies virus gene encoding glycoprotein gIII is not essential for growth in tissue culture. J Virol 59: 635–645
Roizman B, Sears AE (1990) Herpes simplex virus and their replication. Virology. Raven, New York, pp 1795–1841
Sarmiento M, Haffey M, Spear PG (1979) Membrane proteins specified by herpes simplex virus. III. Role of glycoprotein VP7 (B2) in virion infectivity. J Virol 29: 1149–1158
Sastrasinh M, Knauss TC, Weinberg JM, Humes HD (1982) Identification of the aminoglycoside binding site in rat renal brush border membranes. J Pharmacol Exp Ther 222: 350–358
Schreurs C, Mettenleiter TC, Zuckermann F, Sugg N, Ben-Porat T (1988) Glycoprotein gIII of pseudorabies virus is multifunctional. J Virol 62: 2251–2257
Seidel-Dugan C, Ponce de Leon M, Friedman HM, Eisenberg RJ, Cohen GH (1990) Identification of C3b-binding regions of herpes simplex virus type 2 glycoprotein C. J Virol 64: 1897–1906
Shieh M-T, WuDunn D, Montgomery RI, Esko JD, Spear PG (1992) Cell surface receptors for herpes simplex virus are heparan sulfate proteoglycans. J Cell Biol 116: 1273–1281
Spivack JG, Prusoff WH, Tritton TR (1982) A study of the antiviral mechanism of action of 2-deoxy-D-glucose: normally glycosylated proteins are not strictly required for herpes simplex virus attachment but increase viral viral penetration and infectivity. Virology 123: 123–138
Staunton DE, Merluzzi VJ, Rothlein R, Barton R, Marlin SD, Springer TA (1989) A cell adhesion molecule, ICAM-1, is the cell surface receptor for rhinoviruses. Cell 56: 849–853
Svennerholm B, Olofsson S, Lunden R, Vahlne A, Lycke E (1982) Adsorption and penetration of enveloped herpes simplex virus particles modified by tunicamycin or 2-deoxy-D-glucose. J Gen Virol 63: 343–349
Swain MA, Peet RW, Galloway DA (1985) Characterization of the gene encoding herpes simplex virus type 2 glycoprotein C and comparison with the type 1 counterpart. J Virol 53: 561–569
Tanner J, Weis J, Fearson D, Whang Y, Kieff E (1987) Epstein-Barr virus gp 350/220 binding to the B lymphocyte C3b receptor mediates adsorption, capping and endocytosis. Cell 50: 203–213
Vahlne A, Svennerholm B, Lycke E (1979) Evidence for herpes simplex virus type-selective receptors on cellular plasma membranes. J Gen Virol 44: 217–225
Vassbotn FS, Østman A, Siegbahn A, Holmsen H, Heldin C-H (1992) Neomycin is a PDGF antagonist which allows discrimination of PDGF α- and β-receptor signals on cells expressing both receptor types. J Biol Chem 267: 15635–15641

Whitaker M, Aitchison M (1985) Calsium-dependent pholyphosphoinositide hydrolysis is associated with exocytosis in vitro. FEBS Lett 182: 119–124

WuDunn D, Spear PG (1989) Initial interaction of herpes simplex virus with cells is binding to heparan sulfate. J Virol 63: 52–58

Zezulak KM, Spear PG (1984) Mapping of the structural gene for the herpes simplex virus type 2 counterparts of herpes simplex virus type 1 glycoprotein C and identification of a type 2 mutant which does not express this glycoprotein. J Virol 49: 741–747

Zsak L, Sugg N, Ben-Porat T, Robbins AK, Whealy ME, Enquist LW (1991) The gIII glycoprotein of pseudorabies virus is involved in two distinct steps of virus attachment. J Virol 65: 4317–4324

Zuckermann F, Zsak L, Reilly L, Sugg N, Ben-Porat T (1989) Early interactions of pseudorabies virus with host cells: functions of glycoprotein gIII. J Virol 63: 3323–3329

Øyan AM, Langeland N, Dolter KE, Glorioso JC, Haarr L, Crumpacker CS (1991) Neomycin resistance maps to the amino terminal region of HSV gC-2. 16th international herpesvirus workshop, Asilomar, USA, 2B–22: 51

Øyan AM, Dolter KE, Langeland N, Goins WF Glorioso JC, Haarr L, Crumpacker CS (1993) Resistance to herpes simplex virus type 2 to neomycin maps to the N-terminal portion of glycoprotein C. J Virol 67: 2434–2441

Chapter 3 Pathogenicity of Glycoprotein C-Negative Herpes Simplex Virus Type 1 in Herpetic Keratitis

Yasushi Toh, Yasufumi Hidaka, Yuji Kumano, Hiroko Minagawa, and Ryoichi Mori

Summary

Recurrent stromal keratitis induced by herpes simplex virus (HSV) type 1 is one of the most important problems in herpetic ocular diseases and has been considered to represent the host's immunopathologic response to the virus. In the pathogenicity of this disease, glycoprotein C (gC) of HSV-1 has been shown to be essential, because it is reported as immunodominant antigen to induce HSV-specific memory cytotoxic T lymphocytes. gC also functions as a receptor of complement C3b and has been thought to play an important role in natural infection of HSV. Recently, we isolated HSV-1 strains from a patient with recurrent herpetic keratitis. Interestingly, these viruses were shown to be gC-negative by using two different anti-HSV-1, gC-specific monoclonal antibodies. The clinical isolation of the gC-negative HSV strain is an extremely rare event. Molecular analyses of the gC-negative phenotype of these HSV-1 strains revealed that a premature termination of the translation due to a point mutation of the nucleotide occurred in the mid portion of the gC open reading frame and consequently, a truncated, immature gC molecule lacking the transmembrane domain was secreted to the extracellular fluid. We also demonstrated the important role of gC in protection of the virion envelope against complement-mediated damage, by using the gC-positive recombinant viruses derived from one of these clinical isolates. It is thus expected that these rare HSV-1 strains will provide us with valuable information concerning the in vivo functions of gC, especially in ocular disease. The requirement of gC in herpetic stromal keratitis is also discussed.

Introduction

Recurrent stromal keratitis induced by herpes simplex virus type 1 (HSV-1) is currently one of the most important problems in herpetic ocular diseases. Although epithelial keratitis, another type of herpetic keratitis, is caused by the replication of HSV-1 on the cornea, stromal herpetic keratitis is thought to be

---

Department of Virology, Faculty of Medicine, Kyushu University, Fukuoka, 812 Japan caused by the host's immune response to the viral antigens, especially by the cell-mediated immune response (Metcalf et al. 1979; Russel et al. 1984). However, the exact mechanism of this disease remains unclear and is still subject to debate. A major part of the immune response to HSV-1 is directed against viral glycoproteins (Spear 1985), which are incorporated into the virion envelope as well as into the membranes of the infected cells. HSV-1 encodes eight glycoproteins (g); gB, gC, gD, gE, gG, gH, gI, and gJ (Roizman and Sears 1990). Among them, the gC of HSV-1 is one of the major dominant immunogens. gC elicits high-titer, virus-neutralizing antibodies (Glorioso et al. 1984) and plays a dominant role in the induction of HSV-1 specific cytotoxic T lymphocytes (CTL; Glorioso et al. 1985; Rosenthal et al. 1987). Hendricks et al. (1989) reported that these CTL responses to HSV-1 contributed to induce the destruction of the cornea, i.e., severe stromal keratitis, and gC-negative HSV-1, a poor inducer of CTL responses, was also a poor inducer of stromal disease. In addition, gC functions as a receptor for the complement component C3b (Friedman et al. 1986) and inhibits both the complement-mediated cytolysis and complement-mediated neutralization in the presence of antibody (Harris et al. 1990). These findings suggested that gC may play an important role in natural infection in vivo.

Despite these findings, we recently isolated gC-negative HSV-1 strains from a patient with recurrent keratitis (Kumano et al. 1990). These viruses were among the first clinical isolates of gC-negative HSV (Kawana et al. 1991) and are expected to provide us with useful information for studying the roles of gC in the viral pathogenesis in human hosts. In this paper, we will describe the isolation and characterization of these viruses and the results of the study on the interaction between the complement and gC on the viral envelope; we will also discuss the role of gC in the pathogenesis of herpetic ocular diseases.

Experimental Approaches and Results

Clinical Isolation and Characterization of Glycoprotein C-Negative Herpes Simplex Virus Type 1 Strains

Isolation of Herpes Simplex Virus Type 1 TN Strains from a Patient with Recurrent Herpetic Keratitis

From 1987 to 1989, we repeatedly isolated HSV-1 strains 8724TN, 8802TN, and 8904TN (abbreviated as TN1, TN2, and TN3, respectively) from one patient (a 35-year-old male patient in 1987) with recurrent herpetic keratitis of his right eye at intervals of about 1 year by inoculation of Vero cells with lesion scrapings (Kumano et al. 1990). This patient showed a severe stromal keratitis at each episode. This stromal keratitis did not respond to topical treatment with acyclovir, but responded well to the systemic administration of corticosteroids.

The cytopathic effect in Vero cells of TN1, TN2, and TN3 resembled that caused by the HSV-1 KOS strain, and these viruses did not produce any plaque on chick embryonal fibroblasts (data not shown). Judging from these data, the TN strains were thought to be HSV-1. Furthermore, these viruses were proven to be HSV-1 by a direct immunofluorescence test on the infected Vero cells using anti-HSV-1 monoclonal antibody (MAb) conjugated with fluorescein isothiocyanate of BioMerieux Herpes Monoclonal Kit (BioMerieux Co., France), which could recognize the HSV-1 KOS strain, but not the HSV-2 YS-4 strain as controls (Table 1; Hidaka et al. 1990). A restriction endonuclease analysis of the viral DNAs of TN strains also revealed that these were HSV-1 (data not shown).

Herpes Simplex Virus Type 1 TN Strains Are Deficient in Glycoprotein C

To our surprise, an immunofluorescence test using the Syva MicroTrak Herpes anti-HSV-1 MAb (Syva Co., USA) did not recognize any of these TN strains. As this MAb was known to be specific to HSV-1 gC (gC-1), these TN strains were thought to lack gC-1 or to have an alternate epitope of gC-1 (Hidaka et al. 1990). To further confirm the gC-negative phenotype of the TN strains, we performed an immunoreactive plaque assay, the so-called black plaque assay, as previously described by Holland et al. (1983), by using anti-gC-1-specific MAb HC1, which was kindly provided by Dr. Lenore Pereira. The plaques of the HSV-1 KOS strain were all clearly immunostained by HC1, whereas those of the TN strains were never stained by the same MAb (Table 1). These results suggested that the TN strains did not express gC-1 reactive with at least two different MAbs against gC-1 on their infected cell surfaces.

Table1. Direct immunofluorescence test and immunoreactive plaque assay on Vero cells infected with herpes simplex virus (HSV)

HSV strains	MicroTrak Herpes[a]		Herpes monoclonal kit[b]		Anti-gC-1[c] HC1
	HSV-1	HSV-2	HSV-1	HSV-2	
TN1	–	–	+	–	–
TN2	–	–	+	–	–
TN3	–	–	+	–	–
KOS	+	–	+	–	+
YS-4	–	+	–	+	–

gC, glycoprotein C; HC1, anti-gC-1-specific monoclonal antibody from Dr. Pereira.
[a] Syva Kit.
[b] BioMerieux Kit.
[c] Immunoreactive plaque assay (black plaque assay).

Molecular Analysis of Glycoprotein C-Negative Phenotype of Herpes Simplex Virus Type 1 TN Strains

Nucleotide Sequencing of Glycoprotein C Genes of TN Strains

To analyze the molecular basis for the inability of these clinical isolates to express gC, we determined the nucleotide sequences of gC genes of TN strains. A SalI-BamHI fragment R-I of HSV-1 TN1 strain which contained the whole open reading frame of the gC gene was subcloned (pTN1-gC; Fig. 1). Then a deletion mutant containing PstI-BamHI region of pTN1-gC was constructed (pTN1-gC-Pdel). The nucleotide sequence of this PstI-BamHI region was determined for both strands using the dideoxynucleotide chain termination method after the construction of nested sets of the deletion mutants of pTN1-gC-Pdel with exonuclease III and S1 nuclease (Toh et al. 1993). The nucleotide sequence of the TN1-gC gene region thus determined was then compared with that of the KOS strain (Frink et al. 1983). The sequence of the putative promoter region of the TN1 gC gene was very well conserved. In accordance with this, an RNA blot analysis revealed that the gC gene of TN1 strain was transcribed as efficiently as the KOS-gC gene (data not shown). In the open reading frame, there were 15 differences in the 1533 base pairs, ten of which changed the amino acid sequence. The most remarkable change was observed at the codon of the 280th amino acid of TN1-gC, where the codon CAG for glutamine was converted to TAG for the termination codon. This change

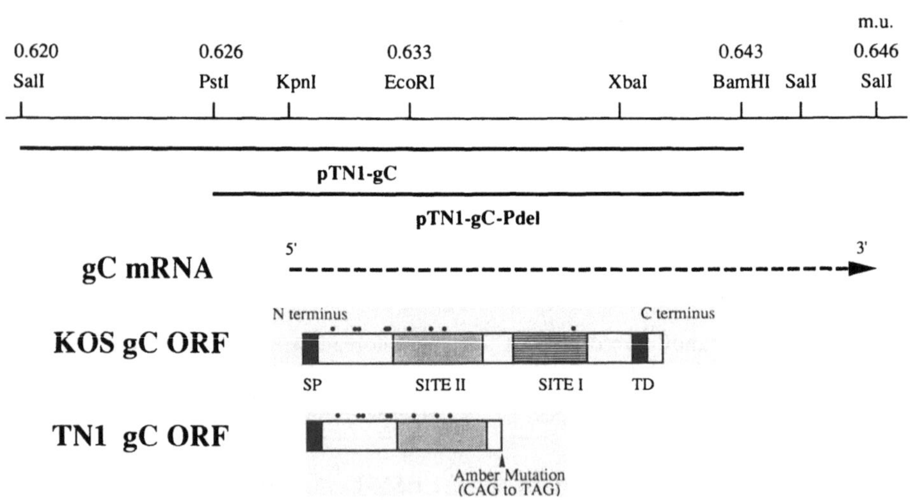

Fig. 1. The glycoprotein C (*gC*)-negative phenotype of HSV-1 TN strains. The location of the plasmid clones, *pTN1-gC* and *pTN1-gC-Pdel*, are indicated with a part of the restriction endonuclease cleavage map and the map units (mu) of HSV-1 genome. A *broken line* shows the major gC-specific mRNA. The open reading frames (*ORF*) of the gC of KOS and TN1 are shown. The regions for the signal peptide (*SP*) and the transmembrane domain (*TD*) are shown as well as the two major antigenic sites. (*site I*, amino acid residues 297–383; *site II*, amino acid residues 127–249; Wu et al. 1990). *Dots* above ORFs indicate the sites for N-glycosylation (Frink et al. 1983). TN1 strain produces a truncated gC molecule caused by an amber mutation (CAG to TAG) at the 280th codon

was also observed in the TN2- and TN3-gC genes. The appearance of a termination codon in the mid-portion of the open reading frame would result in a premature termination of gC translation (Fig.1).

Immunoprecipitation Assay of Glycoprotein C Molecule Produced by KOS and TN1

To confirm the results deduced from the nucleotide sequence, a radioimmunoprecipitation analysis was done by using anti-gC-1 MAb HC1 (data not shown). After the infection of Vero cells with HSV-1 KOS or TN1 at a multiplicity of infection of 5, the cells were incubated in the medium containing [^{14}C]glucosamine hydrochloride to label the carbohydrate moieties of the viral glycoproteins. After a 16-h incubation, an immunoprecipitation assay was performed using anti-gC-1 specific MAb HC1 against the cell media and the cell lysates (Toh et al. 1993). The immune precipitates were subjected to sodium dodecyl sulfate (SDS) polyacrylamide gel, and the gel was autoradiographed on X-ray film. KOS strain produced a mature gC with a relative molecular weight of approximately 120 kDa. It was cell-associated, but rarely recognized in the cell medium. In contrast, a distinct product was precipitated in the medium for TN1 infection at around 80–90 kDa, whereas it was rarely detected in the cell lysate. This smaller molecular weight was compatible with the predicted one of the TN1-gC molecule, when a premature termination and conservation of the N-glycosylation sites (eight out of nine sites; Frink et al. 1993) were taken into consideration (Fig. 1). These results suggested that the TN1 strain produced a truncated gC molecule, which was secreted to the medium because of a lack of the transmembrane domain.

Herpes Simplex Virus Type 1 Glycoprotein C is Essential for the Virus to Evade Antibody-Independent, Complement-Mediated Virus Inactivation

Construction and Isolation of Glycoprotein C-Positive Recombinant TN Strains

gC has been considered to be essential for HSV-1 in vivo, because it functions as a receptor for the complement component C3b and confers on the virus a means of modifying the activity of certain human complement components, thereby evading the immune recognition by this mechanism. However, HSV-1 TN strains isolated from clinical specimens lacked gC on the infected cells and also presumably on the virion envelope. This led us to examine whether an intact gC modifies the interaction between the TN strains and complement in the absence of antibody. We constructed gC-positive recombinant viruses in which the intact gC gene of KOS strain was inserted into the TN1 virus genome (Hidaka et al. 1991). Briefly, the gC gene of the KOS strain was inserted into 5′ noncoding region of the thymidine kinase gene of the KOS strain and the resultant recombinant plasmid clone was designated as pTK1.0-gC. Whole viral DNA of TN1 strain and pTK1.0-gC DNA were introduced into the Vero cells with the calcium phosphate

Table 2. Summary of antibody-independent, complement-mediated virus inactivation assays

HSV strains	Source of sera[a] and treatments					
	Guinea pig	Guinea pig +heat	Rat	Rat + heat	Rat + anti-C3 (1:1 dilution)	Human
KOS	0.19 ± 0.05[b]	/[c]	/	/	/	/
TN 1	1.05 ± 0.12	0.07 ± 0.02	2.70 ± 0.05	0.06 ± 0.05	0.24 ± 0.14	0.61 ± 0.18
C-1	0.20 ± 0.06	/	/	/	/	/
C-2	0.19 ± 0.01	/	/	/	/	/
C-3	0.22 ± 0.06	0.11 ± 0.01	0.93 ± 0.00	0.02 ± 0.11	/	0.05 ± 0.08
C-5	0.15 ± 0.03	/	/	/	/	/

HSV, herpes simplex virus.
[a] Sera from guinea pig (Hartley), rat (Wistar), and sero-negative human.
[b] The rate of reduction in virus titer: $-\log_{10}$ (titer of virus surviving/virus titer of control).
[c] Not determined

coprecipitation method. A selection of the acyclovir-resistant mutants from the progenies of these transfections was performed, and five clones, designated C-1, C-2, C-3, C-4, and C-5 were thus obtained. Successful recombination was confirmed by restriction endonuclease digestion pattern followed by Southern blotting analysis and also by an immunofluorescence test of the recombinant-infected Vero cells using Syva's anti-gC-1 MAb.

Complement-Mediated Virus Inactivation Assay

Using the resultant recombinant viruses C-1, C-2, C-3, and C-5, we carried out a complement-mediated virus inactivation assay to determine the sensitivity of cell-free virus to the complement with sera from various species either with or without pretreatments (Table 2). As a source of complement, sera were collected from guinea pig (Hartley), rat (Wistar), and human (sero-negative healthy male). The method was fully described in our previous report (Hidaka et al. 1991). The rate of reduction of virus titer after incubation of the viruses with serum was calculated as $-\log_{10}$ (titer of virus surviving/virus titer of control).

The reduction rates in KOS and TN1 virus titers were 0.19 ± 0.05 and 1.05 ± 0.12, respectively, which showed that KOS was resistant and TN1 was sensitive to the action of the complement. The rates of reduction in titers of C-1, C-2, C-3, and C-5 were 0.20 ± 0.06, 0.19 ± 0.01, 0.22 ± 0.06, and 0.15 ± 0.03, respectively, which were significantly lower than that of TN1 ($p < 0.01$ for C-2 and C-5; $p < 0.05$ for C-1 and C-3 by Student's *t*-test). Inactivation of the C-3 by complement was not observed using human serum either. Although the rat serum inactivated C-3 despite the expression of gC, the rate of inactivation of C-3 was far less than that of TN1. We then examined whether the inactivation of the viruses by the sera was directly mediated by the complement (Table 2). The guinea pig and rat sera lost their ability to inactivate when they were heated at 56°C for 30 min. The rat serum pretreated with anti-rat complement C3 goat serum also lost its inactivation activity. Furthermore, TN-1-infected Vero cells were lysed by rat complement, while the C-3-infected ones were not, suggesting that gC protected the virus from damage by complement not only on the virus envelope, but also the infected cell surface (data not shown).

Discussion

The gC of HSV-1 functions as a receptor for the complement component C3b (Friedman et al. 1986). gC has a marked protective effect on the complement inactivation of free virus and also blocks the effect of complement of the infected cells (Harris et al. 1990; Hidaka et al. 1991). These findings suggest that protection by gC against complement-mediated damage is most important for in vivo natural infection of HSV-1. HSV-1 TN strains are extremely rare clinical isolates (Kumano et al. 1990) and were isolated from a patient with recurrent herpetic keratitis. These viruses also showed virulence toward mice (Hidaka et al. 1990).

The reason why these strains were isolated from a human is still unknown. The presence of a number of functional complement components within the cornea has been demonstrated by Mondino et al. (1980) and Brawman-Mintzer et al. (1988). TN strains and the cells infected with them will be susceptible to complement-mediated damage, because little or no gC is retained on the infected cell membrane and possibly on the virion envelope (Table 1, Fig. 1), although it is possible that the secreted gC binds to the complement component C3b topically around the infected cells.

Herpetic stromal keratitis by HSV-1 is a common cause of blindness. This disease is generally considered to represent an immunopathologic response. Clinically, corneal stromal lesions respond more favorably to corticosteroids and immunosuppressive therapy than to antiviral drugs that inhibit HSV replication (Boisjoly et al. 1984). Experimentally, athymic nude mice failed to develop stromal lesions following a corneal infection of HSV-1 (Metcalf et al. 1979), while nude mice could be made susceptible to stromal lesions by the adoptive transfer of HSV-specific CTLs (Russel et al. 1984). Furthermore, Hendricks et al. (1989) showed in a mouse model that HSV-1-specific CTLs contributed to the destruction of corneal tissue associated with herpetic stromal keratitis, although delayed-type hypersensitivity did not play an essential role. The gC of HSV-1 is thought to play an important role in the inductive phase as well as the recognition phase as target antigen of the CTL response (Glorioso et al. 1985; Rosenthal et al. 1987). Martin et al. (1989) also reported that gC, but not gB or gD, served as a class I-restricted target antigen for inducing HSV-specific CTL response in mice. These findings were reinforced by Hendricks et al. (1989), who demonstrated that the infection of the corneas of mice with gC-negative HSV-1 resulted in a reduced CTL response and a concomitant reduction in susceptibility to severe stromal keratitis.

In contrast to the facts mentioned above, gC-negative TN strains of HSV-1 were isolated from a patient with recurrent herpetic keratitis (Kumano et al. 1990). This patient developed discform stromal keratitis, which suggested the marked infiltration of T lymphocytes, and the patient responded well to systemic corticosteroids, but not to topical acyclovir treatment. At present, we cannot clearly explain why there was such a discrepancy in the requirement of gC between the previous experimental data in animal models and the actual isolation from a human. Truncated gC molecules of TN strains that are secreted to the extracellular space may still induce HSV-1-specific CTL response, because one of the two major antigenic regions is retained (Fig. 1; Wu et al. 1990). However, the induced CTLs will not be able to recognize the cells infected with TN strains because of the lack of gC on the cell membrane. Our results bring into question the requirement of gC for the induction of herpetic stromal keratitis in a human host. Preliminarily, even in mice, the TN1 strain can also cause stromal keratitis with marked infiltration of mononuclear cells (Sakai et al., manuscript in preparation). Recently, in the mouse model of HSV infection, many groups have focused on the potential of particular glycoproteins to induce CTLs, and their results are conflicting. Some groups have detected gB-specific CTL in some, but not all, haplotypes (Witmer et al. 1990). A gD-specific CTL clone was isolated from an H-2^k mouse (Johnson et al. 1990), although several groups have been unable to detect it (Blacklaws et al. 1987; Witmer et al. 1990). Furthermore, Martin et al. (1988) suggested that the immediate-early gene products might represent 20%–35% of the target antigens and that virion components serve as minor target antigens.

HSV can cause acute retinal necrosis, another devastating ocular disease in man (Lewis et al. 1989). A host immune response also seems to be involved in the pathogenesis of this disease (Atherton et al. 1991). Hemady et al. (1989) reported that when HSV-1 was inoculated into the anterior chamber of mice eyes, gC of HSV-1 was required for developing necrotizing chorioretinitis in the contralateral (uninoculated) eyes. In contrast, here again, the HSV-1 TN1 strain could cause necrotizing chorioretinitis in both the inoculated and uninoculated eyes of BALB/c mice in spite of the deficient gC (Liu et al. 1993). Thus, the role of gC in the pathogenesis of herpetic ocular diseases such as stromal keratitis and retinal necrosis remains to be further investigated. For this purpose, the very rare clinical isolates of the HSV-1 TN strains are expected to provide us with valuable information.

Acknowledgements. The authors thank Dr. Ying Liu, Dr. Shinji Tanaka, and Dr. Yuichiro Sakai, Department of Virology, Faculty of Medicine, Kyushu University, for helpful discussions and suggestions and Mr. Brian T. Quinn for his critical comments. This work was supported in part by a grant from the Yakult Honsha Co., Ltd. (Tokyo, Japan).

References

Atherton SS, Kanter MY, Streilein JW (1991) ACAID requires early replication of HSV in the injected eye. Curr Eye Res 10: 75–80

Blacklaws BA, Nash AA, Darby G (1987) Specificity of the immune response of mice to herpes simplex virus glycoprotein G and D constitutively expressed on L cell lines. J Gen Virol 68: 1103–1114

Boisjoly HM, Woog JJ, Pavan-Langston D, Park N-H (1984) Prophylactic topical cyclosporin in experimental herpetic stromal keratitis. Arch Ophthalmol 102: 1804–1807

Brawman-Mintzer O, Mondino BJ, Mayer FJ (1988) The complement system in sclera. Invest Ophthalmol Vis Sci 29: 1756–1758

Friedman HM, Glorioso JC, Cohen GH, Hastings JC, Harris SL, Eisenberg RJ (1986) Binding of complement component C3b to glycoprotein gC of herpes simplex virus type 1: mapping of gC-binding sites and demonstration of conserved C3b binding in low-passage clinical isolates. J Virol 60: 470–475

Frink RJ, Eisenberg R, Cohen G, Wagner EK (1983) Detailed analysis of the portion of the herpes simplex virus type 1 genome encoding glycoprotein C. J Virol 45: 634–647

Glorioso J, Schroder CH, Kumel G, Szczesiul M, Levine M (1984) Immunogenicity of herpes simplex virus glycoprotein gC and gB and their role in protective immunity. J Virol 50: 805–812

Glorioso J, Kees U, Kumel G, Kirschner H, Krammer PH (1985) Identification of herpes simplex virus type 1 (HSV-1) glycoprotein gC as the immunodominant antigen for HSV-1-specific memory cytotoxic T lymphocytes. J Immunol 135: 575–582

Harris SL, Frank I, Yee A, Cohen GH, Eisenberg RJ, Friedman HM (1990) Glycoprotein C of herpes simplex virus type 1 prevents complement-mediated cell lysis and virus neutralization. J Infect Dis 162: 331–337

Hemady R, Opremcak EM, Zaltas M, Berger A, Foster CS (1989) Herpes simplex virus type-1 strain influence on chorioretinal disease patterns following intracameral inoculation in Igh-1 disparate mice. Invest Ophthalmol Vis Sci 30: 1750–1757

Hendricks RL, Tao MSP, Glorioso JP (1989) Alteration in the antigenic structure of two major HSV-1 glycoproteins, gC and gB, influence immune regulation and susceptibility to murine herpes keratitis. J Immunol 142: 263–269

Hidaka Y, Sakuma S, Kumano Y, Minagawa H, Mori R (1990) Characterization of glycorprotein C-negative mutants of herpes simplex virus type 1 isolated from a patient with keratitis. Arch Virol 113: 195–207

Hidaka Y, Sakai Y, Toh Y, Mori R (1991) Glycoprotein C of herpes simplex virus type 1 is essential for the virus to evade antibody-independent complement-mediated virus inactivation and lysis of virus-infected cells. J Gen Virol 72: 915–921

Holland TC, Sandri-Goldin RM, Holland LE, Marlin SD, Levine M, Glorioso JC (1983) Physical mapping of the mutation in an antigenic variant of herpes simplex virus type 1 by use of an immunoreactive plaque assay. J Virol 46: 649–652

Johnson RM, Lancki KW, Fitch FW, Spear PG (1990) Herpes simplex virus glycoprotein D is recognized as antigen by $CD4^+$ and $CD8^+$ T lymphocytes from infected mice: characterization of T cell clones. J Immunol 145: 702–710

Kawana T, Suzuki R, Hashido M, Sugase M, Sakaoka H, Kurata T (1991) Detection rate of herpes simplex virus type 1 unreactive to the monoclonal antibody (MicroTrak) among genital herpes patients (in Japanese). Clin Virol 19: 376–380

Kumano Y, Yamamoto M, Inomata H, Sakuma S, Hidaka Y, Minagawa H, Mori R (1990) Recurrent herpetic keratitis: failure to detect herpes simplex virus infection using the Syva MicroTrak™ HSV1/HSV2 direct specimen identification/typing test. Ophthalmolgia 201: 169–173

Lewis ML, Culbertson WW, Post MJD, Miller D, Kokame GT, Dix RD (1989) Herpes simplex virus type 1. A cause of the acute retinal necrosis syndrome. Ophthalmolgia 96: 875–878

Liu Y, Sakai Y, Minagawa H, Toh Y, Ishibashi T, Inomata H, Mori R (1993) Induction of bilateral retinal necrosis in mice by unilateral intracameral inoculation of a glycoprotein-C deficient clinical isolate of herpes simplex virus type 1. Arch Virol 129: 105–118

Martin S, Courtney RJ, Fowler G, Rouse BT (1988) Herpes simplex virus type 1-specific cytotoxic T lymphocytes recognize virus nonstructural proteins. J Virol 62: 2265–2273

Martin S, Cantin E, Rouse BT (1989) Evaluation of antiviral immunity using vaccinia virus recombinants expressing cloned genes for herpes simplex virus type 1 glycoproteins. J Gen Virol 70: 1359–1370

Metcalf JF, Hamilton DS, Reichert RW (1979) Herpetic keratitis in athymic (nude) mice. Infect Immun 26: 1164–1171

Mondiano BJ, Ratajczak HV, Goldberg DB, Schanzlin DJ, Brown SI (1980) Alternate and classical pathway components of complements in the normal cornea. Arch Ophthalmol 98: 346–349

Roizman G, Sears AE (1990) Herpes simplex viruses and their replication. In: Fields BN, Knipe DM, Chanock RM, Hirsch MS, Melnick JL, Monath TP, Roizman B (eds) Virology 2nd edn. Raven, New York, pp 1795–1841

Rosenthal KL, Smiley JR, South S, Johnson DC (1987) Cells expressing herpes simplex virus glycoprotein gC but not gB, gD, gE are recognized by murine virus-specific cytotoxic T lymphocytes. J Virol 61: 2438–2447

Russell PG, Nasisse MP, Larsen HS, Rouse BT (1984) Role of T-lymphocytes in the pathogenesis of herpetic stromal keratitis. Invest Ophthalmol Vis Sci 25: 938–944

Spear PG (1985) Glycoproteins specified by herpes simplex virus. In: Roizman B (ed) The herpesviruses 3. Plenum, New York, pp 315–356

Toh Y, Tanaka S, Liu Y, Hidaka Y, Mori R (1993) Molecular characterization of naturally occurring glycoprotein C-negative herpes simplex virus type 1. Arch Virol 129: 119–130

Witmer LA, Rosenthal KL, Graham FL, Friedman HM, Yee A, Johnson DC (1990) Cytotoxic T lymphocytes specific for herpes simplex virus (HSV) studied using adenovirus vectors expressing HSV glycoproteins. J Virol 71: 387–396

Wu C-TB, Levine M, Homa F, Highlander SL, Glorioso JC (1990) Characterization of the antigenic structure of herpes simplex virus type 1 glycoprotein C through DNA sequence analysis of monoclonal antibody-resistant mutants. J Virol 64: 856–863

Chapter 4 Structure and Function of Glycoprotein D of Herpes Simplex Virus

R. J. Eisenberg[1,3], D. Long[2,3], D. L. Sodora[2,3], H.-Y. Chiang [2,3], W. C. Wilcox[1,3], W. R. Abrams[2,3], M. I. Muggeridge[2,3], and G. H. Cohen[2,3]

Summary

Glycoprotein D (gD) is a structural component of the herpes simplex virus (HSV) envelope which is essential for virus penetration into host cells. The function of this protein is highly dependent on its structure, and its structure, in turn, is dependent on maintenance of three intact disulfide bonds. Our goal has been to relate structural features of this glycoprotein to its function in virus infection. We have constructed a detailed antigenic map of the protein. A large panel of gD mutants was assayed for gD function using a complementation assay. At least one antigenic site, which binds antibodies with high titers of virus-neutralizing activity, overlaps a portion of the functional domain of gD. In addition, we constructed mutants which lacked one or all three sites for addition of N-linked oligosaccharides (N-CHO). We found that a triple mutant lacking all three N-CHO sites is able to function normally in virus infection and when recombined back into the virus genome had very little effect on the phenotype of the resulting recombinant. Thus, it appears that N-CHO are not necessary for gD function, though there were some changes in gD structure. In contrast, removal of any of the cysteine residues which participate in disulfide bonding has drastic effects on gD structure and function. Using a combination of genetic and biochemical approaches, we showed that the disulfide bond pattern of gD is Cys 1 (aa 66) to Cys 5 (aa 189), Cys 2 (aa 106) to Cys 6 (aa 202), and Cys 3 (aa 118) to Cys 4 (aa 127). We postulate that a similar disulfide bond arrangement exists in other gD homologs and that gD is part of a superfamily of herpes proteins that includes HSV gI (and its homologs) and HSV gG (and its homologs), all with similar disulfide bonding to gD.

Introduction

Apologia

This review summarizes current studies concerning the herpes simplex virus (HSV) envelope glycoprotein D (gD) in terms of its structure and function. gD is

[1] School of Veterinary Medicine, University of Pennsylvania, Philadelphia, PA 19104, USA
[2] School of Dental Medicine, University of Pennsylvania, Philadelphia, PA 19104, USA
[3] Center for Oral Health Research, University of Pennsylvania, Philadelphia, PA 19104, USA

essential for virus infection and cell–cell spread, but its mechanism of action remains elusive. Since many of the references apply to work carried out in our laboratory, we ask the indulgence of others in the field who have contributed to the overall understanding of gD and other HSV glycoproteins. We emphasize that this article is not a literature review of HSV glycoproteins in general, nor of gD function, nor its use as a vaccine, and direct the reader to excellent reviews on these subjects (Burke 1992, 1993; Dix 1987; Meignier 1985; Muggeridge et al. 1990a; Spear 1985a,b, 1993).

Background

HSV is a major human pathogen which causes a variety of diseases including cold sores, eye and genital infections, neonatal infections, and encephalitis (Hill 1985; Roizman and Sears 1987; Stevens 1989). Most people have antibodies against the oral form of the virus, HSV-1 (Goldstein and Nowinski 1986), and a significant percentage of adults are infected with the genital form, HSV-2 (Breinig et al. 1990; Johnson et al. 1989). Both HSV-1 and HSV-2 establish lifelong latent infections and reactivate periodically to produce recurrent infections (Hill 1985; Meier and Straus 1992; Roizman and Sears 1987; Stevens 1989). Thus, these viruses represent significant morbidity problems.

The HSV virion has a double-stranded DNA genome of 150 kilobases. There are at least 72 open reading frames, of which at least 11 code for glycoproteins (McGeoch et al. 1985,1988; Spear 1993). The DNA genome is encased in an icosahedral protein capsid, which is surrounded by a protein layer called the tegument, and then by a lipid envelope containing the 11 (or more) virion glycoproteins. At time of infection, several of these glycoproteins act singly or in concert to bind HSV to the susceptible cell and then trigger pH-independent fusion between the viral envelope and the cell plasma membrane (Spear 1993). After infection, virus-encoded glycoproteins are expressed on all cellular membranes, including the plasma membrane. Because they are present in the virion envelope and on infected cell plasma membranes, the glycoproteins are major targets of the host's immune response.

gD is an essential protein which is required for virus penetration into susceptible cells (Fuller and Spear 1987; Highlander et al. 1987). It is strongly implicated in receptor binding, cell fusion, and neuroinvasiveness (Fuller and Lee 1992; Fuller et al. 1989; Highlander et al. 1987; Izumi and Stevens 1990; Johnson et al. 1989,1991; Johnson and Ligas 1988; Ligas and Johnson 1988; Minson et al. 1986; Noble et al. 1983). Furthermore, expression of gD on the surface of cells transformed with the gD gene interferes with HSV infection (Arsenakis et al. 1988; Campadelli-Fiume et al. 1988,1990; Johnson and Spear 1989).

Years ago, we speculated that since gD could stimulate virus-neutralizing antibodies in animals, it might represent a viral subunit important for a human vaccine against HSV (Cohen et al. 1972,1978; Eisenberg et al. 1982b). This, in part, drove our efforts to understand the synthesis and processing of gD, to evaluate the properties of a variety of immunological reagents, and to develop methods that could be adapted for large-scale purification of gD (Cohen et al. 1978,1980; Eisenberg et al. 1979,1980,1982b,1987; Matthews et al. 1983; Muggeridge et al. 1990a). As mentioned above, gD, when injected into test animals, induced virus-neutralizing antibodies (Berman et al. 1983; Cohen et al. 1972,1978; Eisenberg et al. 1982a,b; Lasky et al. 1984). The vaccinated animals are protected from HSV infection and establishment of latency (Burke et al. 1989; Chan 1983; Dix and Mills 1985; Eisenberg et al. 1985a; Freifeld et al. 1990; Lasky et al. 1984; Long et al. 1984; Martin et al. 1992; Mishkin et al. 1991; Paoletti et al. 1984; Rooney et al. 1989; Wachsman et al. 1987). At present, two human viral glycoprotein subunit vaccines, both of which contains gD, are in clinical trials in the United States (Burke 1992,1993; Burke et al. 1989; Freifeld et al. 1990; Mertz et al. 1990; Mishkin et al. 1991; Straus et al. 1991,1993). This, in turn, has spurred a "cottage" industry searching for the mechanisms of protection induced by this protein (Lausch et al. 1989,1991; Martin and Rouse 1987; Mishkin et al. 1991; Torseth et al. 1987; Wyckoff et al. 1988; Zarling et al. 1986a,b).

The Goal of this Paper

In this article, we will outline biochemical, genetic, and immunological studies which address basic questions about the structure–function relationship of gD. It is our hope that what we have learned will have an impact on: (a) the successful use of gD as a vaccine and (b) understanding how the interaction of gD with other viral and cell components enables the virus to enter susceptible cells. Here, we will consider four questions that have been the focus of our most recent work on gD. (1) What is the antigenic profile of gD, and in particular, which epitopes are involved in virus neutralization and what structural information is provided from locating epitopes? (2) Which sites on gD are important for its function, and how do these sites relate to its antigenic structure? (3) What is the contribution of the N-linked oligosaccharides (N-CHO) and cysteine residues to the structure and function of gD? (4) What is the disulfide bond structure of this protein, and what implications does this structure have for gD, its homologs, and for other HSV glycoproteins? Answers to these questions have led us to develop a model of gD structure, which is periodically revised as we gain further insight. Clearly, our ultimate modeling goal is to know the structure of gD as solved by X-ray crystallographic means.

Structural Properties of Glycoprotein D

General Structural Features

Figure 1 details some structural features of gD-1 and gD-2. After removal of the 25-amino acid signal sequence (Eisenberg et al. 1984), gD-1 has 369 amino acids (Watson et al. 1982) and gD-2 has 368 (Lasky and Dowbenko 1984; Watson 1983). The two proteins are 85% homologous, and both have three N-CHO

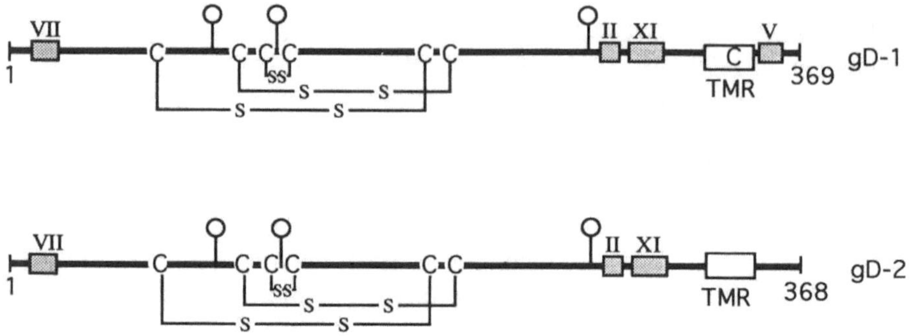

Fig. 1. Stick models of glycoprotein D (*gD*)-*1* and *gD-2*. gD-*1* and gD-2 contain 369 and 368 amino acids, respectively. N-linked glycosylation sites (N-CHO) are shown as *balloons*, and cysteine residues are indicated with a *C*. The seventh cysteine is present only in gD-1 in the predicted transmembrane region (*TMR; empty box*). Also indicated are the positions of four continuous antigenic sites: *VII*, *II*, and *XI* are type common; site *V* is type-1 specific

(Cohen et al. 1983; balloons), two or three O-linked oligosaccharides (Serafini-Cessi et al. 1988), and six cysteine residues (Lasky and Dowbenko 1984; Watson 1983; Watson et al. 1982) in their extracellular domain. A seventh cysteine residue is located in the hydrophobic membrane-anchoring domain of gD-1 (but not gD-2) near the carboxy terminus of the protein (Watson et al. 1982). The first six cysteines are strictly conserved in both number and spacing in gD-1 and gD-2 (lasky and Dowbenko 1984; Watson 1983; Watson et al. 1982). A very similar pattern of cysteine spacing is also found in gD homologs of several α-herpes viruses as well as Marek's disease virus (Bennett et al. 1992; Flowers et al. 1991; Petrovskis et al. 1986a; Ross and Binns 1991; Ross et al. 1991; Tikoo et al. 1990). It is of interest that at least one α-herpesvirus, varicella zoster (VZV) lacks a gD homolog (Davison et al 1986; Davison and McGeoch 1986; Davison and Wilkie 1983), even though this glycoprotein is essential for virus entry of other α-herpesviruses. This implies that gD's function has been assumed by another glycoprotein or that the mechanism of entry of this virus proceeds in a fundamentally different manner than that of other α-herpesviruses. Perhaps some of the differences in pathogenesis of the two viruses relate to the absence of gD in VZV (Straus et al. 1993).

Synthesis and Processing Glycoprotein D

gD is maximally synthesized between 4 and 6 h postinfection (Cohen et al. 1980; Eisenberg et al. 1980; Johnson and Spear 1982; Pizer et al. 1980; Spear 1976). Both precursor and product forms are present in infected cell extracts; gD-1 is first synthesized as a lower molecular weight precursor, pgD, which is post-translationally modified to the product form, gD (Cohen et al 1978; Eisenberg et al. 1979). Processing of gD-1 involves the addition of a high-mannose core oligosaccharide to the nascent polypeptide (Cohen et al. 1980; Eisenberg et al.

1979; Matthews et al. 1983; Pizer et al. 1980). This is followed by another step of processing in which the core oligosaccharide is modified by the addition of other carbohydrates, as well as sialic acid to form a complex-type oligosaccharide (Eisenberg et al. 1979). gD-2 is processed similarly to gD-1 (Eisenberg et al. 1980). Extracellular virions contain only the mature form of gD, whereas intracellular virions contain a mixture of precursor and product forms (H.-Y. Chiang, G. H. Cohen and R. J. Eisenberg, unpublished data). Preparations of immunoaffinity-purified gD-1, but not gD-2, contain a dimer (resistant to sodium dodecyl sulfate, SDS, and reducing conditions) with an apparent relative molecular mass of 128 kDa. This molecule is not present in infected cell extracts (Eisenberg et al. 1982; Spear 1985a), and it is still not known whether gD function involves a monomeric or oligomeric form of the glycoprotein. Other gD homologs have also been observed to exist as homodimers in infected cells (Babiuk et al. 1987; Hughes et al. 1988; Petrovskis et al. 1986a; Van Drunen Littel-Van Den Hurk et al. 1984).

Grouping of Monoclonal Antibodies

The antigenic structure of proteins has been studied for several reasons: (a) in the absence of knowledge of the three-dimensional structure of the protein, it can help determine those amino acids which are located on the surface of the protein: (b) it defines areas involved in the immune response; and (c) it can relate the function of the protein to its physical structure. Function is initially determined by showing that a particular monoclonal antibody (MAb) blocks a step in viral entry; if this occurs in the absence of complement, it is thought that neutralization involves antibody binding at or near the functional site of the protein. Prior work using a panel of MAb has provided a detailed picture of the major antigenic sites of gD and the epitopes within those sites (Eisenberg et al. 1982; Highlander et al. 1987; Muggeridge et al. 1990a; Showalter et al. 1981). Our grouping of MAb utilizes a hierarchical tree, the top-most branch dividing MAb that recognize continuous versus discontinuous epitopes (Cohen et al. 1984, 1985,1986; Dietzschold et al. 1984; Eisenberg and Cohen 1986; Eisenberg et al. 1982a,1985b; Isola et al. 1989; Muggeridge et al. 1988,1990a). Continuous epitopes consist of a linear array of amino acids (those which are insensitive to denaturation), whereas discontinuous epitopes are those which are brought together as a result of more complex protein folding (those which are sensitive to protein denaturation). MAb which recognize discontinuous epitopes are particularly suitable probes for detecting subtle conformational changes in the protein brought about by either mutational alterations or variations in physicochemical conditions. A key technical development for studying and locating discontinuous epitopes on gD was the development of a "native" SDS-polyacrylamide gel system (Cohen et al. 1986). This separation method avoids the use of high SDS concentrations, heat, or reducing agents such as β-mercaptoethanol and separates proteins by their native molecular size (by use of a low concentration of SDS). The proteins in the gel are then transferred to nitrocellulose by the conventional techniques of western blotting (Burnette 1981) and probed with MAb to either continuous or discontinuous epitopes.

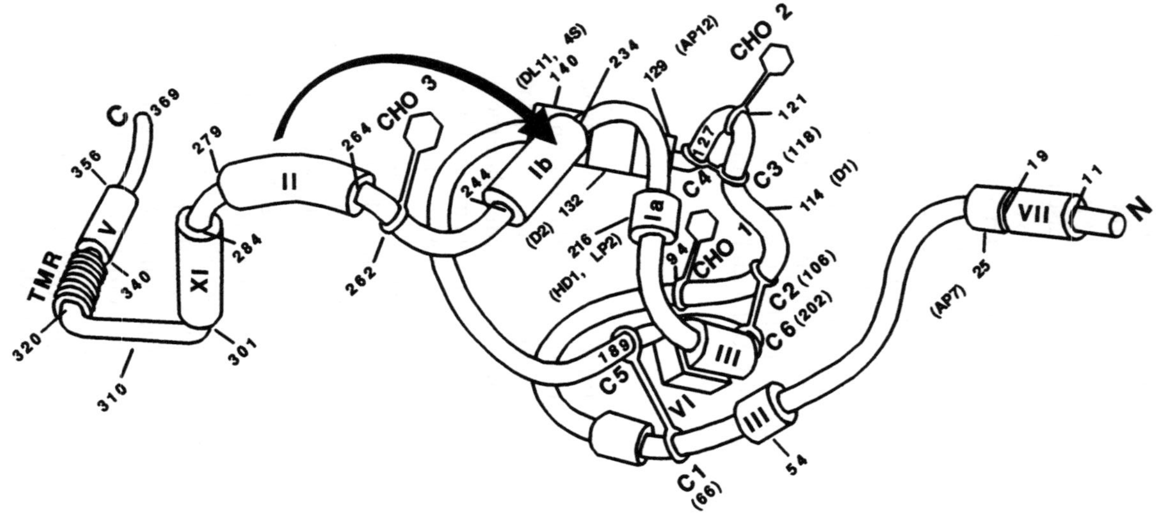

Fig. 2. Structural model of glycoprotein D (gD)-1. This model, adapted from Long et al. (1992), depicts our current concept of the folded form of gD. Both the N terminus and C terminus are depicted as *extended structures* and contain the continuous epitopes VII (residues 11-19), II (residues 264–279), XI (residues 284–301), and *V* (residues 340–356) mapped so far. The first six cysteines (C1–C6) form three disulfide bridges: *C1* (aa 66) bonded to *C5* (aa 189, *C2* (aa 106) bonded to *C6* (aa 202), and *C3* (aa118) bonded to *C4* (aa 127). These bonds are critical for maintaining a complex folded structure containing the discontinuous epitopes. Some of these are shown as; *Ia*, *Ib*,*III*, and *VI*. N-linked oligosaccharides (N-CHO) are designated as *hexagonal balloons*. *TMR*, transmembrane region

Continuous and Discontinuous Antigenic Sites on Glycoprotein D

Localization of a variety of continuous and discontinuous antigenic sites as well as biochemical studies of the protein, particularly resolving the disulfide bond structure of gD (Long et al. 1992), led to the model of gD structure shown in Fig. 2. There are at least 12 defined antigenic sites on the protein (Highlander et al. 1987; Muggeridge et al. 1990a), and our current understanding of where some of them are located is shown. The continuous epitopes within antigenic sites VII, II, XI, and V have been precisely mapped to residues 11–19, 264–279, 284–301, and 340–356, respectively (Cohen et al. 1984; Dietzschold et al. 1984; Isola et al. 1989; Matthews et al. 1983). Three discontinuous sites are also shown (I, III, and VI; Cohen et al. 1986; Muggeridge et al. 1990a). In some cases, we have indicated the position of an amino acid in an epitope corresponding to a specific (and sometimes ungrouped) MAb, such as AP7, for which a monoclonal antibody-resistant mutation was identified at position 25 (Minson et al. 1986). To position other antigenic sites, we examined the properties of mutant forms of gD that were unable to bind MAb within a particular group. For example, site III includes residues downstream as well as upstream of residue 205. A mutation at residue 54 oblates the binding of two group III MAb, but binding of other group III MAb is unaffected (D. Sodora, G. Cohen and R. Eisenberg, unpublished data). In the model for gD (Fig. 2), we placed residue 54 in proximity with residues just downstream of aa 205 to indicate that group III MAb could contact both parts of gD.

Assumptions Concerning the Interpretation of Mutagenesis Studies

In assessing the effect of mutations on gD function, we interpret the data based on the following assumptions: (a) if a mutant protein is both structurally and functionally intact, then the site (residue) altered by that particular mutation is not involved in gD function; (b) if a protein has lost its ability to function, but shows global changes in conformation (inability to bind MAb to discontinuous epitopes) or alterations in transport abilities, the importance of this site for gD function cannot be assessed; one cannot tell whether the loss of function is associated with a change at that particular site or is merely the consequence of an overall conformational change; (c) if the mutation does minimal damage or no detectable harm to the overall conformation or processing, yet its function is abolished, then the site of that mutation is deemed important for function.

Detailed Mapping of Antigenic Site I of Glycoprotein D

Antigenic site I (initially subdivided into sites Ia and Ib based on the properties of monoclonal antibody resistant (mar) mutants (Muggeridge et al. 1988; Fig. 2) is recognized by MAb in group I, of which seven have currently been identified (Muggeridge et al. 1988). Two of these, HD1 and LP2, are in group Ia and the other five (DL11, 4S, D2, III-174, and III-114 are in group Ib (Cohen et al. 1986; Fuller and Spear 1985; Minson et al. 1986; Pereira et al. 1980; Showalter et al. 1981). Each MAb blocks the binding of the others within the group, and all seven bind to discontinuous epitopes that are conserved between gD-1 and gD-2 (Muggeridge et al. 1988). Binding of group I MAb results in neutralization of virus infectivity, presumably due to interference with the function of gD in virus penetration (Eisenberg et al. 1982a; Fuller and Spear 1985; Kümel et al. 1985; Minson et al. 1986; Pereira et al. 1980; Showalter et al. 1981). In addition, these MAb block cell-to-cell spread of virus and also cell fusion by syncytial strains (Fuller and Spear 1987; Highlander et al. 1987; Noble et al. 1983). It therefore seemed likely that site I overlaps a functional site on the protein. We have used a variety of truncation, deletion, and mar mutants to characterize this site.

Using the native gel-western blot procedure, we found that group III MAb such as 11S reacted with a truncated form of gD, resides 1-233, whereas DL11, in group I, did not react with this protein, but did react with a truncated form ending at residue 275 (Cohen et al. 1986). Additional truncation and deletion mutants were created in the gD gene, the mutations were cloned into a eukaryotic expression vector, and the corresponding proteins were then produced in transiently transfected cells. Reaction of group I MAb on western blots of native gels suggested they could be divided into the Ia (HD1 and LP2) and the Ib (DL11, 4S, D2, III-114, and III-174), since only the Ia MAb reacted with a deletion mutant lacking amino acids 234–244 (Δ234–244; Muggeridge et al. 1988). Analysis of truncation mutants also confirmed this division, since only the Ia MAb reacted with a mutant ending at residue 233 (unpublished results). In fact, the shortest fragment which reacts with Is MAb ends at residue 226, whereas the Ib MAb require amino acids from 1–258 (W. Wilcox, C. Handler, M. Muggeridge, G. Cohen and R. Eisenberg, unpublished data). Five of the group I MAb were used to select mar mutant viruses (Highlander et al. 1987; Minson et al. 1986; Muggeridge et al. 1990b), and cross-neutralization experiments demonstrated that mutants selected with Ia MAb were still neutralized by Ib MAb and vice versa (Muggeridge et al. 1988). DNA sequencing showed that both Ia MAb selected a Ser to Asn substitution at aa 216 (Minson et al. 1986; Muggeridge et al. 1990c), consistent with results from the analysis of truncation mutants. Substitutions selected by the Ib MAb were Gln-132 to Leu (D2) and Ser-140 to Asn (DL11 and 4S), identifying an upstream portion of site I. Moreover, the change at 140 created a new site for attachment of an N-CHO which is used (Muggeridge et al. 1988). Removal of that carbohydrate restored some binding of DL11 and 4S, indicating that the carbohydrate prevents MAb binding to gD, in part, by steric hindrance.

In Antigenic Site I part of a Functional Site?

Evidence to support the hypothesis that site I overlaps a functional site on the protein was obtained by constructing a gD mutant containing all three mar mutations (132, 140, and 216) that had been selected with group I MAb; despite the fact that the corresponding protein folded correctly when expressed in transfected cells, the gene could not be recombined intact into the virus genome.

Instead, only the change at aa 216 was retained, implying that the combination of mutations was lethal (Muggeridge et al. 1990c). Further characterization of the functional activity of gD mutants relied on development and use of a complementation assay.

This assay takes advantage of a gD-minus virus called F-gDβ, in which the gD gene has been replaced by that for β-galactosidase (Johnson and Ligas 1988; Ligas and Johnson 1988). Because gD is essential for virus replication, the virus can be propagated only in cells that synthesize gD, such as the VD60 cell line (Johnson and Ligas 1988; Ligas and Johnson 1988). When stocks of FgDβ virus produced in VD60 cells and used to infect non-gD-expressing cells (such as Vero cells), virions are produced, but they lack gD protein and are noninfectious (Johnson and Ligas 1988; Ligas Johnson 1988). However, infectious virions can be obtained from Vero cells if the cells are first transfected with a plasmid expressing gD and then infected with F-gDβ (Muggeridge et al. 1990c). Progeny virions produced by these cells are genotypically gD-minus, but phenotypically they are gD-plus pseudotypes, since the virions incorporate the plasmid-encoded gD protein and are therefore able to enter VD60 cells. Subsequent replication in these cells and incorporation of gD form the VD60 cells allows for virus spread and formation of plaques. The number of plaques resulting from transfection with a mutated gD gene is compared with the number obtained after transfection with the wild-type gene, and the results are usually expressed as percentage of wild-type complementation activity. This percentage is related to the functional activity of the mutant protein. Analysis of genes containing mar mutations agreed with earlier results, in that individual mutations had little effect on complementation, whereas the 140–216 double mutant and the 132–140–216 triple mutant had very low activity (Muggeridge et al. 1990c). In addition to Vero cells, a number of other cell lines such as L cells and COS-1 cells have been used and give higher levels of complementation with wild-type gD (Long et al. 1992; H.-Y. Chiang, and M. Muggeridge, unpublished data).

Combining the complementation assay with an antigenic analysis of mutant proteins (as a measure of the effect of the mutation on gD conformation) has been a powerful approach for mapping of sites involved in gD function. Of several in-frame deletion mutants, those lacking discontinuous epitopes had no complementation activity (Muggeridge et al. 1990b). However, the remainder, with only local conformational changes, were split between those with activity and those without. Thus, the N terminus of the protein (aa 7–21) and the cytoplasmic tail (aa 338–369) were dispensable, whereas aa 234–244 were essential (Muggeridge et al. 1990b). Since absence of these latter residues also abolished binding of group Ib MAb, this result suggests that a portion of site I does indeed overlap a portion of the functional site on gD. Using recombination as an assay for function, a similar region, residues 231–235, was implicated in gD function by Feenstra et al. (1990). To further map the functional site, we constructed a series of deletion mutants of gD, each lacking three residues within the region 220–254, and found that aa 222–251 are required for gD function (Muggeridge et al. 1992). Also, insertion of four amino acids at positions 235, 243, or 246, but not at 257, abolished gD function (Chiang et al. 1991; Muggeridge et al. 1992). In addition, the smaller deletion mutants indicated that antigenic site Ib contains at least two discontinuous elements, one of which comprises aa 222–230. This suggests that neutralization by group Ib antibodies is due to the blocking of the functional site. Furthermore, preliminary analysis of other mutants suggests that the functional site may encompass other parts of the protein which are brought together as a result of folding. Thus, function is abolished by insertions between residues 25 and 43 or by deletion of aa 277–310 (Chiang et al. 1991; Muggeridge et al. 1992). It is important to point out that this latter downstream region is missing at least in part from at least one gD truncation proposed for use as a human subunit vaccine candidate (Berman et al. 1985; Lasky et al. 1984). Parallel studies on MAb binding indicate overlap between functional and antigenic sites (Muggeridge et al. 1992) and suggest that the structural model in Fig. 2 should be modified to bring the N and C termini closer to the mid-region of gD.

What is the Contribution of N-Linked Oligosaccharides to Antigenic Structure and Function of Glycoprotein D?

Properties of Mutants lacking One or All Three N-Linked Oligosaccharide Sites on Glycoprotein D

gD contains three predicted recognition sites for the addition of N-CHO, and all are utilized (Cohen et al. 1983; Matthews et al. 1983). We used site-directed mutagenesis to alter each site individually, and a fourth mutant was altered at all three sites (Sodora et al. 1989, 1991a). The mutated genes were inserted into an expression vector and analyzed. We found that the N-CHO moieties individually and in combination are not essential for the proper processing or surface expression of gD in transfected cells. No significant change in protein structure could be detected in the proteins carrying mutations (substituting Ala for Thr or Ser), which eliminated N-CHO site 2 or site 3. However, substitution of Ala for Thr-96 at N-CHO site 1 eliminated the first N-CHO addition signal, but the amino acid change itself adversely affected protein structure. However, substitution of Gln for Asn-94 at N-CHO site 1 had a less deleterious effect. Thus, we prepared a triple mutant (QAA) which eliminated all three signals and tested the effect of these changes on gD function. Despite changes in antigenic conformation, each mutant form of gD complemented the infectivity of FgDβ. Thus, the absence of all N-CHO had no detectable effect on gD function.

Properties of a Mutant Virus Containing Glycoprotein D Lacking All Three N-Linked Oligosaccharides

These studies predicted that viruses containing gD without N-CHO would be able to replicate normally in cell culture. We carried out recombination experiments (Sodora et al. 1991b) and isolated a virus called F-gD(QAA), which contained the triply mutated form of gD lacking all signals for addition of N-CHO. F-gD(QAA) and the wild-type virus F-gD(WT) were remarkably similar in all properties, except that F-gD(QAA) produced plaques that were 50% smaller. In addition, the gD protein that was detergent extracted from cells infected with F-gD(QAA) was antigenically altered. However, flow cytometry analysis showed that the gD protein that was transported to the cell surface during infection was antigenically intact. We concluded from this study that the N-CHO present on gD are not required for its biological role in HSV infection, but contribute to the structure of gD and to efficient plaque formation and cell-to-cell spread. Although this protein lacks N-CHO, it contains O-linked oligosaccharides (Sodora et al. 1991b). Further studies will be necessary to assess the contribution of these modifications to gD structure and function. In preliminary studies, we found that F-gD(QAA) is able to establish latency and is reactivated from mice in a normal fashion (R. Tal-Singer, G. H. Cohen, N. Fraser, and R. J. Eisenberg, manuscript in preparation). Our results predict that the gD isolated and purified from F-gD(QAA)-infected cells may not function as efficiently as wild-type gD as a subunit vaccine due to its altered antigenic structure. However, the protein might be more stable if it is secreted from the cell and recovered in the absence of detergents.

What is the Contribution of Cysteine Residues to the Structure and Function of Glycoprotein D?

Disulfide bond Structure of Glycoprotein D

Earlier, we found that gD conformation and the efficacy of gD as vaccine was strictly dependent on intact disulfide bonds. Reduced and alkylated gD stimulated production of non-neutralizing antibodies and did not afford protection (Cohen et al. 1992 and unpublished data). Furthermore, this molecule no longer reacted with any MAb to discontinuous epitopes. In contrast, boiling gD in SDS under nonreducing conditions did not destroy several discontinuous epitopes. We concluded that the function of gD is highly dependent on its structure, and its structure is absolutely dependent on maintenance of intact disulfide bonds. Thus, it was important to solve its disulfide bond arrangement.

Mutational Approach for Solving Disulfide Bond Structure

We constructed internal deletion mutations of gD. Those which eliminated any one of the first six Cys residues resulted in a protein with profoundly altered structure (Cohen et al. 1988). Several deletions outside of this Cys grouping were less harmful (Cohen et al. 1988; Feenstra et al. 1990) and some resulted in infectious mutant virus (Feenstra et al. 1990; Muggeridge et al. 1990b). To systematically analyze the importance of each Cys, we constructed seven mutations, specifically altering each Cys to Ser (mutants called Cys-1–Cys-7), thereby causing a genetic reduction of disulfide bonds (Wilcox et al. 1988). The mutated genes were each cloned into an expression vector and the proteins transiently expressed in mammalian cells. The protein expressed by Cys-7 was quite similar to wild-type gD in both structure and function. In contrast, alteration of any of the first six Cys residues had profound effects on conformation and carbohydrate processing, suggesting that all six are important for gD structure and all participate in intramolecular disulfide bonds (Cohen et al. 1992; Long et al. 1990). Elimination of disulfide bonds often increases the susceptibility of proteins to heat denaturation, and several Cys mutants were temperature sensitive (ts) for several properties including function (Cohen et al. 1992; Long et al. 1990). In that study, we utilized similarities in phenotypic properties to predict a possible arrangement of disulfide bonds in gD; however, the properties of the mutants were different enough to lead us to question whether predictions based on the properties of single-Cys mutations were valid. For example, it is possible that individual mutants of a suspected pair will exhibit different phenotypes, as was the case with the Cys-3 and Cys-6 mutants, because the free half-cysteines interact aberrantly. Therefore, we re-examined the problem in two ways: (1) by creating double rather than single-Cys mutants and (2) by solving the disulfide bond structure using a biochemical approach (Long et al. 1992). First, we constructed 15 double-Cys mutants of gD-1, which would represent all possible arrangment of six Cys residues into three disulfide bonds. We hypothesized that if the double mutation eliminated cysteines from two different pairs (not normally bonded), the resulting protein could have only one of the three correct disulfide bonds and would probably lack native confomation. Twelve mutants would fall into this category and would serve as negative controls. If, however, the two mutations eliminated both cysteines in a proper pair, the protein would still have two disulfide bridges and might exhibit conformation similar to that of wild-type gD. Only three double mutants would be expected to exhibit this phenotype. Furthermore, these three mutants should account for all six cysteines that normally form the three disulfide bridges. The mutational data confirmed this hypothesis.

Three double mutants of gD-1, Cys-1,5, Cys-2,6, and Cys-3,4 were able to fold into a structure resembling that of the native configuration (based on MAb binding, lack of aggregation, and transport to the cell surface) and these mutants suggested the disulfide bond pattern of gD in Figs. 1 and 2. This pattern differed from that predicted by the single Cys mutations (Cohen et al. 1992; Long et al. 1990) and therefore it was important to confirm the pattern by standard biochemistry. We also found that all three double-Cys mutants were ts for both structure and function, indicating that all three disulfide bonds are needed for gD stability.

Biochemical Approach for Solving Disulfide Bond Structure

Using the biochemical approach outlined in Fig. 3, we showed that the disulfide bonds of gD-2 are those shown in Fig. 1, and thus the two methods for solving disulfide bond structure agreed with each other (Long et al. 1992). The strategy was to fragment purified gD-2 with cyanogen bromide and proteases and then to separate the disulfide-bonded peptides and carry out direct N-terminal sequence analysis. The results were precisely the same as derived by genetic analysis, and this

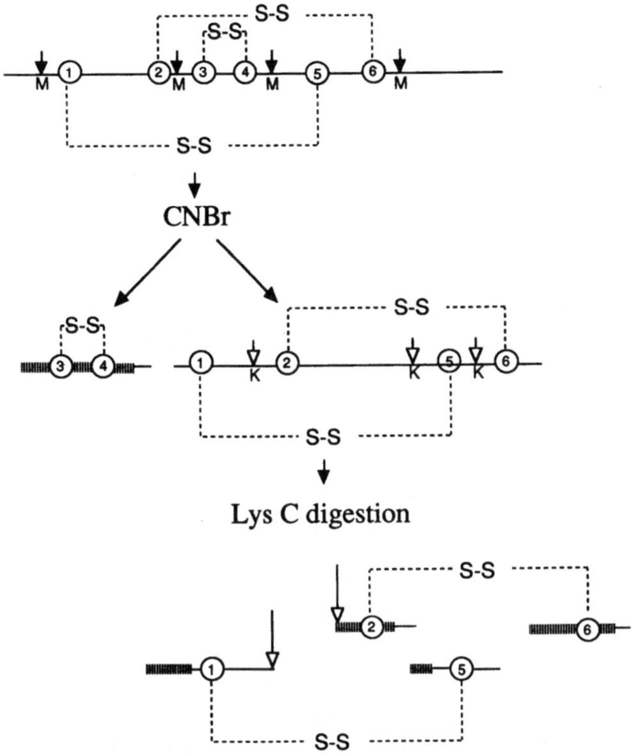

Fig. 3. Biochemical analysis of the disulfide bridges of herpes simplex virus glycoprotein gD-2. The *stick figure* at the top depicts the region of gD-2 that encompasses the six cysteines (numbered *1–6* and *encircled*) that form three disulfide bonds. The arrangement of *S–S* bonds predicted from genetic analysis fo gD-1 is indicated by *dotted lines*. The position of methionine residues (*M*) are indicated. gD-2 labeled with [^{35}S] cysteine was digested with cyanogen bromide (*CNBr*) and the pieces separated by high-pressure liquid chromatography (HPLC). Two fragments containing the cysteine label were separated. One fragment contained cysteines 3 and 4 in a single peptide. The *heavy lines* indicate the residues identified by N-terminal sequencing. A second fragment containing the other four cysteines was digested with lysine C endoproteinease (*Lys C*), and two dipeptides were separated by HPLC. The position of lysine residues (*K*) are indicated. Each dipeptide was sequenced and the *heavy lines* indicate the residues identified. The results indicate that the disulfied bond pattern of gD-2 is: C3 bonded to C4,C1 bonded to C6, and C2 bonded to C5. Thus, this pattern is the same as found by mutational analysis of gD-1 (Long et al. 1992)

gives us confidence that the genetic approach using double-Cys mutants gives Valid results. Furthermore, the biochemical data emphasize that the disulfide bond structure of gD-1 and gD-2 is the same.

Our Newest Concept of Glycoprotein D Structure with Its Three Disulfide Bonds

The study of disulfide bonding of gD provided definitive evidence that the six cysteine residues in the ectodomain of both gD-1 and gD-2 form three disulfide bonds, and the pairing arrangement for both glycoproteins is: Cys 1 (aa 66) bonded to Cys 5 (aa 189), Cys 2 (aa 106) bonded to Cys 6 (aa 202), and Cys 3 (aa 118) bonded to Cys 4 (aa 127). This information allows us to place some "structural" restraints on our current" three-dimensional" model for gD (Fig. 2). Thus, the three disulfide linkages stabilize the secondary and territory structure of the protein and form a complex folded central portion of the protein. Aa 262–315, which are downstream of Cys-6 (aa 202) and bounded on the C-terminal side by the transmembrane region (TMR) anchor (coil, aa 316–337) are portrayed in an extended "stalk" fashion, based on the presence of a string of continuous epitopes in this region (Isola et al. 1989). However, in preliminary studies, we have identified at least one discontinuous epitope downstream of aa 259 (Muggeridge et al. 1992). All of the discontinuous epitopes are stabilized by the presence of the three disulfide bonds; therefore, it is of some interest that residues considerably down stream of Cys 6 at 202 (in the linear sequence) are affected by disulfide bonding. These results suggest that a portion of this downstream stalk region may be folded closer to upstream residues than in shown in the model (Fig. 2).

The three N-CHO are indicated as hexagonal balloons, but the location of the two putative O-linked oligosaccharides are not shown (Serafini-Cessi et al. 1988). Two functional regions of gD are located within aa 231–244 and 277–310 (Feenstra et al. 1990; Muggeridge et al. 1990b, 1992). Again, both regions are outside the central core of disulfide bonds, and one is within the putative stalk region. The significance of this latter finding is now under study in our laboratory. Preliminary data suggests that the region from 277 to 310 is part of the functional site of gD (H.-Y. Chiang, R. J. Eisenberg and G. H. Cohen, manuscript in preparation). Based on the properties of mar mutants, the region from 132 to 140 is important for antigenic site I and is illustrated as overlapping other residues of sites Ia and Ib located further downstream.

The "3-4" Loop of Glycoprotein D Created by One Disulfide Bond

The bonding of Cys-3 (aa 118) to Cys 4 (aa 127) creates a small loop of ten amino acids (Fig. 2). Interestingly, a β-turn is predicated to occur regardless of the presence or absence of the disulfide bond (Chou and Fasman 1974a,b; Long et al. 1992). However, our data would suggest that the stability of this loop is quite different when the disulfide bond is missing, since Cys 3,4 is ts. It is likely that the 3–4 loop and adjacent residues are exposed on the outside of gD. First, the N-CHO at site 2 is within the loop. In addition, several mutations selected with gD-neutralizing MAb (*mar* mutations; Holland et al. 1983) are located immediately adjacent to the loop. These changes include: (a) Gln-114 to Lys obtained using MAb D1 (P. Gage and J. Glorioso, personal communication); (b) Ile-129 to Thr obtained using MAb AP12 (Minson et al. 1986); (c) Gln-132 to Leu obtained using D2 antibody (Highlander et al. 1987; Muggeridge et al. 1990a); and (d) Ser-140 to Asn obtained using 4S or DL11 (Cohen et al. 1986; Muggeridge et al. 1990a; Showalter et al. 1981). The hypothesis that the 3–4 loop has some functional importance is supported by the properties of a deletion mutant d14 (Feenstra et al. 1990). The protein lacked the entire 3–4 loop, i.e., residues 118–127, was unable

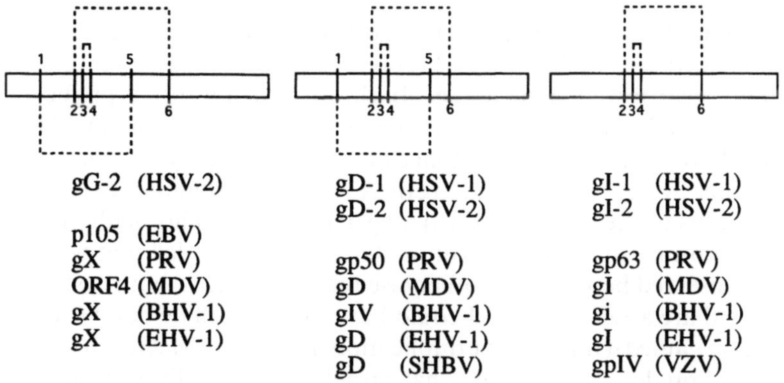

Fig. 4. The glycoprotein D (gD) superfamily of proteins. The *bars* indicate gG-2, gD, and gI of HSV-1 in the order in which they are found in the unique short region of the genome (McGeoch 1990). The disulfide bond arrangement for gD-1 and gD-2 is known. Based on the spacing of the cysteine residues, similar bond arrangement is proposed to exist for gG-2 and gI. In the latter case, the C1–C5 bond is predicted to be missing. Homologs of each of these proteins thus far identified in other herpesviruses are shown under each glycoprotein. We speculate that each will have a similar pattern of disulfide bonding and that the resulting structures will resemble that of gD to form a superfamily of structural homologs

to fold properly, and failed to produce viable recombinants. In view of the ts nature of the Cys-3,4 double mutation, it would be interesting to examine the d14 protein expressed at 31.5°C.

Disulfide Bonding of Glycoprotein D Homologs in Other Herpesviruses

Petrovskis et al. (1986a) first pointed out the homology of sequences between cys-1 and -6 of pseudorabies virus (PRV) gp50 and HSV gD. Since then, other animal herpesviruses were shown to contain a gD homolog with remarkable conservation within this region, including gIV of bovine herpesvirus type 1, Marek's disease virus gD, equine herpesvirus type 1 gD, and simian herpesvirus B gD (Bennett et al. 1992; Flowers et al. 1991; Ross and Binns 1991; Ross et al. 1991; Tikoo et al. 1990). A similar disulfide bond arrangement is likely to be found in all of the gD homologs (Fig. 4), since the spacing of the six cysteine residues is so highly conserved. In support of this proposal, a β-turn is predicted to occur between Cys-3 and 4 of PRV gp50 (Chou and Fasman 1974a,b; Long et al. 1992). However, biochemical analysis will be required to verify this prediction.

Is there a Glycoprotein D "Superfamily" of Proteins with a Similar Pattern of Disulfide Bonding?

At least five HSV glycoproteins are clustered in the unique short region of the HSV genome. McGeoch (1990) noted that the Cys2-Cys-3-Cys-4 region of gD is conserved in HSV gG and gI, and he suggested that these arose by gene duplication. We would note that in each of these other glycoproteins, a fourth Cys is located further downstream. Since Cys-3 is bonded to Cys-4 in gD, we wonder whether this loop is conserved in gG and gI. In this regard, a β-turn is predicted to occur in gI between the two cysteines analogous to Cys-3–Cys-4 of gD (Chou and Fasman 1974a,b; Long et al. 1992). Since gI only has four Cys residues, the bonding of the other two cysteines (assuming there are two disulfide bonds) would have to be analogous to the Cys-2–Cys-6 arrangement of gD. This prediction is supported by Thornton's analysis (Thornton 1981). She concluded that when a disulfide is not conserved between two otherwise related proteins, both cysteines that form the bond are absent. Thus, gI could have lost a disulfide bond (the 1–5) or gD could have gained it during gene duplication. Extending McGeoch's 1990 analysis, we would predict, for example, that proteins in this "gD superfamily" would include HSV gG-2 homologs such as gX of PRV (with the same three disulfides as gD) and HSV gI homologs such as gp63 of PRV and gpIV of VZV (with two possible disulfides, the 3–4 and the 2–6; Davison and McGeoch 1986; McGeoch et al. 1987; Petrovskis et al. 1986b;Rea et al. 1985; Thomsen et al. 1987). However, although these additional proteins may be structural homologs of gD, their functions may be widely divergent. Clearly, biochemical and/or genetic analysis of the disulfide bonds of these proteins is needed to verify our prediction.

Properties of the Functional Double-Cys Mutants of Glycoprotein D

Of the three double-Cys mutants of gD, Cys-1,5 was clearly the least damaged, and this mutant protein had the greatest capacity to rescue F-gDβ infectivity (Long et al. 1992). Thus, the Cys-1–5 bond may be the least important for maintenance of gD structure. interestingly, this may be the same bond that is lacking in HSV gI and other gI homologs. Moreover, since each of the double-Cys mutants was ts for both structure and function, all have the potential for yielding ts recombinants, particularly the Cys-1,5. Such a mutant virus should be ts for virus entry.

Conclusions

We conclude from these studies that the three N-CHO on gD are not necessary for HSV infection, but are critical for the maintenance of the antigenic structure of the molecule when isolated from the cell. Clearly, carbohydrates are important in HSV vaccine development. The N-CHO on gD may also be important for efficient cell-to-cell spread and may be a virulence factor. Secondly, the three disulfide bonds in the extracellular portion of gD are critical for stability of the structure of the protein as well as the function of the virus. Alteration of any one of the three pairs results in global destruction of structure and function when the protein is expressed at 37°C. Function is partially restored by expression at 31.5°C, indicating that each of the double-Cys mutants is ts. Vaccines which incorporate gD must be designed to maintain all three disulfide bonds. Knowledge of disulfide bond structure contributes substantially to our understanding of the antigenic structure of gD and has permitted us to draw our first three-dimensional models of gD (Fig. 2; Long et al. 1992). Finally, we have begun to map the functional sites on gD. Thus far, two regions have been shown to be important for function (Chiang et al. 1991; Muggeridge et al. 1990c,1992). Linker insertion mutagenesis has also been used to extend this map by creating a series of mutations across the genes for gD-1 and gD-2 (Chiang et al. 1991 and H.-Y. Chiang, R. J. Eisenberg and G. H. Cohen, manuscript in preparation). In the folded structure of gD, one site is close to, but distinct from, the major antigenic sites Ia and Ib. Although group I MAb have powerful virus-neutralizing activity, they are not the only MAb to gD with this property. A number of other MAb which have not been placed into any group as yet also exhibit this property (Isola 1988; Muggeridge et al. 1990a). Whether other parts of gD will be found to contribute to its function and whether these will turn out to overlap some of the "ungrouped" MAb remains to be determined. A very important goal is to solve the three-dimensional structure of gD by X-ray crystallographic means. We are now in the process of overexpressing gD in quantities sufficient to carry out crystallization trials. Once the actual structure of gD is known, we will have a much better understanding of the relationship of the physical structure of gD to its antigenic and functional properties.

Acknowledgements. The studies summarized here were supported by Public Health Service Grants DE-08239 from the National Institute of Dental Research and AI-18289 from the National Institute of Allergy and Infectious Diseases and by a grant to R. J. Eisenberg and G. H. Cohen from American Cyanamid Co. D. Long was a predoctoral trainee supported by Public Health Service grant NS-07180 from the National Institute of Neurological Diseases and Stroke, and D. L. Sodora was a predoctoral trainee supported by Public Health Service grant GM-07225 from the National Institute of General Medical Sciences. Special thanks to C. Handler, R.Tal-Singer, and N. Fraser for use of unpublished data.

References

Arsenakis M, Campadelli-Fiume G, Roizman B (1988) Regulation of glycoprotein D synthesis: does α4, the major regulatory protein of herpes simplex virus 1, regulate late genes both positively and negatively? J Virol 62: 148–158

Babiuk LA, L'Italien J, van Drunen Littel-van den Hurk S, Zamb T, Lawman MJP, Hughes G, Gifford GA (1987) Protection of cattle from bovine herpesvirus type 1 (BHV-1) infection by immunization with individual glycoproteins. Virology 159: 57–66

Bennett AM, Harrington L, Kelly DC (1992) Nucleotide sequence analysis of genes encoding glycoproteins D and J in simian herpes B virus. J Gen Virol 73: 2963–2967

Berman PW, Dowbenko D, Lasky LA, Simonsen CC (1983) Detection of antibodies to herpes simplex virus with a continuous cell line expressing cloned glycoprotein D. Science 222: 524–527

Berman PW, Gregory T, Crase D, Lasky LA (1985) Protection from genital herpes simplex virus type 2 infection by vaccination with cloned type 1 glycoprotein D. Science 227: 1490–1492

Breinig MK, Kingsley LA, Armstrong JA, Freeman DJ, Ho M (1990) Epidemiology of genital herpes in Pittsburgh: serologic, sexual, and racial correlates of apparent and inapparent herpes simplex infections. J Infect Dis 162: 299–305

Burke RL (1992) Contemporary approaches to vaccination against herps simplex virus.In: Rouse BT (ed) Herpes simplex virus. Pathogenesis, immunobiology and control. Springer, Berlin Heidelberg New York, pp 137–158 (Current topics in microbiology and immunology, vol 179)
Burke RL (1993) HSV vaccine development. In: Roizman B, Lopez C (eds) The herpesviruses. Raven, New York (in press)
Burke RL, Van Nest G, Carlson J, Gervase B, Goldbeck C, Ng P, Sanchez-Pescador L, Stanberry L, Ott G (1989) Development of herpes simplex virus subunit vaccine. In: Lerner RA, Ginsberg H, Chanock RM, Brown F (eds) Vaccines 89. Modern approaches to new vaccines including prevention of AIDS. Cold Spring Harbor Laboratory New York, pp 377–382
Burnette WN (1981) "Western blotting." Electrophoretic transfer of proteins from SDS-polyacrylamide gels to unmodified nitrocellulose and radiographic detection with antibody and radioiodinated protein A. Anal Biochem 112: 195–203
Campadelli-Fiume G, Arsenakis M, Farabegoli F, Roizman B (1988) Entry of herpes simplex virus 1 in BJ cells that constitutively express viral glycoprotein D is by endocytosis and results in degradation of the virus. J Virol 62: 159–167
Campadelli-Fiume G, Qi S, Avitabile E, Foà-Tomasi L, Brandimarti R, Roizman B (1990) Glycoprotein D of herpes simplex virus encodes a domain which precludes penetration of cells expressing the glycoprotein by superinfecting herpes simplex virus. J Virol 64: 6070–6079
Chan W-L (1983) Protective immunization of mice with specific HSV-1 glycoproteins. Immunology 49: 343–352
Chiang H-Y, Eisenberg RJ, Cohen GH (1991) Linker insertion mutagenesis of HSV-1 and HSV-2 gD. 16th international hespesvirus workshop, 1–5 Aug 1992, Edinburgh, Scotland
Chou PY, Fasman GD (1974a) Conformational parameters for amino acids in helical, β-sheet and random coil regions. Biochemistry 13: 211–222
Chou PY, Fasman GD (1974b) Prediction of protein conformation. Biochemistry 13: 222–245
Cohen GH, Ponce de Leon M, Nichols C (1972) Isolation of a herpes simplex virus-specific antigenic fraction which stimulates the production of neutralizing antibody. J Virol 10: 1021–1030
Cohen GH, Long D, Matthews J, May M, Eisenberg RJ (1983) Glycopeptides of the type-common glycoprotein gD of herpes simplex virus type 1 and 2. J Virol 46: 679–689
Cohen GH, Katze M, Hydrean-Stern C, Eisenberg RJ (1978) Type-common CP-1 antigen of herpes simplex virus is associated with a 59,000-molecular-weight envelope glycoprotein. J Virol 27: 172–181
Cohen GH, Long D, Eisenberg RJ (1980) Synthesis and processing of glycoproteins gD and gC of herpes simplex virus type 1. J Virol 36: 429–439
Cohen GH, Dietzchold B, Long D, Ponce de Leon M, Golub E, Eisenberg RJ (1985) Localization and synthesis of an antigenic determinant of herpes simplex virus glycoprotein D that stimulates neutralizing antibody and protects mice from lethal virus challenge. In: Mergenhangen SE, Rosan B (eds) Molecular basis of oral microbial adhesion. American Society for Microbiology, Washington DC, pp 24–30
Cohen GH, Dietzschold B, Ponce de Leon M, Long D, Golub E, Varrichio A, Pereira L, Eisenberg RJ (1984) Localization and synthesis of an antigenic determinant of herpes of simplex virus glycoprotein D that stimulates production of neutralizing antibody. J Virol 49: 102–108
Cohen GH, Isola VJ, Kuhns, J, Berman PW, Eisenberg RJ (1986) Localization of discontinuous epitopes of herpes simplex virus glycoprotein D: use of a nondenaturing ("native" gel) system of polyacrylamide gel electrophoresis coupled with Western blotting. J Virol 60: 157–166
Cohen GH, Wilcox WC, Sodora DL, Long D, Levin JZ, Eisenberg RJ (1988) Expression of herpes simplex virus type 1 glycoprotein D deletion mutants in mammalian cells. J Virol 62: 1932–1940
Cohen GH, Muggeridge MI, Long D, Sodora DA, Eisenberg RJ (1992) Structural and functional studies of herpes simplex virus glycoprotein D. In: Ciardi (ed) JE Genetically engineered vaccines. Plenum New York, pp 217–228

Davison AJ, McGeoch DJ (1986) Evolutionary comparisons of the S segments in the genomes of herpes simplex virus type 1 and caricella zoster virus. J Gen Virol 67: 597–611
Davison AJ, Wilkie NM (1983) Location and orientation of homologous sequences in the genomes of five herpesviruses. J Gen Virol 64: 1927–1942
Davison AJ, Edson CM, Ellis RW, Forghani B, Gilden D, Grose C, Keller PM, Vafai A, Wroblewska Z, Yamanishi K (1986) New common nomenclature for glycoprotein genes of varicella-zoster virus and their glycosylated products. J Virol 57: 1195–1197
Dietzschold B, Eisenberg RJ, Ponce de Leon M, Golub E, Hudecz F, Varrichio A, Cohen GH (1984) Fine structure analysis of type-specific and type-common antigenic sites of herpes simplex virus glycoprotein D. J Virol 52: 431–435
Dix RD (1987) Prospects for a vaccine against herpes simplex virus types 1 and 2. Prog Med Virol 34: 89–128
Dix RD, Mills J (1985) Acute and latent herpes simplex virus neurological disease in mice immunized with purified virus-specific glycoprotein gB or gD. J Med Virol 17: 9–18
Eisenberg R, Hydrean-Stern C, Cohen G (1979) Structural analysis of precursor and product forms of type-common envelope glycoprotein D (CP-1 antigen) of herpes simplex virus type 1. J Virol 31: 608–620
Eisenberg R, Ponce de Leon M, Cohen G (1980) Comparative structural analysis of glycoprotein gD of herpes simplex virus types 1 and 2. J Virol 35: 428–435
Eisenberg RJ, Cohen GH (1986) Identification and analysis of biologically active sites of herpes simplex virus glycoprotein D. In: Crowell R, Lonberg-Holm K (eds) Virus attachment and entry into cells. American Society for Microbiology, Washington DC, pp 74–84
Eisenberg RJ, Long D, Pereira L, Hampar B, Zweig M, Cohen GH (1982a) Effect of monoclonal antibodies on limited proteolysis of native glycoprotein gD of herpes simplex virus type 1. J Virol 41: 478–488
Eisenberg RJ, Ponce de Leon M, Pereira L, Long D, Cohen GH (1982b) Purification of glycoprotein gD of herpes simplex virus types 1 and 2 by use of monoclonal antibody. J Virol 41: 1099–1104
Eisenberg RJ, Long D, Hogue-Angeletti R, Cohen GH (1984) Amino-terminal sequence of glycoprotein D of herpes simplex virus types 1 and 2. J Virol 49: 265–268
Eisenberg RJ, Cerini CP, Heilman CJ, Joseph AD, Dietzschold B, Golub E, Long D, Ponce de Leon M, Cohen GH (1985a) Synthetic glycoprotein D-related peptides protect mice against herpes simplex virus challenge. J Virol 56: 1014–1017
Eisenberg RJ, Long D, Ponce de Leon M, Matthews JT, Spear PG, Gibson MG, Lasky LA, Berman P, Golub E, Cohen GH (1985b) Localization of epitopes of herpes simplex virus type 1 glycoprotein D. J Virol 53: 634–644
Eisenberg RJ, Ponce de Leon M, Friedman HM, Fries LF, Frank MM, Hastings JC, Cohen GH (1987) Complement component C3b binds directly to purified glycoprotein C of herpes simplex virus types 1 and 2. Microb Pathog 3: 423–435
Feenstra V, Hodaie M, Johnson DC (1990) Deletions in herpes simplex virus glycoprotein D define nonessential and essential domains. J Virol 64: 2096–2102
Flowers CC, Eastman EM, O'Callaghan DJ (1991) Sequence analysis of a glycoprotein D gene homolog within the unique short segment of the EHV-1 genome. Virology 180: 175–184
Freifeld AG, Savarese B, Krause PR, Paar D, Dekker C, Adair S, Tigges M, Burke RL, Strauss WE (1990) Phase 1 testing of a recombinant herpes simplex type 2 glycoprotein D vaccine. 13th interscience conference on antimicrobial agents and chemotherapy, p 206
Fuller AO, Spear PG (1985) Specificities of monoclonal and polyclonal antibodies that inhibit adsorption of herpes simplex virus to cells and lack of inhibition by potent neutralising antibodies. J Virol 55: 475–482
Fuller AO, Spear PG (1987) Anti-glycoprotein D antibodies that permit adsorption but block infection by herpes simplex virus 1 prevent virion-cell fusion at the cell surface. Proc Natl Acad Sci USA 84: 5454–5458
Fuller AO, Lee W-C (1992) Herpes simplex virus type 1 entry through a cascade of virus-cell interactions requires different roles of gD and gH in penetration. J Virol 66: 5002–5012
Fuller AO, Santos RE, Spear PG (1989) Neutralizing antibodies specific for glycoprotein H of herpes simplex virus permit viral attachment to cells but prevent penetration. J Virol 63: 3435–3443

Goldstein LC, Nowinski RC (1986) Diagnosis of herpesviruses with monoclonal antibodies. In: Lopez C, Roizman B (eds) human herpesvirus infections. Raven, New York, pp 195–202

Highlander SL, Sutherland SL, Gage PJ, Johnson DC, Levine M, Glorioso JC (1987) Neutralizing monoclonal antibodies specific for herpes simplex virus glycoprotein D inhibit virus penetration. J Virol 61: 3356–3364

Hill TJ (1985) Herpes simplex virus latency. In: Roizman B (ed) The herpesviruses. Plenum, New York, pp 175–240

Holland TC, Marlin SD, Levine M, Glorioso J (1983) Antigenic variants of herpes simplex virus selected with glycoprotein-specific monoclonal antibodies. J Virol 45: 672–682

Hughes G, Babiuk LA, Van Drunen Littel-Van den Hurk S (1988) Functional and topographical analyses of epitopes on bovine herpesvirus type 1 glycoprotein IV. Arch Virol 103: 47–60

Isola VJ (1988) Studies on the antigenic structure of herps simplex virus glycoprotein D. Thesis, University of Pennsylvania, Philadelphia

Isola VJ, Eisenberg RJ, Siebert GR, Heilman CJ, Wilcox WC, Cohen GH (1989) Fine mapping of antigenic site II of herpes simplex virus glycoprotein D. J Virol 63: 2325–2334

Izumi KM, Stevens JG (1990) Molecular and biological characterization of a herpes simplex virus type 1 (HSV-1) neuroinvasiveness gene. J Exp Med 172: 487–496

Johnson DC, Ligas MW (1988) Herpes simplex viruses lacking glycoprotein D are unable to inhibit virus penetration: quantitative evidence for virus-specific cell surface receptors. J Virol 62: 4605–4612

Johnson DC, Spear PG (1982) Monensin inhibits the processing of herpes simplex virus glycoproteins, their transport to the cell surface, and the egress of virions from infected cells. J Virol 43: 1102–1112

Johnson DC, Gregory T, Burke RL (1989). Inhibition of HSV entry into cells by purified HSV glycoprotein D and characterization of cell surface receptors for HSV. 14th international herpesvirus workshop 1–5 Aug 1989, Nyborg Strand, Denmark, p 237

Johnson DC, Gregory T, Burke R-L (1991) Identification of a receptor for HSV-1 glycoprotein D, a putative receptor for HSV. XVIth international herpesvirus workshop, 7–12 July 1991, Pacific Grove, CA, p 57

Johnson RE, Nahmias AJ, Magder LS, Lee FK, Brooks CA, Snowden CB (1989) A seroepidemologic survey of the prevalence of herpes simplex virus type 2 infection in the United States. N Engl J Med 321: 7–12

Johnson RM, Spear PG (1989) Herpes simplex virus glycoprotein D medicates interference with herpes simplex virus infection. J Virol 63: 819–827

Kümel G, Kaerner HC, Levine M, Schröder CH, Glorioso JC (1985) Passive immune protection by herpes simplex virus-specific monoclonal antibodies and monoclonal antibody-resistant mutants altered in pathogenicity. J Virol 56: 930–937

Lasky LA, Dowbenko DJ (1984) DNA sequence analysis of the type-common glycoprotein D genes of herpes simplex virus types 1 and 2. DNA 3: 23–29

Lasky LA, Dowbenko D, Simonsen CC, Berman PW (1984) Protection of mice from lethal herpes simplex virus infection by vaccination with a secreted form of cloned glycoprotein D. Biotechnology 2: 527–532

Lausch RN, Oakes JE, Metcalf JF, Scimeca JM, Smith LA, Robertson SM (1989) Quantitation of purified monoclonal antibody needed to prevent HSV-1 induced stromal keratitis in mice. Curr Eye Res 8: 499–508

Lausch RN, Oakes JE, Cohen GH, Eisenberg RJ (1991) Prevention of herpes keratitis by monoclonal antibodies specific for discontinuous and continuous epitopes on glycoprotein D. Invest Ophthalmol Vis Sci 32: 2735–2740

Ligas MW, Johnson DC (1988) A herpes simplex virus mutant in which glycoprotein D sequences are replaced by β-galactosiedase sequences binds to but is unable to penetrate into cells. J Virol 62: 1486–1494

Long D, Madara TJ, Ponce de Leon M, Cohen GH, Montgomery PC, Eisenberg RJ (1984) Glycoprotein D protects mice against lethal challenge with herpes simplex virus types 1 and 2. Infect Immun 37: 761–764

Long D, Cohen GH, Muggeridg MI, Eisenberg RJ (1990) Cysteine mutants of herpes simplex virus type 1 glycoprotein D exhibit temperature-sensitive properties in structure and function. J Virol 64: 5542–5552
Long D, Wilcox WC, Abrams WR, Cohen GH, Eisenberg RJ (1992) Disulfide bond structure of glycoprotein D of herpes simplex virus type 1 and 2. J Virol 66: 6668–6685
Martin S, Rouse BT (1987) The mechanisms of antiviral immunity induced by a vaccinia virus recombinant expressing herpes simplex virus type 1 glycoprotein D: clearance of local infection. J Immunol 138: 3431–3437
Martin LB, Montgomery PC, Holland TC (1992) Soluble glycoprotein D blocks herpes simplex virus type 1 infection of rat eyes. J Virol 66: 5183–5189
Matthews JT, Cohen GH, Eisenberg RJ (1983) Synthesis and processing of glycoprotein D of herpes simplex virus types 1 and 2 in an in vitro system. J Virol 48: 521–533
McGeoch DJ (1990) Evolutionary relationships of virion glycoprotein genes in the S regions of alphaherpesvirus genomes. J Gen Virol 71: 2361–2367
McGeoch DJ, Dolan A, Donald S, Rixon FJ (1985) Sequence determination and genetic content of the short unique region in the genome of herpes simplex virus type 1. J Mol Biol 181: 1–13
McGeoch DJ, Moss HW, McNab D, Frame MC (1987) DNA sequence and genetic content of the Hind III l region in the short unique component of the herpes simplex virus type 2 genome: identification of the gene encoding glycoprotein G and evolutionary comparisons. J Gen Virol 68: 19–38
McGeoch DJ, Dalrymple MA, Davison AJ, Dolan A, Frame MC, McNab D, Perry LJ, Scott JE, Taylor P (1988) The complete DNA sequence of the long unique region in the genome of herpes simplex virus type 1. J Gen Virol 69: 1531–1574
Meier JL, Straus SE (1992) Comparative biology of latent varicella-zoster virus and herpes simplex virus infections. J Infect Dis 166 [Suppl 1]: S13–23
Meignier B (1985) Vaccination against herpes simplex virus infections. In: Roizman B, Lopez C (eds) The herpesviruses. Immunobiology and prophylaxis of human herpesvirus infections. Plenum, New York, pp 265–296
Mertz GJ, KAshley R, Burke RL, Benedetti J, Critchlow C, Jones CC, Corey L (1990) Double-blind, placebo-controlled trial of a herps simplex virus type-2 glycoprotein vaccine in persons at high risk for genital herpes infection. J Infect Dis 161: 653–660
Minson AC, Hodgman TC, Digard P, Hanock DC, Bell SE, Buckmaster EA (1986) An analysis of the biological properties of monoclonal antibodies against glycoprotein D of herps simplex virus and identification of amino acid substitutions that confer resistance to neutralization. J Gen Virol 67: 1001–1013
Mishkin EM, Fahey JR, Kino Y, Klein RJ, Abramovitz AS, Mento SJ (1991) Native herpes simplex virus glycoprotein D vaccine: immunogenicity and protection in animal models. Vaccine 9: 147–153
Muggeridge MI, Isola VJ, Byrn RA, Tucker TJ, Minson AC, Glorioso JC, Cohen GH, Eisenberg RJ (1988) Antigenic analysis of a major neutralization site of herpes simplex virus glycoprotein D, using deletion mutants and monoclonal antibody-resistant mutants. J Virol 62: 3274–3280
Muggeridg MI, Roberts SR, Isola VJ, Cohen GH, Eisenberg RJ (1990a) Herpes simplex virus. In: Van Regenmortel MHV, Neurath AR (eds) Immunochemistry of viruses II. The basis for serodiagnosis and vaccines. Elsevier Biochemical, Amsterdam, pp 459–481
Muggeridge MI, Wilcox WC, Cohen GH, Eisenberg RJ (1990b) Identification of a site on herpes simplex virus type 1 glycoprotein D that is essential for infectivity. J Virol 64: 3617–3626
Muggeridge MI, Wu T-T, Johnson DC, Glorioso JC, Eisenberg RJ, Cohen GH (1990c) Antigenic and functional analysis of a nutralization site of HSV-1 glycoprotein D. Virology 174: 375–387
Muggeridge MI, Chiang H-Y, Cohen GH, Eisenberg RJ (1992) Mapping of a functional site on herpes simplex virus glycoprotein D. J Cell Biochem [Suppl]16C: 132
Noble AG, Lee GT-Y, Sprague R, Parish ML, Spear PG (1983) Anti-gD monoclonal antibodies inhibit cell fusion induced by herpes simplex virus type 1. Virology 129: 218–224

Paoletti E, Lipinskas BR, Samsonoff C, Mercer S, Panicali D (1984) Construction of live vaccines using genetically engineered poxviruses: biological activity of vaccinia virus recombinants expressing the hepatitis B virus surface antigen and the herpes simplex virus glycoprotein D. Proc Natal Acad Sci USA 81: 193–197
Pereira L, Klassen T, Baringer JR (1980) Type-common and type-specific monoclonal antibody to herpes simplex virus type 1. Infect Immun 29: 724–732
Petrovskis EA, Timmins JG, Armentrout MA, Marchioli CC, Yancey RJJ, Post LE (1986a) DNA sequences of the gene for pseudorabies virus gp50, a glycoprotein without N-linked glycosylation. J Virol 59: 216–223
Petrovskis EA, Timmins JG, Post LE (1986b) Use of 1gt11 to isolate genes for two pseudorabies virus glycoproteins with homology to herpes simplex virus and varicella-zoster virus glycoproteins. J Virol 60: 185–1193
Pizer LI, Cohen GH, Eisenberg RJ (1980) Effect of tunicamycin on herpes simplex virus glycoproteins and infectious virus production. J Virol 34: 142–153
Rea TJ, Timmons JG, Long GW, Post LE (1985) Mapping and sequence of the gene for the pseudorabies virus glycoprotein which accumulates in the medium of infected cells. J Virol 54: 21–29
Roizman B, Sears AE (1987) An inquiry into the mechanisms of herpes simplex virus latency. Annu Rev Microbiol 41: 543–571
Rooney JF, Wohlenberg C, Cremer KJ, Notkins AL (1989) Immunized mice challenged with herpes simplex virus by the intranasal route show protection against latent infection. J Infect Dis 159: 974–976
Ross LJN, Binns MM (1991) Properties and evolutionary relationships of the Marek's disease virus homologues of protein kinase, glycoprotein D and glycoprotein I of herpes simplex virus. J Gen Virol 72: 939–947
Ross LJN, Binns MM, Pastorek J (1991) DNA sequence and organization of genes in a 5.5 kbp EcoRI fragment mapping in the short unique segment of Marek's disease virus (strain RB1B). J Gen Virol 72: 949–954
Serafini-Cessi F, Dall'Olio F, Malagolini N, Pereira L, Campadelli-Fiume G (1988) Comparative study on O-linked oligosaccharides of glycoprotein D of herpes simplex virus types 1 and 2. J Gen Virol 69: 869–877
Showalter DS, Zweig M, Hampar B (1981) Monoclonal antibodies to herpes simplex virus type 1 proteins, including the immediate-early protein ICP4. Infect Immun 34: 684–692
Sodora DL, Cohen GH, Eisenberg RJ (1989) Influence of asparagine-linked oligosaccharides on antigenicity, processing, and cell surface expression of herpes simplex virus type 1 glycoprotein D. J Virol 63: 5184–5193
Sodora DL, Cohen GH, Muggeridge MI, Eisenberg RJ (1991a) Absence of aspargine-linked oligosaccharides from glycoprotein D of herpes simplex virus type 1 results in a structurally altered but biologically active protein. J Virol 65: 4424–4431
Sodora DL, Eisenberg RJ, Cohen GH (1991b) Characterization of a recombinant herpes simplex virus which expresses a glycoprotein D lacking asparagine-linked oligosaccharides. J Virol 65: 4432–4441
Spear PA (1976) Membrane proteins specified by herpes simplex viruses I. Identification of four glycoprotein precursors and their products in type 1-infected cells. J Virol 17: 991–1008
Spear PG (1985a) Antigenic structure of herpes simplex viruses. In: van Regenmortel MVH, Neurath AR (ed) Immunochemistry of viruses. The basis for serodiagnosis and vaccines. Elsevier Science, Amsterdam, pp 425–446
Spear PG (1985b) Glycoproteins specified by herpes simplex viruses. In: Roizman B (ed) The herpesviruses. plenum, New York, pp 315–356
Spear PG (1993) Membrane fusion induced by herpes simplex virus. In: Bentz J (ed) Viral fusion mechanisms. CRC Press, Boca Raton, pp 201–232
Stevens JG (1989) Human herpesviruses: a consideration of the latent state. Microbiol Rev 53: 318–332
Straus SE, Burke RL, Savarese B, Margolis D, Krause P, Meier J, Adair S, Dekker C (1991) XVI international herpesvirus workshop, 7–12 July, Pacific Grove, CA, p 165

Straus SE, Savarese B, Tigges M, Freifeld A, Krause PR, Margolis DM, Meier JL, Paar DP, Adair SF, Geysen HM, Dina D, Dekker C, Burke RL (1993) Induction and enchancement of immune responses to herpes simplex virus type 2 in humans with a recombinant glycoprotein D vaccine. Science (in press)

Thomsen DR, Marchioli CD, Yancey RJ, Post LE (1987) Replication and virulence of pseudorabies virus mutants lacking glycoprotein gX. J Virol 61: 229–232

Thornton JM (1981) Disulphide bridges in globular proteins. J Mol Biol 151: 261–287

Tikoo SK, Fitzpatrick DR, Babiuk LA, Zamb TJ (1990) Molecular cloning, sequencing, and expression of functional bovine herpesvirus 1 glycoprotein gIV in transfected bovine cells. J Virol 64: 5132–5142

Torseth JW, Cohen GH, Eisenberg RJ, Berman PW, Lasky LA, Cerini CP, Heilman CJ, Kerwar S, Merigan TC (1987) Native and recombinant herpes simplex virus type 1 envelope proteins induce human immune T-lymphocyte responses. J Virol 61: 1532–1539

Van Drunen Littel-Van Den Hurk S, Van Den Hurk J, Gilchrist J, Misra V, Babiuk L (1984) Interactions of monoclonal antibodies and bovine herpesvirs type 1 (BHV-1) glycoproteins: characterization of their biochemical and immunological properties. Virology 135: 466–479

Wachsman M, Aurelian L, Smith CC, Lipinskas BR, Perkus ME, Paoletti E (1987) Protection of guinea pigs from primary and recurrent herpes simplex virus (HSV) type 2 cutaneous disease with vaccinia virus recombinants expressing HSV glycoprotein D. J Infect Dis 155: 1188–1197

Watson RJ (1983) DNA sequence of the herpes simplex virus type 2 glycoprotein D gene. Gene 26: 307–312

Watson RJ, Weis JH, Salstrom JS, Enquist LW (1982) Herpes simplex virus type-1 glycoprotein D gene: nucleotide sequence and expression in Escherichia coli. Science 218: 381–384

Wilcox WC, Long D, Sodora DL, Eisenberg RJ, Cohen GH (1988) The contribution of cysteine residues to antigenicity and extent of processing of herpes simplex virus type 1 glycoprotein D. J Virol 62: 1941–1947

Wyckoff JH III, Osmand AP, Eisenberg RJ, Cohen GH, Rouse BT (1988) Functional T cell recognition of synthetic peptides corresponding to continuous antibody epitopes of herpes simplex virus type 1 glycoprotein D. Immunobiology 177: 134–148

Zarling JM, Moran PA, Burke RL, Pachl C, Berman PW, Lasky LA (1986a) Human cytotoxic T cell clones directed against herpes simplex virus-infected cells. IV. Recognition and activation by cloned glycoproteins gB and gD. J Immunol 136: 4669–4673

Zarling JM, Moran PA, Lasky LA, Moss B (1986b) Herpes simplex virus (HSV)-specific human T-cell clones recognize HSV glycoprotein D expressed by a recombinant vaccinia virus. J Virol 59: 506–509

II. Genes Coding for Enzymes

Chapter 5 The Role of Herpes Simplex Virus Thymidine Kinase Expression in Pathogenesis and Latency

Richard B. Tenser

Summary

The role of herpes simplex virus (HSV) thymidine kinase (TK) expression for HSV neurovirulence and latency has been evaluated by many investigators. Although neurovirulence is complex and involves multiple host and viral factors, in studies with HSV TK-negative (TK^-) mutants, it seemed that TK expression was an important factor. This was apparent in experimental animal studies, as indicated by decreased replication of TK^- HSV mutants in neural tissue and by decreased mortality. Although some virulent TK^- HSV mutants have been reported, most investigators, including studies with HSV deletion mutants, have supported an important role for HSV TK expression and neurovirulence.

In early experimental animal studies in which HSV latency was defined by the ability to recover HSV from explants of latently infected ganglia, HSV TK expression also seemed important for latency. In more recent studies in which molecularly constructed TK^- HSV mutants were utilized, it was evident that ganglia could be infected with TK^- HSV so that HSV latency-associated transcript (LAT) was detected in ganglia during the period of latency. As in most previous studies, reactivation of TK^- HSV was not usually detected in ganglion explants. TK^- HSV could be rescued from such ganglia, however, by superinfection–complementation with TK^+ HSV. These observations suggested that HSV TK expression was important for the reactivation of HSV from latent sensory ganglion neuron infection.

In recent, more speculative studies, the HSV *tk* gene and HSV TK^- mutants have been utilized therapeutically in experimental animal studies. In some of these studies, HSV TK expression was induced in neoplastic cells, which were then destroyed by TK-mediated antivirals. In other studies, TK^- HSV was used to destroy rapidly growing tumor cells or, alternatively, TK^- HSV was used to incorporate foreign genes into the nervous system.

Department of Medicine (Neurology) and Microbiology and Immunology, Pennsylvania State University College of Medicine, Hershey, PA 17033, USA

Introduction

Herpesvirus thymidine kinase has been one of the most discussed virus-coded products in recent years. This has been particularly true for the thymidine kinase (*tk*) gene and the TK activity of herpes simplex virus (HSV). The reasons for this are threefold.

First, in molecular studies, the *tk* gene has proven to be a useful vehicle to introduce viral nucleic acid in studies of genetic control and expression (Cremer et al. 1979; McKnight and Gavis 1980; Smiley 1980; Shih et al. 1984; EI Kareh et al. 1985). For these and other reasons, the HSV *tk* gene have been of interest and has received widespread use in molecular studies of cell biology, without specific interest in HSV per se. HSV deletion and insertion mutants have been created and used to study HSV control mechanisms and, as will be discussed below, these have also been useful in considerations of the role of HSV expression in pathogenesis (Post et al. 1981; Efstathiou et al. 1989; Coen et al. 1989a; Leist et al. 1989).

Second, HSV TK activity was shown to be the means by which the antiviral drug acyclovir is made effective. Phosphorylation of acyclovir by HSV TK is necessary for the drug to reach the active antiviral state. Following HSV TK phosphorylation, acyclovir inhibits HSV replication by inhibition of the viral DNA polymerase and by chain termination of HSV DNA (Elion et al. 1977; Fyfe et al. 1978; Derse et al. 1981). Acyclovir has been recognized as possibly the first rational and selective antiviral. Antiviral activity of acyclovir after phosphorylation by HSV TK is one of the reasons why, in molecular studies, insertion of viral nucleic acid into the HSV *tk* gene has been useful. HSV constructs containing inserts in the viral *tk* gene are converted to the TK phenotype, and since acyclovir is not phosphorylated by these viral mutants, the virus is not inhibited by acyclovir. Therefore, acyclovir can be used to select for such molecularly constructed virus.

Although TK mutants have been useful in experimental studies, the occurrence of such mutants in clinical medicine has been a problem. HSV TK mutants normally occur in clinical isolates of HSV (Parris and Harrington 1982), and there is an expansion of the mutant population in immunosuppressed individuals, particularly those treated with acyclovir (Crumpacker et al. 1982; Burns et al. 1982; Erlich et al. 1989; Birch et al. 1992). In such individuals, standard TK^{+} HSV is inhibited by acyclovir, allowing the selection of and overgrowth by TK^{-} mutants. Several other antivirals which inhibit TK^{+} HSV have been investigated; cross-resistance by HSV TK^{-} mutants is a common occurrence, although resistance to the drugs varies (Darby et al. 1984; Coen et al. 1985, 1989b; Nutter et al. 1987). As in the laboratory, most clinical acyclovir-resistant mutants are TK^{-} and do not significantly phosphorylate thymidine or acyclovir. Some mutants are TK^{+}, but express DNA polymerase activity that is not inhibited by phosphorylated acyclovir (Coen and Schaffer 1980; Darby et al. 1984; Crumpacker et al. 1984). Therefore, while all HSV TK^{-} mutants are resistant to acyclovir, all acyclovir-resistant mutants are not TK^{-}. Lastly, some mutants have been reported which are TK^{+}, but which have altered substrate phosphorylation capacity and do not phosphorylate significant amounts of acyclovir or other antiviral drugs (Larder and Darby 1982; Veerisetty et al. 1983; Hill et al. 1991). Two distinct loci are responsible for the development of HSV TK and polymerase mutants, and in cell culture the mutants can complement each other so that sensitivity to acyclovir results (Coen and Schaffer 1980; Darby et al. 1984). In recent years, in part because of the spread of AIDS, increased numbers of immunosuppressed patients infected with TK⁻ HSV have been described, and altered treatment modalities have been instituted (Erlich et al. 1989; Chatis et al. 1989; Birch et al. 1992). In such patients, cutaneous HSV lesions persist in the presence of acyclovir, although neurological involvement is not common.

In experimental animal studies, there has been recent interest in the use of molecularly constructed TK⁻ virus as a therapeutic agent. In one study, TK⁻ HSV was used to inhibit the progression of human glioma transplanted into nude mice (Martuza et al. 1991); TK⁻ HSV is usually of decreased neurovirulence, as discussed below, and would not be expected to cause severe disease in normal brain tissue of infected animals, but may destroy more rapidly growing glioma cells. In other studies, rather than use TK⁻ HSV as a therapeutic entity, the HSV *tk* gene has been used. The object of these studies has been to use a virus vector to introduce the HSV *tk* gene into neoplastic cells. Following this procedure and the expression of HSV *tk* activity by the neoplastic cells, appropriate TK-mediated antivirals were used to destroy the neoplastic cells (Moolten and Wells 1990; Ezzeddine et al. 1991). In neurobiological investigations, TK⁻ HSV, which is of decreased neurovirulence, as noted above, has been used as a vector to incorporate foreign genes into the nervous system (Ho and Mocarski 1988; Friedman 1989; Palella et al. 1989; Chiocca et al. 1990). This approach may prove to be of value in the treatment of neurological illnesses due to enzyme deficiency.

The third reason for interest in HSV TK has been the evidence that HSV TK expression is important for neurovirulence and particularly for latency in experimental viral pathogenesis. These subjects will be the emphasis of the present review.

Herpesvirus TK expression was first reported in 1963. In that year, Kit and Dubbs (1963) described the induction of TK activity in cells lytically infected with HSV, and Nohara and Kaplan (1963) reported similar results in pseudorabies virus-infected cells. The latter group was the first to report that the TK activity induced was virus specific. Shortly thereafter, it was evident that TK expressed in HSV-infected cells was virus specific (Dubbs and Kit 1964; Klemperer et al. 1967; Thouless and Skinner 1971). In subsequent years, TK activity has been reported for many herpesviruses (Table 1). Most emphasis has been placed on HSV TK, and studies of this protein will be emphasized.

The HSV *tk* gene is a delayed early or β-gene and has been localized to approximately 0.3 map units, in the long unique component, of the HSV type 1 (HSV-1) and HSV-2 genomes (Wigler et al. 1977; Halliburton et al. 1980; McDougall et al. 1980; Reyes et al. 1982). The genes have been sequenced (McKnight 1980; Wagner et al. 1981; Swain and Galloway 1983; Darby et al. 1986), and methylation of the HSV *tk* gene has been investigated (Graessmann and Graessmann 1980; deJong et al. 1989). There is 80%–90% DNA sequence homology between the *tk* of HSV-1 and HSV-2, and although there is little DNA sequence homology with the *tk* of other herpesviruses, there is more amino acid homology among herpesvirus TK proteins (Kit 1985). Transgenic mice contain-

Table 1. Thymidine kinase identified herpesviruses

Virus	References
Human herpesviruses	
HSV-1	Dubbs and Kit (1964)
HSV-2	Thouless and Skinner (1971)
Varicella zoster virus	Doberson et al. (1976); Roberts et al. (1991)
Epstein-Barr virus	Roubal and Klein (1981); Little and Arrand (1988)
Animal herpesviruses	
Bovine herpesvirus type 1	Weinmaster et al. (1982); Miller et al. (1991)
Bovine herpesvirus type 2	May et al. (1990)
Bovine herpesvirus type 4	Kit et al. (1986)
Channel catfish virus	Harrison et al. (1991)
Equine herpesvirus type 1	Allen et al. (1978)
Equine herpesvirus type 3	McGowan et al. (1980)
Equine herpesvirus type 4	Nicolson et al. (1990)
Feline herpesvirus	Nunberg et al. (1989)
Herpesvirus ateles	Honess et al. (1982)
Herpesvirus saimiri	Honess et al. (1989)
Herpesvirus of turkeys	Martin et al. (1989)
Infectious laryngotracheitis virus	Griffen and Boursnell (1990)
Marmoset herpesvirus	Otsuka and Kit (1984)
Marek's disease virus	Scott et al. (1989)
Pseudorabies virus	Nohara and Kaplan (1963); Prieto et al. (1991)

ing the HSV *tk* gene have been constructed (Heyman et al. 1989; Borrelli et al. 1989; Al-Shawi et al. 1991). Cytoplasmic mRNA which codes for HSV *tk* was characterized (Cremer et al. 1979). In some studies, HSV *tk* was used as a marker in the biochemical transformation of cells (Munyon et al. 1971; Bacchetti and Graham 1977; Maitland and McDougall 1977; Pellicer et al. 1978).

HSV TK has been reported to have a relative molecular mass of 40–45 kDa in the denatured form (Honess and Watson 1974; Summers et al. 1975) and approximately 82 kDa (Cheng and Ostrander 1976) in the nondenatured, biologically active dimer form. Antigenic regions of HSV TK have been investigated in epitope-mapping studies (Zimmerman et al. 1991). Cell-free synthesis of active enzyme has been described (Cremer et al. 1978; Mahalingham et al. 1990), and HSV-1 TK enzyme has been purified and crystallized (Sanderson et al. 1988). Unlike cellular TK, it has been noted that HSV TK will phosphorylate various pyrimidine deoxynucleotide substrates (Cooper 1973; Jamieson et al. 1974; Cheng 1977; Coen et al. 1985). This relatively unselective promiscuous behavior also results in the phosphorylation of acyclovir, which permits the synthesis of the active antiviral form of the drug. HSV TK also possesses thymidylate kinase activity (Chen and Prusoff 1978; Fyfe et al. 1983).

Detection of HSV TK activity has most commonly been determined by phosphorylation of substrate. After HSV infection of monolayer cells deficient in cellular TK activity, the level of thymidine phosphorylation by infected cell extracts is taken to indicate the TK activity of the test virus (Maitland and McDougall 1977; Gronowitz and Kallander 1980; Veerisetty and Gentry 1983). Phosphorylation typically requires binding of thymidine and adenosine triphosphate (ATP; Darby et al. 1986; Liu and Summers 1988). Phosphorylation may be significantly altered in some mutants by salt concentration, and this may differentiate between the TK of standard HSV and the enzyme of mutant virus, which otherwise seems to be similar to that of wild-type virus in phosphorylation assays (Larder et al. 1983; Darby et al. 1984). In addition, HSV TK may be temperature sensitive (Larder and Darby 1982; Coen et al. 1989b); this might be of importance in vivo where temperature may approach nonpermissive levels. Inhibition assays, in which drugs such as arabinosylthymine or acyclovir are added to the medium of monolayer cells infected with HSV, have also been used to test for TK phenotype (Miller et al. 1977; Tenser et al. 1981; Cheng et al. 1982). In these assays, the antiviral drug inhibits TK^+, but not TK^- HSV. Substrate phosphorylation and drug inhibition assays require HSV replication and are relative insensitive to HSV mixtures containing both TK^+ and TK^- HSV. For this reason, particularly in the determination of TK phenotype of HSV isolates from experimentally infected animals, we have used a plaque autoradiographic assay. This method permits the graphic detection of amounts of TK^+ and TK^- HSV in TK^+–TK^- HSV mixtures (Summers and Summers 1977; Martin et al. 1985; Tenser and Edris 1986; Westheim et al. 1987). In our use of this procedure, monolayer cells were infected with HSV as in routine plaque assays. After 4 days, ^{14}[C]thymidine was added to the medium for several hours. Monolayers were then washed and fixed and exposed to X-ray film. After exposure, the film was removed and the number of black plaques on the film (which have incorporated thymidine and are therefore TK^+) was compared with total plaques in the cell monolayer (which are TK^+ and TK^-). An example of plaque autoradiographic testing is shown in Fig. 1.

The accurate determination of HSV TK phenotype has become particularly important in biological studies because of two occurrences. First, HSV expressing

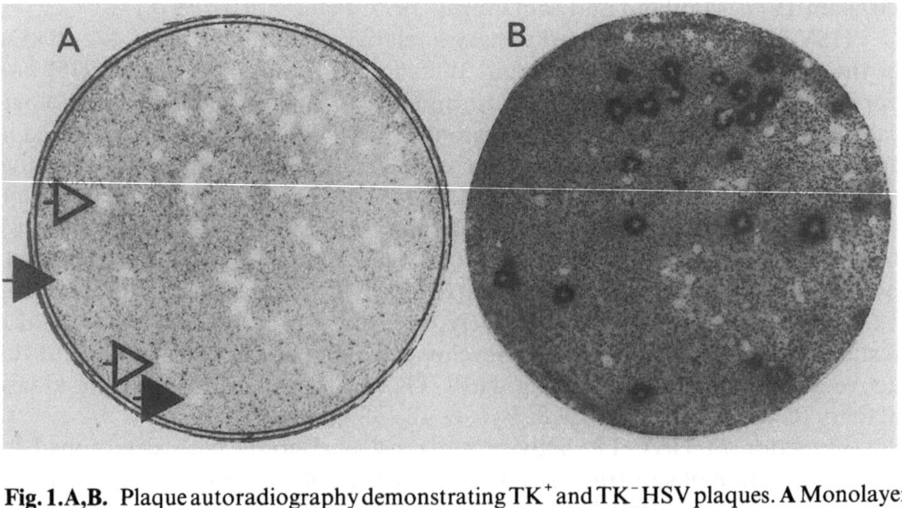

Fig. 1.A,B. Plaque autoradiography demonstrating TK^+ and TK^- HSV plaques. **A** Monolayer culture of primary rabbit kidney cells infected with a TK^+–TK^- HSV mixture. **B** Autoradiograph of the monolayer culture seen in **A** after ^{14}[C] thymidine treatment and X-ray film exposure. *Closed arrows* in **A** indicate plaques which are seen in **B** to be TK^+, and *open arrows* indicate TK^- plaques

an intermediate level of TK activity may occur (Summers et al. 1975; Darby et al. 1981; Dundarov et al. 1978; Martin et al. 1985; Tenser and Edris 1986) and may behave biologically more like TK^+ HSV than TK^- HSV; TK activity as little as 5%–10% of parental levels may be significant in this regard (Darby et al. 1984; Coen et al. 1989b). Second, if mixtures of TK^+ and TK^- HSV are present, the former may complement the latter. Intra-allelic complementation between HSV TK mutants was reported in cell culture experiments by Jamieson and Subak-Sharpe (1978) and is based on the occurrence of TK mutants with defects at different sites. In vivo complementation of TK^- HSV by TK^+ HSV (Field and Lay 1984; Tenser and Edris 1987; Ellis et al. 1989) and in vivo complementation between TK^- HSV mutants (Tenser et al. 1981) have also been demonstrated. Investigators have used HSV TK mutants which have been characterized by one or more of the above methods in studies of the role of HSV TK in pathogenesis, including neurovirulence and latency. The importance of appropriate HSV TK testing is emphasized, since the accurate interpretation of the biological results is predicated on accurate determination of TK phenotype.

The possible biological function(s) of HSV TK were not apparent in initial studies, largely because HSV TK expression did not appear to be necessary for virus replication in cell culture (Dubbs and Kit 1964). It is this feature which has been utilized in the selection of molecularly constructed mutants in some studies. For example, DNA insertion into the HSV *tk* gene converts the virus to TK^-. In the presence of acyclovir, nonmutagenized wild-type HSV is inhibited since it is TK^+, while mutagenized HSV carrying the inserted DNA of interest is TK^- and is not inhibited by acyclovir. The first suggestion of a possible biological role for HSV TK was reported in 1974, when Jamieson et al. (1974) noted that although HSV TK expression was dispensable for HSV replication in dividing cells in culture, TK expression was important for HSV replication in nondividing cells.

In view of the importance of HSV TK expression in nondividing cells, the neuronal locus of HSV latency (reviewed in Stevens 1989; Ho 1992) and the nondividing, TK-deficient nature of neurons, we and others have investigated the role of HSV TK expression in latency. Neurons express undetectable levels of cytoplasmic cellular TK. While mitochondrial TK activity persists in nondividing cells, cytoplasmic TK activity correlates with cell replication (Gross and Merrill 1989). Cellular TKs have been characterized in many studies with emphasis on cytoplasmic TK activity in neoplastic conditions (Munch-Peterson and Tyrsted 1988; Yusa et al. 1988). In studies of the developing nervous system, cellular cytoplasmic TK was reported to closely correlate with neuronal proliferation (Yamada et al. 1980), and in studies of embryonic sensory ganglia, thymidine uptake has been used as a marker of the last neuronal division (Lawson and Biscoe 1979).

Neurovirulence

The in vivo study of presumptive HSV TK mutants was reported in 1975 by Marcialis et al. They noted that iododeoxyuridine-resistant HSV-1 mutants were much less neurovirulent in rabbits and mice than were parental viruses; lethal infection was reduced. In addition, infection of animals with these mutants protected against subsequent challenge with the virulent parental viruses. Protection, presumably immunologically mediated, suggests the possibility that replication of the mutant viruses occurred in vivo. Immunological protection by initially administered TK mutant HSV would be more likely if that immunizing virus replicated in vivo than if it did not. From the pathogenesis standpoint, in vivo replication is needed prior to concluding that HSV TK expression is important, in this case, for neurovirulence. Without in vivo replication, it could not be concluded from this study that HSV TK expression is a neurovirulence factor. If a test virus does not replicate in vivo, it is probable that it will be of decreased pathogenicity. Decreased neurovirulence would be expected in that case, but it would probably not be appropriate to conclude that an identified defective function in the virus is a specific neurovirulence factor. However, if a mutant virus does replicate in vivo, it indicates a certain robustness of the virus, and if decreased lethal infection is noted, it is more appropriate to consider the altered viral function as being a neurovirulence factor.

Table 2. Importance of herpes simplex virus (HSV) thymidine kinase (TK) expression for neurovirulence

Restricted replication of TK^- HSV in neural tissue[a]

HSV inoculated	Days after corneal inoculation of HSV	Isolation of HSV after corneal inoculation (adult mice): PFU/eye	PFU/trigeminal ganglion
TK^+	1	8×10^3	None detected
	3	5×10^4	4×10^3
	5	1×10^3	8×10^0
TK^-	1	5×10^3	None detected
	3	2×10^4	None detected
	5	5×10^3	None detected

Decreased Mortality of TK^- HSV after peripheral inoculation[b]

HSV inoculated	Percent mortality after corneal or nasal inoculation: Adult mice	Newborn mice
TK^+	30%–45%	100%
TK^-	0%	20%–30%

Decreased mortality of TK^- HSV after interacerebral inoculation[a]

HSV inoculated	LD_{50} (PFU) after intracerebral inoculation: Adult mice	Newborn mice
TK^+	27.5	2.6
TK^-	57,000–500,000	1.7–2.2

LD_{50} Median lethal dose.
[a] Data from Tenser et al. (1979).
[b] Unpublished results. TK^+ KOS and TK^- KOS (Tenser et al. 1979) or TK^- 3B (J. Smiley, McMaster University, Hamilton, Ontario, Canada; Wilcox et al. 1992). Bilateral corneal or nasal inoculation (adult mice) or nasal inoculation (newborn mice) was performed (10^6 PFU).
[c] Data from Tenser (1983).

TK^- HSV replication has been well demonstrated in nonneural tissues, but replication is usually restricted in neural tissues (Table 2). Less information is available pertaining to expression of specific HSV gene products in neural tissues. After infection of mice with TK^- HSV, decreased in vivo expression of HSV genes was noted (Kosz-Vnenchak et al. 1990). In another study, the *Escherichia coli lacZ* gene was incorporated into the *tk* gene of HSV, resulting in TK^-HSV, and the gene product was expressed in neurons after animal infection (Ho and Mocarski 1988). Additional study of gene expression by TK^- HSV in neural tissues remains to be performed and would likely further the understanding of HSV TK expression and neurovirulence.

Decreased neurovirulence by HSV TK mutants has been reported by several groups (Field and Wildly 1978; Tenser et al. 1979; Tenser and Dunstan 1979; Klein et al. 1980; Darby et al. 1981; Tenser 1983; Gordon et al. 1983; Halliburton et al. 1987; Hill et al. 1991), although studies of virulent TK mutants has also been reported (Sakuma et al. 1988; lzumi and Stevens 1988). It is likely that HSV neurovirulence involves multiple host factors in addition to viral functions and is more complex than HSV latency. Neurovirulent TK mutants may be important in unraveling some of the complexity. It is thought there is probably sufficient evidence to conclude that HSV TK expression is important for both HSV replication in neural tissues and for neurovirulence. The correlation between TK activity and neurovirulence, however, may not be as strong as that for HSV latency.

It is emphasized that virulence is complex, and it is beyond the scope of this review to discuss other HSV factors which have been related to virulence. In the consideration of HSV TK mutants, complexity is suggested by the possible occurrence of more than one mutation in a virus. For example, it has been reported that the open reading frame of the HSV *UL24* gene overlaps and is on the DNA strand opposite to the *tk* gene (Jacobson et al. 1989), and a TK mutant might also have a defect in *UL24*. However, *UL24* mutants show decreased replication, and this would limit the use of a TK^- *UL24* mutant in attempting to study the role of HSV TK and neurovirulence. Impaired replication in vitro and in vivo in nonneural tissues would suggest a generally defective virus. Replication of a mutant in vivo in nonneuronal tissues would suggest robustness of the virus and permit the consideration of neurovirulence.

The specific role of HSV TK expression in neurovirulence is suggested in studies in which replication of HSV TK mutants in neural tissue was decreased, while remaining relatively intact in nonneural tissue. Similarly, mortality usually due to brain infection after infection with HSV TK mutants has often been noted to be decreased. Examples of decreased neurovirulence by HSV TK mutants are provided in Table 2. In Table 2(A), decreased replication of TK mutant virus in trigeminal ganglion tissue, but intact replication in nonneural ocular tissue, is shown. In Table 2(B and C), decreased mortality of HSV TK mutants in adult mice is shown. Mortality in newborn mice was significantly decreased after peripheral inoculation (B), but was very similar to that of TK^+ HSV after direct intracerebral inoculation (C). While it is well known that immunological reactivity to HSV in newborn mice is impaired, these results also suggested that TK^- HSV might be more virulent because of the presence of dividing cells. The latter situation would exist in newborn mice. The possible importance of cell replication in addition to an immature immune system in newborn mice is suggested from a study of several nude mice infected with TK^- HSV. Little or no TK^- HSV was found in the ganglia of these adult mice, despite the immune impairment present (Field and Wildy 1978). TK^- HSV replication in dividing cells in vivo was directly investigated after partial hepatectomy in adult mice, a situation in which the remaining liver undergoes rapid proliferation. In that study, TK^- HSV replicated to a much greater degree in the presence of dividing liver cells (Fang et al. 1983).

Evidence of decreased neurovirulence by TK^- HSV mutants led to investigations of other herpesviruses. Decreased virulence of TK^- pseudorabies virus was reported (Tenser et al. 1983; Kit et al. 1985), and TK^- pseudorabies virus has been used as a vaccine (van Oirschot et al. 1990; Volz et al. 1992). Reactivation of TK^- pseudorabies virus in the natural pig host after steroid treatment was detected only occasionally (Mengeling 1991), while evidence of viral DNA in trigeminal ganglia was very frequent (Volz et al. 1992). Occasional reactivation of TK^- bovine herpesvirus 1 after steroid treatment of cattle was also reported (Whetstone et al. 1992).

Latency

In the late 1970s, decreased HSV latency after infection of mice and guinea pigs was reported by Field and Wildy (1978) and by our group (Tenser et al. 1979; Tenser and Dunstan 1979). In these studies, TK^- mutants were obtained by HSV selection in the presence of drugs, and characterization of mutants was not complete. We relied upon the presence of adequate viral replication in vivo in ocular tissues to minimize the likelihood that a second mutation in addition to that of TK activity was present. In addition, multiple independently selected mutants were used. Latency was defined by the isolation of HSV after cultivation of ganglion explants. Therefore, these studies did not differentiate among the establishment or maintenance of latency or reactivation. In the following discussion, the term "latency" will be generally considered to include the establishment and maintenance of latent infections as well as the capacity for reactivation. In some instances, reactivation from latency will be specifically considered.

These initial in vivo studies were generally supported by other investigations (Field and Darby 1980; Klein et al. 1980; Darby et al. 1981; Gordon et al. 1983; Tenser et al. 1985), although some reports noted that HSV TK expression was not an absolute requirement for latency (Gordon et al. 1984; Stanberry et al. 1985; Sears et al. 1985). This included studies with HSV-2 (Stanberry et al. 1985), which is usually of increased virulence and may be less dependent on TK expression (Tenser et al. 1985; Sakuma et al. 1988), and an investigation in which immunosuppression appeared to increase TK mutant latency (Gordon et al. 1984). The latter observation may be related to the occurrence of increased latency by TK mutants in newborn mice (Table 3). In addition, it has been considered that the importance of HSV TK expression might vary in different species. Generally, however, an important role for HSV TK expression and latency was supported.

Table 3. Importance of herpes simplex virus (HSV) thymidine kinase (TK) expression for latency

LAT is readily detected after TK^- HSV inoculation[a]

HSV inoculated	Detection of LAT during latency by in situ hybridization: Adult mice, Percentage of ganglia LAT-positive	Adult mice, Percentage of neurons LAT-positive	Newborn mice, Percentage of ganglia LAT-positive	Newborn mice, Percentage of neurons LAT-positive
TK^+	96	0.7	ND	ND
TK^-	95–100	0.2–0.3	100	10.9

Reactivation is restricted after TK-HSV HSV inoculation[b]

HSV inoculated	Reactivation of HSV from latency-infected ganglia: Adult mice, Percentage of ganglia positive	Newborn mice, Percentage of ganglia positive
TK^+	83	ND
TK^-	0	40

LAT, latency-associated transcript; ND, not done.

[a] Studies of dorsal root ganglia after footpad inoculation of HSV-1. Data for adult mice from Tenser et al. (1989). Data for newborns, unpublished (3B TK^- HSV, J. Smiley, McMaster University, Hamilton, Ontario, Canada; 5×105 PFU). Newborn mice do not survive even small amounts of TK^+ HSV, and TK^+ HSV latency could not be produced.

[b] Studies of dorsal root ganglia after footpad inoculation of HSV-1. Data for adult mice from Tenser et al. (1989). Data for newborns, unpublished (3B TK^- HSV as above).

This included more recent studies with defined insertion and deletion HSV TK mutants (Efstathiou et al. 1989; Leist et al. 1989; Coen et al. 1989a; Tenser et al. 1989). Decreased infection of neurons in vitro by TK^- HSV was also reported (Price and Khan 1981), although at a high multiplicity of infection, frequent neuronal infection was observed (Wilcox et al. 1992).

In complementation studies, it was possible to establish in vivo latent infection by TK^- HSV. Complementation of TK^- HSV by TK^+ HSV was detected, following the infection of experimental animals with HSV mixtures (Field and Lay 1984; Tenser and Edris 1987; Ellis et al. 1989; Efstathiou et al. 1989). In addition to complementation between HSV-1 isolates in these studies, complementation of TK^- HSV-2 by HSV-1 was reported (Nishiyama et al. 1991). The occurrence of TK complementation, so that TK^- HSV is isolated during latency, suggested a role for HSV TK expression in latency. In one study, after the inoculation of a TK^+– HSV mixture, both viruses were isolated from latently infected explants, but if arabinosylthymine, which inhibits TK^+ HSV, was present in the explant medium, neither virus was isolated (Tenser and Edris 1987). This suggests that both viruses were present in the same neurons and were, therefore, both inhibited following phosphorylation of acyclovir by TK^+ HSV, or alternatively, that TK^- HSV was reactivation defective. Of potential clinical importance is the possible complementation by wild-type TK^+ HSV of TK^- virus which has been selected by antiviral usage. It was initially though that althought TK^- HSV might be selected for and might replicate in cutaneous tissue of immunosuppressed patients, it would be unlikely that TK^- HSV would establish latent ganglion infection (Klein 1982).

However, if TK^- HSV is complemented by TK^+ HSV, TK^- latency might occur in humans, as has been noted in experimental animals.

In early studies of HSV TK mutants and latency, latency was defined by the ability to isolate HSV from ganglion explants. In 1989, however, Coen et al. (1989a), Leist et al. (1989), and Tenser et al. (1989) reported results of investigations in which, instead of attempting the isolation of HSV from putatively infected ganglia, evidence of latency-associated transcript (LAT) in ganglia was determined by in situ hybridization. LAT was readily detected. In situ hybridization to detect HSV RNA in neurons of latently infected experimental animals was first noted in 1982 (Tenser et al. 1982), and determination of the RNA species expressed was initially described in 1987 by Stevens and coworkers (1987). LAT is the only HSV RNA detected in neurons of latently infected ganglia (reviewed in Stevens 1989; Ho 1992). The detection of LAT in ganglion neurons after infection by TK^- HSV, in conjunction with the general inability to detect reactivated virus in ganglion explants, led to the conclusion that HSV TK was important for the reactivation of latent HSV (Coen et al. 1989b; Leist et al. 1989; Tenser et al. 1989). This conclusion was supported by the report that TK^- HSV in ganglion tissues could be recovered by complementation rescue by TK^+ HSV (Efstathiou et al. 1989). In that study, ganglia from mice infected with TK^- HSV were removed and superinfected with TK^+ HSV in vitro, and the isolation of TK^- HSV from ganglia after long-term infection was demonstrated. In addition, drugs which inhibit HSV TK expression, but which do not or only slightly inhibit HSV replication, were reported to diminish reactivation of TK^+ HSV from ganglion explants (Leib et al. 1990), further evidence for the role of HSV TK expression in reactivation. The observation that similar amounts of HSV DNA were present in ganglia during TK^+ and TK^- latency suggested that HSV TK expression was not important for the establishment of latency (Leist et al. 1989). The expression of LAT after TK^- HSV infection of mice despite the restricted ganglion expression of several HSV genes indicated the establishment of latency in the presence of restricted lytic gene expression (Kosz-Vnenchak et al. 1990). The use of additional drugs which inhibit HSV TK, but not viral replication (Nutter et al. 1987; Martin et al. 1989), and antisense RNA inhibition of HSV TK (Izant and Weintraub 1984; Kim and Wold 1985) may permit further definition of the role of HSV TK expression in latency.

Some have emphasized that HSV TK is not important for latency but only reactivation, using the term “latency” to indicate the establishment of the latent infection. These and other observations raise questions of definitions of latency. Initially, latency was defined by the isolation of HSV in explants (Stevens and Cook 1971). This continues to be our preference, although the observation that exogenous agents may enhance reactivation in explants (Whitby et al. 1987; Leib et al. 1989b) suggests the importance of methodology. If emphasis is placed on reactivation capacity, possibly TK^- HSV latency should be termed “abortive” or “incomplete” latency. In addition, it is evident that the presence of HSV RNA does not define latency, since LAT-defective mutants can establish latent infections (reviewed in Ho 1992). Reactivation from ganglia latently infected with LAT mutants occurs, albeit possibly with reduced kinetics, and reactivation in vivo may be impaired. The presence of HSV DNA is a logical requirement for latency, but a requirement for reactivation capacity remains uncertain. Definitions of latency remain to be clarified, and the role of HSV TK expression in latency will in part based on definitions of latency. At the present time, it is concluded that HSV TK expression is important for latency, using the term to include establishment, maintenance, and reactivation.

Evidence of decreased latency after TK^- HSV inoculation of mice is presented in Table 3 (A), where latency was defined by the decreased isolation of HSV from ganglion explants of adult mice. LAT was readily detected. In newborn mice, TK^- HSV was reactivated from explants during latency from 40% of mice. In addition, the proportion of LAT-positive neurons was almost 50-fold higher than in adult mice. These results suggest that in newborn mice, increased latency may result with TK^- HSV, both in terms of numbers of LAT- positive neurons and HSV isolation. Reactivation from explants after newborn infection may be related to a dosage effect, similar to the observation that while for in vitro latency, TK^- HSV is latency negative at low multiplicity of infection, at high multiplicity, reactivation from latency frequently occurred (Wilcox et al. 1992). TK^- HSV latency in the newborn model may provide a useful method to study the role of HSV TK expression in latency. Increased TK^- HSV latency in newborn mice may also be related to host factors, which are also important for increased virulence (Table 2). It is hypothesized that HSV TK expression is important in the reactivation of HSV from latency, and this is related to the nondividing and generally TK^- status of neurons. Neurons of some species or of immature animals may be more permissive, possibly related to cellular TK or related enzymes.

The conclusion that HSV TK expression is important for HSV reactivation in the "establishment-of-latency–maintenance–reactivation" model is supported by evidence of LAT after TK^- HSV infection (Coen et al. 1989b; Leist et al. 1989; Tenser et al. 1989; Kosz-Vnenchak et al. 1990), by rescue of latent TK^- HSV (Efstathiou et al. 1989), and by the use of drugs which inhibit HSV TK, but not HSV replication (Leib et al. 1990). In addition, evidence that HSV replication is not necessary for the establishment of latency (Steiner et al. 1990; Kosz-Vnenchak et al. 1990) minimizes the possibility that TK expression is important for the initial establishment of latency. It remains probable, however, that limited replication of TK^- HSV in vivo in neural tissues relates to the decreased virulence of these mutants.

Acknowledgements. This work was supported in part by Public Health Service Award NS20684. Helpful comments by Dr Kathleen A. Hay and Dr. M. Judith Tevethia are gratefully acknowledged, as is the secretarial assistance of Mrs. Tricia R. Spitler.

References

Allen GP, McGowan JJ, Gentry Ga, Randall CC (1978) Biochemical transformation of deoxythymidine kinase-deficient mouse cells with UV-irradiated equine herpesvirus type 1. J Virol 28: 361–367

Al-Shawi R, Burke J, Wallace H, Jones C, Harrison S, Buxton D, Maley S, Chandley A, Bishop JO (1991) The herpes simplex virus type 1 thymidine kinase is expressed in the testes of transgenic mice under the control of a cryptic promoter. Mol Cell Biol 11: 4207–4216

Bacchetti S, Graham FL (1977) Transfer of the gene for thymidine kinase to thymidine kinase-deficient human cells by purified herpes simplex viral DNA. Proc Natl Acad Sci USA 74: 1590–1594

Birch CJ, Tyssen DP, Tachedjian G, Doherty R, Hayes K, Mijch A, Lucas CR (1992) Clinical effects and in vitro studies of trifluorothymidine combined with interferon-α for treatment of drug-resistant and -sensitive herpes simplex virus infections. J Infect Dis 166: 108–112

Borrelli E, Heyman RA, Arias C, Sawchenko PE, Evans RM (1989) Transgenic mice with inducible dwarfism. Nature 339: 538–541

Burns WH, Saral R, Santos GW, Laskin OL, Lietman PS, McLaren C, Barry DW (1982) Isolation and characterization of resistant herpes simplex virus after acyclovir therapy. Lancet 1: 421–423

Chatis PA, Miller CH, Schrager LE, Crumpacker CS (1989) Successful treatment with foscarnet of an acyclovir-resistant mucocutaneous infection with herpes simplex virus in a patient with acquired immunodeficiency syndrome. N Engl J Med 320: 297–300

Chen MS, Prusoff WH (1978) Association of thymidylate kinase activity with pyrimidine deoxyribonucleoside kinase induced by herpes simplex virus. J Biol Chem 253: 1325–1327

Cheng Y-C (1977) A rational approach to the development of antiviral chemotherapy: alternative substrates of herpes simplex virus type 1 (HSV-1) and type 2 (HSV-2) thymidine kinase (TK). Ann NY Acad Sci 284: 594–598

Cheng Y-C, Ostrander M (1976) Deoxythymidine kinase induced in HeLa TK^- cells by herpes simplex virus type I and type II. J Biol Chem 251: 2605–2610

Cheng Y–C, Schinazi RF, Dutschman GE, Tan RS, Grill SP (1982) Virus-induced thymidine kinases as markers for typing herpes simplex viruses and for drug sensitivity assays. J Virol Methods 5: 209–217

Chiocca EA, Choi BB, Cai W, DeLuca NA, Schaffer PA, DiFiglia M, Breakefield OX, Martuza RL (1990) Transfer and expression of the lac Z gene in rat brain neurons mediated by herpes simplex virus mutants. New Biol 2: 739–745

Coen DM, Schaffer PA (1980) Two distinct loci confer resistance to acycloguanosine in herpes simplex virus type 1. Proc Natl Acad Sci USA 77: 2265–2269

Coen DM, Fleming HE Jr, Leslie LK, Retondo MJ (1985) Sensitivity of arabinosyladenine-resistant mutants of herpes simplex virus to other antiviral drugs and mapping of drug hypersensitivity mutations to the DNA polymerase locus. J Virol 53: 477–488

Coen DM, Kosz-Vnenchak M, Jacobson JG, Leib DA, Bogard CL, Schaffer PA, Tyler KL, Knipe DM (1989a) Thymidine Kinase-Negative herpes simplex virus mutants establish latency in mouse trigeminal ganglia but do not reactivate. Proc Natl Acad Sci USA 86: 4736–4740

Coen DM, Irmiere AF, Jacobson JG, Kerns KM (1989b) Low levels of herpes simplex virus thymidine-thymidylate kinase are not limiting for sensitivity to certain antiviral drugs or for latency in a mouse model. Virology 168: 221–231

Cooper GM (1973) Phosphorylation of 5-bromodeoxycytidine in cells infected with herpes simplex virus. Proc Natl Acad Sci USA 70: 3788–3792

Cremer KJ, Bodemer M, Summers WP, Summers WC, Gesteland RF (1979) In vitro suppression of UAG and UGA mutants in the thymidine kinase gene of herpes simplex virus. Proc Natl Acad Sci USA 76: 430–434

Crumpacker CS, Schnipper LE, Marlowe SI, Kowalsky PN, Hershey BJ, Levin MJ (1982) Resistance to antiviral drugs of herpes simplex virus isolated from a patient treated with acyclovir. N Engl J Med 306: 343–346

Crumpacker CS, Kowalsky PN, Oliver SA, Schnipper LE, Field AK (1984) Resistance of herpes simplex virus to a 9-[2-hydroxy-1-(hydroxymethy) ethoxy]methyl guanine: physical mapping of drug synergism within the viral DNA polymerase locus. Proc Natl Acad Sci USA 81: 1556–1560

Darby G, Field HJ, Salisbury SA (1981) Altered substrate specificity of herpes simplex virus thymidine kinase confers acyclovir-resistance. Nature 289: 81–83

Darby G, Churcher MJ, Larder BA (1984) Cooperative effects between two acyclovir resistance loci in herpes simplex virus. J Virol 50: 838–846

Darby G, Larder BA, Inglis MM (1986) Evidence that the "active centre" of the herpes simplex virus thymidine kinase involves an interaction between three distinct regions of the polypeptide. J Gen Virol 67: 753–758
de Jong JT, Aker J, den Dulk H, van de Putte P, Giphart-Gassler M (1989) Cytosine methylation in the EcoR 1 site of active and inactive herpesvirus thymidine kinase promoters. Biochim Biophys Acta 1008: 62–70
Derse D, Cheng Y-C, Furman PA, St. Clair MH, Elion GB (1981) Inhibition of purified human and herpes simplex virus-induced DNA polymerases by 9-(2-hydroxyethoxymethyl) guanine triphosphate. Effects on primer-template function. J Biol Chem 256: 11447–11451
Dobersen MJ, Jerkofsky M, Greer S (1976) Enzymatic basis for the selective inhibition of varicella-zoster virus by 5-halogenated analogues of de oxycytidine. J Virol 20: 478–486
Dubbs DR, Kit S (1964) Mutant strains of herpes simplex deficient in thymidine kinase-inducing activity. Virology 22: 493–502
Dundarov S, Dundarova D, Todorov S, Kavaklova L, Falke D (1978) Induction capacity and influence of dThdMP on thymidine kinase activity of type 1 and 2 strains of herpes simplex virus. Arch Virol 56: 243–249
Efstathiou S, Kemp S, Darby G, Minson AC (1989) The role of herpes simplex virus type 1 thymidine kinase in pathogenesis. J Gen Virol 70: 869–879
Elion GB, Furman PA, Fyfe JA, deMiranda P, Beauchamp L, Schaeffer HJ (1977) Selectivity of action of an antiherpetic agent 9-(2-hydroxyethoxymethyl) guanine. Proc Natl Acad Sci USA 74: 5716–5720
El Kareh A, Murphy AJM, Fichter T, Efstratiadis A (1985) "Transactivation" control signals in the promoter of the herpesvirus thymidine kinase gene. Proc Natl Acad Sci USA 82: 1002–1006
Ellis MN, Waters R, Hill EL, Lobe DC, Selleseth DW, Barry DW (1989) Orofacial infection of athymic mice with defined mixtures of acyclovir-susceptible and acyclovir-resistant herpes simplex virus type 1. Antimicrob Agents Chemother 33: 304–310
Erlich KS, Mills J, Chatis P, Mertz GJ, Busch DF, Follansbee SE, Grant RM, Crumpacker CS (1989) Acyclovir-resistant herpes simplex virus infections in patients with the acquired immunodeficiency syndrome. N Engl J Med 320: 293–296
Ezzeddine ZD, Martuza RL, Platika D, Short MP, Malick A, Choi B, Breakefield XO (1991) Selective killing of glioma cells in culture and in vivo by retrovirus transfer of the herpes simplex virus thymidine kinase gene. New Biol 3: 608–614
Fang Z-Y, Tenser RB, Rapp F (1983) Hepatic infection by thymidine kinase-positive and thymidine kinase-negative herpes simplex virus after partial hepatectomy. Infect Immun 42: 402–408
Field HJ, Darby G (1980) Pathogenicity in mice of strains of herpes simplex virus which are resistance to acyclovir and vitro and in vivo. Antimicrob Agents Chemother 17: 209–216
Field HJ, Lay E (1984) Characterization of latent infections in mice inoculated with herpes simplex virus which is clinically resistant to acyclovir. Antiviral Res 4: 43–52
Field HJ, Wildly P (1978) The pathogencity of thymidine kinase-dificient mutants of herpes simplex in mice. J Hyg 81: 267–277
Friedmann T (1989) Progress toward human gene therapy. Science 244: 1275–1281
Fyfe JA, Kellèr PM, Furman PA, Miller RL, Elion GB (1978) Thymidine kinase from herpes simplex virus phosphorylates the new antiviral compound 9-(2-hydroxyethoxymethyl) guanine. J Biol Chem 253: 8721–8727
Fyfe JA, McKee SA, Keller PM (1983) Altered thymidine-thymidylate kinases from strains of herpes simplex virus with modified drug sensitivities to acyclovir and (E)-5-(2 bromovinyl)-2′-deoxyuridine. Mol Pharmacod 24: 316–323
Gordon YJ, Gilden DM, Becher Y (1983) HSV-1 thymidine kinase promotes virulence and latency in the mouse. Invest Ophthalmol Vis Sci 24: 599–602
Gordon YJ, Rao H, Arullo-Cruz T (1984) Immunosuppression promotes ocular virus replication and CNS neurovirulence following corneal inoculation with an avirulent herpes simplex type 1 thymidine kinase negative mutant. Curr Eye Res 3: 651–657
Graessmann A, Graessmann M (1988) DNA methylation, chromotin structure and regulation of herpes simplex virus tk gene expression. Gene 74: 135–137

Griffin AM, Boursnell MEG (1990) Analysis of the nucleotide sequence of DNA from the region of the thymidine kinase gene of infectious laryngotracheitis virus: potential evolutionary relationships between the herpes virus subfamilies. J Gen Virol 71: 841–850
Gronowitz JS, Kallander CFR (1980) Optimized assay for thymidine kinase and its applications to the detection of antibodies against herpes simplex virus type 1- and 2-induced thymidine kinase. Infect Immun 29: 425–434
Gross MK, Merrill GF (1989) Thymidine kinase synthesis is repressed in nonreplicating muscle cells by a translational mechanism that does not affect the polysomal distribution of thymidine kinase mRNA. Proc Natl Acad Sci USA 86: 4987–4991
Halliburton IW, Morse LS, Roizman B, Quinn KE (1980) Mapping of the thymidine kinase genes of type 1 and type 2 herpes simplex viruses using intertypic recombinants. J Gen Virol 49: 235–253
Halliburton IW, Honess RW, Killington RA (1987) Virulence is not conserved in recombinants between herpes simplex virus types 1 and 2. J Gen Virol 68: 1435–1440
Harrison PT, Thompson R, Davison AJ (1991) Evolution of herpesvirus thymidine kinase from cellular deoxycytidine kinase. J Gen Virol 72: 2583–2586
Heyman RA, Borrelli E, Lesley J, Anderson D, Richman DD, Baird SM, Hyman R, Evans RM (1989) Thymidine kinase obliteration: creation of transgenic mice with controlled immune deficiency. Proc Natl Acad Sci USA 86: 2698–2702
Hill EL, Hunter GA, Ellis MN (1991) In vitro and in vivo characterization of herpes simplex virus clinical isolates recovered from patients infected with human immunodeficiency virus. Antimicrob Agents Chemother 35: 2322–2328
Ho DY (1992) Herpes simplex virus latency: molecular aspects. Prog Med Virol 39: 76–115
Ho DY, Mocarski ES (1988) β-Galactosidase as a marker in the peripheral and neural tissues of the herpes simplex virus-infected mouse. Virology 167: 279–283
Honess RW, Watson DN (1974) Herpes simplex virus-specific polypeptides studies by polyacrylamide gel electrophoresis of immune precipitates. J Gen Virol 22: 171–185
Honess RW, O'Hare PO, Young D (1982) Comparison of thymidine kinase activities induced in cells productively infected with herpesvirus saimiri and herpes simplex virus. J Gen Virol 58: 237–249
Honess RW, Craxton MA, Williams L, Gompels UA (1989) A comparative analysis of the sequence of the thymidine kinase gene of a gammaherpesvirus, herpesvirus saimiri. J Gen Virol 70: 3003–3013
Izant JG, Weintraub H (1984) Inhibition of thymidine kinase gene expression by anti-sense RNA: a molecular approach to genetic analysis. Cell 36: 1007–1015
Izumi KM, Stevens JG (1988) Two thymidine kinase deficient herpes simplex viruses exhibit unexpected virulence properties. Microbiol Pathogen 4: 145–153
Jacobson JG, Martin SL, Coen DM (1989) A conserved open reading frame that overlaps the herpes simplex virus thymidine kinase gene is important for viral growth in cells culture. J Virol 63: 1839–1843
Jamieson AT, Subak-Sharpe JH (1978) Interallelic complementation of mutants of herpes simplex virus deficient in deoxypyrimidine kinase activity. Virology 85: 109–117
Jamieson AT, Gentry GA, Subak-Sharpe JH (1974) Induction of both thymidine and deoxycytidine kinase activity by herpes virus. J Gen Virol 24: 465–480
Kim SK, Wold BJ (1985) Stable reduction of thymidine kinase activity in cells expressing high levels of anti-sense RNA. Cell 42: 129–138
Kit S (1985) Thymidine kinase. Microbiol Sci 2: 369–375
Kit S, Dubbs DR (1963) Acquisition of thymidine kinase activity by herpes simplex infected mouse fibroblast cells. Biochem Biophys Res Commun 11: 55–59
Kit S, Kit M, Pirtle EC (1985) Attenuated properties of thymidine kinase-negative deletion mutant of pseudorabies virus. Am J Vet Res 46: 1359–1367
Kit S, Kit M, Ichimura H, Crandell R, McConnell S (1986) Induction of thymidine kinase activity by viruses with group B DNA genomes: bovine cytomegalovirus (bovine herpesvirus 4). Virus Res 4: 197–212
Klein RJ (1982) Acyclovir–resistant herpes simplex virus. N Engl J Med 307: 681–682

Klein RJ, DeStefano E, Brady E, Friedman-Kien AE (1980) Experimental skin infection with an acyclovir resistant herpes simplex virus mutant: response to antiviral treatment and protection against reinfection. Arch Virol 65: 237–246

Klemperer HG, Haynes GR, Shedden WIH, Watson DH (1967) A virus-specific thymidine kinase in BHK21 cells infected with herpes simplex virus. Virology 31: 120–128

Kosz-Vnenchak M, Coen DM, knipe DM (1990) Restricted expression of herpes simplex virus lytic genes during establishment of latent infection by thymidine kinase-negative mutant viruses. J Virol 64: 5396–5402

Larder BA, Darby G (1982) Properties of a novel thymidine kinase induced by an acyclovir-resistant herpes simplex virus type 1 mutant. J Virol 42: 649–658

Larder BA, Cheng Y-C, Darby G (1983) Characterization of abnormal thymidine kinases induced by drug-resistant strains of herpes simplex virus type 1. J Gen Virol 64: 523–532

Lawson SN, Biscoe TJ (1979) Development of mouse dorsal root ganglia: an autoradiographic and quantitative study. J Neurocytol 8: 265–274

Leib DA, Coen DM, Bogard CL, Hicks KA, Yager DR, Knipe DM, Tyler KL, Schaffer PA (1989a) Immediate-early regulatory gene mutants define different stages in the establishment and reactivation of herpes simplex virus latency. J Virol 63: 759–768

Leib DA, Bogard CL, Kosz-Vnenchak M, Hicks KA, Coen DM, Knipe DM, Schaffer PA (1989b) A deletion mutant of the latency-associated transcript of herpes simplex virus type 1 reactivates from the latent state with reduced frequency. J Virol 63: 2893–2900

Leib DA, Ruffner KL, Hildebrand C, Schaffer PA, Wright GE, Coen DM (1990) Specific inhibitors of herpes simplex virus thymidine kinase diminish reactivation of latent virus from explanted murine ganglia. Antimicrob Agents Chemother 34: 1285–1286

Leist TP, Sandri-Goldin RM, Stevens JG (1989) Latent infections in spinal ganglia with thymidine kinase–deficient herpes simplex virus. J Virol 63: 4976–4978

Lettler E, Arrand JR (1988) Characterization of the Epstein-Barr virus-encoded thymidine kinase expressed in heterologous eucaryotic and procaryotic systems. J Virol 62: 3892–3895

Liu Q, Summers WC (1988) Site-directed mutagenesis of a nucleotide-binding domain in HSV-1 thymidine kinase: effects on catalytic activity. Virology 163: 638–642

Mahalingham R, Cabirac G, Wellish M, Gilden D (1990) In-vitro synthesis of functional varicella zoster and herpes simplex viral thymidine kinase. Virus Genes 4: 105–120

Maitland NJ, McDougall JK (1977) Biochemical transformation of mouse cells by fragments of herpes simplex virus DNA. Cell 11: 233–241

Marcialis MA, LaColla P, Schivo ML, Flore O, Firinu A, Loddo B (1975) Low virulence and immunogenicity in mice and in rabbits of variants of herpes simplex virus resistant to 5-iodo-2-deoxyuridine. Experientia 31: 502–503

Martin JA, Duncan IB, Hall MJ, Wong-Kai-In P, Lambert RW, Thomas GJ (1989) New potent and selective inhibitors of herpes simplex virus thymidine kinase. Nucleosides nucleotides 8: 753–764

Martin JL, Ellis MN, Keller PM, Biron KK, Lehrman SN, Barry DW, Furman PA (1985) Plaque autoradiography assay for the detection and quatitation of thymidine kinase-deficient and thymidine kinase—altered mutants of herpes simplex virus in clinical isolates. Antimicrob Agents Chemother 28: 181–187

Martin SL, Aparisio DI, Bandyopadhyay PK (1989) Genetic and biochemical characterization of the thymidine kinase gene from herpesvirus of turkeys. J Virol 63: 2847–2852

Martuza RL, Malick A, Markert JM, Ruffner KL, Coen DM (1991) Experimental therapy of human glioma by means of a genetically engineered virus mutant. Science 252: 854–856

May JT, Awad M, Reum A, Sheppard M (1990) Bovine herpes pammilitis virus thymidine kinase. Acta Virol 34: 188–192

McDougall JK, Masse TH, Galloway DA (1980) Location and cloning of the herpes simplex virus type 2 thymidine kinase gene. J Virol 33: 1221–1224

McGowan JJ, Allen GP, Barnett JM, Gentry GA (1980) Deoxythymidine kinase metabolism in equine herpesvirus type 3 infected horse embryo dermal fibroblasts. Virology 101: 516–519

McKnight SL (1980) The nucleotide sequence and transcript map of the herpes simplex virus thymidine kinase gene. Nucleic Acids Res 8: 5949–5964

McKnight SL, Gavis ER (1980) Expression of the herpes thymidine kinase gene in Xenopus laevis oocytes: an assay for the study of deletion mutants constructed in vitro. Nucleic Acids Res 8: 5931–5948
Mengeling WL (1991) Virus reactivation in pigs latently infected with a thymidine kinase negative vaccine strain of pseudorabies virus. Arch Virol 120: 57–70
Miller JM, Whetstone CA, Bello LJ, Lawrence WC (1991) Determination of ability of a thymidine kinase–negative deletion mutant of bovine herpesvirus–1 to cause abortion in cattle. Am J Vet Res 52: 1038–1043
Miller RL, Iltis JP, Rapp F (1977) Differential effect of arabinofuranosylthymine on the replication of human herpesviruses. J Virol 23: 679–684
Moolten FL, Wells JM (1990) Curability of tumors bearing herpes thymidine kinase genes transferred by retroviral vectors. J Natl Cancer Inst 82: 297–300
Munch–Peterson B, Tyrsted G (1988) Thymidine kinase in human leukemia. Expression of the lymphoblastic isoenzyme in three patients with acute myelocytic leukemia. Leuk Res 12: 173–178
Munyon W, Kraiselburd E, Davis D, Mann J (1971) Transfer of thymidine kinase to thymidine kinaseless L cells by infection with ultraviolet-irradiated herpes simplex virus. J Virol 7: 813–820
Nicolson L, Cullinane AA, Onions DE (1990) The nucleotide sequence of the equine herpesvirus 4 thymidine kinase gene. J Gen Virol 71: 1801–1805
Nishiyama Y, Kimura H, Daikoku T (1991) Complementary lethal invasion of the central nervous system by nonneuroinvasive herpes simplex virus types 1 and 2. J Virol 65: 4520–4524
Nohara H, Kaplan AS (1963) Induction of a new enzyme in rabbit kidney cells by pseudorabies virus. Biochem Biophys Res Commun 12: 189–193
Nunberg JH, Wright DK, Cole GE, Petrovskis EA, Post LE, Compton T, Gilbert JH (1989) Identification of the thymidine kinase gene of feline herpesvirus: use of degenerate oligonucleotides in the polymerase chain reaction to isolate herpesvirus gene homologs. J Virol 63: 3240–3249
Nutter LM, Grill SP, Dutschman GE, Sharma RA, Bobek M, Cheng Y-C (1987) Demonstration of viral thymidine kinase inhibitor and its effect on deoxynucleotide metabolism in cells infected with herpes simplex virus. Antimicrob Agents Chemother 31: 368–374
Otsuka H, Kit S (1984) Nucleotide sequence of the marmoset herpesvirus thymidine kinase gene and predicted amino acid sequence of thymidine kinase polypeptide. Virology 135: 316–330
Palella TD, Hidaka Y, Silverman LJ, Levine M, Glorioso J, Kelley WN (1989) Expression of human HPRT mRNA in brains of mice infected with a recombinant herpes simplex virus-1 vector. Gene 80: 137–144
Parris DS, Harrington JE (1982) Herpes simplex virus variants resistant to high concentrations of acyclovir exist in clinical isolates. Antimicrob Agents Chemother 22: 71–77
Pellicer A, Wigler M, Axel R, Silverstein S (1978) The transfer and stable integration of the HSV thymidine kinase gene into mouse cells. Cell 14: 133–141
Post LE, Mackem S, Roizman B (1981) Regulation of α genes of herpes simplex virus: expression of chimeric genes produced by fusion of thymidine kinase with α gene promoters. Cell 24: 555–565
Price RW, Khan A (1981) Resistance of peripheral autonomic neurons to in vivo productive infection by herpes simplex virus mutants deficient in thymidine kinase activity. Infect Immun 34: 571–580
Prieto J, Martin Hernandez AM, Taborés E (1991) Loss of pseudorabies virus thymidine kinase activity due to a single base mutation and amino acid substitution. J Gen Virol 72: 1435–1439
Reyes GR, Jeang K-T, Hayward GS (1982) Transfection with the isolated herpes simplex virus thymidine kinase genes. I. Minimal size of the active fragments from HSV-1 and HSV-2. J Gen Virol 62: 191–206
Roberts GB, Fyfe JA, Gaillard RK, Short SA (1991) Mutant varicella-zoster virus thymidine kinase: correlation of clinical resistance and enzyme impairment. J Virol 65: 6407–6413

Roubal J, Klein G (1981) Synthesis of thymidine kinase (TK) in Epstein-Barr virus-superinfected Raji TK-negative cells. Intervirology 15: 43–48
Sakuma S, Yamamoto M, Kumano Y, Mori R (1988) An acyclovir-resistant strain of herpes simplex virus type 2 which is highly virulent for mice. Arch Virol 101: 169–182
Sanderson MR, Freemont PS, Murthy HMK, Krane JF, Summers WC, Steitz TA (1988) Purification and crystallization of thymidine kinase from herpes simplex virus type 1. J Mol Biol 202: 917–919
Scott SD, Ross NLJ, Binns MM (1989) Nucleotide and predicted amino acid sequences of the Marek's disease virus and turkey herpesvirus thymidine kinase genes; comparison with thymidine kinase genes of other herpesviruses. J Gen Virol 70: 3055–3065
Sears AE, Meignier B, Roizman B (1985) Establishment of latency in mice by herpes simplex virus 1 recombinants that carry insertions affecting regulation of the thymidine kinase gene. J Virol 55: 410–416
Shih M-F, Arsenakis M, Tiollais P, Roizman B (1984) Expression of hepatitis B virus S gene by herpes simplex virus type 1 vectors carrying α and β regulated gene chimeras. Proc Natl Acad Sci USA 81: 5867–5870
Smiley JR (1980) Construction in vitro and rescue of a thymidine kinase-deficient deletion mutation of herpes simplex virus. Nature 285: 333–335
Stanberry LR, Kit S, Myers MG (1985) Thymidine-kinase deficient herpes simplex virus type 2 genital infection in guinea pigs. J Virol 55: 322–328
Steiner I, Spivack JG, Deshmane SL, Ace CI, Preston CM, Fraser NW (1990) A herpes simplex virus type 1 mutant containing a nontransinducing Vmw65 protein establishes latent infection in vivo in the absence of viral replication and reactivates efficiently from explanted trigeminal ganglia. J Virol 64: 1630–1638
Stevens JG (1989) Human herpesviruses: a consideration of the latent state. Microbiol Rev 53: 318–332
Stevens JG, Cook ML (1971) Latent herpes simplex virus in spinal ganglia of mice. Science 173: 843–845
Stevens JG, Wagner EK, Devi-Rao GB, Cook ML, Feldman LT (1987) RNA complementary to a herpesvirus alpha gene mRNA is prominent in latently infected neurons. Science 235: 1056–1059
Summers WC, Summers WP (1977) [^{125}I]Deoxycytidine used in a rapid, sensitive, and specific assay for herpes simplex virus type 1 thymidine kinase. J Virol 24: 314–318
Summers WP, Wagner M, Summers WC (1975) Possible peptide chain termination mutants in thymidine kinase gene of a mammalian virus, herpes simplex virus. Proc Natl Acad Sci USA 72: 4081–4084
Swain MA, Galloway DA (1983) Nucleotide sequence of the herpes simplex virus type 2 thymidine kinase gene. J Viral 46: 1045–1050
Tenser RB (1983) Intracerebral inoculation of newborn and adult mice with thymidine kinase-deficient mutants of herpes simplex virus type 1. J Infect Dis 147: 956
Tenser RB, Dunstan ME (1979) Herpes simplex virus thymidine kinase expression in infection of the trigeminal ganglion. Virology 99: 417–422
Tenser RB, Edris WA (1986) Thymidine kinase (TK) activity in herpes simplex virus type 1 recombinants that carry insertions affecting regulation of the TK gene. Virology 155: 257–261
Tenser RB, Edris WA (1987) Trigeminal ganglion infection by thymidine kinase-negative mutants of herpes simplex virus after in vivo complementation. J Virol 61: 2171–2174
Tenser RB, Miller RL, Rapp F (1979) Trigeminal ganglion infection by thymidine kinase-negative mutants of herpes simplex virus. Science 205: 915–917
Tenser RB, Ressel S, Dunstan ME (1981) Herpes simplex virus thymidine kinase expression in trigeminal ganglion infection: correlation of enzyme activity with ganglion virus titer and evidence of in vivo complementation. Virology 112: 328–341
Tenser RB, Dawson M, Ressel SJ, Dunstan ME (1982) Detection of herpes simplex virus mRNA in latently infected trigeminal ganglion neurons by in situ hybridization. Ann Neurol 11: 285–291
Tenser RB, Ressel SJ, Fralish FA, Jones JC (1983) The role of pseudorabies virus thymidine kinase expression in trigeminal infection. J Gen Virol 64: 1369–1373

Tenser RB, Jones JC, Ressel SJ (1985) Acute and latent infection by thymidine kinase mutants of herpes simplex virus type 2. J Infect Dis 151: 548–550

Tenser RB, Hay KA, Edris WA (1989) Latency-associated transcript but not reactivatable virus is present in sensory ganglion neurons after inoculation of thymidine kinase-negative mutants of herpes simplex virus type 1. J Virol 63: 2861–2865

Thouless ME, Skinner GRB (1971) Differences in the properties of thymidine kinase produced by cells infected with type 1 and type 2 herpes virus. J Gen Virol 12: 195–197

van Oirschot JT, Terpstra C, Moorman RJM, Berns AJM, Gielkens ALJ (1990) Safety of an Aujeszky's disease vaccine based on deletion mutant strain 783 which does not express thymidine kinase and glycoprotein 1. Vet Rec 127: 443–446

Veerisetty V, Gentry GA (1983) Alterations in substrate specificity and physiochemical properties of deoxythymidine kinase of a drug-resistant herpes simplex virus type 1 mutant. J Virol 46: 901–908

Veerisetty V, Veerisetty IK, Gentry GA (1983) Alterations in the recognition of nucleoside analogues as substrates by the deoxythymidine kinase of a 5-methoxymethyldeoxyuridine-resistant mutant of herpes simplex virus type 1 mutant. J Gen Virol 64: 2767–2770

Volz DM, Lager LM, Mengeling WL (1992) Latency of a thymidine kinase-negative pseudorabies vaccine virus detected by the polymerase chain reaction. Arch Virol 122: 341–348

Wagner MJ, Sharp JA, Summers WC (1981) Nucleotide sequence of the thymidine kinase gene of herpes simplex virus type 1. Proc Natl Acad Sci USA 78: 1441–1445

Weinmaster GA, Misra V, McGuire R, Babiuk LA, DeClercq E (1982) Bovid herpesvirus type-1 (Infectious bovine rhinotracheitis virus)-induced thymidine kinase. Virology 118: 191–201

Westheim AI, Tenser RB, Marks JG (1987) Acyclovir resistance in a patient with chronic mucocutaneous herpes simplex infection. J Am Acad Dermatol 17: 875–880

Whetstone CA, Miller JM, Seal BS, Bello LJ, Lawrence WC (1992) Latency and reactivation of a thymidine kinase-negative bovine herpesvirus 1 deletion mutant. Arch Viro 122: 207–214

Whitby AJ, Blyth WA, Hill TJ (1987) The effect of DNA hypomethylating agents on the reactivation of herpes simplex virus from latently infected mouse ganglia in vitro. Arch Virol 97: 137–144

Wigler M, Silverstein S, Lee L-S, Pellicer A, Cheng Y-C, Axel R (1977) Transfer of purified herpes virus thymidine kinase gene to cultured mouse cells. Cell 11: 223–232

Wilcox CL, Crnic LS, Pizer LI (1992) Replication, latent infection, and reactivation in neuronal culture with a herpes simplex virus thymidine kinase-negative mutant. Virology 187: 348–352

Yamada N, Sawaski Y, Nakajima (1980) Thymidine kinase isozymes in rat cerebellum. Correlation of the activities of specific isozymes with DNA synthesis. Brain Res 195: 485–488

Yusa T, Yamaguchi Y, Ohwada H, Hayashi Y, Kuroiwa N, Morita T, Asanagi M, Moriyama Y, Fujimura S (1988) Activity of the cytosolic isozyme of thymidine kinase in human primary lung tumors with reference to malignancy. Cancer Res 48:5001–5006

Zimmerman N, Beck-Sickinger AG, Folkens G, Krickl S, Müller I (1991) Conformational and epitope mapping of herpes simplex virus type 1 thymidine kinase using synthetic peptide segments. Eur J Biochem 200: 519–528

Chapter 6 Herpes Simplex Virus Type 1 DNA Polymerase: Eukaryotic Model Enzyme and Principal Target of Antiviral Therapy

Charles W. Knopf and Reiner Strick

Summary

As the major essential function of the viral DNA replication machinery, the herpes simplex virus (HSV) DNA polymerase (Pol) represents the ultimate target of many currently used antiherpetic drugs. Since development of antiviral drug resistance may occur for all of these drugs and has already become an important clinical issue, there is an apparent need for other targets which allow drugs to be designed which are to a lesser extent affected by drug resistance. Potential new targets are the recently identified novel interactions of HSV Pol and the viral replication proteins, which have been shown to be essential for viral DNA replication. This review tries to provide an insight into our present knowledge of the structural and functional analysis of HSV Pol, which has become an attractive model enzyme for studies of eukaryotic Pol. The extensive computer analysis carried out by several laboratories on the primary protein sequence of prokaryotic and eukaryotic Pol reveals that DNA-dependent and RNA-dependent Pol are similarly organized. Major replicative enzymes possess both a proofreading exonuclease and a polymerization activity, and according to the sequence comparison these catalytic functions are organized in analogous and separate protein entities, designated exonuclease and polymerization domain, as observed with the prokaryotic *Escherichia coli* Pol I. An updated sequence alignment indicates that amongst the cellular polymerase species, the HSV Pol is more related to the eukaryotic Pol δ, suggesting that the viral enzyme derives from an ancestor of the latter polymerase. Like this major DNA replicase, the HSV Pol has a proofreading function and is associated with an additional subunit. Furthermore, the close relationship is documented by the identical behavior to common replication inhibitors.

Genetic studies and the biochemical analysis of the HSV Pol have emphasized that the herpesviral enzyme may specify at least four distinct functions that are required for the faithful replication of herpesvirus DNA in vivo. Three of these functions can be correlated with known properties of the HSV enzyme such as polymerization activity, proofreading exonuclease, and association of a viral replication protein. The fourth function, which could correspond to a recently identified RNase H activity, is controversial.

Deutsches Krebsforschungszentrum, Angewandte Tumorvirologie, Im Neuenheimer Feld 280, 69120 Heidelberg, Germany

Recent progress has been made in the analysis of the DNA template interaction of HSV Pol. Using a DNA-binding assay, it will be shown that many compounds known to inhibit the Pol activity exert a pronounced effect on the DNA-binding property of the enzyme. By employing a set of monospecific antibodies, directed against partially overlapping peptide domains of HSV-1 Pol of strain Angelotti, two antibodies which neutralized both catalytic activities of the enzyme were shown to interfere with DNA binding. The currently available information from structural modeling of *E. coli* Pol I and human immunodeficiency type 1 (HIV-1) reverse transcriptase, biochemical analysis, and antibody supershift experiments were combined into a working model for the HSV Pol holoenzyme. The usefulness of the identified protein–protein and protein–DNA interactions of HSV Pol and an auxiliary virus protein as possible novel targets for antiherpetic therapy are discussed.

Introduction

The genome of the neurotropic herpes simplex virus (HSV) encodes a complete set of protein functions that are not available, or only present at a very low concentration, in the neuronal cell and which enable the virus to be rescued from the apparently latent state of infection. To these protein functions, required for viral recurrency, belong enzymes of the DNA biosynthesis pathway such as thymidine kinase (TK), ribonucleotide reductase, uracil glycosylase, deoxyuridine triphosphatase, and in particular a complete set of replication proteins which accomplish the replication of the viral DNA. Thus, it is understandable that proteins which play an essential role in the lytic as well as latent type of viral infection represent favored targets of antiherpesvirus therapy. It was one of the fortunate circumstances for antiviral chemotherapy that base analogs such as acyclovir (ACV; Elion et al. 1977) turned out to be very effective drugs in defeating herpesviral infections where HSV DNA polymerase (Pol) represented the final target. In order to become a potent inhibitor of the HSV Pol (Furman et al. 1979), acyclovir has to be converted to the phosphomonoester, which is accomplished exclusively by the viral phosphotransferase activity of HSV TK within the infected cell (Fyfe et al. 1978). The further phosphorylation up to the effective triphosphate derivative and competitor of natural deoxyguanosine triphosphate (dGTP) occurs via cellular enzymes of the salvage pathway of DNA metabolism (Miller and Miller 1980, 1982). Not all base analogs, for example vidarabine (AraA) and acyclic nucleoside phosphonates, depend for their antiherpesvirus activity on phosphorylation by the virus-specified TK. They are phosphorylated exclusively by cellular kinases (Bennet et al. 1978; De Clercq et al. 1987)

Resistance to ACV as well as for most of the currently clinically applied base analogs is readily obtained in vitro by serial passage of HSV in the presence of the drug (Coen 1986). According to the two target genes involved in drug derivatization and utilization within the HSV-infected cell, HSV can acquire resistance to ACV by alterations in the viral TK and the Pol (Coen and Schaffer 1980; Schnipper and

Crumpacker 1980; Furman et al. 1981). Due to the advent of AIDS and the increasing number of immunocompromised individuals, it became apparent that development of drug resistance is also a serious clinical problem (for review, see Coen 1991). Although most of the clinically isolated ACV-resistant mutants were found to exhibit TK deficiency, in a few cases double mutants were isolated with alterations in both TK and Pol. A further problem in the treatment of HSV infection caused by drug-resistant viruses is that these viruses are usually also resistant to another class of commonly used inhibitors whose target is the HSV Pol, i.e., the pyrophosphate analogs. The inhibitory mechanism of these drugs, which include phosphonoacetic acid (PAA) and phosphonoformic acid (foscarnet, PFA), is thought to be based on a competition of the reaction product inorganic pyrophosphate (PPi) during the polymerization reaction, which results in a block of the forward polymerase reaction (Leinbach et al. 1976; Crumpacker 1992). Because of the resistance problem involved with these base and pyrophosphate analogs, a search is going on to find other compounds which do not require TK activation. A comparison with respect to potency and selectivity against clinical HSV strains with a new set of antiherpes compounds is presented by Andrei et al. (1992).

In the past, the study of drug-resistant HSV strains has enabled the gene for HSV Pol to be mapped as the site of mutations conferring resistance to base and pyrophosphate analogs (Purifoy et al. 1977; Chartrand et al. 1979). With the aid of drug-resistant mutants as well as intra- and intertypic recombinants, the polymerase gene mutations could be correlated with altered enzymatic functions (Furman et al. 1981; Knopf et al. 1981; Coen et al. 1982). At present, the polymerase gene sequence of four different HSV-1 strains are available (Quinn and McGeoch 1985; Gibbs et al. 1985; Knopf 1986; Larder et al. 1987). The major polymerase transcripts have been mapped; they are unspliced and include sequences both of a short open reading frame (SORF) and the polymerase coding sequences in a large open reading frame (LORF) of 3705 bp, capable of encoding a 136-kDa polypeptide (Holland et al. 1984; Yager and Coen 1988; Bludau and Freese 1991). The role of SORF is presently unclear. In infected primary human cells, a spliced polymerase transcript has been identified in which the SORF is fused to the N terminus of LORF, giving rise to a potential 133-kDa polypeptide (Bludau and Freese 1991) that has not yet been detected in infected cells. Expression of the polymerase gene is finely regulated at the transcriptional as well as at the post-transcriptional level, which results in an early shutoff of polymerase synthesis during HSV infection (Yager and Coen 1988; Dorsky and Crumpacker 1988; Yager et al. 1990; Bludau and Freese 1991).

Since the initial reports of the virus-specific Pol (Keir and Gold 1963), detailed biochemical characterization has revealed that the viral enzyme can be distinguished from cellular polymerases by ionic strength and DNA template requirements (Weissbach et al. 1973; Powell and Purifoy 1977; Knopf 1979). Like true DNA replicases, the HSV Pol has a proofreading 3′–5′ exonuclease (Knopf 1979; O'Donnell et al. 1987). Recently, other activities, an intrinsic RNase H activity, a novel topoisomerase, and a double-stranded DNA (dsDNA) binding protein have been reported to be associated with the DNA polymerase (Crute and Lehman 1989; Biswal et al. 1983; Powell and Purifoy 1977). From these, the

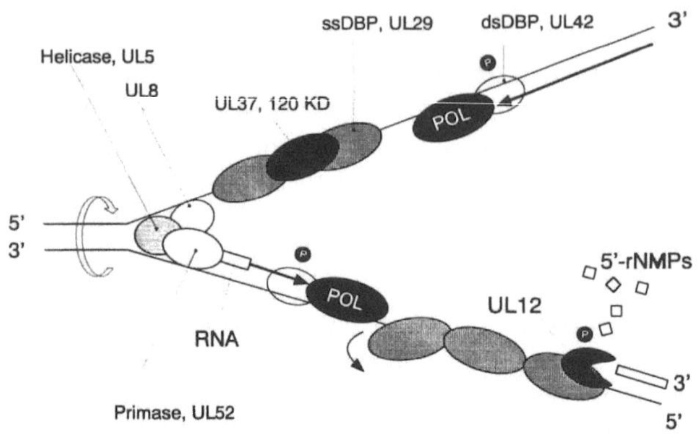

Fig. 1. Protein–protein and protein–DNA interactions of herpes simplex virus type 1 (HSV-1) replication proteins and enzymes involved in fork elongation during viral DNA replication. A schematic replication fork is shown, with the leading strand *above* and lagging strand *below*, each with one HSV DNA polymerase (*Pol*) holoenzyme, composed of HSV Pol and dsDNA-binding protein (UL42), at a 3′–OH primer terminus. The fork is unwound by a complex of three proteins with helicase and primase activity. The primase is shown to produce RNA primers, which are extended by the HSV DNA Pol holoenzyme. Single-stranded (*ss*) DNA is protected against degradation by the concerted action of ssDNA-binding protein and UL37 protein. Hydrolysis of RNA primers in 5′–3′ direction is shown as one of the possible interactions of the HSV DNase (UL12) during DNA synthesis. Proteins are labeled according to the gene order on the HSV-1 genome of strain 17 (McGeoch et al 1988), and phosphoproteins are marked by a circled P. *5′-rNMPS*, ribonucleoside 5′-monophosphates; *ssDBP*, single-stranded DNA-binding protein; *P*, phosphate group (s)

dsDNA binding protein has now been identified as a protein subunit of the enzyme that functions to increase processivity (Gottlieb et al. 1990)

The dsDNA-binding protein performs one of the seven viral gene functions that were recently identified to be essential for viral origin-dependent DNA replication (Wu et al. 1988). In analogy to the better studied T4 bacteriophage system (for review, see Young et al. 1992), the herpesviral genome contains a complete set of enzymes and proteins that are essential and sufficient for autonomously replicating its own viral DNA. Figure 1 shows an illustration of a schematic replication fork for documenting the interaction of viral replication proteins and other proteins, currently in discussion, during DNA replication (for review, see Challberg and Kelly 1989; Matthews et al. 1993). So far, nine viral proteins are reported to be directly involved in DNA synthesis at the replication fork. One function, a sequence-specific origin-binding protein encoded by the *UL9* gene (Olivo et al. 1988), together with cellular factors (Dabrowski and Schaffer 1991), accomplishes initiation of viral DNA replication at specific viral origin sequences. The 94-kDa origin-binding protein was shown to have helicase activity, catalyzing the unwinding of DNA in the 3′–5′ direction, and to require a DNA-dependent assembly of a multimeric UL9 protein complex for helicase activity (Fierer and Challberg 1992; Boehmer et al. 1993; Dodson and Lehman 1993). The unwinding activity is stoichiometric, optimal at high pH and at low salt, and specifically stimulated by the major DNA-binding protein, or infected-cell protein

8 (ICP8), which represents a single-stranded DNA (ssDNA)-binding protein and is encoded by the *UL29* gene. The UL29 protein has itself helix-destabilizing properties (Boehmer and Lehman 1993). At the onset of DNA replication, the UL29 protein moves into the nucleus, where it seems to be responsible for the organization of the HSV DNA replication complex (de Bruyn Kops and Knipe 1988). The C-terminal 28 residues of UL29 protein were shown to be required for the nuclear localization, and in addition these C-terminal sequences may interact with a component of the cell nucleus to initiate viral DNA replication (Gao and Knipe 1993). The 128-kDa UL29 protein binds preferentially and cooperatively to ssDNA (Ruyechan 1983), and residues between 564 and 849 as well as N-terminal residues contribute to the ssDNA-binding activity (Leinbach and Heath 1988; Gao and Knipe 1989; Wang and Hall 1990). The UL29 protein functionally interacts with the Pol (O'Donnell et al. 1987). Recent data suggest that the UL29 protein undergoes a physical association with the HSV alkaline DNase activity encoded by the *UL12* gene (Thomas et al. 1992). The HSV alkaline DNase, a 85-kDa phosphoprotein (Banks et al. 1985), is not essential for viral DNA synthesis, but is required for the efficient production of infectious virus (Weller et al. 1990). The enzyme is an endoexonuclease (Hoffmann and Cheng 1979) with a strong 5′–3′ exonucleolytic activity (Knopf and Weisshart 1990) that is capable of hydrolyzing preferentially ssDNA, but also dsDNA, ssRNA, and RNA of RNA–DNA hybrids under optimal reaction conditions for the viral Pol. Its activity could play a role in DNA recombination events for removing ssDNA extensions or RNA and DNA from ssDNA regions in 5′–3′ direction as emphasized in Fig. 1. Whereas the significance of the association between UL29 and UL12 protein is presently less clear, another newly discovered (W.T. Ruyechan, personal communication) interaction seems to be more relevant. The UL29 protein was found to copurify with a 120-kDa polypeptide identified as the product of the *UL37* gene. This latter protein is believed to balance the otherwise very rigid binding of UL29 to ssDNA. The fork movement is accomplished by a helicase–primase complex formed by the three essential viral replication proteins UL5, UL8, and UL52 (Crute et al. 1989). Subassembly of the UL5 protein (97 kDa) and the UL52 protein (120 kDa) is sufficient for retaining the helicase acitivity, which unwinds DNA in the 5′–3′ direction in the presence of adenosine triphosphate (ATP) or GTP (Crute et al. 1988), and for partially restoring the primase activity (Calder and Stow 1990; Dodson and Lehman 1991). The 70-kDa UL8 subunit in this three-protein complex has been shown to be required for efficient primer utilization, probably by stabilizing the association between nascent oligoribonucleotide primers and the DNA template (Sherman et al. 1992). The helicase activity of this complex is stimulated by the UL29 protein. In Fig. 1, the HSV Pol is tentatively shown as replicating both leading and lagging DNA strands, although the independence from additional cellular polymerase functions has never been demonstrated. The HSV Pol (136 kDa) forms a heterodimeric complex with the phosphorylated 65-kDa dsDNA-binding protein, encoded by the *UL42* gene, that was identified as a processivity factor and shown to be essential for virus replication (Gallo et al. 1989; Gottlieb et al. 1990; Hernandez and Lehman 1990; Digard et al. 1993b). To date, except for the UL37 protein, all other replication proteins have been expressed in insect cells by recombinant baculovirus (Dodson et al. 1989; Olivo

et al. 1989; Calder and Stow 1990; Gottlieb et al. 1990; Marcy et al. 1990c; Stow 1992). They can, therefore, be studied individually and in combination by biochemical means as well as by using the recently developed systems for viral origin-dependent DNA replication (Challberg 1986; Stow 1992).

From this brief overview, it is clear that the autonomous HSV DNA replication machinery has become a valuable model system for the study of eukaryotic DNA replication, and in this context it may play a similar role as the T4 bacteriophage system does for prokaryotic DNA replication. Moreover, since there are several essential and virus-specific interactions required for successful viral DNA replication, such as protein–protein and protein–DNA interactions, the HSV system provides a strong impetus for new developments in antiviral therapy and reasonable prospects of achieving antiviral compounds with a diminished potential for developing drug resistancy.

Materials and Methods

Materials

Radiochemicals were purchased from Amersham Buchler (Braunschweig). Nucleotides and synthetic polymers were obtained from Pharmacia LKB (Freiburg) and Boehringer Mannheim. Aphidicolin was from Boehringer Mannheim. The disodium salt of PAA was obtained from Abbott Laboratories. N^2-(*p*-*n*-Butylphenyl) dGTP (BuPdGTP) and activated calf thymus DNA were a kind gift from Dr.George E. Wright, University of Massachusetts. Enzymes were routinely obtained from GIBCO/BRL (Eggenstein), Pharmacia LKB (Freiburg), and Boehringer Mannheim.

Growth of Cells and Virus

African green monkey kidney monolayer cells (Rita clone, RC-37; Italdiagnostic Products, Rome) were cultivated in Dulbecco's modified Eagle medium (DMEM; Biochrom KG, Berlin), and virus stocks of standard HSV-1 strain Angelotti were prepared as described (Weisshart and Knopf 1988; Knopf and Weisshart 1990). For preparation of infected cells, confluent monolayers were infected at the permissive temperature for a given virus with ten plaque-forming units per cell. At 8 h after infection, cells were collected by scraping and centrifugation, washed twice with phosphate-buffered saline, and the cell pellets were stored at –70°C before use.

Antisera

Monoclonal antibody Q1 raised against purified HSV-2 alkaline DNase was a generous gift from Dr. Kenneth L. Powell, Wellcome Research Laboratories, Beckenham. The preparation and characterization of polyclonal anti-(HSV Pol) rabbit sera have been described elsewhere (Weisshart and Knopf 1988). Purification of antibodies was achieved by incubating antigen-bound nitrocellulose membranes with rabbit serum overnight at 4°C. Nitrocellulose membranes were pretreated and washed with phosphate-buffered saline containing 0.1% Tween. After washing, antibodies were eluted with 0.1 M glycine, pH 2.5, neutralized with NaOH, and dialyzed against phosphate-buffered saline, pH 7.2. To reduce cross-reactivity, a second incubation of the anti-(HSV Pol) rabbit sera was routinely carried out with nitrocellulose membranes containing the heterologous protein portions of the utilized fusion proteins. Antibody preparations were stored at 4°C in the presence of 0.01% sodium azide. Neutralization of the HSV Pol-associated activities by the purified antibodies was performed as previously described (Weisshart and Knopf 1988).

Purification of Herpes Simplex Virus DNA Polymerase

A total of 7×10^8 HSV-1 ANG-infected RC-37 cells were incubated for 10 min in buffer A (10 m*M* Hepes/KOH, pH 7.9, 1.5 m*M* $MgCl_2$, 10 m*M* KCl, 0.5 m*M* dithiothreitol), centrifuged at 1500 g. After resuspension in two volumes, buffer A cells were lyzed with a Dounce glass homogenizer and centrifuged at 2000 g. Nuclei were suspended in 8 ml buffer B (20 m*M* Hepes/KOH, pH 7.9, 10% glycerol, 2 m*M* β-mercaptoethanol, 0.2 m*M* ethylenediaminetetraacetic acid, EDTA, 1 m*M* phenylmethylsulfonyl fluoride, 5 µg/ml leupeptin, 0.7 µg/ml pepstatin) containing 1 *M* NaCl, and disrupted by Dounce homogenization. Nuclear debris was sedimented at 10 000 g and the supernatant cleared from nucleic acids by passing through a 6 ml diethylaminoethanol (DEAE)-cellulose column (DE52-cellulose, Whatman, Maidstone) equilibrated with buffer B containing 1 *M* NaCl. Flow and wash fractions were combined and dialyzed against buffer B containing 0.2 *M* NaCl. After a further clearance at 10 000 g, the dialysate was loaded on a 3 ml phosphocellulose column (P11-cellulose, Whatman, Maidstone) previously equilibrated with buffer B/0.2 *M* NaCl. The column was washed with ten volumes of the same buffer, and the adsorbed proteins were eluted in 1-ml fractions at a flow rate of 0.2 ml min^{-1} with a 50-ml linear gradient from 0.2 to 1 *M* NaCl in buffer B. Peak fractions of HSV Pol activity were pooled, diluted to a NaCl concentration of 0.2 *M* with buffer B, and loaded on a 2-ml heparin-sepharose CL-6B column (Pharmacia, Freiburg) equilibrated with buffer B/0.2 *M* NaCl; a linear gradient (40 ml) from 0.2 to 1 *M* NaCl in buffer B was applied. Peak fractions of HSV Pol eluting at 0.5 *M* NaCl were combined, and the proteins precipitated with 60% (mass/vol.) ammonium sulfate, collected by centrifugation at 12 000 g, and resuspended in buffer B adjusted to 0.1 *M* NaCl. The concentrated protein was applied to a 3.6-ml denatured DNA-cellulose column (Pharmacia, Freiburg) equilibrated with buffer B/0.1 m*M* NaCl, and protein elution was performed with a linear gradient (40 ml) from 0.1 to 1 *M* NaCl in buffer B. Peak fractions of HSV Pol eluting at 0.3 *M* NaCl were concentrated to 250 µl on a Centricon 30 microconcentration (Amicon, Danvers), adjusted to 0.2 *M* NaCl in buffer B, and loaded on a glycerol density gradient (polyallomer tubes; 15%–30% (by vol.) glycerol in buffer B). Velocity sedimentation was performed in a Spinco SW 40.2 rotor at 36 000 rpm at 64 h at 2°C. The following sedimentation standards were used: catalase (244 kDa, 11.3 S; Serva, Heidelberg), lactate dehydrogenase (136 kDa; 6.93 S; Worthington Biochemical Corp.), and bovine serum albumin (68 kDa, 4.4 S; Sigma, Deisenhofen). Peak fractions of HSV Pol were dialyzed against 50 m*M* Tris/Cl, pH 7.5, 1 m*M* dithiothreitol, 0.1 m*M* phenylmethylsulfonyl fluoride, and 40% (by vol.) ethylene glycol, concentrated with the Centricon 30 microconcentrator (Amicon, Danvers), and stored at –20°C before use.

Standard Assay for Herpes Simplex Virus DNA Polymerase

Reaction mixtures contained in a total volume of 100 µl: 50 m*M* Tris/Cl, pH 8.0, 50 µg bovine serum albumin (DNase-free, Gibco BRL, Eggenstein), 0.5 m*M* dithiothreitol, 7.5 m*M* $MgCl_2$, 100 m*M* ammonium sulfate, 0.1 m*M* each of dATP, deoxycytidine triphosphate (dCTP), dGTP, deoxythymidine triphosphate (dTTP) including one [α-^{32}P]dNTP (800 cpm/pmol), 25 µg activated salmon sperm DNA, and enzyme as indicated. After incubation at 37°C for 20 min, aliquots of the reaction mixture were spotted onto Whatman GF/C glass filters and treated to determine acid-insoluble radioactivity as described (Knopf 1979).

Exonuclease Assay

Exonucleolytic activites were determined with the substrates as indicated below under the conditions of the standard assay for HSV Pol, except that deoxynucleoside triphosphates (dNTP) were omitted unless otherwise stated. In order to neutralize HSV alkaline DNase activity, enzyme fractions were preincubated for 15 min in ice with one fifth of the reaction volume of the monoclonal antibody Q1. Incubations were performed at 37°C, and at the indicated times aliquots of the reaction mixtures were analyzed by thin-layer chromatography as described (Knopf and Weisshart 1990).

Preparation of Substrates

Unless otherwise stated, the reaction mixtures were incubated for 30 min at 37°C and the substrates were extracted with phenol, chloroform-isoamylalcohol, purified by gel filtration, precipitated with ethanol, and resuspended in Tris/EDTA buffer (10 m*M* Tris/Cl, pH 8.0, 1 m*M* EDTA) at a template concentration of 0.5 mg/ml.

5′-^{32}P-Labeled poly(dT) and Poly(A)

Reaction mixtures (50 µl) containing 10 m*M* Tris/Cl, pH 9, 20 µg/ml polydeoxythymidine (poly(dT)) or polyadenylate (poly(A)) (Pharmacia, Freiburg) and 25 units bacterial alkaline phosphatase/ml (Gibco BRL, Eggenstein) were incubated at 45°C for 30 min. After repeated extraction with phenol/chloroform and ether, the lyophilized polynucleotides were resuspended in 20 µl reaction mixture containing 50 m*M* Tris/Cl, pH 7.6, 10 m*M* $MgCl_2$, 5 m*M* dithiothreitol, 0.1 m*M* spermidine, 0.1 m*M* EDTA, 75 µCi [γ-^{32}P]ATP (3000 Ci/mmol), and 10 units T4 polynucleotide kinase (Boehringer, Mannheim).

Poly ([^{32}P]dT)·Poly(dA)

Reaction mixtures contained in 100 µl: 50 m*M* Tris/Cl, pH 7.8, 5 m*M* $MgCl_2$, 1 m*M* dithiothreitol, 50 µg bovine serum albumin, 0.1 m*M* [α-^{32}P]dTTP (2000 cpm/pmol), 12.5 µg oligo $(dT)_{12-18}$·poly(dA) (ratio 3:7 by mass), and five units of Klenow enzyme (Boehringer Mannheim).

Poly ([^{32}P]A)·Poly(dT)

Reaction mixture contained in 100 µl: 50 m*M* Tris/Cl, pH 8.0, 5 m*M* $MgCl_2$, 1 m*M* $MnCl_2$, 1 m*M* dithiothreitol, 0.1 m*M* [α-^{32}P]ATP (2000 cpm/pmol), 12.5 µg poly(dT), 5% glycerol and five units of *E. coli* RNA polymerase (Boehringer Mannheim).

DNA Polymerase–DNA Binding Assay

Pol reaction mixtures contained in a final volume of 15 µl: 0.1 µg ^{32}P-labeled EcoRI/BamHI fragment (12 000 cpm Cerenkov), 10 m*M* Hepes/KOH, pH 7.9, 3 m*M* $MgCl_2$, 50 m*M* ammonium sulfate, 0.25 m*M* dithiothreitol, 0.1 m*M* EDTA, 10% (mass/vol.) glycerol, 100 µg/ml bovine serum albumin, and 0.6 µg HSV Pol purified by glycerol gradient sedimentation. Incubation was at 25°C for 10 min. To determine the effect on DNA binding of antibodies directed to the HSV Pol, the polymerase was preincubated with 0.5 µg purified antibodies for 15 min at 25°C. For supershift experiments antibodies were added after the DNA-binding step, and incubations continued for 15 min at 25°C.

The 157-bp EcoRI/BamHI fragment derived from a NaeI fragment of HSV-1 ANG genome corresponding to nucleotides 555–691 of the short repeat region of strain 17 (McGeoch et al. 1986). The NaeI fragment was subcloned into the SmaI site of pUC19, and 5′- and 3′-^{32}P-labeled with T4 polynucleotide kinase and Klenow fragment, respectively, as described above. To ensure that 3′-^{32}P-labeled fragments contained recessed 3′ end, a partial filling reaction was performed with Klenow fragment in the presence of 0.5 m*M* dGTP and [α-^{32}P]dATP.

Band Shift Electrophoresis

After adding 3.8 µl loading buffer (0.5 *M* Tris/borate, pH 8.3, 40% (mass/vol.) glycerol, 50 m*M* EDTA, and 0.25% (mass/vol.) bromophenol blue), reaction mixtures were loaded onto a 5% (30% acrylamide/0.8% bisacrylamide by mass) native polyacrylamide gel or a step-gradient gel, containing a 4-cm bottom layer of 15% polyacrylamide, followed by two layers of 6 and 14 cm in height with 10% and 5% polyacrylamide, respectively. Before loading, gels were prerun for 30 min. Electrophoresis was carried out in 0.1 *M* Tris/borate, pH 8.3, 1 m*M* EDTA at a constant current of 12 mÅ at 4°C, until the marker dye reached the bottom layer. After electrophoresis, gels were dried on Whatman paper and exposed to Kodak XAR film overnight.

Computer Analysis

Sequence alignments were performed with the programs GAP, BESTFIT, and CLUSTAL from University of Wisconsin Genetic Computer Group (Devereux et al. 1984) using the HUSAR program library at the German Cancer Research Center.

Results and Discussion

Evolutionary Aspects

Primary Structure of Herpes Simplex Virus DNA Polymerase Reveals Close Relatedness to Cellular DNA δ-Polymerase

Apart from evolutionary considerations, alignment studies have in the past proven to be very valuable tools for identifying regions essential for polymerase function by means of sequence conservation. In the absence of available data for a direct three-dimensional determination of enzyme structure, this is still the method of choice. Once polymerase sequence motifs have been detected in an unknown sequence, a functional examination can follow. The strategy for a successful structure–function correlation is to perform alignments with sequences of functionally related enzymes which derived from sufficiently closely related as well as unrelated organisms. Using this approach (Doolittle 1981), according to their amino acid sequence homologies with *E. coli* Pol I (family A Pol) and human Pol α (family B Pol), herpesviral Pol were grouped to the family of Pol α (Jung et al. 1987; Leavitt and Ito 1989). Consequently, in recent reports this relatedness is expressed by the term "α-like" (Gibbs et al. 1991). Ito and Braithwaite (1991) have proposed an extension of this classification of Pol into families A, B, C, D, etc., according to the amino acid sequence homologies with *E. coli* Pol I, II, and III. Following this classification, the herpesviral Pol belong to the subfamily of viral Pol within the family B Pol. Unfortunately, this novel classification does not fully consider the evolutionary relatedness of a viral Pol with respect to the host organisms. Thus, a higher degree of sequence conservation may be observed between Pol from disparate subfamilies than from the same subfamily.

Owing to the steadily increasing number of Pol sequences available, it has now been possible to compare the primary sequences of the herpesviral enzyme with those of the known five types of cellular DNA-dependent Pol (Burgers et al. 1990)

Table 1. Amino acid similarities of herpes simplex virus type 1 DNA polymerase to herpesviral and cellular DNA polymerases

Polymerase	Similarities %	Identities
hsv 2	95.6	1119
ehv 1	72.4	666
vzv	70.5	652
hvs	58.8	373
ebv	58.7	386
mcmv	55.2	342
hhv 6	55.2	335
hcmv	53.7	362
ccv	33.2	74
α, *hu*	40.1	219
α, *sc*	37.8	199
α, *dm*	39.3	197
ß, *hu*	33.1	47
ß, *rat*	38.7	41
γ, *sc*	32.5	135
γ, *mz*	34.1	105
γ, *ni*	32.6	107
δ, *hu*	47.4	266
δ, *sc*	43.3	225
δ, *bo*	47.9	267
δ, *pf*	44.2	242
ε, *sc*	34.0	175

Polymerase sequences were retrieved from the EMBL Data Library, and amino acid similarities were determined with the alignment program GAP (Devereux et al. 1984; gap, 5.0; weight, 0.3). Individual alignments were performed with HSV-1 DNA polymerase of strain ANG (Knopf 1986) and the following polymerase sequences abbreviated according to the revised nomenclature for eukaryotic DNA polymerases (Burgers et al. 1990): *hsv* 2, HSV-2 strain 186 (Tsurumi et al. 1987); *ehv* 1, equine herpesvirus type I (Telford et al. 1992); *vzv*, varicella zoster virus (Davison and Scott 1986); *hvs*, herpesvirus saimiri (Albrecht and Fleckenstein 1990); *ebv*, Epstein-Barr virus (Baer et al. 1984); *mcmv*, murine cytomegalovirus (Elliot et al. 1991); *hhv* 6, human herpesvirus type 6 (Teo et al. 1991); *hcmv*, human cytomegalovirus (Kouzarides et al. 1987); *ccv*, channel catfish virus (Davison 1991); α-polymerases: *hu*, human (Wong et al. 1988); *sc*, *Saccharomyces cerevisiae* (Pizzagalli et al. 1988); *dm*, *Drosophila melanogaster* (Hirose et al. 1991); β-polymerases; *hu*, human (SenGupta et al. 1986); *rat* (Zmudzka et al. 1986); γ-polymerases; *sc*, *Saccharomyces cerevisiae* (Foury 1989); *mz*, maize (Paillard et al. 1985); *ni*, *Neurospora intermedia* (Chan et al. 1991); δ-polymerases: *hu*, human (Chung et al. 1991); *sc*, *Saccharomyces cerevisiae* (Boulet et al. 1989; Morrison and Sugino 1991); *bo*, bovine (Zhang et al. 1991); *pf*, *Plasmodium falciparum* (Ridley et al. 1991); ε-polymerases: *sc*, *Saccharomyces cerevisiae* (Morrison et al. 1990).

from different eukaryotic species. Table 1 shows the results of a sequence alignment of HSV-1 Pol of strain ANG with herpesviral and cellular Pol. Under the herpesviral enzymes, the HSV polymerases share the greatest overall similarities with those from equine herpesvirus type I (EHV-1) and varicella zoster virus (VZV), and the least similarity with the sequence from channel catfish virus. This result correlates well with the biology of these viruses. HSV-1, HSV-2, EHV-1, and VZV exhibit neurotropic growth characteristics, and they belong to the family of α-herpesvirinae (Roizman 1982).

Amongst the cellular Pol, comparing the nuclear Pol α, β, δ, and ε, and the mitochondrial Pol γ from different organisms, Pol δ turns out to be more closely related to HSV, and the relatedness with the cellular enzymes decreases in the order δ, α, ε, β, and γ. A similarity homologous to β was obtained for the terminal deoxynucleotidyltransferase. This result is in accordance with the evolutionary relationship as well as with the newer classification of these enzymes (Ito and Braithwaite 1991). HSV-1 Pol and human and yeast Pol α, δ, and ε classify to the family B, Pol γ to the family A, and Pol β and terminal deoxynucleotidyltransferase to the family X Pol, respectively. The evolution of Pol is an important issue because of their essential role in the transmission of genetic information from each generation to the next. From the data gathered so far, it is generally believed that all Pol share some common ancestral gene (Wang 1991). From the sequence similarities shown in Table 1 and previously noted (Boulet et al. 1989; Yang et al. 1992), it can be concluded that the herpesviral enzyme and Pol δ may have the same common ancestor.

Striking Conservation of Functional Domains in the Sequences of Herpesviral and δ-Polymerases

In the past, many investigators have found that Pol share a related structural and functional organization, and different laboratories have used different nomenclatures for the identified homologous regions (Ollis et al. 1985b; Larder et al. 1987; Wong et al. 1988; Spicer et al. 1988; Knopf and Weisshart 1988; Leavitt and Ito 1989). Wong et al. (1988) initially numbered the regions according to their degree of conservation, beginning with the highly conserved amino acid sequence motif YGDTD as region I. This kind of nomenclature is now generally accepted (Coen 1991; Wang 1991) and only varied by additions that are typical for a certain class of Pol. At present, seven conserved sequence elements have been identified in all family B Pol (Hwang et al. 1992). The corresponding regions of HSV Pol are: IV (residues 437–479), II (residues 694–736), VI (residues 772–791), III (residues 805–845), I (residues 881–896) VII (residues 938–946), and V (residues 953–963). An additional region, designated region A, has been recently added which is conserved only in Pol sequences of viruses that are sensitive to certain antiviral drugs (Gibbs et al. 1988).

Three-dimensional structural studies of the Klenow fragment of *E. coli* Pol I, together with biochemical and site-directed mutagenesis analyses, have led to the identification of the catalytic residues for the enzymatic functions, polymerization, and 3′–5′ exonucleolytic activity of this enzyme (Ollis et al. 1985a,b; Joyce and Steitz 1987; Mildvan 1989; Polesky et al. 1990; Derbyshire et al. 1991; Beese and Steitz 1991). Corresponding amino acids involved in catalysis of the Pol I exonuclease were identified by sequence similarities and site-directed mutagenesis using the ϕ29 Pol, and the three protein domains containing the residues critical for metal and 3′-OH binding (D355, E357, D424, D501) and catalytic activity (Y497) in Pol I (Freemont et al. 1988; Derbyshire et al. 1988, 1991) were named Exo I, Exo II, and Exo III (Bernad et al. 1989; Blanco et al. 1991). The location of Exo II and Exo III site corresponds to the above-mentioned conserved regions IV and A (Fig. 2).

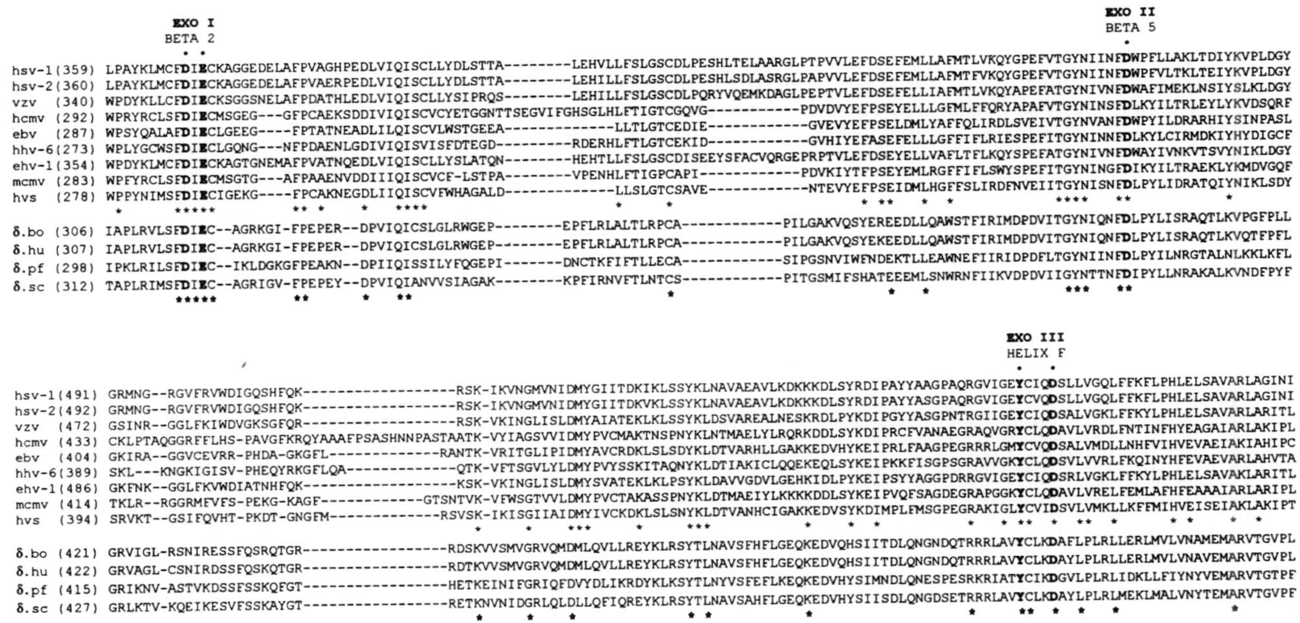

Fig. 2. Exonuclease domain. Multiple alignments were performed with herpesviral and δ-polymerases using the program CLUSTAL and the sequences as stated in Table 1. The single-letter amino acid code is used. The residue number of the first amino acid in each sequence is given in *brackets*. Conserved residues of the Exo I-III segments critical for metal binding and catalytic activity of *E. coli* DNA polymerase (Pol) I are printed in *boldface* as well as marked by a *dot*, and the relevant secondary structures determined in the tertiary structure of *E. coli* Klenow fragment (Ollis et al. 1985a) are stated. Homologous residues among the herpesviral sequences are marked by an *asterisk*, and among all aligned sequences by bold *asterisks*

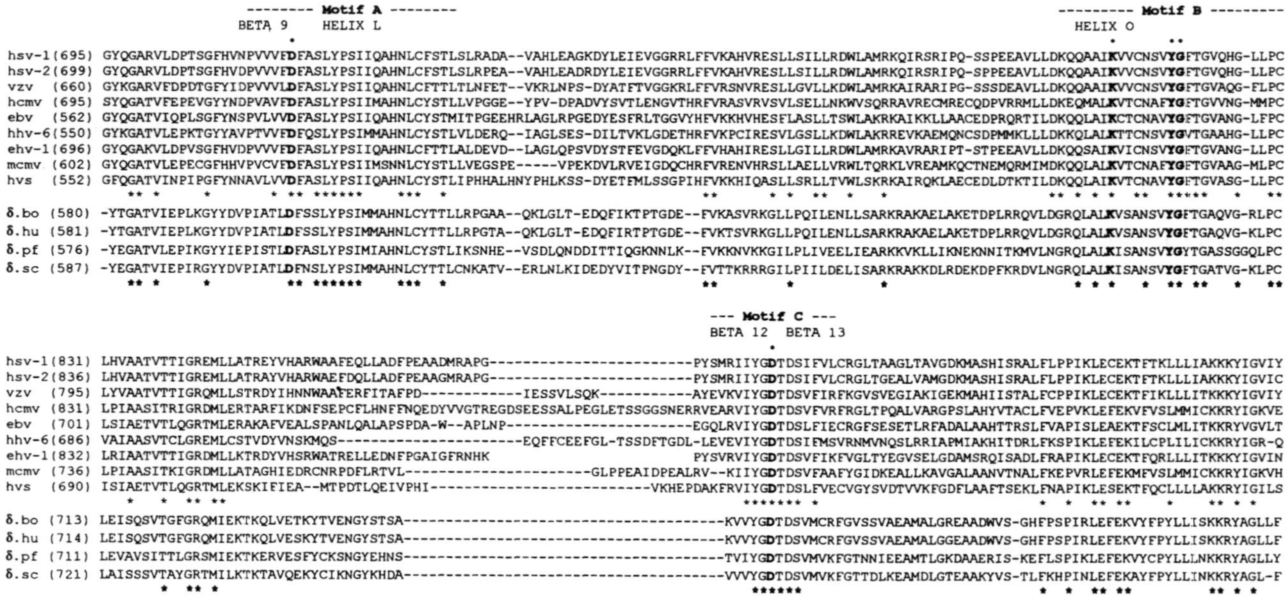

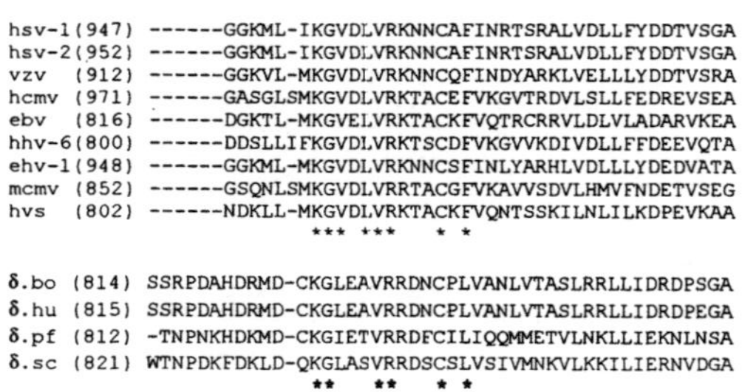

Fig. 3. Polymerization domain. Alignments were performed as in Fig. 1 and the sequence region comprising the previously identified motifs A, B, and C (Delarue et al. 1990) is presented. Secondary structure and conserved residues from *E. coli* DNA polymerase (Pol) I, and homologous residues among the aligned sequences are indicated as in the legend of Fig. 1

According to its X-ray crystal structure, the polymerization function of Pol I resides in the C-terminal residues 515–928, which form a large cleft suitable for accommodating a DNA template primer. Alignment studies (Delarue et al. 1990) demonstrated that five amino acid residues (D705, K758, Y766, G767, D882) of the polymerase domain of *E. coli* Pol I are conserved in three sequence stretches, dubbed motifs A, B, and C, in prokaryotic and eukaryotic DNA-dependent DNA and RNA polymerases. The location of the motifs A, B, and C (Fig. 3) corresponds to the conserved regions II, III, and I, respectively. By contrast, Pol β and terminal deoxynucleotidyltransferase, together with RNA-dependent DNA and RNA polymerases, exhibit similarities only to motifs A and C containing the Pol I homologs D705 and D882. This finding fits well with the restricted properties of the β-polymerase, which is merely a gap-filling enzyme (Chang 1975), and of the terminal deoxynucleotidyltransferase, which adds nucleotides to any 3′-OH-terminated ds- or ssDNA (Chang and Bollum 1971; Baltimore 1974). There is accumulating evidence that motif B is involved in base pair recognition. A synthesized Pol I oligopeptide corresponding to residues 728–777 was shown by circular dichroism spectroscopy (CD) to tightly bind dNTP substrates and also duplex DNA (Mildvan 1989).

In conclusion, prokaryotic and eukaryotic major DNA replicases seem to share an identical structural and functional organization: an N-terminal protein domain is responsible for the exonuclease function, and a C-terminal domain for the polymerase function.

As a reference for, and further documentation of, the close relatedness of herpesviral Pol and δ-polymerases, a recent alignment of the homologous exonuclease and polymerase domains is presented in Figs. 2 and 3. The conserved Pol I residues, displayed in boldface, are found in highly conserved sequence clusters. Striking similarities between herpesviral and δ-polymerases exist not only within the previously identified motifs of the exonuclease and polymerase domains, but also in the more length-variable sequence segments flanking these motifs. Altogether, 64% and 78%, respectively, of the identities detected in the exonuclease and polymerase domains of herpesviral polymerases are conserved in δ-polymerases (Figs. 2, 3).

The Exonuclease Domain of Herpes Simplex Virus Type 1 DNA Polymerase

From our current knowledge based on site-directed mutagenesis, studies of the exonuclease domain with different prokaryotic (ϕ29, Bernad et al. 1989; Soengas et al. 1992; *E. coli* Pol I, Mullen et al. 1990; Derbyshire et al. 1991; Bacillus subtilis Pol III, Barnes et al. 1992; T4, Reha-Krantz et al. 1991; T7, Patel et al. 1991) and eukaryotic pol (*S. cerevisiae* δ, Simon et al. 1991; *S. cerevisiae* ε, Morrison et al. 1991), it is evident that the putative exonuclease domain embodies the proofreading function (for review, see Blanco et al. 1992). This function seems to be organized in a quite separate domain in prokaryotic enzymes and in yeast, since exo-minus mutants are readily obtained which display fully functioning polymerase activity. In addition, the structural and functional independence of these two domains in Klenow fragment was demonstrated by the ability of the polymerase domain to retain activity when cloned separately (Freemont et al. 1986) and by antibody neutralization studies (Ruscitti et al. 1992).

In the HSV-1 Pol molecule, the highly conserved region IV (437–479) correlates with the ExoII segment (464–471) of the predicted exonuclease domain (Fig. 2). It is quite surprising that the earlier alignments (Bernad et al. 1989) did not detect the motif FDIEC (367–371) of the ExoI segment of HSV Pol, which is present in all major Pol that are equipped with a proofreading exonuclease activity (Morrison et al. 1991). The putative ExoIII segment of the herpesviral polymerase maps at residues 577–581. So far, within the exonuclease domain of eukaryotic Pol no homologs have been identified for the Pol I residues L361 and F473, which are essential for the base positioning within the exonuclease domain. From the presented alignment, possible candidates could be residue F381 near the ExoI motif and residue Y539 or L540 in front of the ExoIII motif of the HSV-1 Pol (Fig. 2).

To functionally test the alignment results, Gibbs et al. (1991) introduced mutations at the ExoII site (E460D; V462A; G464V) of the herpesviral polymerase and investigated the viability of virus recombinants carrying the respective mutations. Only the recombinant with the mutation V462A was viable, and its polymerase activity, determined from infected cell extracts, was about one half of the wild-type activity. According to the alignment in Fig. 2, these mutations affected or were near residues highly conserved between herpesviruses and δ-polymerases. Further investigations are necessary in order to prove the possible structural and functional interconnection of exonuclease and polymerization activities of HSV Pol.

The Polymerization Domain of the Herpes Simplex Virus Type 1 DNA Polymerase

All of the known naturally selected drug resistance mutations of HSV-1 Pol map within residues 597 and 961, which comprise seven of the known eight regions (A, II, VI, III, I, VII, and V) that are highly conserved among viral enzymes (Larder et al. 1987; Knopf and Weisshart 1988; Gibbs et al. 1988; Hall et al. 1989; Hwang et al. 1992). According to the nature of the resistance developed towards base and pyrophosphate analogs, it is expected that these mutations effect the recognition of dNTP and the pyrophosphate exchange reaction, both properties of the polymerization function in the better studies *E. coli* Pol I enzyme (for review see: Kornberg 1980). Site-directed mutagenesis and functional analysis of this enzyme (Polesky et al. 1990) suggested that Y766, located near the C terminus of helix O, which corresponds to the HSV-1 polymerase residue Y818 of motif B (Fig. 3), interacts with the incoming dNTP. Furthermore, Y766 could be affinity labeled with the photoactivatable dNTP analog 8-azido-dATP (Rush and Konigsberg 1990). Similar cross-linking with the natural substrate dTTP led to the labeling of residue H881 (Pandey et al. 1987) in motif C (Fig. 3). Mutations of the neighboring residue D882 indicated that this residue could be involved in catalysis of the polymerase reaction. In a more recent study, additional mutations were examined (Polesky et al. 1992), and the results indicate that D705, E710 (motif A), and E883 (motif C) are critical for binding and positioning DNA within the active site. To gain insight into the interactions with the dNTP substrate and template primer, a working model for the Klenow fragment of Pol I was recently elaborated (Yadav et al. 1992). In agreement with the data acquired from functional studied (Englund et al. 1969), this model predicts that the specificity of nucleotide selection by the enzyme is based on a single dNTP pocket developed by the template base bound to E883, D882, H881, N845, and Y766. Ribonucleotides cannot enter this pocket, and in the presence of the template nucleotide, binding is restricted to a complementary nucleotide which binds stably in the pocket.

By analogy to the *E. coli* Pol I model, from the high conservation of the putative catalytic residues as shown in Fig. 3, and from the map limits of drug resistance mutations of HSV Pol, it can be concluded that the polymerization function of the HSV Pol is mainly organized by the residues 597–961. It should be noted that this result is in agreement with a first-generation molecular model of the HSV-1 Pol catalytic core based on the available structure of the *E. coli* Pol I Klenow fragment (Haffey et al. 1990; Matthews et al. 1993).

Attempts to functionally dissect the polymerization domains of HSV Pol according to the different drug resistance targets, base and pyrophosphate analogs, have indicated that no one region is solely involved in the recognition of these analogs. Instead, by folding, several of the conserved sequences may participate in the building of the active center of the polymerization function (Coen 1991). Consequently, mutations at an apparently remote position can confer resistance to a drug. For instance, PAA resistance mutations were mapped in regions A, II, and III (Knopf 1987; Larder et al. 1987; Gibbs et al. 1988), spanning residues 597–842. Site-directed mutagenesis of motif C (equivalent to conserved region I) in HSV-1 polymerase indicated that any mutation of the conserved residues GDTD completely eradicates polymerase activity (Dorsky and Crumpacker 1990; Marcy et al. 1990b). A similar lethal effect exerted the mutation G896V (Dorsky and Crumpacker 1990), whereas mutations F891C, F891Y and V892M conferred resistance towards ACV and PAA (Marcy et al. 1990b).

From Sequence to Functional Predictions

According to the sequence alignment in Figs. 2 and 3, herpesviral and δ-polymerases exhibit a similar organization in distinct domains for exonuclease and polymerase function. On examining residues indicative for protein folding more closely, it is interesting to detect that both domains are flanked by conserved cysteines at position 345, about 30 residues before the ExoI segment, and at position 963, just two residues after the last identified drug resistance mutation located in the conserved region V (Gibbs et al. 1988). In addition, a unique feature of HSV and δ-polymerase, two cysteine residues are found in each of the functional domains and, strikingly, within or immediately near conserved regions: ExoI (C371), ExoIII (C578), motif A (C732), and motif B (C830). Possible protein folding via disulfide bridging would stably link a peptide domain which is reasonably large for harboring the exonucleolytic functions: 3′-OH primer terminus binding and phosphodiester cleavage. The prediction is that a polymerase peptide truncated after the ExoIII segment would still function. In contrast, a similar internal folding at the cysteines in the polymerase domain would not comprise all of the residues identified in *E. coli* Pol I as being critical for the polymerization function (Polesky et al. 1990). Motif C could only be contained inside the active center if folding occurs via the cysteines C963 and any one of the N-terminal cysteines. As a consequence, the disruption of the exonuclease domain would negatively influence the polymerase function, as recently shown (Gibbs et al. 1991). Up to now, there is no evidence that protein folding via disulfide bonding plays a role in the protein structure of the herpesviral enzyme, but this should be examined. In contrast, only two cysteine residues exist in the *E. coli* Pol I holoenzyme, one at the C terminus of the small fragment (C262) and the other in the C-terminal sequences of the large fragment (C907). The ease with which the *E. coli* enzyme can be proteolytically cleaved seems to exclude a vital role of the two cysteines. However, in contrast to the *E. coli* Klenow fragment (Kornberg 1980), cysteine residues are involved in the catalytic activity of the HSV-1 Pol, which is strongly dependent on dithiothreitol and very sensitive to sulfhydril-group blocking agents such as *N*-ethylmaleimide (Knopf 1979) and p-chloromercuribenzenesulfonate (see Fig. 12). A similar sensitivity towards sulfhydril-group blockers is observed for DNA polymerases α and δ (Table 2). More information about a structural and functional relationship between the herpesviral and δ-polymerase can be gained from a comparison of the properties of HSV-1, α-, and δ-polymerases, as illustrated in Table 2. Both, HSV-1 and δ-polymerase are heterodimers, possess an associated proofreading activity, depend on an auxiliary protein for their processivity, and exhibit an identical sensitivity towards aphidicolin, butylphenyl-dGTP, and N-ethylmaleimide. HSV and δ-polymerase can be discerned by their different sensitivity to high ionic strength. The association with a proofreading exonuclease activity is a major criterion for distinguishing HSV and δ-polymerase from Pol α. The latter enzyme from various organisms examined has been found to lack 3′–5′ exonuclease activity (Fry and

Table 2. Properties of herpes simplex virus type 1, α, and δ-polymerases

	α	HSV-1	δ
Catalytic polypeptide	165 kDa	136 kDa	125 kDa
Associated subunits	70, 58, 48 kDa	65 kDa,	48 kDa
Cellular location	Nuclear	Nuclear	Nuclear
Associated enzyme activities:			
3′-5′ Exonuclease	No	Yes	Yes
Primase	Yes	No	No
Processivity	Medium	Low	Low
(With auxiliary factor)	Medium	High	High
Inhibitors			
Aphidicolin	Sensitive	Sensitive	Sensitive
Butylphenyl-dGTP	Sensitive	Relatively resistant	Relatively resistant
N-Ethylmaleimide	Sensitive	Sensitive	Sensitive
Salt stimulation	No	Yes	No

Data were compiled from Downey et al. (1990), Wang (1991), Knopf (1979), and Strick (1993). dGTP, deoxyguanosine triphosphate; HSV, herpes simplex virus.

Loeb 1986). The functional analysis correlates with amino acid alignments, indicating that the exonuclease domain of Pol α is less well preserved (Blanco et al. 1992). However, interestingly enough, the Pol α of Drosophila melanogaster was found to have cryptic proofreading 3′–5′ exonuclease activity (Cotteril et al. 1987).

In conclusion, it is generally assumed that viral DNA replicases were evolutionary derived from a primordial sequence with polymerase function of cellular origin. In view of their sequence similarity (Table 1), the high conservation of functional domains (Figs. 2, 3), and their properties (Table 2), we propose that both HSV Pol and Pol δ, the major cellular replicase, which is responsible for the leading strand synthesis (Downey et al. 1988), are derived from the same ancestral gene.

Structural and Functional Aspects

Genetic Analyses of Herpes Simplex Virus Type 1 DNA Polymerase

Since its discovery by Keir and Gold (1963), the HSV-1 Pol has been intensively studied both genetically as well as biochemically. The first reasonable purification was achieved by Weissbach et al. (1973), who showed that the enzyme copurifies with a deoxyribonuclease activity. Schaffer and colleagues (Schaffer et al. 1973; Aron et al. 1975) provided the first clear evidence for the existence of a distinct viral replicase gene function by examining temperature-sensitive (ts) virus mutants which failed to synthesize viral DNA at the nonpermissive temperature. Two complementation groups, C and D, were found, comprising ts mutants which induced a ts polymerase activity in vivo, and it was emphasized that there may be two subunits involved in the formation of functional HSV-induced polymerase (Aron et al. 1975). Further genetic analysis of polymerase ts mutants has identified three complementation groups, which were named 1–3, 1–4, and 1–14 (Weller et al. 1983). All ts mutations map within the previously identified polymerase coding sequences (Purifoy et al. 1977; Holland et al. 1984; Coen et al. 1984). Quite recently, a polymerase mutant, 6C4, was described, obtained by forced recombination, which apparently belongs to a novel complementation group (Marcy et al. 1990a), but whose mutation has not been identified. From the genetic analyses, which revealed complex overlapping patterns of complementation, one should expect that the HSV-1 Pol has at least four distinct functions.

General Aspects of DNA Polymerase Function

Based upon the detailed functional analysis of *E. coli* Pol I (for review, see Kornberg 1980), the following minimal requirements have to be met by a DNA-dependent DNA replicase: (a) in order to replicate, the enzyme has to bind to a template primer; the primer terminus must carry a 3′-hydroxyl group at the ribose moiety; (b) for a simple filling reaction, the enzyme requires a single pocket for the uptake of all four dNTPs (Englund et al. 1969), which in addition holds the nucleotide of the ssDNA matrix; to enable the correct incorporation of the dNTPs according to Watson-Crick base pairing, the shape of the pocket has to be such that only the right dNTP can enter; according to the working model for *E. coli* Pol I (Yadav et al. 1992), the inlying nucleotide of the ssDNA matrix changes the pocket size and thus determines the uptake of the correct dNTP; and (C) the base-paired dNTP is incorporated at the 3′-OH primer terminus via phosphodiester linkage; this step is the catalysis of the polymerization reaction; pyrophosphate and H_2O are liberated, and the gained energy is used for the translocation of the enzyme on the template primer, thus moving the protein along the DNA matrix by one base in distance and in the 5′–3′ direction relative to the primer terminus. The described mechanism is adequate for very simple DNA polymerases such as the eukaryotic Pol β. It is more complicated in the case of true replicative enzymes, which possess an additional proofreading 3′–5′ exonuclease and which experience specific protein interactions during the DNA replication event in vivo. Again, the reaction pathway of the Klenow enzyme, as shown in Fig. 4, can serve as a valuable example for the complex interactions of an exonuclease-proficient Pol during polymerization. In the reaction pathway of processive polymerization shown (Dahlberg and Benkovic 1991; Kunkel 1992), the polymerase binds DNA first (Step 1) and this complex then binds dNTP. At this level (step 2), discrimination against errors can take place, as discussed above. In step 3, a conformational change of the ternary complex occurs by which the dNTP is positioned near the catalytic residues involved in phosphodiester bond formation (step 4). Prior to the release of pyrophosphate (step 6), another slow step has been detected with the Klenow enzyme that is believed to be caused by another conformational change of the enzyme–template primer complex (step 5). At this level of the reaction, a polymerase-associated proofreading activity could remove errors (step 5a). After incorporation of dNMP into the DNA and translocation of the enzyme, the polymerase can dissociate from the DNA or continue with the next cycle of processive polymerization by accepting another dNTP (step 2 and so forth). At this state, there is another opportunity for proofreading by polymerases (step 7), resulting from the slower rate of incorporating the next correct nucleotide onto a mispaired, rather than a paired, 3′-OH primer terminus. In compliance with the described scheme, in the presence of pyrophosphate and without dNTP, Pol carry out the reversal of the polymerization reaction, which is called the pyrophosphorolysis reaction (Deutscher and Kornberg 1969). The enzyme–template primer complex binds the pyrophosphate, and the reaction proceeds from step 5 to step 3, liberating dNTP as a reaction product, which is then available again for being incorporated in the forward reaction of the proposed pathway (Fig. 4). Thus, it is clear that in excess of template primer at a given concentration of pyrophosphate, and after a sufficiently long reaction time, an equilibrium will be reached for the generation and incorporation of dNTP. The reaction intermediates of steps 3–5 can be monitored by incubating the enzyme in the presence of dNTP and radiolabeled pyrophosphate. Under this condition, the radiolabel of pyrophosphate is transferred to the dNTP, indicating that the enzyme at this conformational stage is able to exchange the pyrophosphate moiety of the correctly bound dNTP with the new incoming pyrophosphate. We call this DNA-dependent and strictly base-specific reaction the "pyrophosphate exchange reaction" (Deutscher and Kornberg 1969).

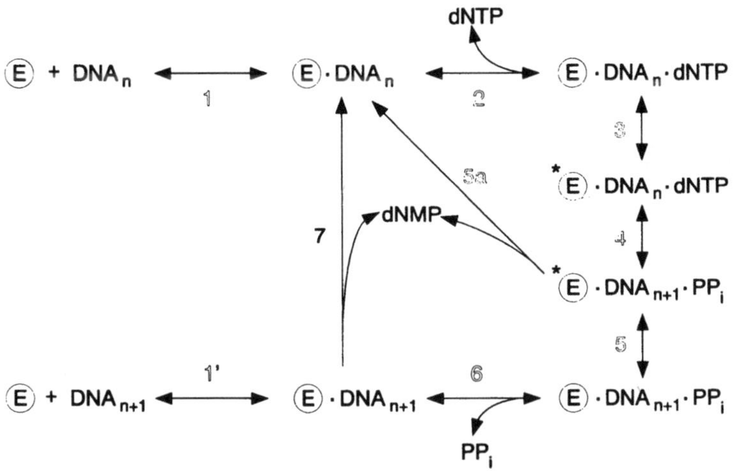

Fig. 4. Reaction pathway of Klenow fragment according to Dahlberg and Benkovic (1991) and Kunkel (1992). In the forward reaction, the polymerase (*circled E*) binds to DNA (*1*), accepts dNTP (*2*), undergoes different conformational changes (*circled E with asterisk*; *3–5*), and forms the phosphodiester bond by releasing inorganic pyrophosphate (*PP_i*) (*6*). The enzyme may dissociate from the DNA (*1′*) or translocate and consequently remain bound for another cycle of deoxynucleoside 5′-triphosphate (dNTP) addition (*2*). Translocation that must take place during processive polymerization has not been assigned to a particular step in the reaction

After discovering the general properties of a proofreading exonuclease-associated DNA polymerase, it is obvious that, with HSV Pol as target enzyme, the success of antiherpetic therapy based on the use of dNTP and PP_i analogs strongly depends on the specificity of the inhibitor–enzyme interaction. Figure 5 summarizes the essentials of the kinetic reactions for an exonuclease-proficient DNA polymerase in a more mechanistic fashion. By analogy to a recent model proposed for processive proofreading of T4 Pol that derived from experiments performed under single-turnover conditions by using a heparin trap (Reddy et al. 1992), a similar orientation relative to the template primer was adopted for the catalytic domains of exonuclease and polymerase. Primer terminus binding as the first step in recognition of DNA in the absence of nucleotides can be observed with a template primer having a perfectly matched or frayed base at the 3′-OH primer terminus. It is proposed that the exonuclease site (Fig. 5A) is in closer contact with the 3′-OH terminus than the polymerase site. Without dNTPs, both the frayed as well as the paired bases will be translocated again by the exonuclease site (Fig. 5B). In the T4 enzyme, the exonuclease site can uptake at most two unpaired nucleotides at a time (Reddy et al. 1992). In the presence of low levels of dNTP, the polymerase pocket accepts nucleotides and the DNA polymerization reaction starts at a limited rate of synthesis (Fig. 5C). Frayed primer termini are mostly degraded. Eventually, the exonuclease and polymerase activities, monitored by excision of dNMP and incorporation of dNTP, attain a steady state plateau. At higher levels of dNTP, the polymerization function is the overwhelming activity which makes it less likely to remove a mismatched nucleotide, once erroneously incorporated, during polymerization. Depending on the type of polymerase, therefore, distinct proofreading qualities can be found.

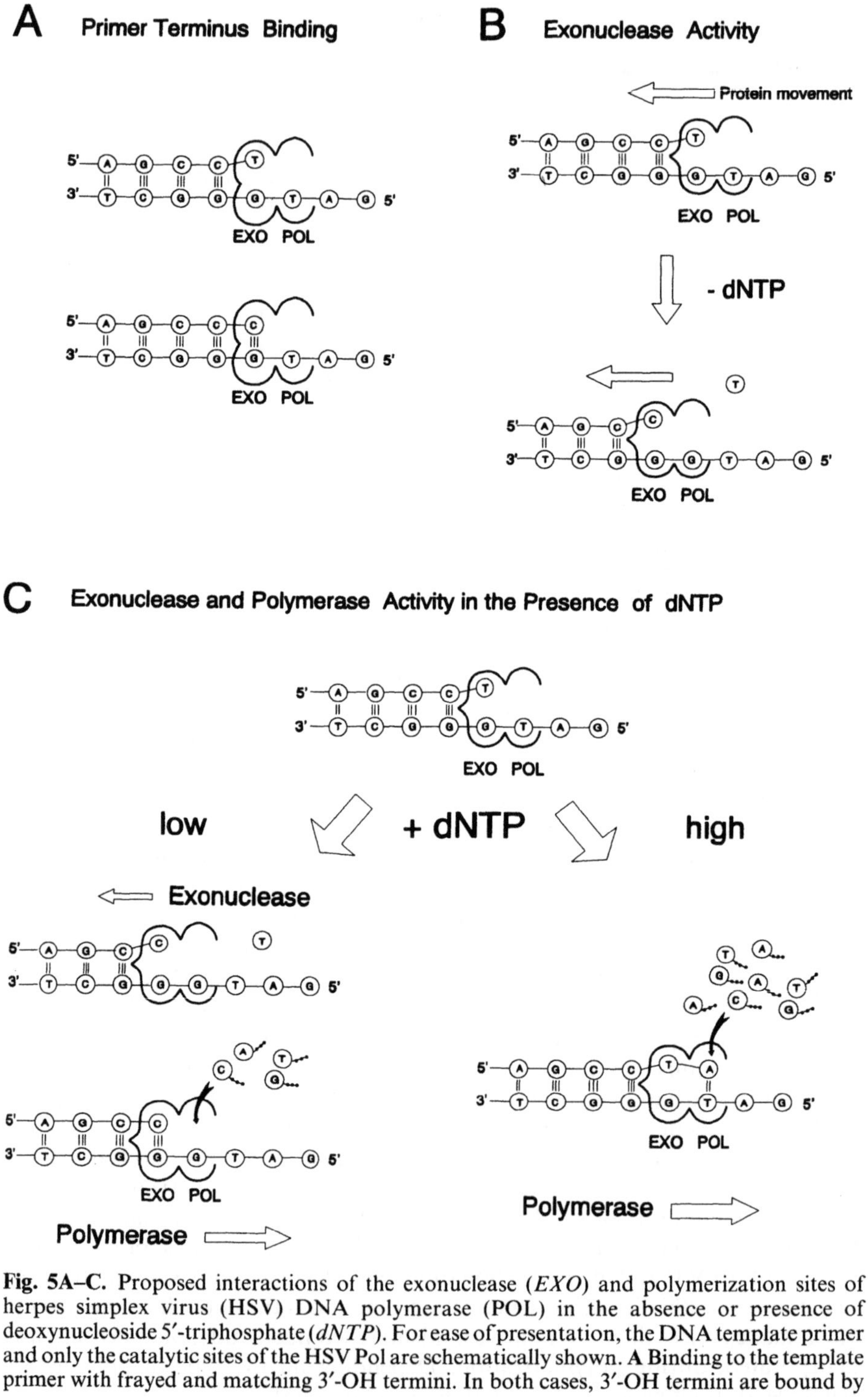

Fig. 5A–C. Proposed interactions of the exonuclease (*EXO*) and polymerization sites of herpes simplex virus (HSV) DNA polymerase (POL) in the absence or presence of deoxynucleoside 5′-triphosphate (*dNTP*). For ease of presentation, the DNA template primer and only the catalytic sites of the HSV Pol are schematically shown. **A** Binding to the template primer with frayed and matching 3′-OH termini. In both cases, 3′-OH termini are bound by

Enzymatic Functions of Herpes Simplex Virus Type 1 DNA Polymerase

The HSV Pol from both serotypes has been extensively purifed and biochemically studied in many laboratories (Weissbach et al. 1973; Powell and Purifoy 1977; Knopf 1979; Ostrander and Cheng 1980; O'Donnell et al. 1987). The biochemical analysis has been facilitated by recent achievements in recombinant technology. Functionally active HSV Pol has been expressed in vitro (Dorsky and Crumpacker 1988), in COS-1 cells (Dorsky et al. 1987), in yeast (Haffey et al. 1988), in insect cells using recombinant baculovirus (Marcy et al. 1990c; Stow 1992), and in African green monkey kidney cells by recombinant vaccinia virus (Kronenwett et al. 1990). Expression of recombinant HSV Pol in a catalytically active form in prokaryotic cells has not yet been achieved. Since the similarly sized *S. cerevisiae* pol δ, whose expression is toxic to bacteria, has now been functionally expressed in *E. coli* by using a new T7 expression vector (Brown et al. 1993), it is to be expected that the HSV polymerase can be overexpressed by this procedure.

The Polymerization Activity One of the important criteria for distinguishing the herpesviral enzyme from cellular Pol is its peculiar requirement for salt. As already shown by the earlier studies (Keir et al. 1966; Hay et al. 1971), HSV-induced Pol activity from soluble extracts is stimulated four- to six-fold by 100–200 m*M* potassium or ammonium sulfate. The purified enzyme exhibits a similar salt preference when activated DNA is used as template. Maximal incorporation of nucleotides with this template occurs at 100 m*M* ammonium sulfate and 150 m*M* potassium sulfate, respectively. At this salt concentration, the polymerization activity of human Pol α, δ (Table 2), and ε (Kesti and Syväoja 1991) are inhibited by more than tenfold. In addition to activated DNA, synthetic templates such as poly(dA) · oligo(dT) and poly(dC) · oligo(dG) can be used, the latter being the prefered one (Weissbach et al. 1973). The enzyme accepts oligoribonucleotides as primers (Müller et al. 1979). Depending on the synthetic template primer used, Mg^{2+} can be replaced by Mn^{2+}. The herpesviral polymerase is sensitive to 0.1 m*M* Zn^{2+} and 2.5 m*M* *N*-ethylmaleimide.

The second important criterion for distinguishing between HSV and cellular polymerases derives from previous studies on antiherpetic chemotherapy using base and pyrophosphate analogs to inhibit herpesvirus DNA replication with the herpesviral enzyme as ultimate target. During propagation of herpesvirus in tissue culture in the presence of these compounds, drug-resistant HSV polymerases have been easily generated, and it has been shown that mutations within the polymerase domain (residues 597–961) confer resistance to a variety of different drugs (Coen

the exonuclease site, and the polymerase site forms a pocket containing the single-stranded DNA template. **B** In the absence of dNTP, hydrolysis of the frayed primer termini by the exonuclease activity occurs in 3′–5′ direction. With the matching terminus melting of the duplex DNA is required prior to hydrolysis. This melting may destabilize the enzyme–DNA template primer complex, and thus the enzyme frequently dissociates from the template. **C** In the presence of low concentrations of dNTP, both proofreading exonuclease as well as polymerization activity can be observed. At high concentrations of deoxynucleoside 5′-triphosphates (dNTP), the polymerase forward reaction is favored, and errors, once incorporated, are no removed by the proofreading activity

1991). Biochemical analysis of the HSV Pol from a PAA-resistant variant of strain KOS (PAA^r-5) has revealed that the apparent antimutator phenotype of this strain correlates with the altered polymerase function. In addition, it was shown that not the proofreading, but the altered polymerization function was responsible for the higher replication fidelity observed in vivo and in vitro and that this resulted from an improved nucleotide selection of the enzyme (Hall et al. 1985). Frank and Cheng (1985) examined the interaction of acyclic NTP, PFA, and aphidicolin on the HSV polymerase activity by dual inhibitor studies (Segel 1975) and made the interesting observation that inhibition by one agent prevented the simultaneous inhibition by a second agent. This indicates a similar target of interaction for all three different classes of compounds: (1) the acyclic NTP, which like acyclovir triphosphate competes for dGTP (Furman et al. 1979); (2) aphidicolin, which was found to be a competitive inhibitor for dATP, dCTP, and dTTP, but noncompetitive for dGTP (Frank et al. 1984; Pedrali-Noy and Spadari 1980); and (3) PFA, which belongs to the class of pyrophosphate analogs. As discussed before, according to the inhibitor specification, the likely target is the polymerization function, and specifically the dNTP pocket, formed by the enzyme–DNA complex, which accomplishes dNTP binding as well as the pyrophosphate interaction (Fig. 4).

When assayed with a singly primed ϕX ssDNA template, the processivity of the wild-type HSV polymerase is low to moderate under optimal reaction conditions and at high salt. At low salt concentrations, the enzyme becomes highly processive (O'Donnell et al. 1987). ssDNA-binding proteins from *E. coli*, T4, and human origin (O'Donnell et al. 1987; Weisshart 1989), and polyamines such as spermidine and spermine (Ostrander and Cheng 1980; Wallace et al. 1980), also stimulate the HSV polymerase activity.

The herpesviral enzyme, like *E. coli* Pol I, performs a DNA-dependent pyrophosphate exchange reaction that proceeds linearly up to 45 min (Knopf 1979; Leinbach et al. 1976). Recent data (Strick and Knopf 1992) demonstrate that the herpesviral enzyme is further able to carry out the pyrophosphorolysis reaction in the presence of PP_i and absence of dNTP in the reaction mixture. Finally, the polymerization function of the herpesviral enzyme displays each of the major catalytic activities described for the forward and backward polymerase reaction pathway of the *E. coli* Pol I, as shown in Fig. 4.

The 3′–5′ Exonucleolytic Activity HSV Pol, like every major replicative DNA polymerase, has associated proofreading 3′–5′ exonucleolytic activity (Knopf 1979; Ostrander and Cheng 1980; O'Donnell et al. 1987). This activity was shown to exhibit the properties of the polymerization function, as well as its sensitivities towards the inhibitors PAA, dideoxyguanosine 5′-triphosphate (didGTP), and Zn^{2+} with poly(dC) · oligo(dG) as template primer (Knopf 1979; Knopf and Weisshart 1990), and to degrade ssDNA twofold better than dsDNA (Strick 1993). The proofreading capability is quite apparent at limiting concentrations of dNTPs, where up to 50% of the incorporated dNTPs are excised by the exonuclease at ongoing polymerization (Knopf 1979). The proofreading activity hydrolyzed an unpaired 3′-terminal nucleotide at about twice the rate than a paired 3′-terminal nucleotide on hairpin DNA templates in the presence of dNTP (O'Donnell et al. 1987). The 3′–5′ exonuclease can be inhibited in a template-dependent fashion by rNMP (Knopf 1979). Dual inhibitor studies on the 3′–5′ exonuclease with aphidicolin and PFA revealed the same mutually exclusive inhibition as for the polymerase function (Frank and Cheng 1985) when 3′-end labeled activated DNA was used. This could indicate that the binding of either one of these compounds to the dNTP pocket, as discussed before (and illustrated in Figs. 4, 5), prevents the translocation to the 3′-OH primer terminus. An alternative explanation is that binding to any site by one compound alone would produce such an inhibitory effect. A clue to this may be given by recent studies (Strick 1993) demonstrating that PAA and didNTP do not inhibit the hydrolysis of 3′-terminally labeled ssDNA by the 3′-5′ exonuclease activity, suggesting that these compounds are not bound by the exonuclease active site or by the polymerase active site, or, if binding to the dNTP pocket occurred, this did not affect the translocation step required for hydrolysis. On ssDNA, the HSV Pol-associated exonuclease behaves quite similarly to PAA as the 3′–5′ exonuclease activity of the ϕ29 DNA polymerase (Bernad et al. 1987). In addition, it has been shown that the polymerization activity of the PAA-resistant (PAA^r) polymerase of strain ANG (Knopf 1987) with a mutation at residue 719 in motif A (Fig. 3) was resistant against PAA, PFA, and aphidicolin, but its exonuclease activity was still sensitive towards aphidicolin on a dsDNA template. Thus, the PAA^r mutation prevents PAA, PFA, and aphidicolin from entering the dNTP pocket. The observed inhibitory effect of aphidicolin could then be related to its binding to the exonuclease site, with the consequence that either recognition of the 3′-OH primer terminus or the translocation step is affected. In this context, we believe that the alternate binding capacity of aphidicolin to both the exonuclease and the polymerase site, and its interference with DNA binding, play a key role for some of the drug resistance mutations reported. One example is the observed suppression of mutation Aph^{r}10, rendering the virus resistant to aphidicolin and hypersensitive to PAA by a second mutation (Su1, Su2) in the N-terminal portion of the HSV-1 polymerase gene of double mutants from strain KOS (Hall et al. 1989). The genetic analysis of the mutants carrying only a single suppressor mutation, Su1 (D531N) or Su2 (V258A), has now revealed that the Su mutants were hypersensitive to aphidicolin and resistant to ACV and arabinosylthymine (Wang et al. 1992). This is an intriguing finding that is reminiscent of earlier reports of certain ts mutations of the ssDNA-binding protein, which also affected viral sensitivity against PAA and aphidicolin (Chiou et al. 1985) and which provided strong evidence that the major DNA-binding protein and DNA polymerase closely interact in infected cells. Thus, in addition to the DNA-binding, protein interactions with the HSV polymerase may also influence drug sensitivity. A prerequisite for the better understanding of the very complex enzyme–drug interactions is the biochemical analysis of the purified enzymes and the study of possible interacting proteins in vitro. As the first elementary step, a combination of site-directed mutagenesis and biochemical analysis of selected mutant enzymes is recommended; interactions of the mutated proteins can then be followed for their capabilities to replicate DNA using HSV origin-containing plasmids in reconstituted cellular systems, as described (Stow 1992; Challberg 1986). Future work should focus on the aspect of the structural organization of polymerization and exonuclease function and on the question as to whether it is possible to achieve an active exonuclease-negative polymerase mutant, and if so, what role this mutation plays in the fidelity of viral DNA replication in vivo. From previous in vitro studies, it seems to be apparent that, depending on the reaction conditions, different error frequencies can be achieved. Using a ϕX174 natural DNA template under nonoptimal assay conditions, where exonuclease excision is three- to tenfold higher than incorporation, i.e., at limiting dNTP concentration and without salt, the error rates of HSV-2 Pol have been reported to be 1/650 000, reaching those of the *E. coli* Pol I (Abbotts et al. 1987). Using synthetic template primers under optimal assay conditions, a considerably lower maximum rate of 1/7 750 has been reported for the natural substrates, whereas for the incorporation of base analogs by a drug-resistant polymerase, error rates of more than 1/100 000 were found (Hall et al. 1985). These earlier comparative studies of wild-type and drug-resistant HSV polymerases show that the accuracy of the enzyme is primarily ensured by the base recognition of the polymerization function and not dependent on the proofreading exonuclease activity.

The 5′–3′ Exonucleolytic Activity Crute and Lehman (1989) have reported that HSV-1 Pol contains an intrinsic 5'–3′ exonuclease with ribonuclease H activity. The authors identified this activity in highly purified enzyme fractions and showed that the 5′–3′ exonuclease was optimally active at alkaline pH. By an in situ assay for RNase H activity, using a synthetic RNA–DNA hybrid with a continously radiolabeled RNA strand, the polymerase peptide, as well as a 30-kDa polypeptide of unknown origin, was found to hydrolyze the RNA strand of the hybrid. This was obviously a key experiment for correlating the polymerase polypeptide with the so-called RNase H activity. From the moment this report was published, many investigators quoted this original finding without an additional examination, and in order to document the structural correspondence with *E. coli* Pol I-associated 5′–3′ exonuclease domain, attempts have been made to identify the site for this activity in the very N-terminal residues of T4 and HSV Pol (Reha-Krantz 1990). Our laboratory, too, reported at the Eighth International Herpesvirus Workshop in Oxford that highly purified Pol preparations contained both 3′–5′ and 5′–3′ directed exonucleolytic activities (Knopf 1983). Further purification experiments indicated that the Pol-associated 5′–3′ exonuclease activity was a purification artifact. Therefore, a detailed examination of the nucleolytic activities associated with the purified polymerase fraction was undertaken (Knopf and Weisshart 1990), which provided evidence for the fact that highly purified polymerase preparations contain a 5′–3′ exonuclease activity, which was identified by antibody studies and physical properties to derive from the HSV alkaline DNase. In agreement with previous data, the HSV alkaline DNase was characterized as a 5′–3′ endoexonuclease (Hoffmann and Cheng 1979). For the alkaline HSV DNase, a pH-optimum of 9 has been reported, and at this pH the endonucleolytic activity of this enzyme, which shows a preference for ssDNA, is considerably increased, so that both ends of a DNA fragment are degraded in a very processive manner (Hoffmann 1981). Presently unclear is whether the 30-kDa polypeptide, which we also detected to be associated with the glycerol gradient enzyme (Knopf 1979), exerts the 5′–3′ exonucleolytic activity as a possible proteolytic cleavage product of the alkaline DNase. The important finding of

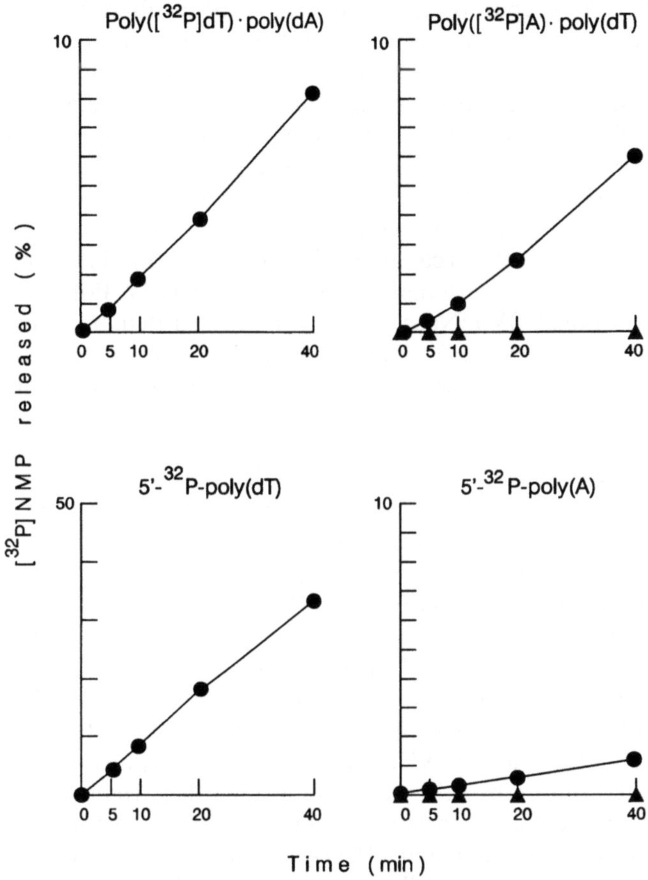

Fig. 6. Exonucleolytic activities of herpes simplex virus (HSV) DNA polymerase (Pol). Kinetics of hydrolysis of the indicated synthetic templates were performed with the glycerol gradient fraction of wild-type HSV-1 ANG Pol, as described in "Materials and Methods." Degradation of 5′-^{32}P-poly(A) and poly([^{32}P]A) · poly(dT) could be completely prevented in the presence of dNTPs as well as monoclonal antibody Q1 directed against HSV-2 alkaline DNase

biochemical analysis has been that the associated 3′–5′ exonuclease was able to hydrolyze both RNA as well as DNA of a synthetic RNA–DNA hybrid and that hydrolysis could be prevented by the presence of dNTP. This property of the HSV Pol proofreading activity explains the results of the in situ assay for RNase H activity (Crute and Lehman 1989). According to the template used in this assay, hydrolysis occurring in 3′–5′ and /or 5′–3′ direction would generate an identical autoradiographic image. Figure 6 illustrates the hydrolysis of different synthetic RNA and DNA templates by the 3′–5′ exonuclease of a HSV-1 ANG Pol prepared as described in "Materials and Methods." The polymerase-associated exonuclease degrades DNA strands from DNA hybrids slightly faster than RNA strands from RNA–DNA hybrids. Poor hydrolysis is observed with ssRNA, and hydrolysis is best with ssDNA. Hydrolysis of RNA–DNA hybrids is inhibited by the presence of dNTP. Complete inhibition of hydrolysis of RNA strands by a standard wild-type HSV-1 Pol preparation, even for extended incubation times (more than 30 min), requires both the presence of dNTP, in order to diminish the 3′–5′-directed activity, and the monoclonal DNase antibody Q1 (Banks et al. 1983), which neutralizes the template primer degradation in 5′–3′ direction by traces of HSV-1 DNase. We have also examined recombinant HSV-1 Pol purified from insect cells as well as from African green monkey kidney cells using recombinant baculovirus and vaccinia virus, respectively. Evidence for a proposed RNase H activity could not be provided. In conclusion, according to our analysis, the proofreading activity of HSV-1 Pol is capable of hydrolyzing both RNA and DNA in 3′–5′ direction. In glycerol gradient preparations of HSV Pol from infected cells, the majority of the HSV DNase activity can be separated, but some fraction sticks to the polymerase and its 5′–3′ exonucleolytic activity is detected only with a 5′-labeled DNA template and under specific reaction conditions (Knopf and Weisshart 1990).

The Auxiliary Protein UL42 Powell and Purifoy (1977) found in HSV Pol preparations two polypeptides with molecular masses of 150 kDa and 54 kDa. The latter polypeptide, present at a significant amount in HSV-2 Pol preparations, was identified as the DNA-binding protein ICSP34/35 (Vaughan et al. 1985). After extensive purification, the peak fraction of HSV-1 Pol activity of strain ANG from dsDNA-cellulose chromatography was shown to copurify with three polypeptides having molecular masses of 144 kDa, 74 kDa, and 29 KDa and to sediment as two fractions with 7.5 S and 6.2 S at high ionic strength (Knopf 1979). Biswal et al. (1983) found that the DNA-cellulose enzyme copurifies with three polypeptides of 145 kDa, 78 kDa, and 45 kDa and with a DNA topoisomerase activity, which could be neutralized by rabbit antisera against the virus-infected cells. Contradictory to this are the results of a detailed study on DNA topoisomerases I and II in HSV-2 infected cells by Bapat et al. (1987), who were able to separate the DNA topoisomerase activities from HSV Pol and further showed that the DNA topoisomerases were of cellular origin. For HSV-1 Pol, Gallo et al. (1988,1989) have demonstrated the close association with the 65-kDa dsDNA-binding protein UL42 and its relatedness to the HSV-2 counterpart ICSP34/35. Crute and Lehman (1989) found four polypeptides in the HSV Pol peak fraction after sucrose density gradient sedimentation (136 kDa, 60–64 kDa, 40 kDa, 30 kDa). From the sedimentation coefficient (7.3 S) and Stokes radius (59 Å), they calculated a molecular weight of 190 000, assuming a partial specific volume of 0.725, and concluded that HSV Pol is a heterodimer consisting of the polymerase (136 kDa) and the DNA-binding protein UL42 (65 kDa). Previous studies have indicated that the DNA-binding protein UL42 is one of only seven HSV-1 genes required to replicate plasmids containing an HSV origin of replication (Wu et al. 1988) and is essential for viral DNA replication in vivo (Marchetti et al. 1988; Schaffer et al. 1973). The polymerase–UL42 protein interaction is very tight and ressembles that of cellular Pol δ and its associated 48-kDa subunit (Table 2). The two subunits can be partially separated by centrifugation through a glycerol gradient (Gallo et al. 1988). The identification of the functional role of the UL42 protein has been greatly facilitated by the recent application of the baculovirus for individually expressing the HSV-1 replication proteins. It was found that the UL42 protein increases the processivity of the polymerization

(Hernandez and Lehman 1990; Gottlieb et al. 1990) and that an additional stimulatory effect of the UL42 protein on the Pol activity can only be obtained at high salt concentrations (i.e., >150 m*M* KCl), which is an apparent consequence of the different affinities towards DNA of the polymerase in the presence and in the absence of UL42 protein (Hart and Boehme 1992). Since the interaction between the herpesviral polymerase and the UL42 protein is essential for viral DNA replication (Marchetti et al. 1988; Johnson et al. 1991), it could serve as a target for antiviral therapy. Progress has been made in the exploration of the possible protein-binding sites between the two proteins. The UL42-binding site on the polymerase has been characterized in detail. Previous work has shown that the 227 C-terminal amino acids of HSV-1 Pol are sufficient to bind the UL42 protein (Digard and Coen 1990) and that the final 27–40 residues are crucial for complex formation (Digard et al. 1993a; Stow 1993; Tenney et al. 1993). The studies with yeast HSV Pol suggest that residues 1195–1216 appear to be particularly important for the stability of the complex (Tenney et al. 1993). In a recent functional analysis of the UL42 protein, it was shown that the 340 N-terminal residues are sufficient for all functional activities associated with this protein, e.g., dsDNA-binding, polymerase binding, and processivity (Digard et al. 1993b). Interestingly, an insertion mutation at position 160 destroyed the ability of UL42 protein to bind the polymerase in vitro. By complementation analysis, it was demonstrated that this mutation likewise affected the protein binding in vivo, confirming that the specific interaction between the polymerase and the DNA-binding protein is essential for virus replication and may represent a valid target for the design of novel therapeutic agents based on small peptides and nonpeptide analogs, e.g., peptidomimetica. A preliminary study on the inhibitory effect of 96 overlapping pentadecapeptides spanning the entire 488 amino acids of the UL42 protein on the HSV-1 Pol activity revealed that peptides from five noncontiguous regions inhibited the polymerase in both the presence and absence of UL42 protein (Owsianka et al. 1993). Four of these peptides were found to bind DNA. Although this study seemed to indicate that the inhibitory effect of the peptides was largely nonspecific, the DNA-binding properties of some of these peptides correlates quite nicely with the involvement of the N-terminal sequences of UL42 protein in DNA binding as identified by Digard et al. (1993b). It is to be expected that further progress can be achieved by directly examining the inhibitory effect of peptides from UL42 protein or HSV Pol on protein–protein interaction. At the moment, the interaction between the polymerase and its auxiliary protein offers a promising target for antiviral drug design.

Recent Approaches to Dissect Structural and Functional Domains of Herpes Simplex Virus Type 1 DNA Polymerase

Use of Herpes Simplex Virus Type 1 DNA Polymerase Antibodies

To achieve better insight into the structural and functional organization of the herpes polymerase, we and others have used an immunobiochemical approach (Thomas et al. 1988; Matthews et al. 1990; Weisshart and Knopf 1988). Our laboratory has previously generated polyclonal antibodies which were raised against selected N-terminal, central, and C-terminal domains of the polymerase-coding sequences using MS2 RNA polymerase expression vectors. These antibodies were successfully employed for probing the HSV-1 ANG Pol. Analysis of the capacities for binding the native enzyme and for neutralizing the enzymatic activities indicated that antibodies raised to the central (residues 597–685) and C-terminal domains (residues 1072–1235) of the protein could specifically neutralize both 3′–5′ exonuclease and polymerization activities of HSV Pol (Knopf and Weisshart 1988, 1990). Matthews et al. (1990) initially prepared antisera against seven peptide sequences present in HSV-1 Pol of strain KOS and found, in agreement with our data, that antipeptide sera, produced against C-terminal residues 1100–1108 and 1216–1224, could specifically neutralize HSV-1 Pol activity in a dose-dependent manner. Whereas the neutralization potential of the antibodies raised against residues 597–685 correlated with the protein region harboring the domains for exonuclease and polymerase function, the inhibitory effect of the C terminus-directed antibodies was puzzling, and it was suggested that these antibodies most likely interfere with DNA binding of the enzyme (Knopf and Weisshart 1988). Thomas et al. (1988) also used antipeptide sera to localize the polymerase within the infected cell and to determine the time of synthesis during productive infection. They found that the very N-terminal residues 1–15 of the enzyme are available for antibody binding. In order to identify possible surface regions of the enzyme and regions involved in DNA binding, we have raised additional antibodies and currently have a set of monospecific antibodies against the complete coding sequence of the enzyme, consisting of 12 partly overlapping subdomains of the HSV-1 ANG Pol. Figure 7 shows a summary of the antibody characterization with respect to polymerase binding, neutralization of 3′–5′ exonuclease, and polymerase activity of the enzyme, which will be described in detail elsewhere (Strick, Weisshart and Knopf, unpublished results). The tenor of this study is that antibodies (EX1051, EX3) which display stronger binding to the enzyme exert a greater inhibitory effect on both enzymatic activities. An interesting and reproducible finding is that antibodies raised against residues 685–924 of the polymerase domain, and which displayed significant, above-average polymerase binding, exclusively inhibit the polymerization activity. This result suggests that 3′–5′ exonuclease and polymerase functions are organized in separate domains. On the other hand, antibodies directed to residues 297–536 and spanning the proposed exonuclease domain, which also exerted strong polymerase binding, inhibited exonuclease and polymerase to the same extent. It is very possible that this inhibition is a consequence of protein folding, which brings certain sequences of the exonuclease site into contact with essential regions of the polymerase site, but it could also be that these antibodies inhibit the DNA-binding step of HSV-1 Pol.

Analysis of DNA-Binding Property

To further investigate the role of the DNA binding in the antibody–polymerase interaction and in order to gain an insight into interactions of HSV Pol with the

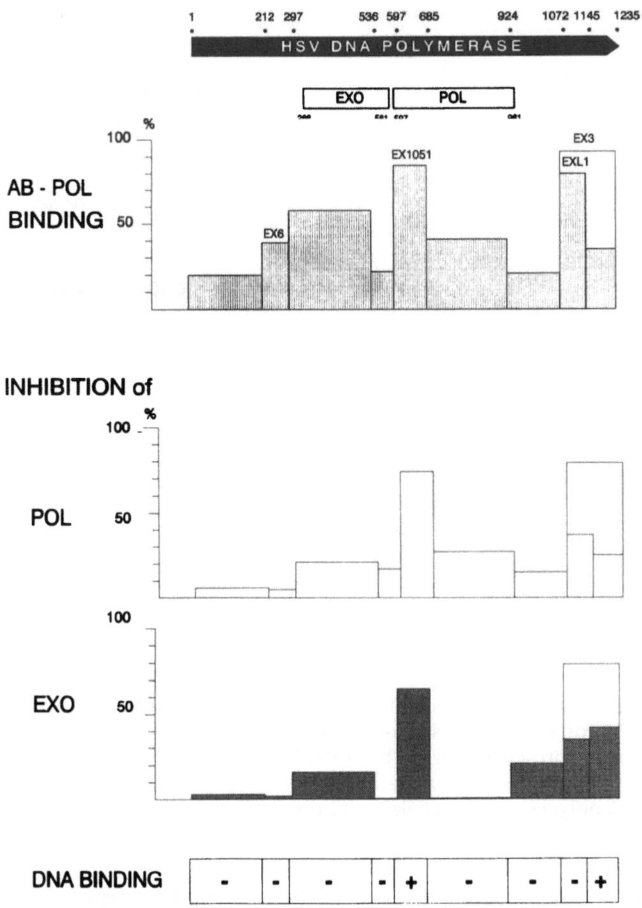

Fig. 7. Summary of the analysis of polymerase antibodies. Monospecific antibodies were raised against portions of the herpes simplex virus (HSV) DNA polymerase (Pol), as indicated by the residue number on the linear polypeptide map of HSV Pol in the *top panel*. Presented are the binding potential to native HSV Pol and the inhibitory activity of the purified polymerase antibodies (1.5 mg/ml) on polymerase (*POL*), exonuclease (*EXO*) and DNA-binding functions of the HSV Pol as described

DNA template, we have employed the previously described band shift assay, which allows the effect of individual reaction components on both DNA and enzyme to be monitored during the Pol–template interaction (Strick and Knopf 1992). Figure 8 shows the polycrylamide gel electrophoresis (PAGE) analysis of polymerase–DNA complexes by the described method. Prior to gel analysis, an excess of a 157-bp 5′-labeled DNA fragment was incubated with the indicated enzymes in the presence of glutaraldehyde, which generated a single band for a given enzyme-DNA complex. Distinct mobilities are received for polymerase-DNA complexes of *E. coli* Pol I and Klenow enzyme, T4 Pol, and HSV-1 ANG Pol, and the mobilities of the enzyme–DNA complexes correspond quite well with the different masses of the tested enzymes.

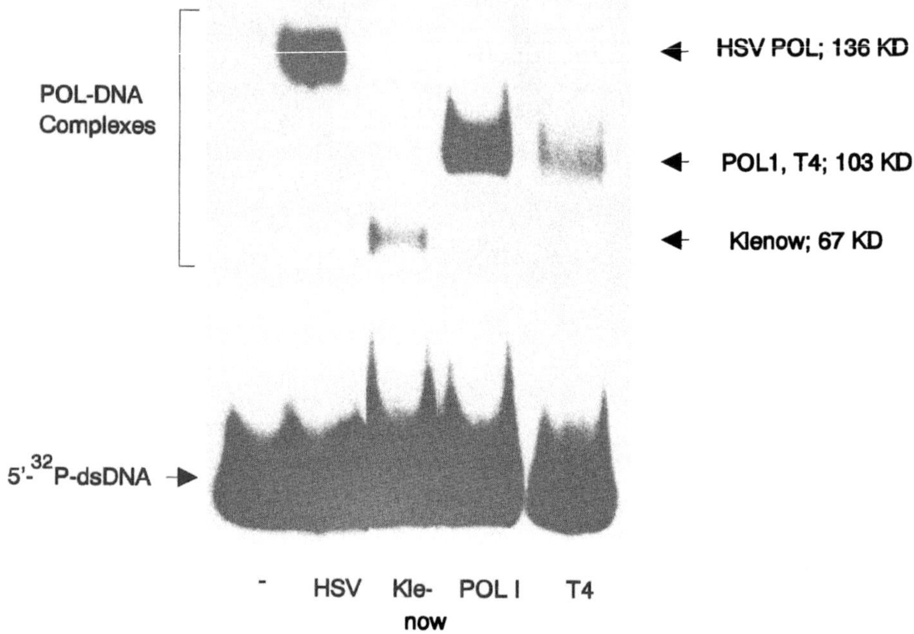

Fig. 8. DNA-binding properties of *E. coli* DNA polymerase (*Pol*)*I*, *Klenow* fragment, *T4* Pol, and herpes simplex virus (*HSV*) Pol. Polymerases (1 unit each) were incubated with a 157-bp, 5′ end-labeled DNA fragment in the presence of 0.1% (by vol.) glutaraldehyde as described for the DNA polymerase–DNA binding assay. Autoradiography shows the analysis of the reaction products by native 5% polycrylamide gel electrophoresis

In order to determine the role of DNA binding in the observed neutralization of the catalytic activities of HSV Pol by antibodies, and to identify polypeptide domains at the surface of the protein still accessible once the enzyme had bound DNA, we used the band shift analysis and performed supershift experiments with the polymerase antibodies. HSV Pol was incubated with antibodies prior to and after adding the DNA template, and the mobilities of the complexes were analyzed by native PAGE. When antibodies were added after enzyme–DNA complex formation, supershifted bands, but no inhibition of DNA binding, was observed. Addition of antibodies prior to enzyme–DNA complex formation revealed that only antibodies EX1051 and EX3, which are directed to the central and C-terminal sequences, completely inhibited the enzyme–DNA complex formation (Fig. 7). This result further indicated that the strong neutralization of the catalytic activities of HSV Pol by these antibodies demonstrated before (Weisshart and Knopf 1988) was a consequence of an impairment of the enzyme–DNA binding interaction and did not necessarily mean that the target sequences were involved in the catalytic function.

Using this assay and composite native polyacrylamide gels for analysis, we have been able to assess DNA binding as well as the exonuclease and polymerase activities taking place during the DNA-binding reaction, as shown in Fig. 9. Using

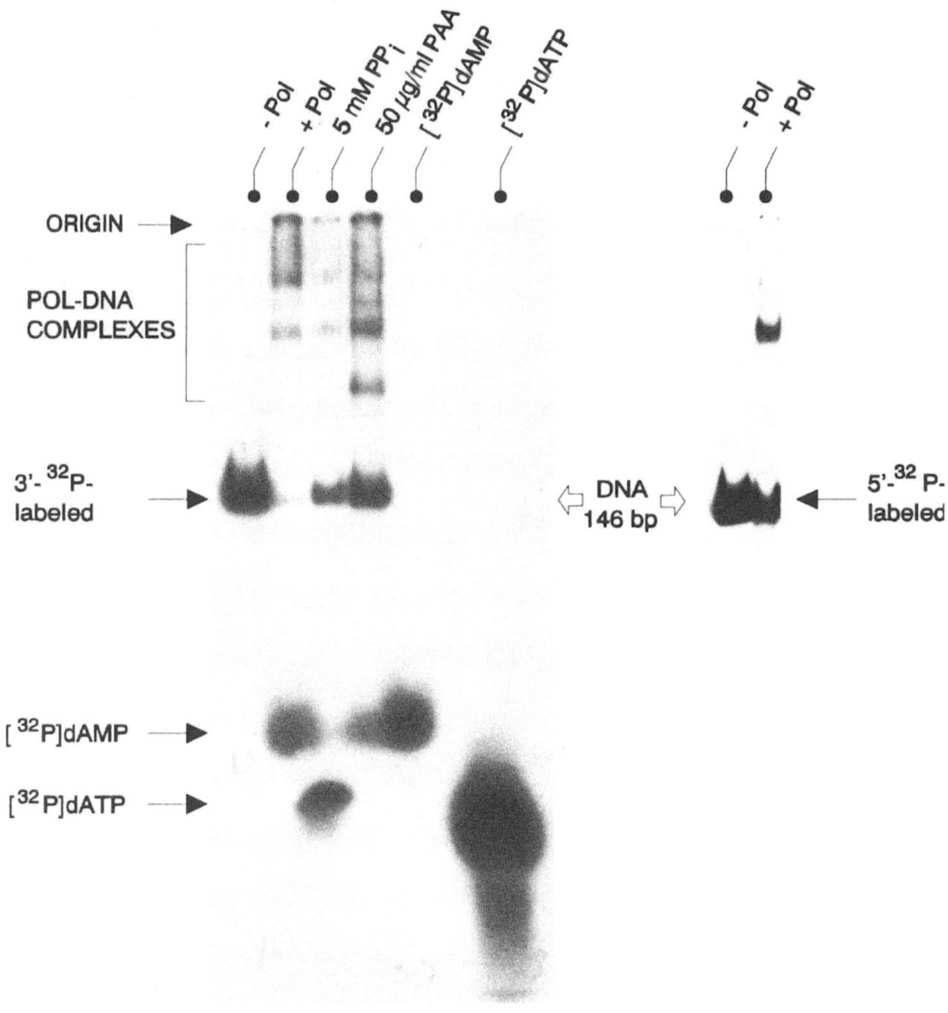

Fig. 9. Band shift electrophoresis in a composite native polyacrylamide gel of DNA polymerase (Pol)-DNA binding reactions containing herpes simplex virus (HSV) Pol, differently labeled DNA, inorganic pyrophosphate (PP_i), or phosphonoacetic acid (PAA). [^{32}P] Deoxyadenosine monophosphate (*[^{32}P]dAMP*) and [^{32}P] deoxyadenosine triphosphate (*[^{32}P]dATP*) were used as size reference

differently end-labeled DNA, it was demonstrated that radiolabeled dNMP is removed only from the 3′-labeled DNA, which indicated that the HSV polymerase preparation used lacked both 5′–3′ exonuclease and endonuclease activities. The reverse reaction of DNA polymerization, the pyrophosphorolysis reaction, can be examined by the generation of dNTP as a result of adding PP_i to the DNA-binding reaction. At 5 m*M* PP_i, the pyrophosphorolysis reaction is favored, and the proofreading activity apparently inhibited. We have further exploited DNA-

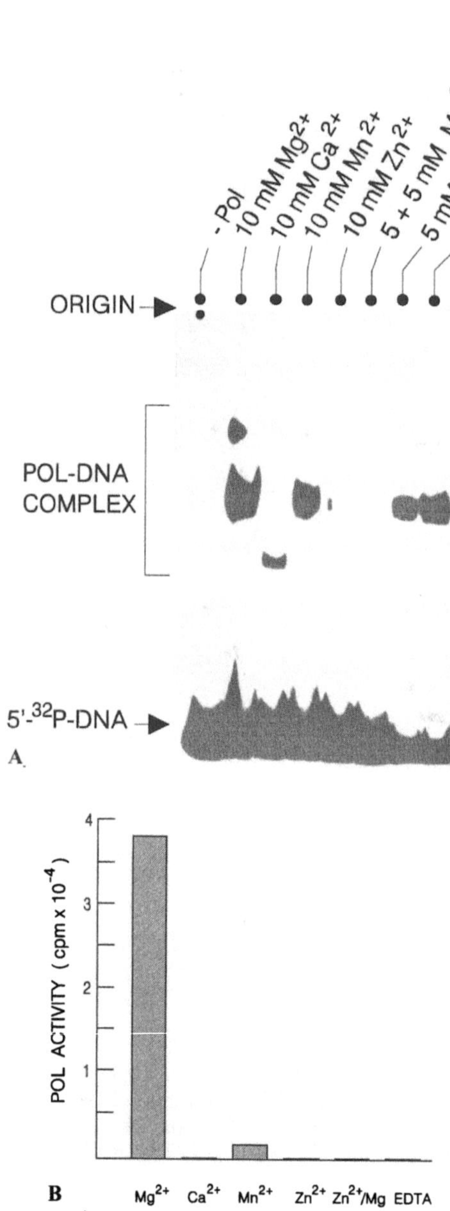

Fig. 10 Effect of divalent cations on herpes simplex virus (HSV) DNA polymerase (Pol)–DNA binding (**A**) and HSV Pol activity (**B**). A Pol–DNA binding reactions were performed with the indicated additions and subjected to band shift electrophoresis in native 5% polyacrylamide gels, as described. **B** HSV Pol activity was determined with 10 m*M* each of the stated ingredients under standard reaction conditions. Zn^{2+} and Mg^{2+} together were present at 5 m*M* each. *EDTA*, ethylenediaminetetraacetic acid

binding analysis to study other components known to interfere with the catalytic functions of the enzyme such as divalent cations, salt, and replication inhibitors. We were able to demonstrate that Ca^{2+} and Zn^{2+} ions have a specific effect on DNA binding of HSV Pol. Ca^{2+} ions produce to a lesser extent enzyme-DNA complexes of apparently faster mobility (Fig. 10A). Zn^{2+} ions, even in the presence of equimolar amounts of Mg^{2+}, prevent DNA binding totally. This result correlates with the polymerase activity when assayed in the presence of these cations (Fig.

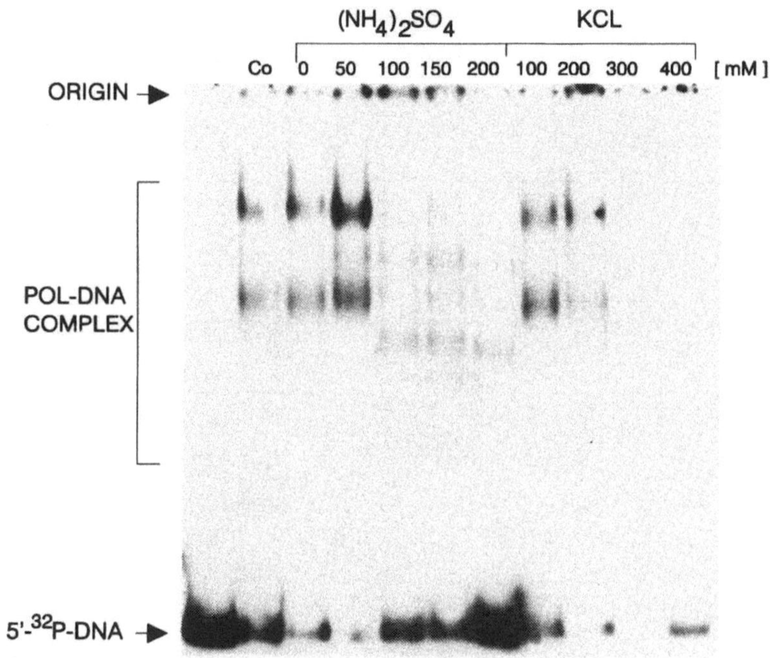

A

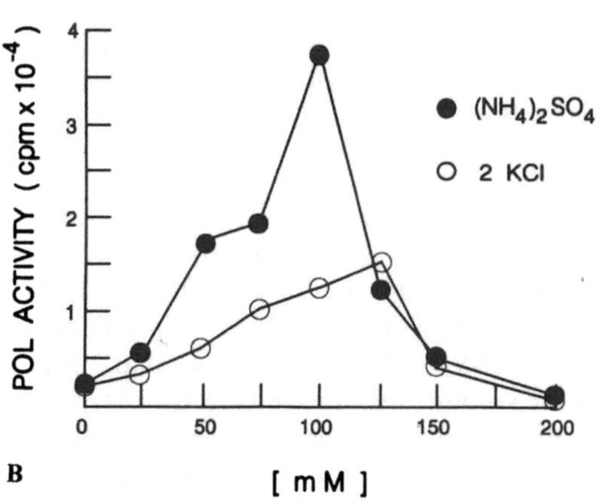

B

Fig. 11. Influence of salt on herpes simplex virus (HSV) DNA polymerase (Pol)–DNA binding (**A**) and HSV Pol activity (**B**). A DNA Pol–DNA binding reactions were performed in the presence of the indicated concentrations of ammonium sulfate and potassium chloride and subjected to band shift electrophoresis in native 5% polyacrylamide gels, as described. **B** HSV Pol activity was determined in the presence of different salt concentrations under standard reaction conditions

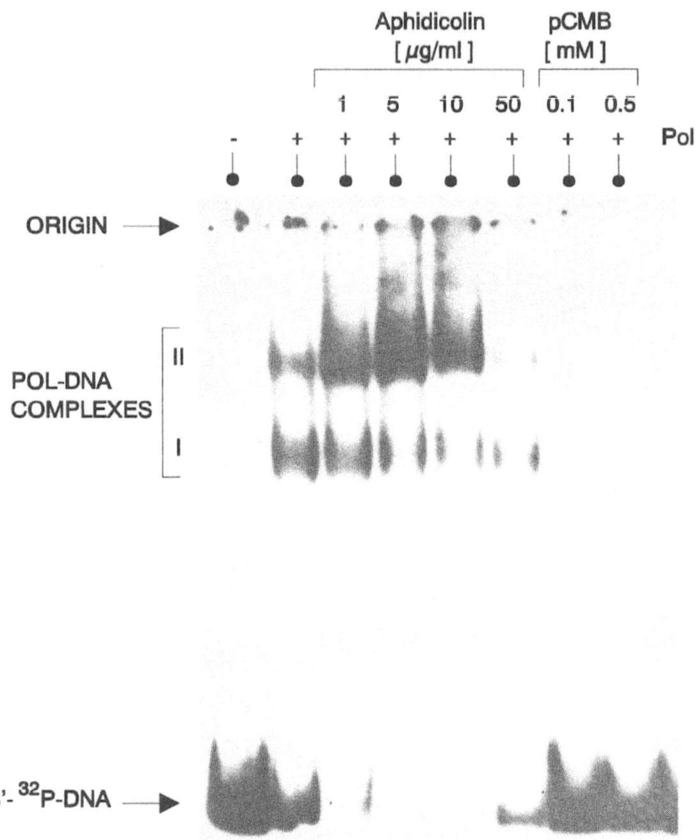

Fig. 12. Effect of aphidicolin and *p*-chloromercuribenzenesulfonate (*pCMB*) on herpes simplex virus (HSV) DNA polymerase (Pol)–DNA binding. Pol–DNA binding reactions were performed with the stated concentrations of aphidicolin and pCMB and subjected to band shift electrophoresis in native 5% polyacrylamide gels, as described

10B) and shows that the enzyme preferentially binds Zn^{2+} ions. Zn^{2+} ions, known to selectively inhibit HSV Pol activity (Fridlender et al. 1978), interfere with DNA binding, an observation which is reminiscent of the role of Zn^{2+} ions in dNMP recognition of the *E. coli* Klenow fragment (Ollis et al. 1985a). Analysis of the effect of salt on DNA binding of wild-type HSV Pol (Fig. 11A) shows that optimal binding occurs at 50 m*M* ammonium sulfate and 100 m*M* KCl. At higher ammonium sulfate concentrations, DNA binding decreases and several intermediate bands of faster mobility can be detected, suggesting that, like the *E. coli* Klenow fragment (Freemont et al. 1988), the enzyme binds DNA in an alternate conformation. This alternate and less tight binding is apparently important for optimal polymerase activity, which is achieved only at higher salt concentrations (100 m*M* ammonium sulfate, 200 m*M* KCl; Fig. 11B).

The DNA-binding analysis has proven to be useful to explore molecular mechanisms of drug interactions. A few examples are presented in Fig. 9 and 12.

Addition of the pyrophosphate analog PAA, a potent inhibitor of HSV Pol, to the DNA-binding reaction enhances the enzyme complex formation and produces strong intermediate bands, indicating a stable conformational change of the enzyme in DNA binding in the presence of the inhibitor. In addition, analogous to the generation of dATP in the presence of PP_i, dAMP–PAA complexes are formed which, on native PAGE, exhibit a mobility similar to dAMP, but they are clearly resolved by thin-layer chromatography (Strick 1993). The inhibitory effect of PAA is consequently based upon maintaining the enzyme in reaction steps 4 and 5 of the polymerization pathway as illustrated in Fig. 4, favoring the reverse reaction of polymerization and keeping the enzyme in an alternate enzyme–DNA-binding conformation. The effect is much stronger than with the natural PP_i, since the presence of the newly made wrong triphosphate dAMP–PAA can additionally block the pyrophosphate exchange reaction and hence the forward polymerization reaction. Analysis of aphidicolin, a potent replication inhibitor of herpes, α, δ-, and ε-polymerases (Wang 1991), demonstrated that this compound also exerted a pronounced effect on the DNA-binding interaction of the herpes polymerase. At concentrations commonly used to inhibit eukaryotic DNA polymerases, aphidicolin greatly enhances DNA binding of HSV Pol (Fig. 12). Indicative of tight binding is the finding that predominantly enzyme–DNA complexes of higher molecular weight (from II) are observed. These complexes consist of DNA fragments with polymerase molecules bound at each free 3′ terminus (Strick and Knopf 1992). Another inhibitor of HSV Pol activity, shown in Figure 12, is the sulfhydril-group blocking agent *p*-chloromercuribenzene sulfonate, which inhibited the enzyme–DNA complex formation as well, suggesting that cysteine residues are most likely critical for the DNA-binding interaction.

In conclusion, analysis of the DNA-binding interaction of HSV Pol has revealed that many commonly used inhibitors affect the DNA-binding property of the enzyme in a specific manner, providing clues for the molecular mechanisms of these inhibitors, which should aid in identifying the right targets for future antiherpetic drug design. The present study and the analysis of the UL42 protein demonstrate how important it is to consider that any subtle effect on the protein–DNA interaction of the enzyme may have consequences for any DNA-dependent activity within the ternary polymerase–UL42 protein–DNA complex.

Working Model for Herpes Simplex Virus DNA Polymerase–DNA Interaction

In consideration of both the structural and functional similarities to *E. coli* Pol I already discussed (Ollis et al. 1985a,b; Joyce and Steitz 1987; Joyce 1991) and our present knowledge about the crystal structure of the HIV-1 reverse transcriptase (Kohlstaedt et al. 1992), we propose a model for the DNA interaction of the HSV Pol holoenzyme, as illustrated in Fig. 13. As shown for the HIV-1 reverse transcriptase (Fig. 13A) and as previously emphasized for the proper orientation of exonuclease and polymerase sites relative to the primer terminus of *E. coli* Pol I (Lecomte and Ninio 1988), the 3′–5′ exonuclease site of HSV Pol is assumed to be located near the 3′-OH primer terminus. The polymerase site follows, and it embodies the ssDNA matrix. The polymerase domain (residues 597–961) com-

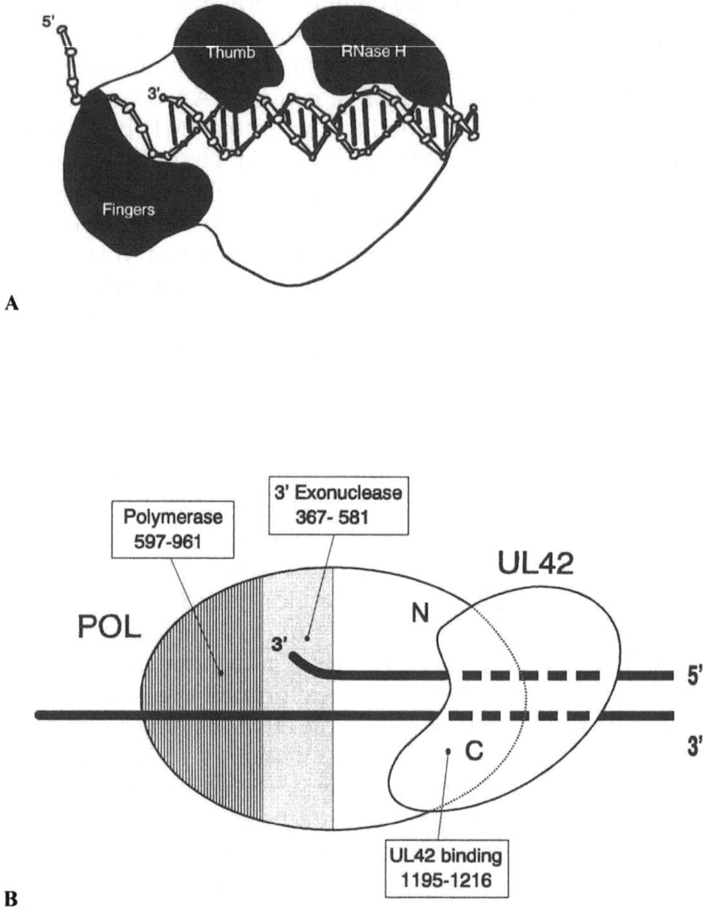

Fig. 13. Model of the human immunodeficiency virus type 1 (HIV-1) reverse transcriptase (**A**) compared to the enzyme–DNA complex of herpes simplex virus type1 (HSV-1) DNA polymerase (Pol) (**B**). A DNA template primer binding of HIV-1 reverse transcriptase according to Kohlstaedt et al. (1992). **B** HSV-1 Pol holoenzyme with subunit UL42. The 3′-OH primer terminus is displaced and located in the exonuclease site. The polymerase domain covers mainly the single strand of the template, and UL42 protein as well as C-terminal sequences interact with the duplex DNA segment

prises the protein regions designated "fingers" and "thumb" in the "right hand" polymerase model (Branden and Tooze 1991) and as indicated in Fig. 13A. No information is available at present as to how close the exonuclease and polymerase sites are situated in the HSV Pol molecule. In ultraviolet cross-linking experiments using UL42 protein-free HSV-1 Pol, the size of a protected DNA primer was found to be 8–10 bases (Strick 1993). According to the mapping of the UL42 protein-binding site (Digard et al. 1993a; Tenney et al. 1993), the C terminus (residues 1195–1216) contains the UL42 protein, which itself has a great affinity

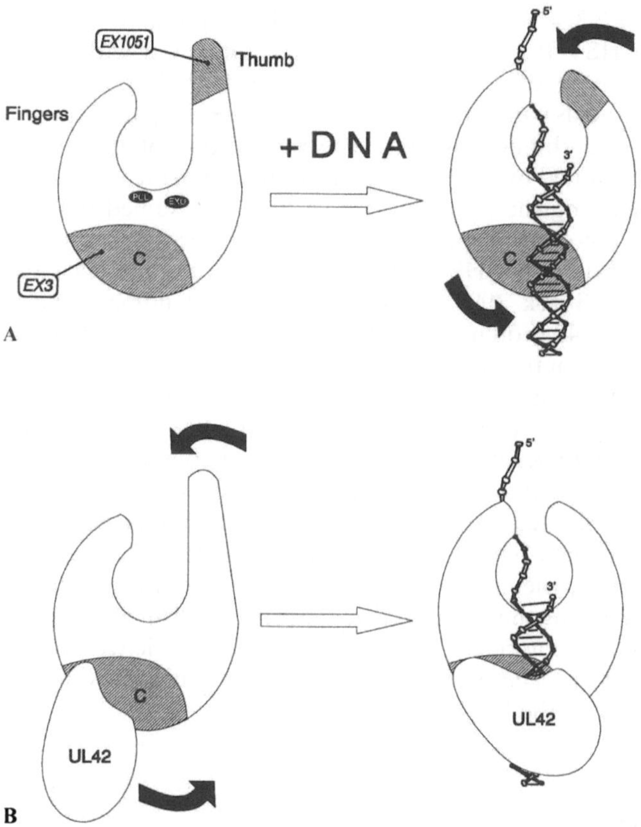

A

B

Fig. 14. DNA interaction of herpes simplex virus type 1 (HSV-1) DNA polymerase (Pol) in the absence (**A**) and in the presence (**B**) of UL42 protein. The drawing depicts a simpler version of the "right-hand" polymerase model (Branden and Tooze 1991) with "*fingers*" and "*thumb*" regions. Target regions for the polymerase antibodies are *shadowed*. The sites for exonuclease and polymerase domain are shown. **A** In the absence of UL42 protein, the template primer is held primarily by fingers and thumb, but in addition binding to C-terminal (*C*) residues is possible. **B** In the HSV-1 Pol holoenzyme, the UL42 protein ensures a further contact with the duplex DNA portion. *Bent arrows* indicate a conformational change of protein regions as result of DNA binding

for dsDNA. In line with the mapping data of Digard et al. (1993b), about two thirds of the mass of the UL42 protein is involved in contact to DNA, and a defined smaller region (around residues 160) binds the polymerase.

A more mechanistic view is provided in Fig. 14, which presents a possible explanation for the inhibitory effect on the DNA binding of the enzyme by the antibodies EX1051 and EX3 (Fig. 7). From sequence alignments, as well as from the strong hydrophilic nature of the target residues of antibody EX1051 (597–685), it can be concluded that these sequences are part of the thumb, which apparently consists of a four-helix bundle (Kohlstaedt et al. 1992). In the absence of DNA, the target sites for both antibodies, i.e., the thumb for EX1051 and the

C-terminal sequences for EX3, are accessible for the antibody interaction. This is true for both situations, HSV Pol alone (Fig. 14A) or complexed with UL42 protein (Fig. 14B). In the absence of UL42 protein, DNA is held by the polymerase through contacts to amino acids within the so-called cleft comprising the catalytic domains and by fingers and thumb domains. The right fitting of the primer terminus is most likely warranted by the 3′-OH group of the primer terminus and further by the ssDNA entering the polymerase pocket. It is assumed that the thumb closes down to the DNA, as proposed for the HIV-1 reverse transcriptase model. It should be added here that HSV Pol binds DNA in the presence and absence of UL42 protein (Strick 1993). When the HSV Pol holoenzyme is incubated with DNA, the UL42 protein can make an additional contact with the ds portion of DNA. By means of this interaction, the C-terminal sequences of the polymerase come in to contact with the dsDNA. The thumb region and the UL42 protein wrap around the DNA, forming a sliding clamp. Preincubating the enzyme with antibodies, either EX1051 or EX3, will block the uptake of DNA by the enzyme. In agreement with the presented model are recent studies on the recombinant HSV Pol using limited proteolysis, which showed that the enzyme is in a different confirmation when bound to DNA (Weisshart et al. 1993). Once bound to the DNA template, the enzyme exposes additional cleavage sites at residues 620 and 1060, suggesting that nearby regions are involved in a conformational change. It is probably no accident that these cleavage sites lie within (for EX1051) and near (for EX3) target sites of the antibodies which prevent the interaction of the enzyme with the DNA template.

Conclusions and Perspectives

HSV Pol remains a major target of antiherpesvirus drug therapy. Its multiple functions and novel interactions with replication proteins in particular offer a wide range of opportunities for interesting novel targets of drug interaction. An important aid for selecting drugs with the target Pol is the analysis of the DNA-binding property, which helps to sort out inhibitors according to their interactions with the enzyme alone or in conjunction with a DNA template. Adapted for routine screening, the DNA-binding analysis may facilitate the discovery of novel suicide inhibitors. The protein interaction between the polymerase and its processivity factor recently emerged as a major promising target, and it has almost been characterized completely. It should be expected that, based on this specific and essential protein-binding, peptide analogs can be developed that successfully inhibit heterodimerization and consequently viral replication. Further, since protein binding is specified by two different and highly adapted replication proteins, antiherpetic therapeutics directed against this interaction should be much less afflicted with the problem of resistance. A promising aspect is that with the identification of novel drug targets, aimed at the HSV Pol holoenzyme and its interactions with replication proteins, a combined antiherpetic drug therapy can be realized in the near future.

References

Abbotts J, Nishiyama Y, Yoshida S, Loeb LA (1987) On the fidelity of DNA replication: herpes DNA polymerase and its associated exonuclease. Nucleic Acids Res 15: 1185–1198

Albrecht JC, Fleckenstein B (1990) Structural organization of the conserved gene block of herpesvirus saimiri coding for DNA polymerase, glycoprotein B, and major DNA binding protein. Virology 174: 533–542

Andrei G, Snoeck R, Goubau P, Desmyter J, De Clercq E (1992) Comparative activity of various compounds against clinical strains of herpes simplex virus. Eur J Clin Microbiol Infect Dis 11: 143–151

Aron GM, Purifoy DJM, Schaffer PA (1975) DNA synthesis and DNA polymerase activity of herpes simplex virus type 1 temperature-sensitive mutants. J Virol 16: 498–507

Baer R, Bankier AT, Beggin MD, Deininger PL, Farrell PL, Gibson TJ, Hatfull G, Hudson GS, Satchwell SC, Seguin C, Tuffnell PS, Barrell BG (1984) DNA sequence and expression of the B95-8 Epstein-Barr virus genome. Nature 310: 207–211

Baltimore D (1974) Is terminal deoxynucleotidyl transferase a somatic mutagen in lymphocytes? Nature 248: 409–411

Banks L, Purifoy DJM, Hurst P-F, Killington RA, Powell KL (1983) Herpes simplex virus non-structural proteins. IV. Purification of the virus-induced deoxyribonuclease and characterization of the enzyme using monoclonal antibodies. J Gen Virol 64: 2249–2260

Banks LM, Halliburton IW, Purifoy DJM, Killington RA, Powell KL (1985) Studies on the herpes simplex virus alkaline nuclease: detection of type-common and type-specific epitopes on the enzyme. J Gen Virol 66: 1–14

Bapat AR, Han F-S, Liu Z, Zhou B-S, Cheng Y-C (1987) Studies on DNA topoisomerases I and II in herpes simplex virus type 2-infected cells. J Gen Virol 68: 2231–2237

Barnes MH, Hammond RA, Kennedy CC, Mack SL, Brown NC (1992) Localization of the exonuclease and polymerase domains of *Bacillus subtilis* DNA polymerase III. Gene 111: 43–49

Beese LS, Steitz TA (1991) Structural basis for the 3′–5′ exonuclease activity of *Escherichia coli* DNA polymerase I: a two metal ion mechanism. EMBO J 10: 25–33

Bennet LL, Shannon WM, Allen PW, Arnett G (1978) Studies on the biochemical basis for the antiviral activities of some nucleoside analogs. Ann N Y Acad Sci 255: 342–352

Bernad A, Zaballos A, Salas M, Blanco L (1987) Structural and functional relationships between prokaryotic and eukaryotic DNA polymerases. EMBO J 6: 4219–4225

Bernad A, Blanco L, Lázaro JM, Martin G, Salas M (1989) A conserved 3′ to 5′ exonuclease active site in prokaryotic and eukaryotic DNA polymerases. Cell 59: 219–228

Biswal N, Feldan P, Levy CC (1983) A DNA topoisomerase activity copurifies with the DNA polymerase induced by herpes simplex virus. Biochim Biophys Acta 740: 379–389

Blanco L, Bernad A, Blasco MA, Salas M (1990) A general structure for DNA-dependent DNA polymerases. Gene 100: 27–38

Blanco L, Bernad A, Salas M (1992) Evidence favouring the hypothesis of a conserved 3′–5′ exonuclease active site in DNA-dependent DNA polymerases. Gene 112: 139–144

Bludau H, Freese UK (1991) Analysis of the HSV-1 strain 17 DNA polymerase gene reveals the expression of four different classes of pol transcripts. Virology 183: 505–518

Boehmer PE, Lehman IR (1993) Herpes simplex virus type 1 ICP8: helix-destabilizing properties. J Virol 67: 711–715

Boehmer PE, Dodson MS, Lehman IR (1993) The herpes simplex virus type-1 origin binding protein. DNA helicase activity. J Biol Chem 268: 1220–1225

Boulet A, Simon M, Faye G, Bauer GA, Burgers PMJ (1989) Structure and function of the *Saccharomyces cerevisiae CDC2* gene encoding the large subunit of DNA polymerase III. EMBO J 8: 1849–1854

Branden C, Tooze J (1991) Introduction to protein structure. Garland, New York

Brown WC, Duncan JA, Campbell JL (1993) Purification and characterization of the *Saccharomyces cerevisiae* DNA polymerase δ overproduced in *Escherichia coli*. J Biol Chem 268: 982–990

Burgers PMJ, Bambara RA, Campbell JL, Chang LMS, Downey KM, Hübscher U, Lee MYWT, Linn SM, So AG, Spadari S (1990) Revised nomenclature for eukaryotic DNA polymerases. Eur J Biochem 191: 617–618
Calder JM, Stow ND (1990) Herpes simplex virus helicase-primase: the UL8 protein is not required for DNA-dependent ATPase and DNA helicase activities. Nucleic Acids Res 18: 3573–3578
Challberg MD (1986) A method for identifying the viral genes required for herpes DNA replication. Proc Natl Acad Sci USA 83: 9094–9098
Challberg MD, Kelly TJ (1989) Animal virus DNA replication. Annu Rev Biochem 58: 671–718
Chan BSS, Court DA, Verula PJ, Bertrand H (1991) The kalilo linear senescence-inducing plasmid of neurospora is an invertron and encodes DNA and RNA polymerases. Curr Genet 20: 225–237
Chang LMS (1975) The distributive nature of enzymatic DNA synthesis. J Mol Biol 93: 219–235
Chang LMS, Bollum FJ (1971) Deoxynucleotide-polymerizing enzymes of calf thymus gland. J Biol Chem 246: 909–916
Chartrand P, Stow ND, Timbury MC, Wilkie NM (1979) Physical mapping of PAA^r mutations of herpes simplex virus type 1 and type 2 by intertypic marker rescue. J Virol 31: 265–276
Chiou HC, Weller SK, Coen DM (1985) Mutations in the herpes simplex virus major DNA-binding protein gene leading to altered sensitivity to DNA polymerase inhibitors. Virology 145: 213–226
Chung DW, Zhang J, Tan C-K, Davie EW, So AG, Downey KM (1991) Primary structure of the catalytic subunit of human DNA polymerase δ and chromosomal location of the gene. Proc Natl Acad Sci USA 88: 11197–11201
Coen DM (1986) General aspects of virus drug resistance with special reference to herpes simplex virus. J Antimicrob Agents Chemother 18[Suppl B]: 1–10
Coen DM (1991) The implications of resistance to antiviral agents for herpesvirus drug targets and drug therapy. Antiviral Res 15: 287–300
Coen DM, Schaffer PA (1980) Two distinct loci can confer resistance to acycloguanosine in herpes simplex virus type 1. Proc Natl Acad Sci USA 77: 2265–2269
Coen DM, Furman PA, Gelep PT, Schaffer PA (1982) Mutations in the herpes simplex virus DNA polymerase gene can confer resistance to 9-β-arabinosyl adenine. J Virol 41: 909–918
Coen DM, Aschman DP, Gelep PT, Retondo MJ, Weller SK, Schaffer PA (1984) Fine mapping and molecular cloning of mutations in the herpes simplex virus DNA polymerase locus. J Virol 49: 236–247
Cotteril SM, Reyland ME, Loeb LA, Lehman IR (1987) A cryptic proofreading 3′ to 5′ exonuclease associated with the polymerase subunit of the DNA polymerase-primase from *Drosophila melanogaster*. Proc Natl Acad Sci USA 84: 5635–5639
Crumpacker CS (1992) Mechanism of action of foscarnet against viral polymerases. Am J Med 92[suppl 2A]: 3S–7S
Crute JJ, Lehman IR (1989) Herpes simplex- 1DNA polymerase. Identification of an intrinsic 5′–3′ exonuclease with ribonuclease H activity. J Biol Chem 264: 19266–19270
Crute JJ, Mocarski ES, Lehman IR (1988) A DNA helicase induced by herpes simplex virus type 1. Nucleic Acids Res 16: 6585–6596
Crute JJ, Tsurumi T, Zhu L, Weller SK, Olivo PD, Challberg MD, Mocarski ES, Lehman IR (1989) Herpes simplex virus 1 helicase-primase: a complex of three herpes-encoded gene products. Proc Natl Acad Sci USA 86: 2186–2189
Dabrowski CE, Schaffer PA (1991) Herpes simplex virus type 1 origin-specific binding protein: oriS-binding properties and effects of cellular proteins. J Virol 65: 3140–3150
Dahlberg ME, Benkovic SJ (1991) Kinetic mechanism of DNA polymerase I (Klenow fragment): identification of a second conformational change and evaluation of the internal equilibrium constant. Biochemistry 30: 4835–4843
Davison AJ (199) Channel catfish virus: a new type of herpesvirus. EMBL Data Library, Heidelberg

Davison AJ, Scott JE (1986). The complete DNA sequence of Varicella-zoster virus. J Gen Virol 67: 1759–1816
De Bruyn Kops A, Knipe DM (1988) Formation of DNA replication structures in herpes virus-infected cells requires a viral DNA binding protein. Cell 55: 857–868
De Clercq E, Sakuma T, Baba M, Pauwels R, Balzarini J, Rosenberg I, Holy A (1987) Antiviral activity of phosphonylmethoxyalkyl derivatives of purine and pyrimidines. Antiviral Res 8: 261–272
Delarue M, Poch O, Tordo N, Moras D, Argos P (1990) An attempt to unify the structure of polymerases. Protein Eng 3: 461–467
Derbyshire V, Freemont PS, Sanderson MR, Beese LS, Friedman JM, Joyce CM, Steitz TA (1988) Genetic and crystallographic studies of the 3′,5′-exonucleolytic site of DNA polymerase I. Science 240: 199–201
Derbyshire V, Grindley NDF, Joyce CM (1991) The 3′–5′ exonuclease of DNA polymerase I of *Escherichia coli*: contribution of each amino acid at the active site to the reaction. EMBO J 10: 17–24
Deutscher MP, Kornberg A (1969) Enzymatic synthesis of deoxyribonucleic acid. XXVIII. The pyrophosphate exchange and pyrophosphorolysis reactions of deoxyribonucleic acid polymerase. J Biol Chem 244: 3019–3028
Devereux J, Haeberli P, Smithies O (1984) A comprehensive set of sequence analysis programs for the VAX. Nucleic Acids Res 12: 387–395
Digard P, Coen DM (1990) A novel functional domain of an α-like DNA polymerase. The binding site on the herpes simplex virus polymerase for the viral UL42 protein. J Biol Chem 265: 17393–17396
Digard P, Bebrin WR, Weisshart K, Coen DM (1993a) The extreme C terminus of herpes simplex virus DNA polymerase is crucial for functional interaction with processivity factor UL42 and for viral replication. J Virol 67: 398–406
Digard P, Chow CS, Pirrit L, Coen DM (1993b) Functional analysis of the herpes simplex virus UL42 protein. J Virol 67: 1159–1168
Dodson MS, Lehman IR (1991) Association of DNA helicase and primase activities with a subassembly of the herpes simplex virus 1 helicase-primase composed of the UL5 and UL52 gene products. Proc Natl Acad Sci USA 88: 1105–1109
Dodson MS, Lehman IR (1993) The herpes simplex virus type I origin binding protein. DNA-dependent nucleoside triphosphatase activity. J Biol Chem 268: 1213–1219
Dodson MS, Crute JJ, Bruckner RC, Lehman IR (1989) Overexpression and assembly of the herpes simplex virus type I helicase-primase in insect cells. J Biol Chem 26: 20835–20838
Doolittle RF (1981) Similar amino acid sequences: chance or common ancestry? Science 214: 149–159
Dorsky DI, Crumpacker CS (1988) Expression of herpes simplex virus type 1 DNA polymerase gene by in vitro translation and effects of gene deletions on activity. J Virol 62: 3224–3232
Dorsky DI, Crumpacker CS (1990) Site-specific mutagenesis of a highly conserved region of the herpes simplex virus type 1 DNA polymerase gene. J Virol 64: 1394–1397
Dorsky DI, Chatis P, Crumpacker CS (1987) Functional expression of a cloned herpes simplex virus type 1 DNA polymerase gene. J Virol 61: 1704–1707
Downey KM, Tan C-K, Andrews DM, Li X, So AG (1988) Proposed roles for DNA polymerases alpha and delta at the replication fork. Cancer Cells 6: 403–410
Downey KM, Tan C-K, So AG (1990) DNA polymerase delta: a second eukaryotic DNA replicase. BioEssays 12: 231–236
Elion G, Furman PA, Fyfe JA, De Miranda P, Beauchamp L, Schaeffer HJ (1977) Selectivity of action of an antiherpetic agent, 9-(2-hydroxyethoxymethyl)guanine. Proc Natl Acad Sci USA 74: 5716–5720
Elliot R, Clark C, Jaquish D, Spector DH (1991) Transcription analysis and sequence of the putative murine cytomegalovirus DNA polymerase gene. Virology 185: 169–186
Englund PT, Huberman JA, Jovin TM, Kornberg A (1969) Enzymatic synthesis of deoxyribonucleic acid. XXX. Binding of triphosphates to deoxyribonucleic acid polymerase. J Biol Chem 244: 3038–3044

Fierer DS, Challberg MD (1992) Purification and characterization of UL9, the herpes simplex virus type 1 origin-binding protein. J Virol 66: 3986–3995

Foury F (1989) Cloning and sequencing of the nuclear gene MIP1 encoding the catalytic subunit of the yeast mitochondrial DNA polymerase. J Biol Chem 264: 20552–20560

Frank KB, Cheng Y-C (1985) Mutually exclusive inhibition of herpesvirus DNA polymerase by aphidicolin, phosphonoformate, and acyclic nucleoside triphosphates. Antimicrob Agents Chemother 27: 445–448

Frank KB, Derse DD, Bastow KF, Cheng Y-C (1984) Novel interaction of aphidicolin with herpes simplex virus DNA polymerase and polymerase-associated exonuclease. J Biol Chem 259: 13282–13286

Freemont PS, Ollis DL, Steitz TA, Joyce CM (1986) A domain of the Klenow fragment of *Escherichia coli* DNA polymerase I has polymerase but no exonuclease activity. Proteins Struct Funct Genet 1: 66–73

Freemont PS, Friedman JM, Beese LS, Sanderson MR, Steitz TA (1988) Co-crystal structure of an editing complex of Klenow fragment with DNA. Proc Natl Acad Sci USA 85: 8924–8928

Fridlender B, Chejanovsky N, Becker Y (1978) Selective inhibition of herpes simplex virus type 1 DNA polymerase by zinc ions. Virology 84: 551–554

Fry M, Loeb LA (1986) Animal cell DNA polymerases. CRC Press, Boca Raton

Furman PA, St. Clair MH, Fyfe JA, Rideout JL, Keller PM, Elion GB (1979) Inhibition of herpes simplex virus-induced DNA polymerase activity and viral DNA replication by 9-(2-hydroxyethoxymethyl) guanine and its triphosphate. J Virol 32: 72–77

Furman PA, Coen DM, St Clair MH, Schaffer PA (1981) Acyclovir-resistant mutants of herpes simplex virus type 1 express altered DNA polymerase or reduced acyclovir phosphorylating activities. J Virol 40: 936–941

Fyfe JA, Keller PM, Furman PA, Miller RL, Elion GB (1978) Thymidine kinase herpes simplex virus phosphorylates the new antiviral compound 9-(2-hydroxyethomethyl)guanine. J Biol Chem 253: 8721–8727

Gao M, Knipe DM (1989) Genetic evidence for multiple nuclear function of the herpes simplex virus ICP8 DNA-binding protein. J Virol 63: 5258–5267.

Gao M, Knipe DM (1993) Intragenic complementation of herpes simplex virus ICP8 DNA-binding protein mutants. J Virol 67: 876–885

Gallo ML, Jackwood DH, Murphy M, Marsden HS, Parris DS (1988) Purification of the herpes simplex virus type 1 65 kilodalton protein and evidence for its association with the virus-encoded DNA polymerase. J Virol 62: 2874–2883

Gallo ML, Dorsky DI, Crumpacker CS, Parris DS (1989) The essential 65-kilodalton DNA-binding protein of herpes simplex virus stimulates the virus-encoded DNA polymerase. J Virol 63: 5023–5029

Gibbs JS, Chiou HC, Hall JD, Mount DW, Retondo MJ, Weller SK, Coen DM (1985) Sequence and mapping analyses of the herpes simplex virus DNA polymerase gene predict a C-terminal substrate binding domain. Proc Natl Acad Sci USA 82: 7969–7973

Gibbs JS, Chiou H, Bastow KF, Cheng YC, Coen DM (1988) Identification of amino acids in herpes simplex virus DNA polymerase involved in substrate and drug recognition. Proc Natl Acad Sci USA 85: 6672–6676

Gibbs JS, Weisshart K, Digard P, De Bruyn-Kops A, Knipe DM, Coen DM (1991) Polymerization activity of an α-like DNA polymerase requires a conserved 3′–5′ exonuclease active site. Mol Cell Biol 11: 4786–4795

Gottlieb J, Marcy AI, Coen DM, Challberg MD (1990) The herpes simplex virus type 1 UL42 gene product: a subunit of DNA polymerase that functions to increase processivity. J Virol 64: 5976–5987

Haffey ML, Stevens JT, Terry BJ, Dorsky DI, Crumpacker CS, Wietstock SM, Ruyechan WT, Field AK (1988) Expression of herpes simplex virus type 1 DNA polymerase in *Saccharomyces cerevisiae* and detection of virus-specific enzyme activity in cell free lysates. J Virol 62: 4493–4498

Haffey ML, Novotny J, Bruccoleri RE, Carroll RD, Stevens JT, Matthews JT (1990) Structure–function studies of the herpes simplex virus type 1 DNA polymerase. J Virol 64: 5008–5018

Hall JD, Furman PA, St. Clair MH, Knopf CW (1985) Reduced in vivo mutagenesis by mutant herpes simplex DNA polymerase involves improved nucleotide selection. Proc Natl Acad Sci USA 82: 3889–3893
Hall JD, Wang Y, Pierpont J, Berlin MS, Rundlett SE, Woodward S (1989) Aphidicolin resistance in herpes simplex virus type I reveals features of the DNA polymerase dNTP binding site. Nucleic Acids Res 17: 9231–9244
Hart GJ, Boehme RE (1992) The effect of the UL42 protein on the DNA polymerase activity of the catalytic subunit of the DNA polymerase encoded by herpes simplex virus type 1. FEBS Lett 305: 97–100
Hay J, Moss H, Halliburton IW (1971) Induction of deoxyribonucleic acid polymerase and deoxyribonuclease activities in cells infected with herpes simplex virus type II. Biochem J 124: 64p
Hernandez TR, Lehman IR (1990) Functional interaction between the herpes simplex-1 DNA polymerase and UL42 protein. J Biol Chem 265: 11227–11232
Hirose F, Yamaguchi M, Nishida Y, Masutani M (1991) Structure and expression during development of *Drosophila melanogaster* gene for DNA polymerase alpha. Nucleic Acids Res 19: 4991–4998
Hoffmann PJ (1991) Mechanism of degradation of duplex DNA by the DNase induced by herpes simplex virus. J Virol 38: 1005–1014
Hoffmann PJ, Cheng Y-C (1979) DNase induced after infection of KB cells by herpes simplex virus type 1 or type 2. II. Characterization of an associated endonuclease activity. J Virol 32: 449–457
Holland LE, Sandri-Goldin RM, Goldin AL, Glorioso JC, Levine ML (1984) Transcriptional and genetic analyses of the herpes simplex virus type 1 genome: coordinates 0.29 to 0.45. J Virol 49: 947–955
Hwang CB, Ruffner KL, Coen DM (1992) A point mutation within a distinct conserved region of the herpes simplex virus DNA polymerase gene confers drug resistance. J Virol 66: 1774–1776
Ito J, Braithwaite DK (1991) Compilation and alignment of DNA polymerase sequences. Nucleic Acids Res 19: 4045–4057
Johnson PA, Best MG, Friedman T, Parris DS (1991) Isolation of a herpes simplex virus type 1 mutant deleted for the essential UL42 gene and characterization of its null phenotype. J Virol 65: 700–710
Joyce CM (1991) Can DNA polymerase I (Klenow fragment) serve as a model for other DNA polymerases? Curr Opin Struct Biol 1: 123–129
Joyce CM, Steitz TA (1987) DNA polymerase I: from crystal structure to function via genetics. Curr Opin Struct Biol 1: 123–129
Jung G, Leavitt MC, Hsieh J-C, Ito J (1987) Bacteriophage PRD1 DNA polymerase: evolution of DNA polymerases. Proc Natl Acad Sci USA 8: 8287–8291
Keir HM, Gold E (1963) Deoxyribonucleic acid nucleotidyltransferase and deocyribonuclease from cultured cells infected with herpes simplex virus. Biochim Biophys Acta 72: 263–276
Keir HM, Subak-Sharpe H, Shedden WIH, Watson DH, Wildy P (1966) Immunological evidence for a specific DNA polymerase produced after infection by herpes simplex virus. Virology 30: 154–157
Kesti T, Syväoja (1991) Identification and tryptic cleavage of the catalytic core of Hela and calf thymus DNA polymerase ε. J Biol Chem 266: 6336–6341
Knopf KW (1979) Properties of herpes simplex virus DNA polymerase and characterization of its associated exonuclease activity. Eur J Biochem 98: 231–244
Knopf C (1983) HSV DNA polymerase: highly purified enzyme preparations contain both, 3′ to 5′ and 5′ to 3′ directed exonucleolytic activities. VIIIth international herpesvirus workshop, Oxford, 31 July–5 Aug, p 289
Knopf CW (1986) Nucleotide sequence of the DNA polymerase gene of herpes simplex virus type 1 strain Angelotti. Nucleic Acids Res 14: 8225–8226
Knopf CW (1987) The herpes simplex virus type 1 DNA polymerase gene: site of phosphonoacetic acid resistance mutation in strain Angelotti is highly conserved. J Gen Virol 68: 1429–1433

Knopf CW, Weisshart K (1988) The herpes simplex virus DNA polymerase: analysis of the functional domains. Biochim Biophys Acta 951: 298–314
Knopf CW, Weisshart K (1990) Comparison of exonucleolytic activities of herpes simplex virus type I DNA polymerase and DNase. Eur J Biochem 191: 263–273
Knopf KW, Kaufman ER, Crumpacker C (1981) Physical mapping of drug resistance mutations defines an active center of the herpes simplex virus DNA polymerase enzyme. J Virol 39: 746–757
Kohlstaedt LA, Wang J, Friedman JM, Rice PA, Steitz TA (1992) Crystal structure at 3.5 Å resolution of HIV-1 reverse transcriptase complexed with an inhibitor. Science 256: 1783–1790
Kornberg A (1980) DNA replication. Freeman, San Francisco
Kouzarides T, Bankier AT, Satchwell SC, Weston K, Tomlinson P, Barrell BG (1987) Sequence and transcription analysis of the human cytomegalovirus DNA polymerase gene. J Virol 61: 125–133
Kronenwett R., Weisshart K., Knopf CW (1990) Expression of functional herpes simplex virus type I DNA polymerase using recombinant vaccinia virus. VIIIth international congress of virology, Berlin, 26–31 Aug
Kunkel TA (1992) DNA replication fidelity. J Biol Chem 267: 18251–18254
Larder BA, Kemp SD, Darby G (1987) Related functional domains in virus DNA polymerases. EMBO J 6: 169–175
Leavitt MC, Ito J (1989) T5 DNA polymerase: structural-functional relationships to other DNA polymerases. Proc Natl Acad Sci USA 86: 4465–4469
Lecomte PJ, Ninio J (1988) Nucleotide excision by *E. coli* DNA polymerase I in proofreading and non-proofreading modes. Biochim Biophys Acta 951: 255–260
Leinbach SS, Heath LS (1988) A carboxyl-terminal peptide of the DNA-binding protein ICP8 of herpes simplex virus contains a single-stranded DNA-binding site. Virology 166:10–16
Leinbach SS, Reno JM, Lee LF, Isbell AF, Boezi JA (1976) Mechanism of phosphonoacetic acid inhibition of herpes virus induced DNA polymerase. Biochemistry 15: 426–430
Marchetti ME, Smith CA, Schaffer PA (1988) A temperature-sensitive mutation in a herpes simplex virus type I gene required for viral DNA synthesis maps to coordinates 0.609 through 0.614 in U_L. J Virol 62: 715–721
Marcy AI, Yager DR, Coen DM (1990a) Isolation and characterization of herpes simplex virus mutants containing engineered mutations at the polymerase locus. J Virol 64: 2208–2216
Marcy AI, Hwang CBC, Ruffner KL, Coen DM (1990b) Engineered herpes simplex virus DNA polymerase point mutants: the most highly conserved region shared among α-like DNA polymerases is involved in substrate recognition. J Virol 64: 5883–5890
Marcy AI, Olivo PD, Challberg MD, Coen DM (1990c) Enzymatic activities of overexpressed herpes simplex virus DNA polymerase purified from recombinant baculovirus-infected cells. Nucleic Acids Res 18: 1207–1215
Matthews JT, Stevens JT, Terry BJ, Cianci CW, Haffey ML (1990) Neutralization of purified herpes simplex virus DNA polymerase by two antipeptide sera. Virus Genes 3: 343–354
Matthews JT, Terry BJ, Field AK (1993) The structure and function of the HSV DNA replication proteins: defining novel antiviral targets. Antiviral Res 20: 89–114
McGeoch DJ, Dolan A, Donald S, Brauer DHK (1986) Complete DNA sequence of the short repeat region in the genome of herpes simplex virus type 1. Nucleic Acids Res 14: 1727–1745
McGeoch DJ, Dalrymple MA, Davison AJ, Dolan A, Frame MC, McNab D, Perry LJ, Scott JE, Taylor P (1988) The complete DNA sequence of the long unique region in the genome of herpes simplex virus type 1. J Virol 69: 1531–1574
Mildvan A (1989) NMR studies of the interactions of substrates with enzymes and their peptide fragments. FASEB J 3: 1705–1714
Miller WH, Miller RL (1980) Phosphorylation of acyclovir monophosphate by GMP kinase. J Biol Chem 255: 7204–7207
Miller WH, Miller RL (1982) Phosphorylation of acyclovir diphosphate by cellular enzymes. Biochem Pharmacol 31: 3879–3884
Morrison A, Sugino A (1991) Nucleotide sequence of the POL3 gene encoding DNA polymerase III (δ) of *Saccharomyces cerevisiae.* Nucleic Acids Res 20: 375

Morrison A, Araki H, Clark AB, Hamatake RK, Sugino A (1990) A third essential DNA polymerase in *S. cerevisiae.* Cell 62: 1143–1151
Morrison A, Bell JB, Kunkel TA, Sugino A (1991) Eukaryotic DNA polymerase amino acid sequence required for 3′–5′ exonuclease activity. Proc Natl Acad Sci USA 88: 9473–9477
Müller WEG, Zahn RK, Arendes J, Falke D (1979) Oligoribonucleotide initiators for herpes simplex virus DNA synthesis in vivo and in vitro. Virology 98: 200–210
Mullen GP, Serpersu EH, Ferrin LJ, Loeb LA (1990) Metal binding to DNA polymerase I, its large fragment, and two 3′–5′ exonuclease mutants of the large fragment. J Biol Chem 265: 14327–14334
O'Donnell ME, Elias P, Lehman IR (1987) Processive replication of single-stranded DNA templates by the herpes simplex virus-induced DNA polymerase. J Biol Chem 262: 4252–4259
Olivo PD, Nelson NJ, Challberg MD (1988) Herpes simplex virus DNA replication: the UL9 gene encodes an origin-binding protein. Proc Natl Acad Sci USA 85: 5414–5418
Olivo PD, Nelson NJ, Challberg MD (1989) Herpes simplex virus type 1 gene products required for DNA replication: identification and overexpression. J Virol 63: 196–204
Ollis DL, Brick P, Hamlin R, Xuong NG, Steitz TA (1985a) Structure of large fragment of *Escherichia coli* DNA polymerase I complexed with dTMP. Nature 313: 762–766
Ollis DL, Kline C, Steitz TA (1985b) Domain of *E. coli* DNA polymerase I showing sequence homology to T7 DNA polymerase. Nature 313: 818–819
Ostrander M, Cheng Y-H (1980) Properties of herpes simplex virus type 1 and type 2 DNA polymerase. Biochim Biophys Acta 609: 232–245
Owsianka AM, Hart G, Murphy M, Gottlieb J, Boehme R, Challberg M, Marsden HS (1993) Inhibition of herpes simplex virus type 1 DNA polymerase activity by peptides from the UL42 accessory protein is largely nonspecific. J Virol 67: 258–264
Paillard M, Sederoff RR, Levings III CS (1985) Nucleotide sequence of the S-1 mitochondrial DNA from the S cytoplasm of maize. EMBO J 4: 1125–1128
Pandey VN, Williams KR, Stone KL, Modak MJ (1987) Photoaffinity labeling of the thymidine triphosphate binding domain in *Escherichia coli* DNA polymerase I: Identification of histidine-881 as the site of cross-linking. Biochemistry 26: 7744–7748
Patel SS, Wong I, Johnson KA (1991) Pre-steady-state kinetic analysis of processive DNA replication, including complete characterization of an exonuclease-deficient mutant. Biochemistry 30: 511–525
Pedrali-Noy G, Spadari S (1980) Mechanism of inhibition of herpes simplex virus and vaccinia virus DNA polymerases by aphidicolin, a highly specific inhibitor of DNA replication in eucaryotes. J Virol 36: 457–464
Pizzagalli A, Valsasnini P, Plevani P, Lucchini G (1988) DNA polymerase I gene of *Saccharomyces cerevisiae*: nucleotide sequence, mapping of a temperature-sensitive mutation, and protein homology with other DNA polymerases. Proc Natl Acad Sci USA 85: 3772–3776
Polesky AH, Steitz TA, Grindley NDF, Joyce CM (1990) Identification of residues critical for the polymerase activity of the Klenow fragment of DNA polymerase I from *Escherichia coli.* J Biol Chem 265: 14579–14591
Polesky AH, Dahlberg ME, Benkovic SJ, Grindley NDF, Joyce CM (1992) Side chains involved in catalysis of the polymerase reaction of DNA polymerase I from *Escherichia coli.* J Biol Chem 267: 8417–8428
Powell K, Purifoy (1977) Nonstructural proteins of the herpes simplex virus 1. Purification of the induced DNA polymerase. J Virol 24: 618–626
Purifoy DJM, Lewis RB, Powell KL (1977) Identification of the herpes virus DNA polymerase gene. Nature 269: 621–623
Quinn JP, McGeoch DJ (1985) DNA sequence of the region in the genome of herpes simplex virus type 1 containing the genes for DNA polymerase and the major DNA binding protein. Nucleic Acids Res 13: 8143–8163
Reddy MK, Weitzel SE, von Hippel, PH (1992) Processive proofreading is intrinsic to T4 DNA polymerase. J Biol Chem 267: 14157–14166
Reha-Krantz LJ (1990) Genetic evidence for two protein domains and a potential new activity in bacteriophage T4 DNA polymerase. Genetics 124: 213–220

Reha-Krantz LJ, Stocki S, Nonay RL, Dimayuga E, Goodrich LD, Konigsberg WH, Spicer EK (1991) DNA polymerization in the absence of exonucleolytic proofreading: in vivo and in vitro studies. Proc Natl Acad Sci USA 88: 2417–2421
Ridley RG, White JH, McAleese SM, Goman M, Alano P, de Vries E, Kilbey BJ (1991) DNA polymerase δ: gene sequences from *Plasmodium falciparum* indicate that this enzyme is more highly conserved than DNA polymerase α. Nucleic Acids Res 19: 6731–6736
Roizman B (1982) The family herpesviridae: general description, taxonomy and classification. In: Roizman B (ed) The herpesviruses, vol 1. Plenum, New York
Ruscitti T, Polayes DA, Karu AE, Linn S (1992) Selective immunoneutralization of the multiple activities of *Escherichia coli* DNA polymerase I supports the model for separate active sites and indicates a complex 5′ to 3′ exonuclease. J Biol Chem 267: 16806–16811
Rush J, Konigsberg WH (1990) Photoaffinity labeling of the Klenow fragment with 8-azido-dATP. J Biol Chem 265: 4821–4827
Ruyechan WT (1983) The major herpes simplex virus DNA-binding protein hold single-stranded DNA in an extended configuration. J Virol 46: 661–666
Schaffer PA, Aron GM, Biswal N, Benyesh-Melnick M (1973) Temperature-sensitive mutants of herpes simplex virus type 1: isolation, complementation and partial characterization. Virology 52: 57–71
Schnipper LE, Crumpacker CS (1980) Resistance of herpes simplex virus to acycloguanosine: the role of viral thymidine kinase and DNA polymerase loci. Proc Natl Acad Sci USA 77: 2270–2273
Segel IH (1975) Enzyme kinetics. Wiley-Interscience New York
SenGupta DN, Zmudzka BZ, Kumar P, Cobianchi F, Skowronski J, Wilson SH (1986) Sequence of human DNA polymerase β mRNA obtained through cDNA cloning. Biochem Biophys Res Commun 136: 341–347
Sherman G, Gottlieb J, Challberg MD (1992) The UL8 subunit of the herpes simplex virus helicase–primase complex is required for efficient primer utilization. J Virol 66: 4884–4892
Simon M, Giot L, Faye G (1991) The 3′ to 5′ exonuclease activity located in the DNA polymerase δ subunit of Sacharomyces cerevisiae is required for accurate replication. EMBO J 10: 2165–2170
Soengas MS, Esteban JA, Lázaro JM, Bernad A, Blasco MA, Salas M, Blanco L (1992) Site-directed mutagenesis at the Exo III motif of ϕ29 DNA polymerase; overlapping structural domains for the 3′–5′ exonuclease and strand-displacement activities. EMBO J 11: 4227–4237
Spicer EK, Rush J, Fung C, Reha-Krantz LJ, Karam JD, Konigsberg WH (1988) Primary structure of T4 DNA polymerase: evolutionary relatedness of prokaryotic and eukaryotic DNA polymerases. J Biol Chem 263: 7478–7486
Stow ND (1992) Herpes simplex virus type 1 origin-dependent DNA replication in insect cells using recombinant baculovirus. J Gen Virol 73: 313–321
Stow ND (1993) Sequences at the C-terminus of the herpes simplex virus type 1 UL30 protein are dispensable for DNA polymerase activity but not for viral origin-dependent DNA replication. Nucleic Acids Res 21: 87–92
Strick R (1993) Feinanalyse der katalytischen Proteindomänen des Enzyms HSV DNA-Polymerase. Doctoral thesis, University of Heidelberg, Heidelberg
Strick R, Knopf CW (1992) Improved band shift assay for the simultaneous analysis of protein–DNA interactions and enzymatic functions of DNA polymerases. FEBS Lett 300: 141–144
Telford EA, Watson MS, McBride KE, Davison AJ (1992) The DNA sequence of equine herpesvirus 1. EMBL Data library, Heidelberg
Tenney DJ, Micheletti PA, Stevens JT, Hamatake RK, Matthews JT, Sanchez AR, Hurlburt WW, Bifano M, Cordingley MG (1993) Mutations in the C terminus of herpes simplex virus type 1 DNA polymerase can affect binding and stimulation by its accessory protein UL42 without affecting basal polymerase activity. J Virol 67: 543–547
Teo IA, Griffin BE, Jones MD (1991) Characterization of the DNA polymerase gene of human herpesvirus 6. J Virol 65: 4670–4680
Thomas MS, Banks LM, Purifoy DJM, Powell KL (1988) Production of antibodies of predetermined specificity against herpes simplex virus DNA polymerase and their use in characterization of the enzyme. J Virol 62: 1550–1557

Thomas MS, Gao M, Knipe DM, Powell KL (1992) Association between the herpes simplex virus major DNA-binding protein and alkaline nuclease. J Virol 66: 1152–1161

Tsurumi T, Maeno K, Nishiyama Y (1987) Nucleotide sequence of the DNA polymerase gene of herpes simplex virus type 2 and comparison with the type 1 counterpart. Gene 52: 129–137

Vaughan PJ, Purifoy DJM, Powell KL (1985) DNA-binding protein associated with herpes simplex virus DNA polymerase. J Virol 53: 501–508

Wallace HM, Baybutt HN, Pearson CK, Keir HM (1980) The effect of polyamines on herpes simplex virus type 1 DNA polymerase purified from infected baby hamster kidney cells (BHK-21/C13). J Gen Virol 49: 387–400

Wang TS-F (1991) Eukaryotic DNA polymerases. Annu Rev Biochem 60: 513–552

Wang Y, Hall JD (1990) Characterization of a major DNA-binding domain in the herpes simplex virus type 1 DNA-binding protein (ICP8). J Virol 64: 2082–2089

Wang Y, Woodward S, Hall JD (1992) Use of suppressor analysis to identify DNA polymerase mutations in herpes simplex virus which affect deoxynucleoside triphosphate substrate specificity. J Virol 66: 1814–1816

Weissbach A, Hong A-CL, Aucker J, Muller R (1973) Characterization of herpes simplex virus-induced deoxyribonucleic acid polymerase. J Biol Chem 248: 6270–6277

Weisshart K (1989) Kartierung und Expression der HSV-1 ANG DNA Polymerase. Doctoral thesis, University of Heidelberg, Heidelberg

Weisshart K, Knopf CW (1988) The herpes simplex virus type I DNA polymerase. Polypeptide structure and antigenic domains. Eur J Biochem 174: 707–716

Weisshart K, Kuo AA, Painter GP, Wright LL, Furman PA, Coen DM (1993) Conformational changes induced in herpes simplex virus DNA polymerase upon DNA binding. Proc Natl Acad Sci USA 90: 1028–1032

Weller SK, Aschman DP, Sacks WR, Coen DM, Schaffer PA (1983) Genetic analysis of temperature-sensitive mutants of HSV-1: the combined use of complementation and physical mapping for cistron assignment. Virology 130: 290–305

Weller SK, Seghatoleslami MR, Shao L, Rowse D, Carmicheal EP (1990) The herpes simplex virus type 1 alkaline nuclease is not essential for viral DNA synthesis: isolation and characterization of a *lac Z* insertion mutant. J Gen Virol 71: 2941–2952

Wong SW, Wahl AF, Yuan P-M, Arai N, Pearson BE, Arai K, Korn D, Hunkapiller MW, Wang TS-F (1988) Human DNA polymerase α gene expression is cell proliferation dependent and its primary structure is similar to both prokaryotic and eukaryotic replicative DNA polymerases. EMBO J 7: 37–47

Wu CA, Nelson NJ, McGeoch DJ, Challberg MD (1988) Identification of herpes simplex virus type 1 genes required for origin-dependent DNA synthesis. J Virol 62: 435–443

Yadav PNS, Yadav JS, Modak JM (1992) A molecular model of the complete three-dimensional structure of the Klenow fragment of *Escherichia coli* DNA polymerase I: binding of the dNTP substrate and template-primer. Biochemistry 31: 2879–2886

Yager DR, Coen DM (1988) Analysis of the transcipt of the herpes simplex virus DNA polymerase genes provides evidence that polymerase expression is inefficient at the level of translation. J Virol 62: 2007–2015

Yager DR, Marcy AI, Coen DM (1990) Translational regulation of herpes simplex virus DNA polymerase. J Virol 64: 2217–2225

Yang C-L, Chang L-S, Zhang P, Hao H, Zhu L, Toomey NL, Lee MYWT (1992) Molecular cloning of the cDNA for the catalytic subunit of human DNA polymerase δ. Nucleic Acids Res 20: 735–745

Young MC, Reddy MK, von Hippel PH (1992) Structure and function of the bacteriophage T4 DNA polymerase holoenzyme. Biochemistry 31: 8675–8690

Zhang J, Chung DW, Tan CK, Downey KM, So AG, Davie EW (1991) Primary structure of the catalytic subunit of calf thymus DNA polymerase delta: sequence similarities with other DNA polymerases. Biochemistry 30: 11742–11750

Zmudzka BZ, SenGupta D, Matsukage A, Kumar P, Wilson SH (1986) Structure of rat DNA polymerase β revealed by partial amino acid sequencing and cDNA cloning. Proc Natl Acad Sci USA 83: 5106–5110

Chapter 7 Ribonucleotide Reductase and the Ocular Virulence of Herpes Simplex Virus Type 1

Curtis R. Brandt[1,2], Randall Kintner[2], Robert J. Visalli[2], and Anne M. Pumfery[2]

Summary

Herpes simplex virus (HSV) type 1 infections of the eye affect approximately 500 000 people in the United States of America each year and are the leading cause of blindness due to infectious disease in developed countries. Several factors, including the strain of virus, the host immune response, and innate host resistance affect the severity of the infection. The fact that the strain of virus is important indicates that genes in the virus affect virulence. Little is known about the genetics of ocular HSV virulence. The viral thymidine kinase, ICP0, and possibly glycoprotein C genes have been reported to be involved in ocular virulence. We have been studying the role of the HSV ribonucleotide reductase (RR) gene in ocular infections using a null (deletion) mutant (ICP6Δ). The ICP6Δ deletion mutant was unable to cause blinding eye disease in our murine model of HSV keratitis compared to the parent HSV-1 KOS and a genetically engineered revertant (ICP6Δ$^+$3.1). These results indicate that HSV-1 RR is a virulence determinant. ICP6Δ grew poorly in mouse and human corneal fibroblasts and in the mouse eye when compared to HSV-1 KOS and ICP6Δ$^+$3.1, suggesting that the role of the HSV-1 RR is to enhance growth in the animal. RR may therefore be a valid target for new antiviral drugs. Mutants lacking the viral RR gene may also be useful as vaccine strains and vectors for neuronal gene therapy.

Introduction

Herpes simplex virus (HSV) ocular infections are the leading cause of blindness due to infectious disease in developed countries. In the United States, there are an estimated 500 000 cases each year, and approximately 1% of these individuals will suffer a permanent reduction in visual acuity (Binder 1977; Liesegang 1988; Dawson and Togni 1976). Once this has occurred, restoration of vision requires a corneal transplant.

[1] Department of Ophthalmology, University of Wisconsin, 6630 Medical Sciences Center, 1300 University Avenue, Madison, WI 53706, USA
[2] Department of Medical Microbiology/Immunology, University of Wisconsin, 6630 Medical Sciences Center, 1300 University Avenue, Madison, WI 53706, USA

Herpes Simplex Virus-induced Ocular Disease

Clinically, the disease usually presents as an ulcerative blepharitis (eyelid infection) or acute follicular conjunctivitis with an incubation period of 3–12 days and lasts 2–3 weeks (Fedukowicz 1963; Leopold and Sery 1963). The infection can then proceed to involve the cornea. Epithelial keratitis is the most common manifestation of corneal involvement and results from viral replication in corneal epithelial cells (Gunderson 1936), which causes cell destruction and the appearance of lesions that are visible after staining with dyes such as fluorescein or rose bengal (Feenstra and Tseng 1992). In most cases, the epithelial lesions heal without major damage to the cornea, although latency is established in the trigeminal ganglia. HSV can reactivate from latency and the disease can reappear. In some individuals, the virus is able to penetrate into the corneal stroma. This results in stromal keratitis, which is manifested as a clouding of the cornea (Pavan-Langston 1975). In severe stromal keratitis, the entire cornea becomes opaque and blood vessels can grow into the cornea from the sclera. In humans, the development of stromal keratitis is usually the result of multiple rounds of reactivation of the latent infection.

In animal models, stromal keratitis begins approximately 1 week after corneal involvement, coinciding with the development of a specific immune response (Metcalf and Kaufman 1976; Metcalf et al. 1976). The early phase of corneal clouding is primarily the result of corneal edema, perhaps due to alterations in the function of corneal endothelial cells which remove water from the cornea (O'Brien et al. 1990). As the infection continues, the cornea is invaded by cells of the immune system. Neutrophils and monocytes are the first cells detected (Meyers-Elliot and Chitjian 1981), which are then replaced by lymphocytes (Russell et al. 1984; Hendricks and Tumpey 1990). Both $CD4^+$ and $CD8^+$ T cells can be found in the

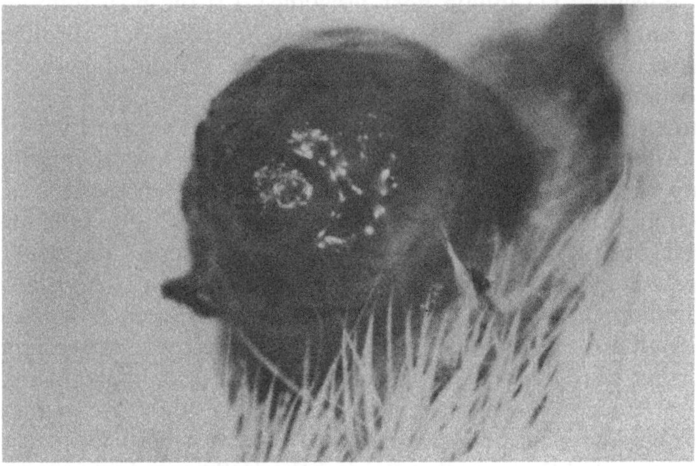

Fig. 1. Typical severe stromal keratitis in a mouse eye following infection with herpes simplex virus type 1 KOS. Note the corneal clouding and blood vessels present in the normally clear cornea

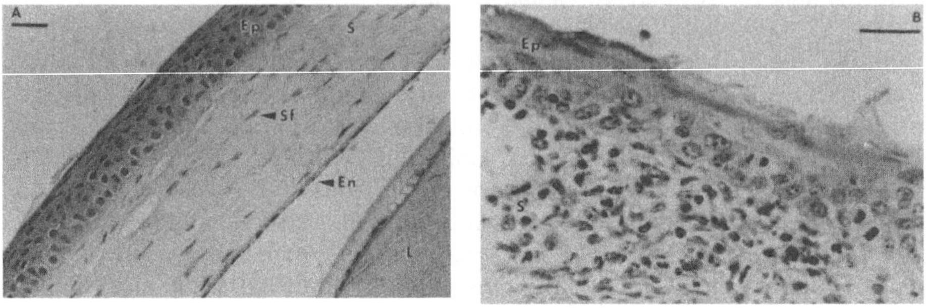

Fig. 2 A,B. Histopathology of herpes simplex virus type 1 (HSV-1)-induced stromal keratitis. **A** Normal mouse cornea (From Brandt et al. 1990, with permission.) **B** HSV-1 infected cornea with typical cellular infiltrate and pathological changes. *Ep*, epithelium; *S*, corneal stroma; *Sf*, stromal fibroblast; *En*, endothelium; *L*, Lens; *bar*, 0.5 mm

cornea. This cellular phase of the pathology leads to further clouding and structural damage to the cornea and may be accompanied by the growth of blood vessels into the normally avascular cornea. Complexes of antibody and antigens are also present in infected corneas (Metcalf and Kaufman 1976). Figure 1 shows severe stromal keratitis in a mouse infected with HSV-1 KOS. Figure 2 shows the typical histopathology in a mouse infected with a strain of HSV-1, DRG1A3, which causes severe stromal keratitis.

The Genetics of Herpes Simplex Virus Ocular Disease

Studies in mice and rabbits have shown that three factors contribute to the severity of an ocular HSV infection: (1) innate host resistance, (2) the host immune response, and (3) the strain of virus. The fact that different strains of virus differ in their ability to cause blinding eye disease indicates that the genetic make-up of the virus is an important factor.

Very little is known about the genes in HSV which influence the severity of an ocular infection. Gordon et al. (1983) reported that mutants lacking thymidine kinase (TK) activity were unable to cause stromal keratitis. Hendricks et al. (1989a,b) reported that viruses unable to express glycoprotein C (gC) failed to induce severe stromal keratitis, suggesting that gC is a virulence gene. Conversely, Centifanto-Fatzgerald et al. (1982) studied a series of recombinant viruses and suggested that gC did not play a role in the type of ocular disease seen in infected rabbits. The reason for these differences are not clear, but these studies suggest the role of gC needs further clarification. The studies of Centifanto-Fitzgerald also implicated a region of the HSV-1 genome extending from 0.7 to 0.83 map units in ocular virulence, but the gene or genes involved have not been identified. Gordon et al. (1990) reported that a null mutant of the *ICP0* gene had reduced ocular virulence. However, in the TK and gC studies, revertant viruses were not tested, and other mutations in the genome may have affected virulence. It is critical to test revertants in virulence studies using mutant viruses.

The identification of virulence genes is important for a number of reasons. There is intellectual interest in advancing our understanding of the virus and its relationship with the host. Also, a better understanding of the pathogenesis of the disease may indicate new ways to treat infections. Since stromal keratitis is an immunopathological disease, interactions between putative virulence factors and the immune system may be identified. Genes that are required for virulence may be excellent targets for antiviral drugs. Therefore, a more rational approach to drug design would be to identify potential target genes, determine their role in virulence, and interrupt the processes involved. Finally, the identification of virulence genes is crucial for the development of vaccines and for the use of HSV as a vector for neuronal gene therapy.

The traditional method of identifying gene functions has been to isolate conditional (e.g., temperature-sensitive) mutants, host range mutants, drug-resistance mutants, and plaque morphology mutants (reviewed in Subak-Sharpe and Timbury 1977). These types of mutants, however, are generally not helpful in virulence studies. The methods that have proven most useful in studying virulence include the isolation of spontaneous avirulent mutants, construction of recombinant viruses, and the direct genetic engineering of mutations in specific genes. Although each of these has problems, virulence genes have been identified with each of these methods.

HSV carries several genes that duplicate the functions of host cell genes. The best studied example is the viral TK gene. Why the virus should duplicate a function provided by the host cell may be explained by the biology of the virus. HSV must successfully replicate in a number of different cell types. Some of these cells types, neurons in particular, are not actively dividing and therefore may have low levels of certain proteins needed by the virus. One mechanism for providing these enzymes is for the virus to encode its own version of the gene. The virus-encoded TK is one of these genes. HSV TK has properties that differ from the host cell enzyme, and this has made the TK protein a primary target for the development of antiviral drugs.

Treatment of Ocular Herpes Simplex Virus

Treatment of ocular HSV infections has presented considerable difficulties and at the same time was the first viral infection for which successful antiviral therapy was licensed. Prior to the successful introduction of antiviral therapy with idoxuridine (IdU) in 1963, treatment was limited. For epithelial keratitis, treatment usually involved mechanical debridement of the infected epithelium (Binder 1977) in the hope of removing the infected epithelial cells. Steroids were sometimes included to suppress the inflammatory response (stromal keratitis), but this sometimes exacerbated the infection.

The first successful use of antiviral therapy came in 1963 when Kaufman and his colleagues reported the successful use of IdU (Kaufman et al. 1962) against HSV keratitis. Since that time, a number of nucleoside analogs, including adenosine arabinoside (ara-A), trifluorothymidine (TFT), and acyclovir (ACV) have been introduced for use in ocular HSV infections. In the United States, both

IdU and TFT are available as topical solutions. ACV is not available in the United States as a topical preparation for HSV keratitis, but oral ACV capsules are being tested in clinical trials. The topical antivirals are effective against epithelial keratitis, but have limited effectiveness against stromal keratitis and do not prevent reactivation from latency. In addition, there are other problems associated with the use of these drugs, particularly TFT and IdU. These include toxicity, selection of resistant viral strains, and the requirement for frequent dosing (Paterson and Jones 1967; Wellings et al. 1972; Coster et al. 1976; Pavan-Langston and Foster 1977; McGill et al. 1988; Young et al. 1982; Hovding 1989). More effective antiviral drugs are therefore needed.

Ribonucleotide Reductase

Another enzyme encoded by HSV-1, HSV-2, and host cells is ribonucleotide reductase (RR). RR reduces ribonucleotides to deoxyribonucleotides and plays a critical role in providing substrates for DNA synthesis in dividing cells. RR is expressed at high levels only in dividing cells, and certain cell types infected by the virus (e.g., neurons) are likely to have low levels of the enzyme. The host shutoff functions of the virus may also reduce the levels of the host cell enzyme. The virus therefore carries its own gene to insure that there are sufficient substrates for replication of DNA. RR is therefore similar to TK in its function in the viral life cycle. Because of the similar roles of RR and TK, attention has been focused on the viral RR gene as a target for antiviral drugs.

The HSV RR is composed of two nonidentical subunits (Ponce de Leon et al. 1977; Huszar and Bacchetti 1981; Dutia 1983; Bacchetti et al. 1984; Preston et al. 1984; Frame et al. 1985). The large subunit (ICP6; M_r 140 000) is encoded by the *UL39* gene, and the small subunit (M_r, 38 000) is encoded by the *UL40* gene (McGeoch et al. 1988). Interestingly, the transcription units for both genes overlap and use the same polyadenylation site, but have separate promoters (Anderson et al. 1981; McLauchlan and Clements 1983). Peptides consisting of the nine amino terminal residues of the small subunit block the association of the large and small subunits and inhibit the activity of the enzyme (Dutia et al. 1986; Cohen et al. 1986), suggesting that RR could be a target for antiviral drugs.

Studies in cell culture have shown that RR is not essential for growth (Goldstein and Weller 1988). If RR is not essential, then it is unlikely that inhibitors of RR would be effective antivirals. Cells in culture are growing rapidly and may have higher levels of host cell-encoded RR than would encountered in the animal host, so it is possible that RR might be essential in the animal. It is therefore necessary to determine whether the virally encoded RR is required in the host before investing a substantial amount of time and money in developing antiviral compounds that act through the viral RR.

Mutants that fail to express RR make it feasible to determine whether viral RR is a virulence factor in vivo and therefore a valid therapeutic target. Recently, Cameron et al. (1988) reported that the mutant *ts1222* (temperature-sensitive mutant) is avirulent in albino Charles River mice following intraperitoneal or intracerebral inoculation. Jacobson et al. (1989) have shown that the RR deletion mutant ICP6Δ (Goldstein and Weller 1988) shows reduced ability to replicate following corneal inoculation in CD-1 mice. Although Jacobson et al. (1989) have shown that the ICP6Δ mutant grows poorly, as measured by eye swabs, the effect of the deletion of the RR gene on keratitis was not examined. Since it is important to know whether RR is a virulence factor in ocular infections, and a valid target for antiviral therapy, we tested the ability of ICP6Δ to cause ocular disease. To rule out the presence of mutations elsewhere in the genome that could affect the virulence, we also tested a revertant (ICP6Δ$^+$3.1) in which the *ICP6* gene had been restored using marker transfer.

Methods

Cell Culture

African green monkey kidney (Vero), BALB/c 3T3 fibroblasts, and human corneal fibroblasts (HCF) were used in these studies. The procedures and media used have been described previously (Brandt et al. 1991).

Viruses

HSV-1 KOS was the parental strain for ICP6Δ and ICP6Δ$^+$3.1. The RR deletion mutant (ICP6Δ) and the revertant (ICP6Δ$^+$3.1) were constructed using marker transfer techniques and have been described previously (Goldstein and Weller 1988; Brandt et al. 1991). The mutation in ICP6Δ is a deletion that removes 90% of the coding sequence, including the initiation codon. High titer stocks of the viruses were prepared as described previously (Grau et al. 1989).

Animal Inoculation

Female BALB/c mice, 4–6 weeks old, were used for all studies and were obtained from Harlan Sprague-Dawley (Indianapolis, IN, USA). The animals were inoculated with 1×10^5 plaque-forming units (PFU) of the appropriate virus in 5 μl of media using the eye scratch method we have described previously (Grau et al. 1989; Brandt et al. 1992). A total of 22, 10, and 11 mice were inoculated with HSV-1 KOS, ICP6Δ$^+$3.1, and ICP6Δ, respectively.

Disease Scoring

The severity of blepharitis (eyelid inflammation), vascularization of the cornea, and stromal keratitis were scored as we have described previously (Grau et al. 1989; Brandt et al. 1992). Briefly, to score for blepharitis a number ranging from 1 to 4 was assigned to each eye based on the swelling and crusting present. To score for vascularization, a number ranging from 1 to 3 was assigned each eye depending on the area of the cornea that contained blood vessels. For stromal keratitis, scores ranging from 1 to 3 were assigned to each eye depending on the area of the cornea involved and the degree of opacity. A score of 4 was assigned if the cornea had perforated.

Viral Growth

To measure viral growth in the eye, mice were killed at various times after infection. The infected eyes were removed, placed in 600 μl of Dulbecco's Modified Eagle Medium (DME) with 2% serum, homogenized, frozen and thawed three times, centrifuged at 1000 rpm for 5 min, and then titered on Vero cell monolayers. To measure growth in cell culture, cells were infected at 2 PFU/cell. At various times, the cells were harvested, frozen and thawed three times, centrifuged at 1000 rpm for 5 min to remove debris, and titered on Vero cells.

Statistics

Statistical methods have been described previously (Grau et al. 1989; Brandt and Grau et al. 1990; Brandt et al. 1991, 1992).

Results

Ocular Disease

The parental wild type HSV-1 KOS caused severe ocular infections (Fig. 3). Blepharitis was visible as early as day 3 postinfection and continued to increase in severity until days 7–9, when it peaked at a mean disease score of 2.75. Blood vessels began infiltrating the cornea as early as day 5 after infection and continued to increase in severity until scoring was halted at day 15. A slight amount of corneal clouding appeared at days 5 and 7 in mice infected with HSV-1 KOS. By day 9, corneal clouding increased substantially to a score of 1.5 and then increased slowly until it reached a score of 1.6 on day 15, when scoring was halted.

The RR deletion mutant, ICP6Δ, caused very little eye disease. A slight but noticeable amount of eyelid inflammation was detected on days 3 and 5, but this resolved completely. ICP6Δ-infected mice showed no evidence of vascularization of the cornea, nor did ICP6Δ-infected mice develop any clouding of the cornea. These results suggest that deletion of the RR gene severely restricts the ability of the virus to cause ocular disease.

Since the techniques used to genetically engineer null mutants (e.g., transfection) can be mutagenic, it is important to rule out the effect of potential mutations elsewhere in the genome of the virus. We therefore tested a revertant of ICP6Δ in

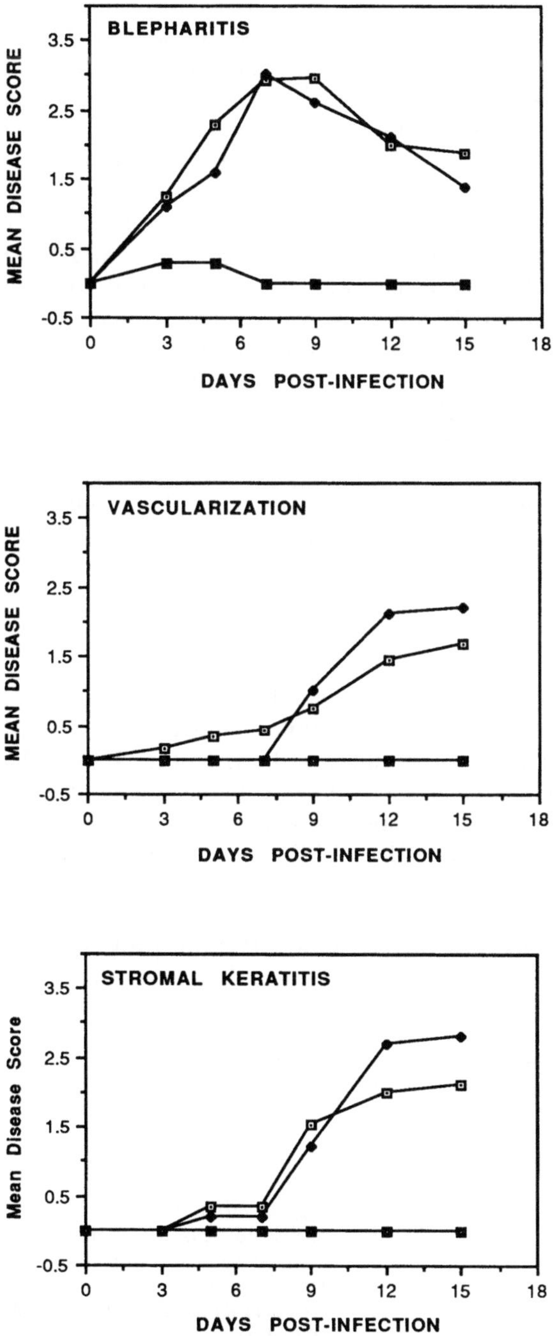

Fig. 3. Time course of ocular disease in infected mice. *Top*, blepharitis; *middle*, corneal vascularization; *bottom*, stromal karatitis. ■ Herpes simplex virus type 1 (HSV-1) KOS; ♦, ICP6Δ$^+$3.1; ■ ICP6Δ. The number of animals used were 22, 10, and 11, for HSV-1 KOS, ICP6Δ$^+$3.1, and ICP6Δ, respectively. From Brandt et al. 1991, with permission)

which the deletion had been rescued using marker transfer. The revertant virus, ICP6Δ^{+}3.1, caused severe ocular disease that was very similar to the HSV-1 KOS parent (Fig. 3). Blepharitis was visible by day 3 and continued to increase in severity until days 7–9, where it peaked at a score of 2.75. The blepharitis then began to heal. Vascularization in ICP6Δ^{+}3.1-infected mice was not seen until day 9. It then increased quickly, peaking at a score of 2.1 on day 15. The vascularization caused by ICP6Δ^{+}3.1 was slightly more severe than that caused by HSV-1 KOS. A slight amount of corneal clouding appeared on days 5 and 7 in ICP6Δ^{+}3.1-infected mice. Corneal clouding increased dramatically between days 7 and 9 and then began to level off at scores of 2.3 – 2.4 on days 12 and 15. The corneal clouding in ICP6Δ^{+}3.1-infected mice appeared to be more severe than in mice infected with HSV-1 KOS. However, the fact that ICP6Δ^{+}3.1 was able to cause ocular disease nearly identical to that caused by the HSV-1 KOS parent indicates that mutations elsewhere in the genome that adversely affect virulence are unlikely and supports the conclusion that the RR gene is required for ocular virulence.

Statistical analysis of the disease scores at times of peak severity indicate that HSV-1 KOS and ICP6Δ^{+}3.1 are not significantly different ($p > 0.05$), but both HSV-1 KOS and ICP6Δ^{+}3.1 are significantly more severe than ICP6Δ ($p < 0.05$).

Viral Growth in the Eye

To determine whether the avirulence of ICP6Δ was related to the ability of the virus to grow in the eye, we infected mice and at various times removed the eyes and measured the amount of virus present. The results are shown in Fig. 4. In HSV-1 KOS-infected mice, we found approximately 1×10^3 PFU in the eyes 1 day after infection. The amount of virus peaked on day 1 and then declined. By day 8 postinfection, there was no virus detected in HSV-1 KOS-infected animals. Very similar results were obtained in mice infected with ICP6Δ^{+}3.1. Eye titers peaked at around 1×10^3 PFU on day 1 and then declined until day 8, when no virus was detected in the eyes of these mice. In mice infected with ICP6Δ, the peak titer was delayed by 24 h (day 2) and was more than ten fold lower than in HSV-1 KOS-

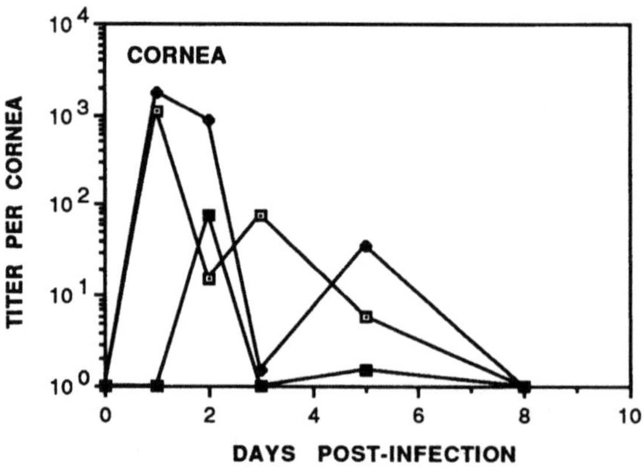

Fig. 4. Virus growth in BALB/c mouse corneas. Three mice were assayed for each time point. ■ Herpes simplex virus type 1, (HSV-1) KOS; ♦, ICP6Δ^{+}3.1; ■ ICP6Δ. (From Brandt et al. 1991, with permission)

or ICP6Δ⁺3.1-infected mice. By day 3, there was no detectable virus in the eyes of ICP6Δ-infected mice. These results show that deletion of the RR gene delays replication of the virus in the eye and inhibits the ability of the virus to grow to high titers; they suggest that the ability to cause severe ocular disease is related to the ability to grow in the host.

Growth in Cell Culture

Although we have shown that the RR gene is required for virulence in our mouse model, this does not necessarily mean it is required for virulence in humans. We cannot repeat these experiments in humans, however, and must therefore rely on indirect means to answer such questions. HCF can be cultured from corneas donated to eye banks, and virus replication can be measured in these cultures. We reasoned that if we could establish a correlation between the ability of viruses to grow in cultured mouse cells and the ability to cause disease and grow in the mouse, then we could obtain some indication of the role of the HSV-1 RR in human infections by measuring viral growth in HCF cells.

One-step growth curves were carried out in Balb/C 3T3 fibroblasts. The results are shown in Fig. 5. HSV-1 KOS and ICP6Δ⁺3.1 grew equally well in

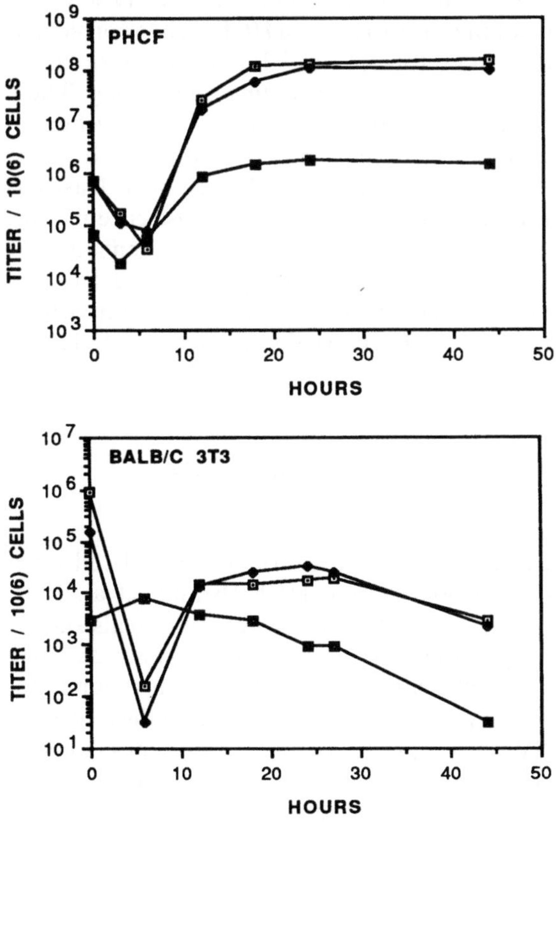

Fig. 5. In vitro growth curves. Top, primary human corneal fibroblasts (*PHCF*); bottom, *BALB/c 3T3* fibroblasts. ■, Herpes simplex virus type 1 (HSV1) KOS; ♦, ICP6Δ⁺3.1; ■, ICP6Δ. (From Brandt et al. 1991, with permission)

BALB/c 3T3 fibroblasts, reaching peak titers of approximately 1×10^4 PFU/10^6 cells. In contrast, ICP6Δ did not grow well. ICP6Δ titers were approximately ten fold lower 24 h after infection and 100-fold lower 48 h after infection. These results are consistent with the results from the in vivo growth experiments and establish a connection between disease severity, growth in the eye, and growth in culture.

When we carried out single-step growth curves in HCF cells (Fig. 5), we found that HSV-1 KOS and ICP6Δ^+3.1 grew equally well, reaching peak titers of approximately $1–2 \times 10^8$ PFU/10^6 cells 18 h after infection. ICP6Δ grew less well, reaching peak titers of $1–2 \times 10^6$ PFU/10^6 cells 18 h after infection, a 100-fold reduction compared to the other viruses. This result suggests that HSV-1 RR is also required for efficient growth in human cells and, when considered with the results from the mouse experiments, is probably required for virulence in human ocular infections.

Discussion

Our studies clearly show that HSV-1 RR is a virulence factor and plays an important role in the ability of the virus to cause blepharitis, vascularization of the cornea, and stromal keratitis in our murine model of ocular disease. Deletion of the RR gene also reduces the ability of the virus to grow in HCF. When considered in total, these studies suggest that inhibitors of viral RR have considerable potential as antiviral drugs and would be worth developing and testing. These results also have important implications for the use of HSV-1 as a vector in neuronal gene therapy and as a vaccine. Any use of HSV-1 as a live viral vector or vaccine requires that the virus either does not cause disease or at least has reduced virulence. It is important that the virus be unable to cause encephalitis, but it must still be capable of entering neurons if it is to be used as a gene therapy vector. The RR gene is an excellent gene to target for these purposes. Deletion of the gene will reduce virulence in vivo yet allow propagation of the vector virus in certain cell lines. In addition, new genes can be inserted into the virus in place of the RR gene.

The Role of Herpes Simplex Virus Type 1 Ribonucleotide Reductase in Ocular Virulence

Deletion of the RR gene reduces the growth of the virus in several cell types including Vero cells, BALB/c 3T3 fibroblasts, and HCF (Brandt et al. 1991; Goldstein and Weller 1988; Jacobson et al. 1989). The deletion mutant also grows less well in the eye. These results suggest that the role of RR is to promote growth of the virus in cells in the eye and that efficient growth of the virus is required for virulence. This conclusion is consistent with previous hypotheses on the role of the enzyme in the viral life cycle. HSV must replicate in several cell types, some of which are not rapidly dividing and likely have low levels of endogenous RR. Corneal epithelial cells, stromal fibroblasts, and endothelial cells in the normal developed eye divide only slowly and are likely to have low RR levels. In this situation, the virus-encoded RR would be required for efficient growth. One function of RR, then, is to provide sufficient substrates for viral DNA synthesis in slowly growing corneal cells infected with the virus. It is interesting to note, however, that the RR deletion mutant was able to replicate in the eye. Peak titers are slightly lower, and the peak titer was delayed by 24 h. The RR mutant was also cleared from the eye faster. These results suggest that rapid clearance of virus from the eye may be important in preventing the immunopathological response that leads to blinding eye disease.

The Role of Ribonucleotide Reductase in Latency

Several studies have shown that the ICP6Δ mutant is defective for either the establishment of or reactivation from latency (Jacobson et al. 1989; Brandt et al. 1990; Katz et al. 1990). More recently, ICP6Δ DNA has been found in trigeminal ganglia, but at considerably lower levels than the HSV-1 KOS parent (Katz et al. 1990). These results suggest that RR is required for reactivation. This conclusion is consistent with RR being needed to provide substrates for DNA replication in a cell type that is not normally dividing (e.g., neurons) and likely has low endogenous RR levels. The fact that RR is required for reactivation of latency has important implications for antiviral drug development, the use of HSV-1 as a vector, and development of vaccines.

Conclusions

In summary, the HSV-1 RR gene is required for ocular virulence in mice and probably in humans. Our studies and the studies of others (Cameron et al. 1988; Idowu et al. 1992; Yamada et al. 1991) show that viral RR is required for neurovirulence and ocular virulence and point to the important role for this enzyme in the life cycle of the virus. Our studies also show that null mutants can be used effectively in animal models to identify and study viral genes that play a role in enhancing the virulence of the virus. Many of these genes may become important targets for antiviral therapy.

Acknowledgements. We would like to thank Dr. Sandra K. Weller, University of Connecticut, for providing the viruses used in these studies and the Wisconsin Eye Bank for providing normal donor corneas. These studies were supported by a grant (EY07336) from the NIH and an unrestricted grant from Research to Prevent Blindness, Inc., to the Department of Ophthalmology, University of Wisconsin-Madison. We would also like to thank Bernadette Bull for typing the manuscript and Dorene Grau for technical assistance.

References

Anderson KP, Frink RJ, Devi GB, Gaylord BH, Costa RH, Wagner EK (1981) Detailed characterization of the mRNA mapping in the HindIII K region of the herpes simplex virus type 1 genome. J Virol 37: 1011–1027

Bacchetti S, Evelegh MJ, Muirhead B, Sartori CS, Huszar D (1984) Immunological characterization of herpes simplex virus type 1 and 2 polypeptide(s) involved in viral ribonucleotide reductase activity. J Virol 49: 591–593
Binder PS (1977) Herpes simplex keratitis. Surv Opathalmol 21: 313–330
Brandt CR, Grau DR (1990) Mixed infection with herpes simplex virus type 1 generates recombinants with increased ocular and neurovirulence. Invest Ophthalmol Vis Sci 31: 2214–2223
Brandt CR, Knupfer PB, Boush GA, Gausas RE, Chandler, JW (1990) In vivo induction of Ia expression in murine cornea after intravitreal injection of interferon gamma. Invest Ophthalmol Vis Sci 31: 2248–2253
Brandt CR, Kintner RL, Pumfery AM, Visalli RJ, Grau DR (1991) The herpes simplex virus ribonucleotide reductase is required for ocular virulence. J Gen Virol 72: 2043–2049
Brandt CR, Coakley LM, Grau DR (1992) A murine model of herpes simplex virus-induced ocular disease for antiviral drug testing. J Virol Methods 36: 209–222
Cameron JM, McDougall I, Marsden HS, Preston VG, Ryan DM, Subak-Sharpe JH (1988) Ribonucleotide reductase encoded by herpes simplex virus is a determinant of pathogenicity of the virus in mice and a valid antiviral target. J Gen Virol 69: 2607–2612
Centifanto-Fitzgerald YM, Yamaguchi T, Kaufman HE, Tognon M, Roizman B (1982) Ocular disease pattern induced by herpes simplex virus is genetically determined by a specific region of viral DNA. J Exp Med 155: 475–489
Cohen EA, Gadreau P, Brazeau P, Langllier Y (1986) Specific inhibition of herpes simplex virus ribonucleotide reductase by a nonapeptide derived from the carboxy terminus of subunit 2. Nature 321: 441–443
Coster DJ, McKinnon JR, McGill JL, Jones BR, Frauenfelder FT (1976) Clinical evaluation of adenine arabinoside and trifluorothymidine in the treatment of corneal ulcers caused by herpes simplex virus. J Infected Dis 133: A173–A177
Dawson CR, Togni B (1976) Herpes simplex eye infections: clinical manifestations, pathogenesis and management. Surv Ophthalmol 21: 121–135
Dutia BM (1983) Ribonucleotide reductase induced by herpes simplex virus has a virus-specified constituent. J Gen Virol 64: 513–521
Dutia BM, Frame MC, Subak-Sharpe JH, Clark WN, Marsden HS (1986) Specific inhibition of herpes virus ribonucleotide reductase by synthetic peptides. Nature 321: 439–441
Fedukowicz HB (1963) External infections of the eye. Appleton Century Crofts, New York, pp 122–126
Feenstra RPG, Tseng SLG (1992) Comparison of fluorescein and rose bengal staining. Ophthalmology 99: 605–617
Frame MC, Marsden HS, Dutia DM (1985) The ribonucleotide reductase induced by herpes simplex virus type I involves minimally a complex of two polypeptides (136 K and 38 K). J Gen Virol 64: 513–521
Goldstein DJ, Weller SK (1988) Factor(s) present in herpes simplex virus type 1 infected cells can compensate for the loss of the large subunit of the viral ribonucleotide reductase: characterization of an ICP6 deletion mutant. Virology 166: 41–51
Gordon YJ, Gilden DM, Backer Y (1983) HSV-1 thymidine kinase promotes virulence and latency in the mouse. Invest Ophthalmol Vis Sci 24: 599–602
Gordon YJ, Romanowski E, Balouris C, Araullo-Cruz T (1990) A herpes simplex type 1 ICPO deletion mutant demonstrates diminished pathogenicity during acute ocular infection in different host animals. Invest Ophthalmol Vis Sci 31: 681–688
Grau DR, Visalli RJ, Brandt CR (1989) Herpes simplex virus stromal keratitis is not titer-dependent and does not correlate with neurovirulence. Invest Ophthalmol Vis Sci 30: 2474–2480
Gunderson T (1936) Herpes cornea: with special references to its treatment with strong solution of iodine. Arch Ophthalmol 15: 225–249
Hendricks RL, Tumpey TM (1990) Contribution of virus and immune factors to herpes simplex virus type I induced corneal pathology. Invest Ophthalmol Vis Sci 31: 1929–1939
Hendricks RL, Tao MSP, Glorioso JC (1989a) Alterations in the antigenic structure of two major HSV-1 glycoproteins, gC and gB influence immune regulation and susceptibility to murine herpes keratitis. J Immunol 142: 263–269

Hendricks RL, Epstein RJ, Tumpey T (1989b) The effect of cellular immune tolerance to HSV-1antigens on the immunopathology of HSV-1 keratitis. Invest Ophthalmol Vis Sci 30: 105–115

Hovding G (1989) A comparison between acyclovir and trifluorothymidine ophthalmic ointment in the treatment of epithelial dendritic keratitis, a double blind randomized parallel group trial. Acta Ophthalmol 67: 51–54

Huszar D, Bacchetti S (1981) Partial purification and characterization of ribonucleotide reductase induced by HSV infection of mammalian cells. J Virol 37: 580–588

Idowu AD, Fraser-Smith EB, Poffenberger KL, Herman RC (1992) Deletion of the herpes simplex virus type I ribonucleotide reductase gene alters virulence and latency in vivo. Antiviral Res 17: 145–156

Jacobson JG, Leib D, Goldstein DJ, Bogard CL, Schaffer PA, Weller SK, Cohen DM (1989) A herpes simplex virus ribonucleotide reductase deletion mutant is defective for productive acute and reactivable latent infections of mice and for replication in mouse cells. Virology 173: 276–283

Katz JP, Bodin ET, Coen DM (1990) Quantitative polymerase chain reaction analysis of herpes simplex virus DNA in ganglia of mice infected with replication-incompetent mutants. J Virol 64: 4288–4295

Kaufman HE, Nesburn AB, Maloney ED (1962) IDU therapy of herpes simplex. Arch Ophthalmol 67: 583–591

Leopold IH, Sery TW (1963) Epidemiology of herpes simplex keratitis. Invest Ophthalmol 2:498–503

Liesegang TJ (1988) Ocular herpes simplex infection: pathogenesis and current therapy. Mayo Clin Proc 63: 1092–1105

McGeoch DJ, Dalrymple MA, Davison AJ, Dolan A, Frame MC, McNab D, Perry LJ, Scott JE, Taylor P (1988) The complete DNA sequence of the long unique region in the genome of herpes simplex virus type 1. J Gen Virol 69: 1531–1574

McGill J, Tormey P, Walker CB (1981) Comparative trial of acyclovir and adenine arabinoside in the treatment of herpes simplex corneal ulcers. Br J Ophthalmol 65: 610–613

McLauchlan J, Clement JB (1983) Organization of the herpes simplex virus type 1 transcription unit encoding two early proteins with molecular weights of 140,000 and 40,000. J Gen Virol 64: 997–1006

Metcalf JF, Kaufman HE (1976) Herpetic stromal keratitis: evidence for cell-mediated immunopathogenesis. Am J Ophthalmol 82: 827–834

Metcalf JF, McNeill JI, Kaufman HE (1976) Experimental disciform edema and necrotizing keratitis in the rabbit. Invest Ophthalmol 15: 979–985

Meyers-Elliott RH, Chitjian PA (1981) Immunopathogenesis of corneal inflammation in herpes simplex virus stromal keratitis: role of the polymorphonuclear leukocyte. Invest Ophthalmol Vis Sci 20: 784–798

Newell FW (1986) Ophthalmology: principles and concepts. Mosby, St Louis, pp 8–13

O'Brien WJ, Guy J, Taylor JL (1990) Pathogenesis of corneal oedema associated with herpetic eye disease. Br J Ophthalmol 74: 723–730

Paterson A, Jones BR (1967) The management of ocular herpes. Trans Ophthalmol Soc UK 87: 59–84

Pavan-Langston D (1975) Diagnosis and management of herpes simplex ocular infection. Int Ophthalmol Clin 15: 19–35

Pavan-Langston D, Foster CS (1977) Trifluorothymidine and idoxuridine therapy of ocular herpes. Am J Ophthalmol 84: 818–825

Ponce de Leon M, Eisenberg RJ, When GH (1977) Ribonucleotide reductase from herpes simplex virus (types 1 and 2) infected and uninfected KB cells: properties of the partially purified enzymes. J Gen Virol 36: 163–173

Preston VG, Palferyman JW, Dutia BM (1984) Identification of a herpes simplex virus type 1 polypeptide which is a component of the virus-induced ribonucleotide reductase. Virology 65: 1457–1466

Russell RG, Nasisse MP, Larson HE, Rouse BT (1984) Role of T-lymphocytes in the pathogenesis of herpetic stromal keratitis. Invest Ophthalmol Vis Sci 25: 938–944

Subak-Sharpe JH, Timbury MC (1977) Genetics of herpes viruses. In: Fraenkel-Contat H, Wagner RR (eds) Comprehensive virology. Academic, New York, pp 89–131

Wellings PC, Awdry PN, Bars FH, Jones BR, Brown DC, Kaufman HE (1972) Clinical evaluation of trifluorothymidine in the treatment of herpes simplex corneal unclers. Am J Ophthalmol 73: 932–942

Yamada Y, Kimura H, Morishima T, Daikoku T, Maeno K, Michiyama Y (1991) The pathogenicity of ribonucleotide reductase-null mutants of herpes simplex virus type 1 in mice. J Infect Dis 164: 1091–1097

Young BJ, Patterson A, Ravenscroft T (1982) A randomized double-blind clinical trial of acyclovir (Zovirax) and adenosine arabinoside in herpes simplex corneal ulceration. Br J Ophthalmol 66: 361–363

Chapter 8 Ribonucleotide Reductase Gene in Herpes Simplex Virus Type 2 and Virus Pathogenicity

Laure Aurelian[1,2]

Summary

The large subunit of herpes simplex virus type 2 (HSV-2) ribonucleotide reductase (RR; ICP10) is a chimera that consists of a Ser/Thr-specific protein kinase (PK) localized at the N terminus and the large subunit of RR (RR1) localized at the C terminus. The product of the PK minigene is myristylated and has features of a signal peptide and a transmembrane (TM) helical segment followed by a basic residue (Arg^{107}) such as described for growth factor receptor kinases (GFR). Membrane immunofluorescence of ICP10-transformed cells with antibodies to synthetic peptides upstream and downstream of the TM indicates that ICP10 is a membrane-spanning protein. PK-catalytic motifs I (nucleotide binding) and II (invariant Lys) are not required for ICP10 kinase activity, but the TM segment is essential. A leucine zipper located at the PK and RR junction and the 50 C-terminal residues are involved in stable complex formation with the small subunit of RR (RR2). The ICP10 promoter contains an Oct-1/TAATGARAT, two AP-1, and two HPV E2 binding motifs. ICP10 expression is regulated as an immediate-early (IE) gene and is enhanced by HPV16 E2. The PK minigene causes neoplastic transformation of immortalized cells, and its expression is required for virus growth and neoplastic transformation. Sequences homologous to a segment of the PK minigene were amplified and cloned from human DNA, and a 60 kDa protein antigenically and structurally similar to ICP10-PK, was identified in human cervical cancer tissues. A model for PK-minigene linked carcinogenesis is proposed.

Introduction

Ribonucleotide reductase (RR) provides a major pathway in the formation of DNA precursors, thereby playing a key role in DNA synthesis. The HSV holoenzyme consists of two nonidentical subunits, termed RR1 (140 kDa; also known as ICP6 for HSV-1 and ICP10 for HSV-2) and RR2 (38 kDa), that are

[1] Virology/Immunology Laboratories, Departments of Pharmacology and Microbiology, The University of Maryland School of Medicine, Baltimore, MD 21201, USA
[2] Departments of Comparative Medicine and Biochemistry, The Johns Hopkins Medical Institutions, Baltimore, MD 21205, USA

encoded by adjacent genes, respectively designated UL39 and UL40 (McGeoch et al. 1988). Sequence comparison of RR1 proteins has identified an additional N-terminal domain unique to ICP6 and ICP10 that is responsible for their 50% greater molecular weight (Nikas et al. 1986; Swain and Galloway 1986) and has only 38% intertypic (HSV-1 vs. HSV-2) homology as compared to 93% for the remaining carboxyl two thirds of the molecule (Nikas et al. 1986). This N-terminal domain is unrelated to RR activity (Conner et al. 1992a), and HSV-2 DNA sequences that encode it cause neoplastic transformation of immortalized cells (Hayashi et al. 1985). RR1 is essential for HSV-1 growth in nondividing cells in vitro at 37°C and in dividing cells growing at higher temperatures (Goldstein and Weller 1988), for HSV-1 pathogenicity in mice and guinea pigs (Jacobson et al. 1989; Cameron et al. 1988; Idowu et al. 1992) and for the reactivation of latent virus (Jacobson et al. 1989; Idowu et al. 1992). Our studies sought to further characterize ICP10 and elucidate its role in HSV-2 pathogenesis.

Materials and Methods

Plasmids

ICP10 plasmids and constructs have already been described (Chung et al. 1989, 1991; Wymer et al. 1989; Luo et al. 1991; Luo and Aurelian 1992). pJW17 contains the entire ICP10 gene. pJW32 and pJW31, respectively, express the ICP10 protein kinase (PK) and RR domains, and pJW30 expresses RR2. pJW33 expresses a 123-kDa protein that lacks the last 50 residues in ICP10. The frameshift mutant pJW21 is expression negative. Site-directed mutants are pJHL2 (Gly^{106} to Pro) and pJHL4 (Lys^{176} to Leu). pJHL15 is deleted in the transmembrane (TM) segment and expresses a 139-kDa protein ($p139^{TM}$). pJHL9 lacks all PK-catalytic motifs (residues 106–445) and expresses a 95-kDa protein (p95). pJHL17 is deleted in PK-catalytic motifs I and II and expresses a 136-kDa protein ($p136^{I/II}$). pICP10-cat contains the ICP10 promoter driving the *CAT* structural gene. pJL11 is an ICP10-PK bacterial expression vector; pJL39 is its invariant Lys mutant. Protein synthesis is regulated by the λ-temperature-sensitive (ts) repressor. Plasmid pGH62 contains the HSV-1 gene for Vmw65 (O'Hare and Hayward 1985), pC59 encodes the BPV-1 E2 protein (Yang et al. 1985), pSV2-350-E2 encodes the HPV16 E2 protein (Wymer and Aurelian 1990), and p1225 encodes the HPV16 E7 protein (Phelps et al. 1988).

Antibodies

Peptide synthesis and antibody production have already been described (Aurelian et al. 1989; Chung et al. 1990,1991).

DNA Transfection

Calcium phosphate precipitation with a 15% glycerol shock was performed as described (Wymer et al. 1989).

Immunoprecipitation and Subcellular Fractionation

Total cell extracts and soluble, cytoskeleton, and nuclear fractions were prepared from cells labeled with [^{35}S]methionine or [^{32}P]orthophosphate at 30–40 and 2–40 h postransfection, respectively. Plasma membranes were purified by centrifugation on a 35% sucrose cushion, and purity was confirmed by glucose-6-phosphatase assays. Immunoprecipitation used antibody and protein A-Sepharose CL4B. Proteins were resolved by Sodium dodecyl sulfate polyacrylamide gel electrophoresis (SDS-PAGE) on 7% or 10% polyacrylamide gels (Chung et al. 1989).

Ribonucleotide Reductase and Protein Kinase Assays

RR and immunocomplex kinase assays were as described (Chung et al. 1990; Smith et al. 1992). For K_m measurements serial dilutions of [γ-^{32}P] adenosine triphosphate ([γ-^{32}P]ATP;0.1–20 μM) were added to each immunoprecipitate. In some experiments, immunoprecipitates were separated by electrophoresis on 8.5% SDS-PAGE and transferred onto Immobilon membranes. Proteins on the blots were renatured and assayed for kinase activity with [γ-^{32}P]ATP (Luo et al. 1992).

Peptide Mapping

Bands obtained on 7% SDS-PAGE were excised and loaded onto 12.5% polyacrylamide gels with 0.1% SDS together with 25 μg of V8 protease (Aurelian et al. 1989; Luo and Aurelian 1992).

DNA Sequencing and Site-Directed Mutagenesis

DNA sequence of fragments cloned into M13 bacteriophage was identified by the Sequenase dideoxynucleotide chain termination methods. Site-directed mutagenesis was performed with the Muta-Gene system (Wymer et al. 1989,1992; Smith et al. 1992; Luo and Aurelian 1992).

Cloning of ts *5–152 ICP10* Gene into Wild-Type Herpes Simplex Virus Type 2

Rabbit kidney cells were cotransfected with the cloned ts *5–152 ICP10* gene (pMK1) and DNA extracted from cytoplasmic virions of wild-type herpes simplex virus type 2 (HSV-2; strain 859). Virus progeny was plaque purified and assayed for growth and ICP10 synthesis at 34° and 39°C (Smith et al. 1992).

Transformation and Anchorage-Independent Growth

Transformation of diploid (SHE) and immortalized (AE) cells was by continuous passage (Hayashi et al. 1985; Smith et al. 1992). A cell line (JHLa1) that constitutively expresses ICP10 was established from 293 cells transfected with SV_2neo containing pJW17 and selected with G418 (Luo and Aurelian 1992). Anchorage-independent growth was assayed as described (Hayashi et al. 1985; Smith et al. 1992). Cloning efficiency is expressed as the number of colonies (×100)/ number of seeded cells.

Polymerase Chain Reaction Analysis

The Perkin Elmer Cetus GeneAmp™ DNA amplification kit was used with AmpliTaq™ recombinant DNA polymerase and primers that bracket a 200-bp segment of HSV-2 DNA that includes the LA-1 determinant in ICP10 (residues 13–26). Reaction products were analyzed by electrophoresis on 8% PAGE and Southern blot hybridization with [^{32}P]-labeled HSV-2 *Sal*I C DNA from pJW17 (Smith et al. 1991).

Results and Discussion

Computer Analysis

ICP10 is a fusion protein consisting of an N-terminal domain (residues 1–411), which has eight PK-catalytic motifs, and a C-terminal domain (residues 412–1144), which contains five RR motifs. PK-catalytic motifs I and II (nucleotide binding and invariant Lys) are not conserved in ICP6, and the first RR motif (around ICP10 sequence G-XX-SALKYEEFYL) is present only in the herpes-simplex viruses RR1 proteins. TM helical segment is located at ICP10 residues 88–105. It is followed by a basic residue (Arg^{107}) that in GFR is thought to anchor the membrane-spanning helix (Yarden and Ullrich 1988). This operationally defines all ICP10 upstream residues as an extracellular domain and all downstream residues as an intracellular domain (Figs. 1, 2). Additional features shared with GFR include: (a) a signal peptide at residues 1–12, (b) a potential N-glycosylation site at residue 69, (c) a Cys-rich region close to the TM (residues 37, 66, and 70), (d) a consensus pattern for myristylation at residue 67, and (e) two highly related acidic peptide regions (at residues 190 and 223) that may be involved in basic protein regulation of PK activity (Morrison et al. 1989). At the junction of the PK and RR domains (residues 409–437), there is a leucine zipper repeat and DNA sequences consistent with a core enhancer element (Chung et al. 1989,1991; Smith

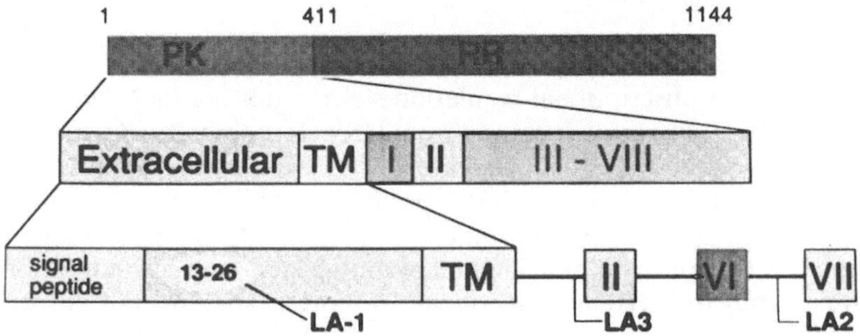

Fig. 1. ICP10 is a chimera consisting of the protein kinase (*PK*; residues 1–411) and ribonucleotide reductase (*RR*; residues 412–1144) domains. Residues upstream of the transmembrane (*TM*; extracellular) include a signal peptide and the kinase modulatory determinant *LA-1* (residues 13–26). PK-catalytic motifs (I-VIII) and the location of the LA-3 and LA-2 determinants are identified

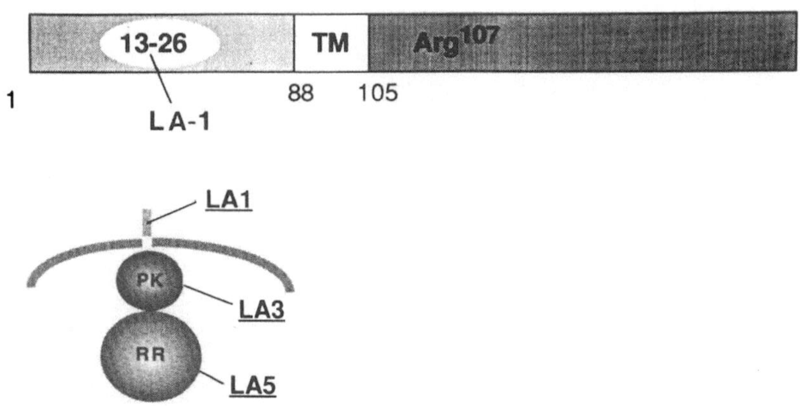

Fig. 2. Representation of ICP10 showing the transmembrane segment (*TM*; residues 88–105) and the anchoring Arg^{107}(top). The polarity of its membrane association defines *LA-1* as an extracellular determinant; *LA-3* and *LA-5* are intracellular (bottom). *PK*, protein kinase; *RR*, ribonucleotide reductase

et al. 1991). Elements in the ICP10 promoter that are required for IE regulation include a binding site for the ubiquitous octamer binding protein (Oct-1) that overlaps a TAATGARAT sequence and two AP-1 elements characteristic of cellular IE genes, but absent in other HSV promoters (Wymer et al. 1989,1992). We propose that the gene coding for ICP10-PK (PK minigene) originated from a cellular GFR gene (LA-1 protooncogene) which fused with an ancestral HSV RR1 gene and was modified in the process. The results of our recent studies support this interpretation.

ICP10 is Regulated as an Immediate-Early Protein

In HSV-2-infected cells, ICP10 expression is temporally regulated as a delayed early gene. However, transcriptional regulation studies indicate that the virion protein Vmw65, which activates expression from HSV IE promoters (Kristie and Sharp 1990), causes a dose-dependent increase in CAT expression from pICP10-cat. Both the POU-homeo (PHB) and POU-specific (PSB) binding motifs within the Oct-1-binding site in the ICP10 promoter are required for Vmw65 complex formation and ICP10 activation, as evidenced by mutagenesis and gel retardation assays. However, only the PHB motif is critical, and correct alignment of Vmw65 about the GARAT motif is essential for activation. CAT expression from pIC10-cat is also enhanced by co-transfection with c-jun and c-fos (Wymer et al. 1989, 1992), these findings indicate that under certain conditions, such as may occur in some cells. ICP10 can be regulated as an IE gene. This is consistent with its potential involvement in reactivation of latent virus and neoplastic transformation.

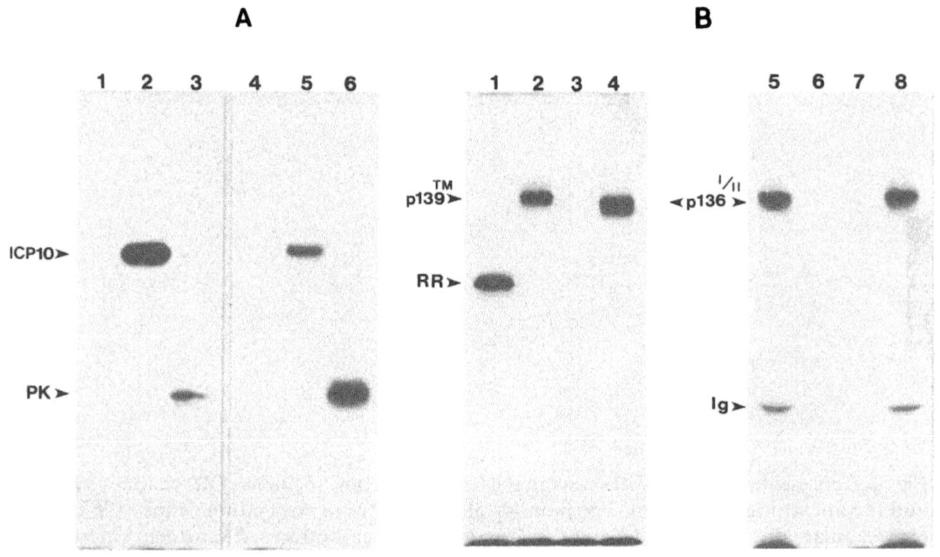

Fig. 3. **A** Immunoprecipitation of $[^{35}S]$ methionine-labeled extracts of cells transfected with pJW21 (*lane 1*), pJW17 (*lane 2*), or pJW32 (*lane 3*) with monoclonal antibody (MAb) that recognizes amino acid residues within the protein kinase (*PK*) domain. Immunocomplex kinase assays of cells transfected with pJW21 (*lane 4*), pJW17 (*lane 5*), or pJW32 (*lane 6*) indicate that kinase activity is associated with ICP10 and the 57-kDa PK. **B** Immunoprecipitates of $[^{35}S]$methionine-labeled extracts of cells transfected with pJHL9 (ribonucleotide reductase, p95, *RR*; *lane 1*), pJHL15 ($p139^{TM}$; *lane 2*), or p JHL17 ($p136^{I/II}$; *lane 4*) with a MAb that recognizes amino acid residues within the RR domain. *Lane 3* is a normal ascites precipitate of pJHL17-transfected cells. PK activity of immunocomplexes from unlabeled duplicates of 293 cells transfected with pJHL17 (*lanes 5, 8*), pJHL9 (*lane 6*), and pJHL15 (*lane 7*). *Ig*, trans-phosphorylated immunoglobulin

ICP10 Has Intrinsic Protein Kinase Activity

Immune complex kinase assays of extracts from HSV-2-infected cells with ICP10-specific MAb indicate that ICP10 has auto- and transphosphorylating activity. The activity is localized in the soluble and cytoskeleton fractions and in purified plasma membranes, and it is Ser/Thr specific. The PK and RR domains are functionally dissociated by endogenous proteolysis (Chung et al. 1989).

The following evidence supports the interpretation that the kinase activity is intrinsic to the ICP10 N-terminal domain. First, PK activity was observed in ICP10-specific immunoprecipitates from cells transfected with pJW17 (entire ICP10) or pJW32 (57-kDa N-terminal domain). PK activity was not seen in precipitates of cells transfected with the expression-negative frameshift mutant or with pJW31 or pJHL9, both of which express the 95-kDa C-terminal domain of ICP10 (Fig. 3). Second, PK activity was evidenced by a 29-kDa protein ($pp29^{la1}$), structurally and antigenically identical to ICP10 residues 13–283, that is expressed by the bacterial expression vector pJL11 (Luo et al. 1991). Since most bacterial PKs have a structure totally unrelated to eukaryotic PKs, it is highly unlikely that a putative contaminant is structurally identical in both eukaryotic and prokaryotic cells. Third, kinase activity was retained by ICP10 after elution from SDS-PAGE and protein renaturation on a membrane filter, and finally, ICP10 specifically bound the [^{14}C]-labeled ATP affinity analog FSBA, and binding was abolished by Adenosyl imidodiphosphate (AMP-PNP) (Luo and Aurelian 1992).

ICP10 is a New Protein Kinase Type

Three lines of evidence support the conclusion that PK- catalytic motifs I/II that are implicated in the activity of most PKs (Hanks et al. 1988) are not essential for the kinase activity of ICP10. First, mutants pJHL2, pJHL4, and pJHL17 have auto- and transphosphorylating activities in vitro and in [^{32}P]-labeled transfected cells. pJHL9, which is deleted in all PK-catalytic motifs, lacks kinase activity (Fig. 3B). Second, $p140^{II}$ (expressed by the invariant Lys mutant pJHL4) specifically binds [^{14}C]FSBA, and kinase activity is evidenced by $p140^{II}$ expressed in Escherichia coli (pJL39; Luo et al. 1991, Luo and Aurelian 1992). Finally, PK activity is retained by the highly purified ICP6 (Paradis et al. 1991), which did not conserve these motifs. However, motifs I and II contribute toward optimization of kinase activity, since: (a) K_m values are increased by mutation of Lys^{176} or deletion of motifs I and II; (b) in vivo kinase activity is lower in cells transfected with mutants pJHL2, pJHL4, or pJHL17 than with pJW17; and (c) ICP6 has lower kinase activity than that of ICP10 (Luo and Aurelian 1992). These findings suggest that ICP10 is a new kinase that functions in phosphotransfer with regions of the molecule distinct from those described previously. Other proteins that have intrinsic kinase activity, while lacking a classical eukaryotic PK sequence, include the hepatitis virus protein hbx (Wu et al. 1990) and the products of the *BCR* (Maru and Witte 1991) and *PC-1* (Oda et al. 1991) genes.

Alternate ATP-binding sites in ICP10 could be Tyr or Cys residues that can also bind FSBA (Esch and Allison 1978) or a sequence (residues 253–280) that encompasses a nucleotide-binding consensus motif (Gly^{273}–X–Gly^{271}–X–X–Gly^{268})

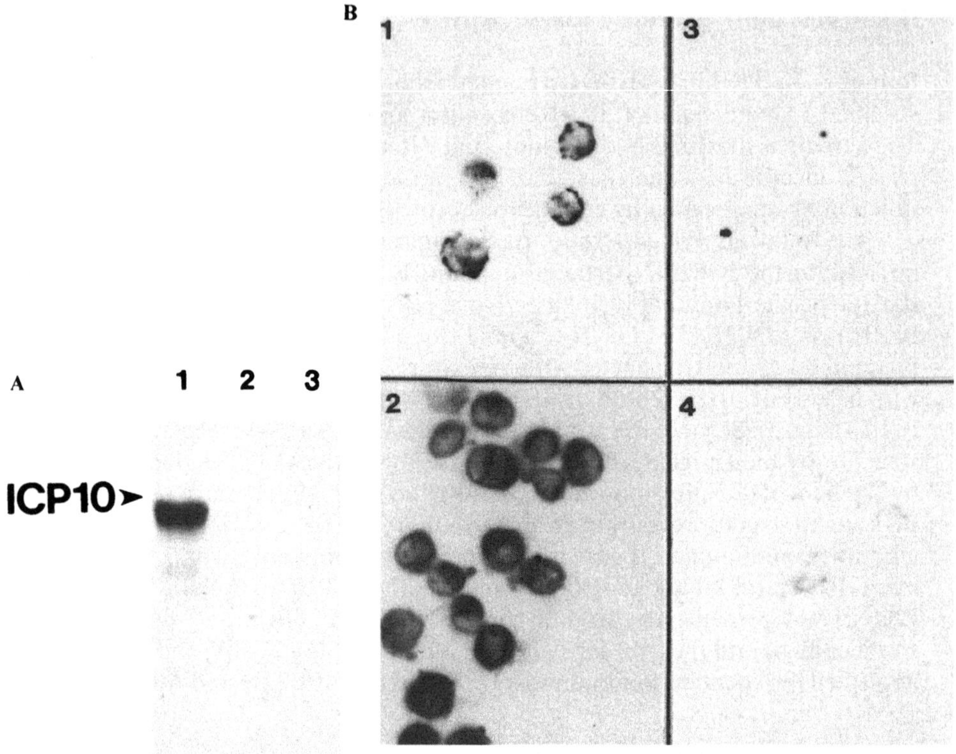

Fig. 4. A Immunoprecipitation of [^{35}S]methionine labeled extracts of JHLa1 (*lanes 1,2*) or 293 (*lane 3*) cells with ICP10-specific monoclonal antibody (MAb; *lanes 1,3*) or preimmune serum (*lane 2*). **B** Membrane immunofluorescent staining of JHLa1 cells with anti-LA-1 (*1*), anti-LA-3 (*3*), or anti-LA-5 (*4*) sera; fixed JHLa1 cells stained with anti-LA-3 serum (*2*)

and Lys residue (Lys^{253}) separated by 12 amino acids that form a short β-strand. Finally, a MgATP complex could be generated by salt bridge formation with the carboxyl group of the Asp residue in PK-catalytic motif VII. Identification of the alternate ATP-binding site must await the results of ongoing sequence analysis of the [^{14}C]FSBA-labeled ICP10 and $p140^{II}$ proteins. The PK activity of the HSV-1 RR1 (ICP6) is only autophosphorylating (Conner et al. 1992b), suggesting that it may be functionally distinct from the ICP10 PK.

ICP10 Protein Kinase is a Growth Factor Receptor

ICP10-PK shares a number of properties with established GFR. It is myristilated, its kinase activity is modulated by an extracellular (LA-1), but not by an intracellular (LA-3), determinant, and it is stimulated by basic proteins that presumably interact with acidic peptide regions in the PK-catalytic domain (Chung et al. 1990). Studies of the ICP10 polarity in plasma membranes from transformed cells (JHLa1) that constitutively express ICP10 (Fig. 4A) indicate that ICP10 is a membrane-spanning protein. Thus, only anti-LA-1 serum, which recognizes ICP10 residues 13–26, stains JHLa1 cells in membrane immunofluorescence. Antibody to intracellular determinants LA-3 (residues 165–179) or LA-5 (residues 918–927) do not strain JH1a1 cells in membrane immunofluorescence, although they stain permeabilized JHLa1 cells (Fig. 4B). It remains to be determined whether biologically relevant basic proteins (such as c-ras) stimulate ICP10-PK activity. Furthermore, the amino acid residue or residues through which myristic acid links to ICP10 have yet to be identified, and the in vivo function of myristylation (e.g., interaction with other cellular proteins) and of the LA-1 determinant are still unclear.

A significant aspect of our findings is the observation that pJHL15, which is deleted in the TM segment, is negative for both auto- and transphosphorylating activities (Fig. 3B). The mechanism whereby the TM segment contributes to ICP10 kinase activity is unclear. It is not due to nonspecific binding of a contaminant protein, since pJHL9, which lacks all eight PK-catalytic motifs but retains the TM, is kinase negative (Fig. 3B). TM is not the autophosphorylation site, since peptides (14 and 21 kDa) that encompass the TM segment are not phosphorylated (Luo and Aurelian 1992). Possibly, TM forms a functional conformation with PK-catalytic motifs. Alternatively, deletion of the TM segment produces an allosteric effect, so that the tertiary structure of the PK-catalytic motifs is detrimentally altered. Finally, the TM deletion mutant may lose its association with the plasma membrane and the polarity of its distribution, thereby abrogating its enzymatic activity. Ongoing studies with pJHL15 and ICP10-PK mutants that contain distinct nuclear or cytoskeleton association signals should elucidate the role of plasma membrane localization in kinase activity and neoplastic transformation.

The Protein Kinase Minigene is a Cellular Gene

PCR-amplified genomic DNA sequences cloned from human cervical tissue (pCSla1) had 68.6% homology with the LA-1 region of the PK minigene. The coding potential of pCS1a1, if any, is unknown. Terminators were not found in reading frame 2. However, there was a potential donor splice site, suggesting that pCS1a1 may have a gene intron. We do not believe that the homology between pCS1a1 and the PK minigene is due to accidental matches between short G-rich tandem repeats, since: (a) sequence analysis did not reveal G-rich tandem repeats; (b) the pCS1a1 sequence is not abundant in mammalian genomic DNA, its detection required amplification by the PCR method; and (c) equally G- and C-rich primers in the HSV DNA polymerase gene did not amplify genomic DNA sequences, while amplification was also observed with primers that bracket the LA-3 determinant of ICP10-PK (Smith et al. 1991; Sharma et al. 1993). Significantly, western blot assays with anti-LA-1 antibody identified the presence in cervical tissues of a 60-kDa protein that shares nine of its 12 peptides with ICP10-PK. The expression of the 60-kDa protein was amplified in cancer as compared to normal cells (Aurelian et al. 1989).

We interpret these findings as indicating that, by analogy to the oncogene–protooncogene relationship (Aaronson 1991), the PK minigene originated from a human cellular gene that encodes an ICP10-PK-like GFR and is henceforth designated LA-1 protooncogene. It was inserted into the polypeptide-coding region of an ancestral HSV RR1 such as suggested for the human somatomammotropin gene (Selby et al. 1984). Presumably, the upstream (5′) recombination site is within the ICP10 promoter region, thereby explaining the presence in the ICP10 promoter of *cis*-response elements such as AP-1 that are characteristic of cellular, but not HSV, genes. The C terminus (3′) recombination site at the junction of PK and RR presumaby occurred within the promoter region of the ancestral HSV-2 RR1, thereby giving rise to an open reading frame that contains enhancer core and functional promoter elements. Indeed, mathematical modeling identified the presence at the PK and RR junction of distortion patterns consistent with an enhancer region, and DNA sequence analysis revealed a consensus core enhancer element (G)-T-G-G-A(T)-(G) (Smith et al. 1991). When cloned into CAT expression vectors, this region had functional promoter elements and bound cellular factors (Jones 1989).

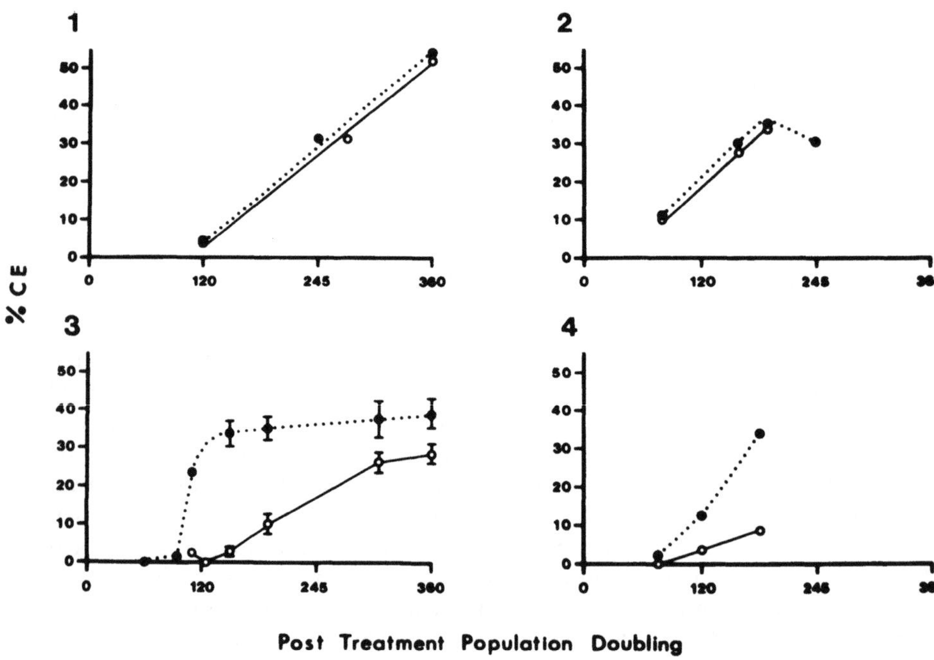

Fig. 5. Anchorage-independent growth of SHE cells transformed by wild-type UV-inactivated herpes simplex virus type 2 (HSV-2; *1*), an unrelated temperature-sensitive mutant *tsH9* (*2*), *ts5-152* (*3*), and 859/152 (*4*). Cell lines were established at 34°C. At increasing population doublings (ptpds), they were assayed for cloning efficiency in agarose at 34°C (•) and 39°C (o). Data are shown as average cloning efficiency of two lines independently established with HSV-2, *tsH9*, and 859/152, and four lines independently established with *ts5–152* ± SEM

The Protein Kinase Minigene Mediates Neoplastic Transformation

Immortalized cells transfected with the PK minigene acquire anchorage independence and neoplastic potential (Hayashi et al. 1985). Expression of the PK minigene is required for transformation. Thus, cells transformed with a ts mutant (*ts5-152*) defective in ICP10 synthesis, or a recombinant (859/152) in which the mutant ICP10 gene was inserted in a wild-type virus, had significantly decreased anchorage-independent growth at the nonpermissive temperature (39°C). Cells transformed by wild-type HSV-2 or an unrelated ts mutant cloned equally well at 34° and 39°C (Fig. 5) The conclusion that expression of the PK minigene is required for transformation is also supported by the finding that cells transfected with pJW17 (ICP10) or pJW32 (PK domain) clone in agarose, while cloning is not observed with cells transfected with pJW21 or the ICP10 RR vector pJW31 (Smith et al. 1992).

The dependence of the *ts5-152*-transformed cells on ICP10-PK was severe only during the first 210–304 population doublings (ptpd), suggesting that beyond this stage the function of the PK minigene is partially replaced by other, as yet unidentified, cellular gene(s). These genes presumably function in the selection of cells with reduced requirements for the GFR kinase function of the PK minigene. Indeed, a 21-kDa protein distinct from the ras oncogene was identified in the transformed cells at ptpd>300, but not earlier. This protein may belong to the growing family of guanosine triphosphate (GTP)-binding proteins involved in signal transduction pathways (Takai et al. 1992). Its increased expression/activation would explain the reduced requirement for ICP10-PK. Ongoing studies in our laboratory are designed to address these questions and identify signal transduction pathways involved in transformation.

A Model for Protein Kinase Minigene-Linked Carcinogenesis

Serologic evidence has associated ICP10 expression with cervical cancer. However, detection of HSV-2 DNA has generally been difficult to document (Aurelian et al. 1981; DiLuca et al. 1989). We considered the possibility that the association of ICP10 with cervical cancer is due to activation of the 60-kDa cellular protein (Aurelian et al. 1989) by the human papillomavirus (HPV) E2 protein. This interpretation is based on previous findings indicating that: (a) cancer is a multifactorial disease, (b) HPV (particularly type 16) is associated with cervical cancer, and expression of the 3′ early genes (including E2) is not disrupted in HPV16 positive tissues (Durst et al. 1992), (c) the PK minigene originated from the *LA-1* protooncogene that presumably codes for the 60-kDa protein, and (d) the ICP10 promoter has two HPV E2 response elements (E2RE) that are absent in the promoter of the HSV-1 RR1 (Wymer et al. 1991), which lacks transforming potential (Camacho and Spear 1978).

Transcriptional regulation studies indicated that HPV-16 E2 DNA enhances CAT expression from pICP10-cat, but not from CAT constructs using other HSV promoters. Activation was significantly reduced by site-directed mutagenesis of the E2RE consensus motifs, and it was not observed with HPV-16 E7 DNA

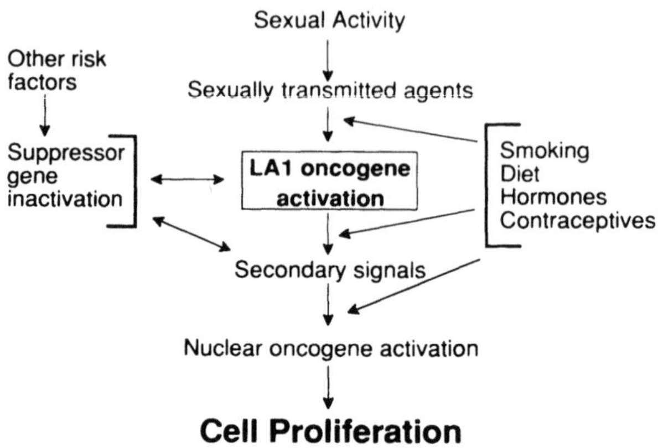

Fig. 6. *LA1* hypothesis of cervical cancer, protein kinase minigene-linked carcinogenesis

(Wymer et al. 1990). In view of the finding that patients exposed to multiple risk factors (most notably ICP10 and HPV) are at higher risk of developing cervical cancer (DiLuca et al. 1989), our studies support a postulated role for the PK minigene and its protein product (henceforth designated *LA-1* oncogene) as a necessary component for neoplastic progression (Aurelian et al. 1990). According to this interpretation (Fig. 6), HSV-2 as well as other infectious (sexually transmitted) and noninfectious risk factors activate the *LA-1* protooncogene, resulting in overexpression/modification of the 60-kDa cellular protein that is homologous to ICP10-PK. HSV-2 can cause neoplastic transformation by activating the *LA-1* protooncogene, under which circumstances HSV-2 DNA would not be maintained, or it can transform by virtue of the maintenance and expression of the PK minigene. Either way, *LA-1* oncogene expression correlates with neoplastic progression, as recently shown (Aurelian et al. 1989; Terzano et al. 1992). The hypothesis visualizes carcinogenesis as involving an oncogene cascade in which the *LA-1* oncogene, which is a GFR, in turn activates nuclear oncogenes through secondary messengers. Among changes that occur in the nucleus are increases in the expression of nuclear protooncogenes (c-myc expression is increased in cells transformed by the PK minigene, Hayashi et al. 1985) and in *LA-1* oncogene-positive cervical cancer tissues, DiLuca et al. 1989 and their interaction (e.g., fos and jun). Nuclear oncogenes are thought to regulate the transcription of genes whose products ultimately disrupt the cellular growth cycle, committing them to enter additional rounds of division. Other risk factors can also contribute, as shown in Fig. 6.

ICP10 Sequences Involved in Complex Formation with Ribonucleotide Reductase 2

The last nine residues in RR2 (H–Tyr–Ala–Gly–Ala–Val–Val–Asn–Asp–Leu–OH) are required for complex formation with RR1 and enzymatic activity (Dutia et al. 1986; McClements et al. 1988). In immunoprecipitation studies designed to

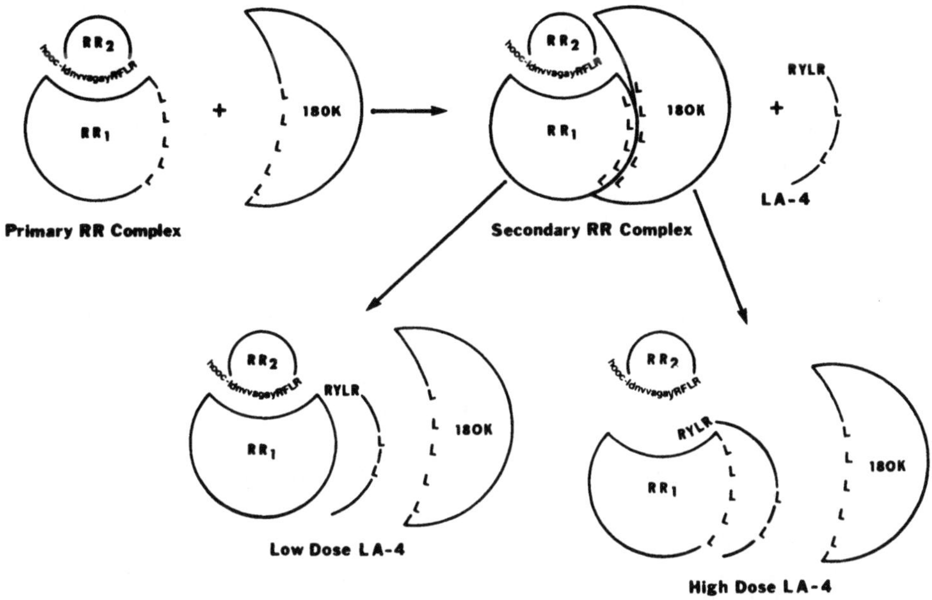

Fig. 7. Multiprotein complex formation involved in ribonucleotide reductase (*RR*) activity highlighting the ICP10 leucine zipper and its interaction with the 180-kDa cellular protein. Low-dose LA-4 causes dissociation of the 180-kDa protein and reduces RR activity. High-dose LA-4 causes RR1–RR2 subunit dissociation and abrogates RR activity

identify RR1 sequences that are required for complex formation, we found that a MAb specific for RR2 coprecipitates ICP10, RR2, (38 kDa), and a 180-kDa cellular protein from cells cotransfected with pJW17 (ICP 10) and pJW30 (RR2), but only RR2 from cells cotransfected with pJW30 and pJW33 (ICP10 deleted in the 50 C-terminal residues) or pJW30 and pJW32 (ICP10-PK). A synthetic peptide (LA-4) that corresponds to the ICP10 leucine zipper caused dose-dependent inhibition of RR activity. This was paralleled by complex dissociation, beginning with a decrease in the amount of coprecipitated 180 kDa protein and reaching total subunit dissociation. However, the 180-kDa protein was not required for RR activity (Fig. 7). Oligopeptides that represent other ICP10 regions (e.g., LA-1) did not inhibit RR activity and failed to cause subunit dissociation. We interpret these findings as indicating that the last 50 residues of ICP10 are involved in complex formation with RR2. A cellular protein (180 kDa) that binds the RR1–RR2 complex via the ICP10 leucine zipper is involved in enzyme stabilization. This is consistent with the results of in vitro reconstitution studies that implicated a cellular protein in optimization of RR activity (Huang et al. 1988).

The Protein Kinase Minigene and Herpes Simplex Virus Growth

It seems plausible that the PK minigene was acquired by an ancestral HSV, because it conferred selective advantage to the virus by virtue of the ability of its product to phosphorylate viral and cellular substrates. Therefore, an obvious question is what role does the PK minigene play in HSV-2 growth in vitro and in vivo? While a definitive answer is still elusive, it is important to point out that the *ts5-152* mutant is defective in ICP10 synthesis, RR and PK activities, and the ability to grow at 39°C. Sequence analysis revealed a pair of frameshift mutations and a downstream single bp mutation in the PK minigene resulting in the substitution of ICP10 residues 275–293 and 309. Reduced kinase activity of *ts5-152 ICP10* is consistent with our previous conclusion that residues 253–280 are the ATP-binding site (Luo and Auretian 1992). However, the effect of this mutation on ICP10 synthesis is more difficult to interpret. Pulse-chase experiments indicate that *ts5-152 ICP10* is stable at 39°C and temperature has no effect on its intracellular localization. Possibly kinase activity is required for optimal ICP10 synthesis. Alternatively, the decreased conformational stability resulting from the unwinding of part of the α-helix interferes with the activity of the enhancer functions at the PK and RR junction that are required for optimal ICP10 synthesis (Smith et al. 1992). In any case, the reduced RR activity that results from decreased ICP10 synthesis is responsible for impaired HSV-2 growth at 39°C. Further studies are needed in order to define the distinct roles of the PK and RR domains in virus growth in dividing and nondividing cells, particularly as pertains to the reactivation of latent virus.

Clinical Applications

The utilization of HSV-1 RR as a target for the design of antiviral drugs has recently attracted great interest (Spector et al. 1985; McClements et al. 1988). Our findings extend this conclusions to HSV-2 and identify additional targets for antiviral activity, including the LA-4 peptide, which inhibits RR activity and causes subunit dissociation, and the last 50 residues of ICP10, which are required for complex formation and could potentiate the antiviral activity of the RR2 nonapeptide. The PK minigene provides an additional target for antiviral activity in that it may be required for optimal ICP10 synthesis (and therefore RR activity), as well as for the phosphorylation of the products of viral (e.g., transactivating IE) and cellular genes, the function of which depends on a phosphorylated state. This is particularly significant within the context of the RR role in reactivation of latent virus.

There is good reason to assume that elimination of viruses associated with human cancer as risk factors would greatly reduce the incidence of those cancers with which they are linked. Our studies indicate that the PK minigene and its product (*LA-1* oncogene) is a necessary component for neoplastic transformation. As such, the *LA-1* oncogene provides a marker for the identification of the patient with an early cervical lesion that is at high risk of progressing to invasive cancer (Aurelian et al. 1989,1990; Terzano et al. 1992). Furthermore, by using antisense oligonucleotides complementary to the PK minigene, it should be possible to specifically inhibit its expression, thereby potentially inhibiting the reactivation of latent virus and the progression of the cervical precursor lesions to invasive cancer.

References

Aaronson SA (1991) Growth factors and cancer. Science 25: 1146–1153

Aurelian L, Kessler II, Rosenshein NB, Barbour G (1981) Viruses and gynecologic cancers: herpesvirus protein (ICP10/AG-4) a cervical tumor antigen that fulfills the criteria for a marker of carcinogenecity. Cancer 48: 455–471

Aurelian L, Terzano P, Smith CC, Chung TD, Shamsuddin A, Costa S, Orlandi, C (1989) Amino-terminal epitope of herpes simplex virus type 2 ICP10 protein as a molecular diagnostic marker for cervical intraepithelial neoplasia. Cancer Cells 7: 187–191

Aurelian L, Costa S, Terzano, P, Orlandi C (1990) Sexually transmitted viruses and cervical cancer: the LA-1 oncogene hypothesis. Reach Gynecol 2: 64–70

Camacho A, Spear PG (1978) Transformation of hamster embryo fibroblasts by a specific fragment of the herpes simplex virus genome. Cell 15: 993–1002

Cameron JM, McDougall I, Marsden HS, Preston VG, Ryan DM, Subak-Sharpe JH (1988) Ribonucleotide reductase encoded by herpes simplex virus is a determinant of the pathogenicity of the virus in mice and a valid antiviral target. J Gene Virol 69: 2607–2612

Chung TD, Wymer JP, Smith CC, Kulka M, Aurelian L (1989) Protein kinase activity associated with large subunit of herpes simplex virus type 2 ribonucleotide reductase (ICP10). J Virol 63: 3389–3398

Chung TD, Wymer JP, Kulka M, Smith CC, Aurelian L (1990) Myristylation and polylysine-mediated activation of the protein kinase domain of the large subunit of herpes simplex virus type 2 ribonucleotide reductase (ICP10). Virology 179: 168–178

Chung TD, Luo JH, Wymer JP, Aurelian L (1991a) Leucine-repeat motif in the large subunit of herpes simplex virus type 2 (HSV-2) ribonucleotide reductase (ICP10) is involved in RR activity and subunit complex formation. J Gen Virol 72: 1139–1144

Conner J, Macfarlane J, Lankinen H, Marsden H (1992a) The unique N terminus of the herpes simplex virus type 1 large subunit is not required for ribonucleotide reductase activity. J Gen Virol 73: 103–112

Conner J, Cooper J, Furlong J, Clements JB (1992b) An autophosphorylating but not transphosphorylating activity is associated with the unique N terminus of the herpes simplex virus type 1 ribonucleotide reductase large subunit. J Virol 66: 7511–7516

DiLuca D, Costa S, Monini P, Rotola A, Terzano P, Savioli A, Grigioni W, Cassai E (1989) Search for human papillomavirus, herpes simplex virus and c-myc oncogene in human genital tumors. Int J Cancer 43: 570–577

Durst M, Glitz D, Schneider A, Zur Hausen H (1992) Human papillomavirus type 16 (HPV16) gene expression and DNA replication in cervical neoplasia: analysis by in situ hybridization. Virology 189: 132–140

Dutia BM, Frame MC, Subak-Sharpe JH, Clark, WN, Marsden HS (1986) Specific inhibition of herpes ribonucleotide reductase by synthetic peptides. Nature 321: 439–441

Esch FS, Allison WS (1978) Identification of a tyrosine residue at a nucleotide binding site in the beta subunit of the mitochondrial ATPase with p-fluorosulfonyl [14C]-benzoyl-5′-adenosine. J Biol Chem 253: 6100–6106

Goldstein DJ, Weller SK (1988) Factor (s) present in herpes simplex virus type 1-infected cells can compensate for the loss of the large subunit of the viral ribonucleotide reductase: characterization of an ICP6 deletion mutant. Virology 166: 41–51

Hanks SK, Quinn AM, Hunter T (1988) The protein kinase family: conserved features and deduced phyogeny of the catalytic domains. Science 241: 42–51

Hayashi Y, Iwasaka T, Smith CC, Aurelian L, Lewis GK, Ts'O, P,O.P (1985) Multistep transformation by defined fragments of herpes simplex virus type 2 DNA: oncogenic region and its gene products. PNAS 82: 8493–8497

Huang A, Jacobi G, Haj-Ahmad Y, Bacchetti S (1988) expression of the HSV-2 rebonucleotide reductase subunits in adenovirus vectors or stably transformed cells: Restoration of enzymatic activity by reassociation of enzyme subunits in the absence of other HSV proteins. Virology 163: 462–470

Idowu AD, Fraser-Smith EB, Poffenberger KL, Herman RC (1992) Deletion of the herpes

simplex virus type 1 ribonucleotide reductase gene alters virulence and latency in vivo. antiviral Res 17: 145–156
Jacobson JG, Leib DA, Goldstein DJ, Bogard CL, Schaffer PA, Weller SK, Coen DM (1989) A herpes simplex virus ribonucleotide reductase deletion nutant is defective for productive acute and reactivatable latent infections of mice and for replication in mouse cells. Virology 173: 276–283
Jones C (1989) The minimal transforming fragment of HSV-2 mtrIII can function as a complex promoter element. Virology 169: 346–353
Kristie TM, Sharp PA (1990) Interactions of the oct-1 POU subdomains with specific DNA sequences and with the HSV a-trans-activator protein. Genes Dev 4: 2383–2396
Luo JH, Aurelian L (1992) The transmembrane helical segment but not the invariant lysine is required for the kinase activity of the large subunit of herpes simplex virus type 2 ribonucleotide reductase (ICP10). J Biol Chem 267: 9645–9653
Luo JH, Smith CC, Kulka M, Aurelian L (1991) A truncated protein kinase domain of the large subunit of herpes simplex virus type 2 ribonucleotide reductase (ICP10) expressed in Escherichia coli. J Biol Chem 266: 20976–20983
Maru Y, Witte ON (1991) The BCR gene encodes a novel serine/threonine kinase activity within a single exon. Cell. 67: 459–468
McClements W, Yamanaka G, Garsky V, Perry H, Bacchetti S, Colonno R, Stein RB (1988) Oligopeptides inhibit the ribonucleotide reductase of herpes simplex virus by causing subunit separation. Virology 102: 270–273
McGeoch DJ, Dalrymple MA, Davison AJ, Dolan A, Frame MC, McNab D, Perry LJ, Scott JE et al. (1988) The complete DNA sequence of the long unique region in the genome of herpes simplex virus type 1. J Gen Virol 69: 1531–1574
Morrison BD, Feltz SM, Pessin JE (1989) Polylysine specifically activates the insulin-dependent insulin receptor protein kinase. J Biol Chem 264: 994–1001
Nikas I, McLauchlan J, Davison AJ, Taylor WR, Clements JB (1986) Structural features of ribonucleotide reductase. Proteins Struct Funct Genet 1: 376–384
Oda Y, Kuo MD, Huang SS, Huang JS (1991) The plasma cell membrane glycoprotein, PC-1, is a threonine-specific protein kinase stimulated by acidic fibroblast growth factor. J Biol Chem 266: 16791–16795
O'Hare P, Hayward GS (1985) Three trans-acting regulatory proteins of herpes simplex virus modulate immediate-early gene expression in a pathway involving positive and negative feedback regulation. J Virol 56: 723–733
Paradis H, Gaudreau P, Massie B, Lamarche N, Guilbault C, Gravel S, Langelier Y (1991) Affinity purification of active subunit 1 of herpes simplex virus type 1 ribonucleotide reductase exhibiting a protein kinase activity. J Biol Chem 266: 9647–9651
Phelps WC, Yee CL, Munger K, Howley PM (1988) The human papillomavirus type 16 E7 gene encodes transactivation and transformation functions similar to those of adenovirus E1A Cell 53: 539–547
Selby MJ, Barta A, Baxter JD, Bell GL, Eberhardt NL (1984) Analysis of a major human chorionic somatomammotropin gene. Evidence for two functional promoter elements. J Biol Chem 259: 13131–13138
Sharma BK, Nelson JW, Smith CC, Aurelian L (1993) Detection of *LA-1* oncogene in paraffin embedded cervical cancer tissues by polymerase chain reaction. Int J Oncol 3 (in press)
Smith CC, Wymer JP, Luo JH, Aurelian L (1991) Genomic sequences homologous to the protein kinase region of the bifunctional herpes simplex virus type 2 protein ICP10. Virus Genes 53: 215–226
Smith CC, Kulka M, Wymer JP, Chung TD, Aurelian L (1992) Expression of the large subunit of herpes simplex virus type 2 ribonucleotide reductase (ICP10) is required for virus growth and neoplastic transformation. J Gen Virol 73: 1417–1428
Spector T, Averett DR, Nelson DJ, Lambe CU, Morrison RW Jr, St Clair MH, Furman PA (1985) Potentiation of antiherpetic activity of acyclovir by ribonucleotide reductase inhibition. PNAS 82: 4254–4257
Swain M, Galloway DA (1986) Herpes simplex virus specifies two subunits of ribonucleotide reductase encoded by 3′-coterminal transcripts. J Virol 57: 802–808

Takai Y, Kaibuchi K, Kikuchi A, Kawata M (1992) Small GTP-binding proteins. Int Rev Cytol 133: 187–230

Terzano P, Martini F, Costa S, Martinelli GN (1993) Immunohistochemistry with antibody to the LA-1 oncogene as a prognostic marker in cervical intraepithelial neoplasia. Gynecol Oncol 48: 317–327

Wu JY, Zhou Z-Y, Judd A, Cartwright CA, Robinson WS (1990) The hepatitis B virus-encoded transcriptional trans-activator hbx appears to be a novel protein serine/threonine kinase. Cell 63: 687–695

Wymer JP, Aurelian L (1990) Papillomavirus trans-activator protein E2 activates expression from the promoter for the ribonucleotide reductase large subunit from herpes simplex virus type 2. J Gen Virol 71: 1817–1821

Wymer JP, Chung TD, Chang Y-N, Hayward GS, Aurelian L (1989) Identification of immediate-early-type cis-response elements in the promoter for the ribonucleotide reductase large subunit from herpes simplex virus type 2. J Virol 63: 2773–2784

Wymer JP, Aprhys CMJ, Chung TD, Feng CP, Kulka M, Aurelian L (1992) Immediate early and functional AP-1 cis-response elements are involved in the transcriptional regulation of the large subunit of herpes simplex virus type 2 ribonucleotide reductase (ICP10). Virus Res 23: 253–270

Yang YC, Okayama H, Howley PM (1985) Bovine papillomavirus contains multiple transforming genes. PNAS 82:1030–1034

Yarden Y, Ullrich A (1988) Growth factor receptor tyrosine kinases. Ann Rev Biochem 57: 443–478

III. Genes Related to Intraperitoneal Pathogenicity

Chapter 9 Effect of Herpes Simplex Virus Type 1 *UL41* Gene Product on *mdr-1* Gene-mRNA in Infected Adrenal Glands

Michal Moyal and Yechiel Becker

Summary

The adrenal glands are the primary site for herpes simplex virus type 1 (HSV-1) replication in the intraperitoneal (IP) route of inoculation. After replication in the adrenal glands, the virus penetrates the neurons that innervate the glands and reaches the spinal cord. HSV-1 infection leads to shutoff of both cellular and viral genes, due to the activity of the UL41 gene product. The mutant vhs-1 expresses an inactive form of UL41 product and does not cause shutoff of gene expression.

The *mdr-1* (multidrug resistance) gene is normally expressed in the adrenal glands and is apparently involved in hormone secretion. The effect of HSV-1 infection of the adrenal glands on the expression of the mdr-1 gene was studied. Mice were infected with several HSV-1 strains, recombinants, and the vhs-1 mutant. Shutoff of the expression of the mdr-1 gene in the adrenal glands was observed in all the infected mice except for those infected with the vhs-1 mutant. The level of *mdr-1*-specific mRNA in HSV-1 vhs-1-infected glands was comparable to that in uninfected glands. The relevance of the shutoff phenomenon to HSV-1 pathogenicity and to future application in gene therapy is discussed in this chapter.

Introduction

Herpes Simplex Virus Type 1 Infection of the Adrenal Glands

The adrenal glands are the primary site of herpes simplex virus type 1 (HSV-1) replication after intraperitoneal (IP) administration of the virus to mice (Peles et al. 1990; Hill et al. 1986; Potratz et al. 1986). The enhanced tropism of HSV-1 for the adrenal glands may be explained by either: (a) the ectodermal embryonic origin of the adrenal medulla (HSV-1 is known to be tropic for ectodermal tissues, e.g., nervous system, skin, and cornea) or (b) the high concentration of steroid

---

Department of Molecular Virology, Faculty of Medicine, The Hebrew University of Jerusalem, Jerusalem, Israel

hormones, which leads to a local immunosuppression (Frenkel 1960; Goodpasture and Teaggue 1923), such that the immune system cannot prevent virus replication in the adrenal glands. It has been shown that hydrocortisone is required to maintain persistent HSV-1 infection (Nishiyama and Rapp 1979), although other studies revealed an inhibitory effect of steroid hormones on HSV-1 replication (Notter and Docherty 1979).

Bilateral paralysis of hind legs is often observed after HSV-1 replication. The paralysis occurs after the penetration of the virus into the spinal cord. Unilateral adrenalectomy was shown to lead to unilateral paralysis after HSV-1 infection (Towensond and Baringer 1981). Adrenalectomy of both glands prevented penetration of the virus into the spinal cord and no paralysis was observed (Peles et al. 1990). These observations suggest that after IP administration, the virus infects the adrenal glands, penetrates the neurons that innervate the glands, and spreads to the central nervous system (Irie et al. 1987).

The Multidrug Resistance Gene (*mdr-1*)

The *mdr-1* (multidrug resistance) gene is known to be expressed at high levels in transformed cells, and its product, P-glycoprotein (gp170), functions as a non-specific, adenosine triphosphate (ATP)-dependent pump that carries the drug out of the cells (DiMarco et al. 1983; Garratt et al. 1989; Rankin et al. 1990; Gros et al. 1986).

The model for the human multidrug transporter suggests that this glycoprotein is membrane anchored, with 12 transmembrane domains. As was shown by the use of anti-*mdr-1*-specific antibodies, only small parts of the protein are located outside the cell, and most of it, including the two ATP-binding sites, is intracellular (Gottesman and Pastan 1988; Kane et al. 1990).

The *mdr-1* gene is clearly homologous to bacterial ATP-dependent transporters (Juranka et al. 1989). The only eukaryotic homolog to *mdr-1* with a known defined function is the yeast *STE* gene, which is involved in secretion of the α-factor mating pheromone (Gros and Shutik 1991; McGrath and Vashavsky 1989; Hochhausen and Harris 1991). The normal function of the *mdr-1* gene product in mammalian cells is obscure. It was suggested that this protein participates in secretion of steroid hormones (Yang et al. 1989) or that it acts in an alternative pathway for secretion of biological substances out of the cell (McGrath and Vashavsky 1989; Kuchler et al. 1989; Genlach et al. 1986). The latter suggestion also explains the absence of specificity and the broad range of substances that are recognized and can be transported by the *mdr-1* product, the P-glycoprotein. The *mdr-1* gene is normally expressed in several tissues, including cells of the cortex of the adrenal glands (Thiebaut et al. 1987; Sawara et al. 1988).

Effect of Herpes Simplex Virus Type 1 on Cellular Genes

Infection of cells with HSV-1 leads to the shutoff of gene expression. The shutoff function is encoded by the viral gene *UL41* (Strom and Frenkel 1987) and is accompanied by degradation of both viral and cellular mRNAs (Fenwick and

Everett 1990; Fenwick and McMenamin 1984; Mayman and Nishioka 1985; Scheck and Bachenheimer 1985; Becker et al. 1993). A mutant with a deletion in the *UL41* gene was isolated. This mutant, vhs-1, is unable to cause shut-off of cellular genes. It was found that the loss of the *UL41* gene affects virus pathogenicity, and the vhs-1 mutant is apathogenic to mice even when a high titer of virus was used for infection (Becker et al. 1993). In the present study, the shutoff of the *mdr-1* gene expression in adrenal glands of mice infected with several strains of HSV-1 was investigated.

Materials and Methods

Virus Strains and Recombinants

Seven HSV-1 strains and recombinants were used: F, HFEM, KOS, MlC1, R19, R15, and the mutant vhs-1. The HSV-1 intratypic recombinant MlC1 was obtained by cotransfection of intact HFEM genomic DNA with an *MluI–MluI* fragment (0.7615–0.796 mu) cloned from the IP virulent HSV-1 strain F (Rösen et al. 1986). HSV-1 intratypic recombinants R15 and R19 were obtained by cotransfection of HFEM DNA with the *BamHI* fragment B (0.744–0.813 mu) cloned from strain F (Rösen et al. 1985). The vhs-1 mutant was obtained from Dr. N. Frenkel, Rockville, MD, USA. The viruses were propagated on BSC-1 cells to titers of 10^7–10^8 plaque-forming units (PFU) ml.

Preparation of mRNA from Infected and Control Mouse Adrenal Glands

Groups of five female Sabra mice were injected IP with HSV-1 strains R15, R19,MlC1, HFEM KOS, F, and vhs-1 (3×10^6 PFU/0.1 ml for each mouse). Mice were sacrificed 72 h postinfection and adrenal glands were removed and pooled. The adrenal glands were homogenized by Dounce homogenizer, and total RNA was isolated using the guanidium–isothiocyanate and cesium–chloride method (Chirgwin et al. 1979).

Northern Blot Analysis and Hybridization Probes

RNA was electrophoresed on 1.5% agarose formaldehyde gels and transferred to nylon filters (Genescreen, DuPont) by the northern blot method (Thomas 1980). Hybridization with a nick-translated DNA probe (BRL kit) was performed in 50% formamide, 5 × Denhardt's solution, 0.05 M tromethamine (TRIS) HCl pH 7.5, 1M NaCl, 0.1% sodium pyrophosphate, 1% $NaDodSO_4$, 10% dextran sulphate, at 42°C for 16 h. The filters were washed by stepwise lowering of salt concentration from 1 × SSC to 0.1 × SSC (1 × SSC = 0.15 M NaCl and 0.015 M sodium citrate) and raising the temperature from room temperature to 65°C followed by exposure to X-ray film.

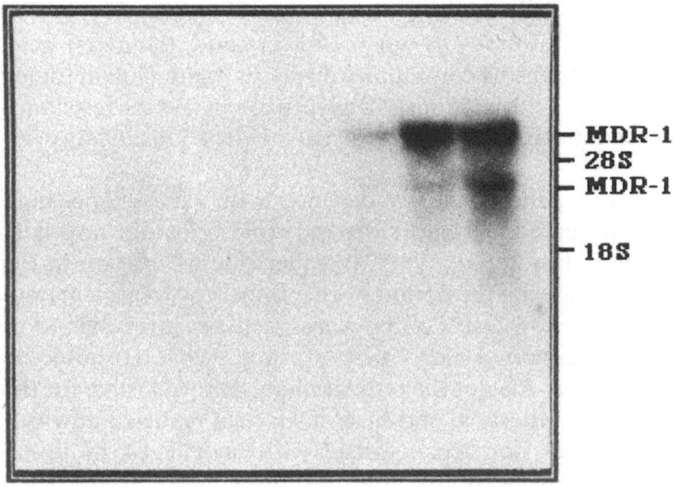

Fig. 1. Northern blot hybridization of RNA extracted from adrenal glands infected with HSV-1 strains: R15 (*lane 1*), R19 (*lane 2*), MIC1 (*lane 3*), HFEM (*lane 4*), F (*lane 5*), KOS (*lane 6*), and vhs-1 (*lane 7*); *lane 8*, uninfected mice. RNA was hybridized with *mdr-1*-specific probe

Results

Mice were infected with the following HSV-1 strains: R15, R19, MIC1, HFEM, F, KOS, and vhs-1. They were killed 3 days postinfection and the adrenal glands were removed. RNA was extracted from the adrenal glands, blotted, and probed with *mdr-1* DNA.

As can be seen in Fig. 1, no trace or a very low level of *mdr-1*-specific RNA was observed when mice were infected with the various HSV-1 strains (lanes 1–6). A significantly higher level of *mdr-1* RNA was detected in adrenal glands of uninfected mice (lane 8). Approximately the same level of *mdr-1* mRNA was detected in the adrenal glands of mice which were infected with the HSV-1 vhs-1 mutant, which does not express a functional shutoff protein (lane 7).

Discussion

The pathogenicity of HSV-1 is determined by many factors, including virus strain, site of infection, virus titer, the strain of the infected animal, and its age. According to our concept, specific genes, encoded by the virus, control virus pathogenicity at a specific site of infection. These genes are designated "pathogenicity genes."

The *UL41* gene may be classified as a pathogenicity gene, since the vhs-1 mutant, lacking the functional UL41 gene product, is apathogenic to mice by the IP route of inoculation (Becker et al. 1993).

HSV-1 HFEM and R15 are not virulent when injected IP, although their shutoff function is normal (Fig. 1, lanes 4 and 1, respectively), suggesting that IP pathogenicity is a multifactorial trait and is not solely determined by one virus gene.

Although the normal role of *mdr-1* is not yet understood, the *mdr-1* gene attracted the attention of the scientific community due to its expression in tumor cells and the correlation between the level of the P-glycoprotein and the resistance of the cells to chemotherapeutic drugs (Hochhausen and Harris 1991; Chan et al. 1991).

The expression of *mdr-1* gene is usually dependent on tumor type; high expression of the gene is observed in tumors derived from cells that normally express high levels of *mdr-1* (Fojo et al. 1987). In other cases, elevation in the expression of *mdr-1* occurred during the period of chemotherapy treatment, and cells which expressed relatively low levels of the *mdr-1* gene became resistant to drugs as a result of over expression of *mdr-1* (Salmon et al. 1989). It should be noted that the expression of *mdr-1* is not the only mechanism responsible for the responsiveness of cells to chemotherapy, and cases have been reported in which resistance to drug therapy was not accompanied with high levels of *mdr-1* expression (Moscow and Cowan 1988).

It could be suggested that improvement of the tumor cells' responsiveness to chemotherapy could be achieved by the inhibition of *mdr-1* gene expression in the tumor cells; therefore, the ability of the *UL41* gene to shut off the *mdr-1* gene in the adrenal glands might be of interest. Although the shutoff is not specific for the *mdr-1* gene only, one might suggest the use of the *UL41* gene to eliminate *mdr-1* activity in tumor cells.

The idea of utilizing HSV-1 genes for gene therapy of tumor cells was recently investigated by Culver et al. (1992). In their study, a retroviral vector was used to introduce the HSV-1 gene thymidine kinase (TK) into replicating cells. The tumor cells which absorbed the viral TK gene cloned in the retroviral vector could then be treated with the anti-HSV-1 drug gancyclovir (GCV), which would be phosphorylated by the TK gene and interfere with cellular DNA synthesis.

We would like to add that if the *mdr-1* gene is expressed in tumor cells, then treatment with GCV might not be possible, since the drug would be pumped out of the cells by the *mdr-1* product, the P-glycoprotein. Therefore, we suggest that the first step in gene therapy of tumor cells should be the shutoff of the *mdr-1* gene, and the HSV-1 *UL41* gene could be used to advantage in reaching this objective.

Acknowledgments. This study was supported by the Foundation for the Study of Molecular Virology and Cell Biology, Tucson, Arizona.

References

Becker Y, Tavor E, Asher Y, Berkowitz C, Moyal M (1993) Effect of herpes simplex virus type 1 UL41 gene on the stability of mRNA from the cellular genes: β-actin, fibronectin, glucose transporter 1 and docking protein and on virus intraperitoneal pathogenicity to newborn mice. Virus Genes 7: 133–143

Chan HSL, Hadad G, Thorner PS, DeBoer G, Lin YP, Ondrusk N, Yager H, Ling V (1991)

P-glycoprotein expression as a prediction of the outcome of therapy for neuroblastoma. N Engl J Med 35: 1608–1614

Chirgwin JM, Przybyla AE, MacDonald RS, Rutter WS (1979) Isolation of biologically active ribonucleic acids from sources enriched in ribonucleases. Biochemistry 18: 5294–5299

Culver KW, Ram Z, Wallbridge S, Ishii H, Oldfield EH, Blaese RM (1992) in vivo gene transfer with retroviral vector-producer cells for treatment of experimental brain tumor. Science 256: 1550–1552

DiMarco JP, Sellers TD, Berne RM, West GA, Belardinelli L (1983) Adenosine: electrophysiological effects and therapeutic use for terminating paroxymal supraventricular tachycardia. Circulation 68: 1254–1263

Fenwick ML, Everett RD (1990) Transfer of UL41 gene controlling virion associated host cell shut-off between different strains of herpes simplex virus. J Gen Virol 71: 411–418

Fenwick ML, McMenamin MM, Jr (1984) Early virion-associated suppression of cellular protein synthesis by herpes virus is accompanied by inactivation of mRNA. J Gen Virol 65: 1225–1228

Fojo AT, Veda K, Salmon DJ, Poplack DG, Gotessman MM, Pastan I (1987) Expression of a multidrug resistance gene in human tumors and tissues. Proc Natl Acad Sci USA 84: 265–269

Frenkel JK (1960) Pathogenesis of infection of the adrenal gland leading to Addision's disease in man: role of corticoids in adrenal and generalized infection. Ann NY Acad Sci 84: 393–440

Garratt CJ, Linker NJ, Griffith MJ, Ward DE (1989) Comparison of adenosine and verapamil for termination of paroxymal junctional tachycardia. Cann AJ Am J Cardiol 64: 1310–1316

Genlach JH, Endicott A, Juranka PF, Herdrson G, Sarangi F, Deuchass KL, Ling V (1986) Homology between P-glycoprotein and a bacterial haemolysis transport protein suggests a model for multidrug resistance. Nature 324: 485–489

Goodpasture EW, Teague O (1923) Experimental production of herpetic lesions in organs and tissues of the rabbit. J Med Res 44: 121–139

Gottesman MM, Pastan I (1988) The multidrug transporter, a double-edged sword. J Biol Chem 263: 12163–12166

Gros P, Shustik C (1991) Multidrug resistance: a novel class of membrane-associated transport proteins is identified. Cancer Invest 9: 563–569

Gros P, Ben Neriyah Y, Croop J, Hausman DE (1986) Isolation and expression of a complementary DNA that confers multidrug resistance. Nature 323: 728–731

Hill TJ, Yirrell DC, Blyth WA (1986) Infection of the adrenal glands as a route to the centeral nervous system after viremia with herpes simplex virus in the mouse. J Gen Virol 67: 309–320

Hochhausen D, Harris AL (1991) Drug resistance. Br Med Bull 47: 178–196

Irie H, Harada Y, Kurokawa E, Saito M, Sugarawa Y, Ohami K, Mori W (1987) Early adrenal infection by herpes simplex virus type-1: special reference to inoculation dose and spread from the adrenal to central nervous system. Virchows Arch [B] 53: 325–331

Juranka PE, Zastawny RL, Ling V (1989) P glycoprotein: multidrug resistance and a superfamily of membrane associated transport proteins. FASEB J 3: 2583–2592.

Kane ES, Pastan I, Gottesman MM (1990) Genetic basis of multidrug resistance of tumor cells. J Bioenerg Biomembr 22: 593–618

Kuchler K, Sterne RE, Thorner J (1989) Saccharomyces cerevisiae STE gene product: a novel pathway of protein export in eukaryotic cells. EMBO J 13, 3973–39484

Mayman BA, Nishioka Y (1985) Differential stability of host mRNA in Friend erythroleukemia cells infected with herpes simplex virus. J Virol 53: 1–6

McGrath JP, Varshavsky A (1989) The yeast STE gene encodes a homologue of the mammalian multidrug resistance p-glycoprotein. Nature 340: 400–404

Moscow JA, Cowan KH (1988) Multidrug resistance J Natl Cancer Inst 80: 14–20

Nishiyama Y, Rapp F (1979) Regulation of persistant infection with herpes simplex virus in vitro by hydrocortisone. J Virol 31: 841–844

Notter MFD, Docherty JJ (1979) Steroid hormone alteration of herpes simplex virus type 1 replication. J Med Virol 2: 247–252

Peles E, Rosen H, Darai G, Rosen Wolff A, Becker Y (1990) Importance of the Hpal-P sequence for herpes simplex virus-1 replication in the adrenal glands. Arch Virol 113: 151–163

Potratz D, Brake B, Dienes HP, Schulz TF, Hosp M, Dierich MP, Falke D (1986) Herpes simplex virus type 1 and 2 in the adrenal glands: replication and histopathology. Arch Virol 90: 207–222

Rankin AC, Rae AP, Oldroyd KG, Cobbe SM (1990) Verapamil or adenosine for the immediate treatment of superventricullar tachycardia. Q J Med 74: 203–208

Rösen A, Gelderblom H, Darai G (1985) Transduction of virulence in herpes simplex virus type 1 from a pathogenic to an apathogenic strain by a cloned DNA fragment. Med Microbiol Immunol 173: 257–278

Rösen A, Ernst E, Koch HG, Gelderblom H, Darai G, Hadar J, Tabor E, Ben Hur T, Becker Y (1986) Replacement of the deletion in the genome (0.762–0/789 mu) of avirulent HSV-1 HFEM using cloned MluI-MluI DNA fragment (0.7615–0.796 mu) of virulent HSV-1 F leads to generation of virulent intratypic recombinants. Virus Res 5: 157–175

Salmon SE, Grogan TM, Miller T, Scheper R, Dalton WS (1989) Prediction of doxorubici resistance in vitro in myeloma, lymphoma and breast cancer by P-glycoprotein staining. J Natl Cancer Inst 81: 696–701

Sawara I, Kataoka I, Morishita Y, Hamada H, Tsurvo T, Itoyama S, Mori S (1988) Tissue distribution of P-glycoprotein encoded by a multidrug resistant gene as revealed by monoclonal antibodies. Cancer Res 48: 1926–1929

Scheck N, Bachenheimer SL (1985) Degradation of cellular mRNAs induced by virion-associated factor during herpes simplex virus infection of Vero cells. J Virol 55: 601–610

Strom T, Frenkel N (1987) Effects of herpes simplex virus on mRNA stability. J Virol 61: 2198–2207

Thiebaut F, Tsurvo T, Hamada H, Gottesman MM, Pastan I, Willingham MC (1987) Cellular localization of multidrug resistance gene product P-glycoprotein in normal human tissues. Proc Natl Acad Sci USA 84: 7735–7738

Thomas PS (1980) Hybridization of denaturated RNA and small DNA fragments transferred to nitrocellulose. Proc Natl Acad Sci USA 71: 5201–5205

Towensond JJ, Baringer JR (1981) The demyelinating effect of corneal HSV infection in normal and nude mice. J Neurosci 50: 435–441

Yang CPH, DePhino SG, Greenberger LM, Arceci RJ, Horwitz SB (1989). Progesterone interacts with P-glycoprotein in multidrug resistant cells in the endometrium of gravid uterus. J Biol Chem 264: 782–788

Chapter 10 Pathogenicity and Latency of Herpes Simplex Virus in the Animal Model System Tree Shrew

Angela Rösen-Wolff, Roland Kehm, Eva Lorentzen, Wolfram Lamade, and Gholamreza Darai

Summary

The high susceptibility of *Tupaia belangeri* (tree shrews) to infection with herpes simplex virus (HSV) type 1 and 2 was used to develop a system for investigation of the molecular mechanisms involved in viral pathogenicity and latency of HSV. All HSV strains tested caused lethal generalized infections with the exception of HSV-1 strain HFEM. Investigation of the state of viral latency in those animals which survived an acute infection revealed that wild-type HSV-1 and -2 colonize the ganglia of latently infected animals and can efficiently be recovered (75%–100%) from cocultivated nervous tissue. In contrast, it was found that avirulent HSV-1 strain HFEM, whose genome harbors a deletion of 4.1 kbp within DNA sequences of the BamHI DNA fragment B, could only be recovered from the spleen of latently infected animals. The virulent phenotype of HSV-1 strain HFEM was restored by replacing the deleted DNA sequences using the corresponding DNA sequences of the BamHI DNA fragment B of virulent HSV-1 strain F. The resulting pathogenic recombinant viruses regained the ability to persist as latent virus in the ganglia of latently infected animals. Whilst testing the transcriptional activity of this particular region of the genome of avirulent and virulent HSV-1 strains, two novel RNA transcripts of 3.5 and 1.5 kb were found. The 1.5-kb RNA was identified as the transcript of the UL56 gene of HSV-1. Further analysis of the *UL56* gene indicated that the lack of *UL56* gene transcription in HSV-1 HFEM-infected cells is due to the fact that the DNA sequences of the promoter region of the *UL56* gene is a part of the natural deletion (4.1 kbp) within the DNA sequences of the BamHI DNA fragment B of HSV-1 strain HFEM. The elimination of the DNA sequences of the *UL56* gene from the genome of pathogenic HSV-1 strain F led to generation of avirulent deletion mutants.

The investigation of the genes involved in the pathogenicity of HSV-2 revealed that the activity of the ribonucleotide reductase gene is one of the factors that influence intraperitoneal virulence in tree shrews. However, it was found that the lack of ribonucleotide reductase expression did not affect the viral capability to colonize the ganglia of latently infected animals.

Institut für Medizinische Virologie der Universität Heidelberg, Im Neuenheimer Feld 324, 69120 Heidelberg, Germany

Introduction

The neurotropic potential of herpes simplex virus (HSV) and the properties of this virus for remaining in a persistent state within the nervous system were first postulated by Goodpasture in 1923 (Goodpasture and Teague 1923). In 1970, Plummer succeeded in recovering infectious HSV from the ganglia of latently infected rabbits using cocultivation of ganglia and spinal cord cells with cells susceptible for HSV in vitro (Plummer et al. 1970). The discovery of latent HSV within the ganglia of the nervous system confirmed Goodpasture's hypothesis, and the neurotropism and latency of HSV have since been well documented. Several investigators developed and established a variety of animal systems for the investigation of the biological properties of HSV in vivo. It was found that the transport of HSV in nerve tracks approximated the speed of retrograde transport of proteins to the nervous cells (Stevens 1975). After intravenous administration of the virus, latent HSV was recovered from cervical and thoracic dorsal ganglia, but much less in the anterior and/or posterior region of the brain and spinal cord (Cook and Stevens 1976). Although more than half a century has passed since Goodpasture's hypothesis was postulated, it is still not known which factors are necessary for establishing the latency of HSV in ganglia (neurotropism of HSV) and for maintaining the infection in the latent state. It is obvious that the viral gene functions and virus–cell interactions are responsible for tropism and latency of HSV. In the last two decades, a variety of experimental approaches have been applied for investigating the basic mechanisms manifesting HSV pathogenicity and latency. Studies of viral pathogenicity and latency in the nervous system were performed using a variety of laboratory animals, e.g., rabbits (Plummer et al. 1970), mice (Stevens and Cook 1971; Stevens 1975; Cook and Stevens 1976), and guinea pigs (Scriba 1977).

The use of tree shrews as a model system for investigation of biological properties of HSV in vivo has been established in our laboratory (Darai et al. 1978, 1979, 1980a,b, 1982, 1983; Darai and Scholz 1984). The choice of this animal for the study of HSV pathogenicity, latency, and tropism was based on the fact that the tree shrew is highly susceptible to HSV (Darai et al. 1978, 1979, 1980a,b, 1982, 1983a; Darai and Scholz 1984) and also that it belongs phylogenetically to a more highly developed species than rodents. Tree shrews, members of the family *Tupaiidae*, have been placed in a phylogenetic position intermediate between prosimians and insectivores (Fiedler 1956; Moore and Goodman 1968). In the present report, we summarize the results of our studies on experimental infections of *Tupaia belangeri* with HSV-1 and -2 with the goal to identify the genetic information involved in viral tropism, pathogenicity, and latency.

Experimental Approach

The experimental approach for this study (e.g., animals, HSV-1 and -2 strains, temperature-sensitive (ts) mutants of HSV-1 and -2, inter- and intratypic recombinant viruses, virus propagation, titration, cells and media, cocultivation assays, transfection assays in vivo and in vitro, viral and cellular DNA and RNA preparations, DNA–DNA and DNA–RNA hybridizations, molecular cloning, recombinant DNA technology, restriction enzyme analysis and physical mapping, DNA nucleotide sequencing, and polymerase chain reaction, PCR) was carried out as described previously (Darai et al. 1983a; Scholz et al. 1983; Rösen et al. 1985; Rösen and Darai 1985a; Rösen et al. 1986; Koch et al. 1987; Rösen-Wolff et al. 1988; Rösen-Wolff et al. 1989; Rösen-Wolff and Darai 1991; Rösen-Wolff et al. 1991, 1992).

Susceptibility of Tree Shrews to Herpes Simplex Virus Type 1 and 2 Infections

Susceptibility of Tree Shrews to Infection with Wild-Type Strains

The susceptibility of *Tupaia belangeri* to infection with HSV-1 and -2 and the pathogenesis of HSV in this animal model system were investigated. Juvenile (28- to 45-day-old) and adult (150-day-old) animals were inoculated intravenously, intraperitoneally, or subcutaneously with different HSV-1 or -2 strains; the results are given in Tables 1 and 2 for HSV-1 and -2, respectively.

As shown in Table 1, all HSV strains tested were highly pathogenic with the exception of HSV-1 strain HFEM. In the case of HSV-2 strains (Table 2), it was found that with the exception of HSV-2 strain HG-52, which showed lower virulence, all strains of HSV-2 were highly pathogenic. The results of these studies demonstrate the high pathogenicity of HSV-1 and -2 in juvenile tree shrews.

Clinical illness usually appeared in juvenile animals on the second day after inoculation, and the animals died between 2 and 14 days after inoculation. The features of the disease and its progression to death in young adult and adult tree shrews differed in no way from the illness appearing in juvenile animals. High titers of infectious HSV were detected in liver, spleen, adrenal gland, kidney, pancreas, brain, and blood (Table 3). The histopathologic examination always showed severe liver changes with numerous necrotic areas. The morphologic events in the liver were designated as herpetic hepatitis. The next most common morphologic findings were encephalitis and fibrosis in the spleen. In contrast, those animals that died after oral infection showed no herpetic hepatitis, but did show necrosis of the stomach and intestine. The highest titers of infectious virus were found in the stomach followed by intestine and adrenal glands, brain, and lung. No virus could be detected in other organs such as liver, spleen, pancreas, and blood of orally infected animals.

State of Viral Latency

As shown in Tables 1 and 2, some adult animals survived an HSV-1 or -2 infection. The possibility that latent infections occur in those animals which survived a viral infection was investigated by cocultivating segments of spinal cord, brain, whole

Table 1. Sensitivity of juvenile (28- to 45-day-old) and adult (150-day-old) *Tupaia belangeri* (tree shrews) to infection with herpes simplex virus type 1 (HSV-1)

Age of animals (days)	Strain of HSV inoculated	Route of injection	PFU of virus/ animal	Time of death after HSV infection (days)	No. animals dead/ no. animals infected
35	HSV-1 F	i.p.	5.0×10^2	5–6	2/2
30	HSV-1 F	s.c.	1.0×10^3	3–4	2/2
548	HSV-1 F	i.p.	1.0×10^5	3–4	3/3
548	HSV-1 F	i.p.	1.0×10^2	7–9	2/8
28–35	HSV-1 17	i.v.	1.0×10^3	4	2/2
28–32	HSV-1 WAL	i.v.	2.5×10^4	2–4	4/4
29	HSV-1 WAL	i.v.	2.5×10^2	2	2/2
34–35	HSV-1 WAL	i.p.	2.5×10^2	6–7	3/3
35	HSV-1 WAL	i.p.	2.5×10^1	10	1/1
30–32	HSV-1 Thea	i.p.	1.0×10^3	4	2/2
30–34	HSV-1 Thea	s.c.	1.0×10^3	4–5	2/2
30–34	HSV-1 Thea	p.o.	1.0×10^4	3	2/2
50	HSV-1 Thea	i.p.	1.0×10^5	3	4/4
150	HSV-1 Thea	i.p.	1.0×10^5	4	1/1
365	HSV-1 Thea	i.p.	4.5×10^4	4	3/3
1035	HSV-1 Thea	i.p.	4.5×10^4	4–5	4/4
720	HSV-1 Thea	i.p.	2.5×10^2	8–12	3/6
28	HSV-1 KOS	i.p.	1.0×10^2	3–4	3/3
28–30	HSV-1 ANG	i.p.	2.5×10^2	3–4	2/2
548	HSV-1 ANG	i.p.	1.0×10^5	4–5	4/4
28–35	HSV-1 HFEM	i.p.	1.4×10^7	–	0/4
28–35	HSV-1 HFEM	i.v.	2.5×10^7	–	0/2
28–35	HSV-1 HFEM	s.c.	5.0×10^7	–	0/4

i.p., Intraperitoneally; i.v., intravenously; s.c., subcutaneously; p.o., orally.

blood, and leukocytes with susceptible monkey kidney cells (RC-37), as described previously (Darai et al. 1980b). The results of recovery of infectious virus from animals latently infected with HSV-1 and -2 are summarized in Table 4. With the exception of HSV-1 strain HFEM, all strains of HSV-1 and -2 tested were recovered from the ganglia of the nervous system of infected animals. This indicates that HSV-1 and -2 are able to colonize the ganglia of the central and peripheral nervous system and to develop viral latency.

The unexpected latency feature of HSV-1 strain HFEM, which can only be recovered from the spleen and not from the ganglia of latently infected animals, was surprising. The altered virulence and latency phenotype of HSV-1 HFEM in the tree shrew model system is of particular interest and led to the development of a system for localizing the corresponding genes involved in the pathogenicity and latency of HSV-1.

DNA of Herpes Simplex Virus Types 1 and 2 Caused Generalized and Lethal Infection in Tree Shrews

The infectivity of the HSV-1 and -2 genome was investigated in vivo using tree shrews as a highly susceptible animal model system for HSV-1 or -2 infection

Table 2. Sensitivity of juvenile (28- to 45-day-old) and adult (150-day-old) *Tupaia belangeri* (tree shrews) to infection with herpes simplex virus types 2 (HSV-2)

Age of animals (days)	Strain of HSV inoculated	Route of injection	PFU of virus/ animal	Time of death after HSV infection (days)	No. animals dead/ no. animals infected
100	HSV-2 G	i.p.	1.0×10^6	2–3	4/4
548	HSV-2 G	i.p.	1.0×10^5	2–3	4/4
28–35	HSV-2 3345	i.p.	1.0×10^6	2–3	2/2
30–32	HSV-2 3345	i.p.	1.0×10^4	3	2/2
165	HSV-2 3345	i.p.	7.5×10^2		0/2
30–34	HSV-2 Müller	i.p.	1.0×10^3	3–6	2/2
30–31	HSV-2 Müller	s.c.	1.0×10^3	4–5	2/2
30–34	HSV-2 Müller	p.o.	1.0×10^4	4	2/2
30	HSV-2 HG-52	i.p.	5.0×10^4	4	1/1
30–32	HSV-2 HG-52	i.p.	5.0×10^2	5–6	4/4
31	HSV-2 HG-52	i.p.	5.0×10^1	14	1/1
40–45	HSV-2 HG-52	i.p.	1.0×10^5	5–7	9/12
60	HSV-2 HG-52	i.p.	3.8×10^5	6	1/6
60	HSV-2 HG-52	i.p.	1.0×10^6	8	1/3
80	HSV-2 HG-52	i.p.	1.8×10^4	–	0/4
150	HSV-2 HG-52	i.p.	1.0×10^5	–	0/2

i.p., Intraperitoneally; i.v., intravenously; s.c., subcutaneously; p.o., orally.

Individual tree shrews were inoculated intrahepatically with a transfection mixture containing 25 µg HSV-1 or -2 DNA, as described elsewhere (Darai et al. 1983b). Under the conditions used, a generalized and lethal herpesvirus infection was generated in tree shrews. The new virus progeny was found in different organs of transfected animals (for details, see Darai et al. 1983b). This experimental approach is a suitable system for testing the infectivity of genetically engineered HSV DNA in vivo.

Table 3. Isolation and titration of herpes simplex virus (HSV) from juvenile, infected *Tupaia belangeri* (tree shrews) found dead or killed when moribund

	Titer of HSV (PFU/g of tissue)[a]			
	Inoculated virus[b]			
Tissue	HSV-1 F	HSV-1 17	HSV-2 G	HSV-2 HG-52
Adrenal gland	2.4×10^5	2.2×10^5	8.8×10^4	1.9×10^5
Blood	1.8×10^4	1.5×10^4	3.2×10^3	2.8×10^3
Brain	3.4×10^4	2.7×10^4	2.6×10^4	1.5×10^4
Kidney	2.2×10^6	1.8×10^6	6.2×10^5	4.3×10^5
Liver	2.8×10^7	3.9×10^7	4.5×10^7	8.6×10^6
Lung	6.5×10^5	3.2×10^5	4.2×10^4	7.1×10^4
Spleen	3.4×10^6	9.2×10^5	1.2×10^5	8.6×10^4

[a] The resulting titers (an average of four animals) of the virus obtained from the plaque assay were normalized to 1.0 g of the corresponding original tissue.
[b] The juvenile (28- to 32-day-old) animals were inoculated intraperitoneally.

Table 4. Recovery of herpes simplex virus types 1 and 2 during chronic latent infection of tree shrews

No. of animals	Strain of HSV inoculated	PFU of virus/ animal	Time when animals were killed after infection (months)	Site of recovery of virus[d]				Rate of recovery (%)
				Brain	Spinal cord	Spleen	Other organs	
6	HSV-1 F[a]	1.0×10^2	8–14	1/6	4/6	0/6	0/6	83
4	HSV-1 17[a]	1.0×10^2	10–18	0/4	3/4	0/4	0/4	75
2	HSV-1 KOS[a]	5.0×10^2	12	1/2	2/2	0/2	0/2	100
12	HSV-1 HFEM[a]	1.0×10^6	6–36	0/12	0/12	9/12	0/12	75
4	HSV-1 HFEM[a]	1.4×10^7	12	0/4	0/4	4/4	0/4	100
2	HSV-1 HFEM[b]	2.5×10^7	16	0/2	0/2	2/2	0/2	100
4	HSV-1 HFEM[c]	5.0×10^7	12–36	0/4	0/4	3/4	0/4	75
2	HSV-2 3345[a]	7.5×10^2	8–10	0/2	2/2	0/2	0/2	100
2	HSV-2 HG-52[a]	1.4×10^6	12	0/2	2/2	0/2	0/2	100
10	HSV-2 HG-52[a]	$1\text{–}3 \times 10^5$	4–24	1/10	8/10	0/10	0/10	90

HSV, herpes simplex virus
Infection route:
[a] Intraperitoneal infection route.
[b] Intravenous infection route.
[c] Subcutaneous infection route.
[d] The identity of the individually isolated virus to its original inoculum was confirmed using restriction endonuclease analysis of the viral DNAs.

Taken together, the following facts justify the further use of tree shrews as a susceptible animal model system for determination of HSV gene functions in vivo: (a) it was shown that juvenile tree shrews are extremely susceptible to HSV infections; the rate of virus recovery from latently infected animals was found to be 75%–100%; (b) the avirulent phenotype of HSV-1 strain HFEM in tree shrews was combined with the recovery of the latent virus from spleen of infected animals. In the light of these observations, it was rational to use this animal system for screening of the virulence and latency phenotype of different temperature-sensitive mutants, deletion mutants, and inter- and intratypic recombinant viruses.

Identification of Herpes Simplex Virus Type 1 Genes Which Influence Viral Pathogenicity

Susceptibility of Tree Shrews to Infection with Temperature-Sensitive Mutants

The susceptibility of juvenile *Tupaia belangeri* to infection with a variety of temperature-sensitive mutants of HSV-1 strain 17 (*tsD+*, *tsG+*, *tsH+*, *tsH*, and *tsHsyn*) or of HSV-2 strain HG-52 (*ts-1*, *ts-2*, *ts-5*, *ts-9*, *ts-10*, *ts-12*, and *ts-13*) was tested (Darai et al. 1978). Although the animals were inoculated with a high dose of infectious virus (5×10^6–1×10^7 plaque-forming units (PFU)/animal), no clinical picture of illness or death was observed. The detection of neutralizing antibodies (Darai et al. 1983a,b; Darai and Scholz 1984) in the surviving animals 15 days after infection indicates that the temperature-sensitive mutants tested are able to replicate in tree shrews, but are unable to generate an acute viral infection.

Those animals which had initially been infected with the temperature-sensitive mutants of HSV-1 and/or -2 were protected against a superinfection of a lethal dose of HSV-1 or HSV-2 (Darai et al. 1980a, 1983a; Darai and Scholz 1984). The investigation of viral latency in surviving animals showed that infectious viruses were recovered only from the spleens of those animals which had initially been infected with temperature-sensitive mutants and superinfected with HSV-1. The characterization of reactivated viruses revealed that recovered viruses from the spleens of tree shrews lost their pathogenicity and tropism in tree shrews. Secondly, that significant changes in the genomes of recovered viruses from the spleens were detected between map coordinates 0.1 and 0.22, 0.7 and 0.8, or 0.94 and 1 (Scholz et al. 1983; Darai and Scholz 1984). The alteration in the genomic region of HSV-1 between map coordinates 0.7 and 0.8 corresponds to the BamHI DNA fragment B (0.738-0.809 mu HSV-1 genome). This finding was in agreement with the observation that the avirulent HSV-1 HFEM (Tables 1 and 4) had been shown to harbor a deletion within the DNA sequences of the BamHI DNA fragment B. Consequently, it was essential to focus our attention on the analysis of this particular region of the genome of HSV-1, which seemed to encode genetic information for the viral pathogenicity and latency.

Conversion of Avirulent Phenotype of Herpes Simplex Virus Type 1 Strain HFEM to Virulent Phenotype by Manipulation of the DNA Sequences of the BamHI DNA Fragment B of the Viral Genome

The hypothesis that the genetic information located within the DNA sequences of the BamHI DNA fragment B of HSV-1 is responsible for the virulent phenotype of the virus was confirmed by replacement of the deletion of the genome of HSV-1 strain HFEM using the DNA sequences of the BamHI DNA fragment B of virulent HSV-1 strain F (Rösen and Darai 1985a; Rösen et al. 1986). Marker rescue experiments were performed by the cotransfection technique using native DNA of HSV-1 strain HFEM and cloned BamHI DNA fragment B (0.738–0.809 mu) derived from the pathogenic HSV-1 strain F. A number of different intratypic recombinants were established in which the deleted region of HSV-1 HFEM had been repaired (Rösen-Wolff et al. 1988). The pathogenicity of these recombinant viruses was examined in vivo. Two of the recombinants (HSV-R-HFehx-C19, R19, and C26, R26) caused generalized and lethal herpesvirus infection in juvenile and adult tree shrews (Table 5), indicating that the virulence of the pathogenic HSV-1 strain F can be transduced by the cloned BamHI DNA fragment B to the apathogenic HSV-1 strain HFEM (for details see Rösen and Darai 1985a and Rösen et al. 1986). This finding was in agreement with the observations of Thompson et al. (1983), in which the involvement of the genetic information located between 0.71 and 0.83 map units of the HSV-1 genome on the neurovirulent phenotype of HSV-1 in mice was described.

In further experiments, it has been shown that the HSV-1 DNA sequences of the *Mlu*I DNA fragment between the map coordinates 0.7615 and 0.789 (HSV-1 nucleotide positions 116030–121753) were sufficient to transduce the virulent phenotype to the avirulent HSV-1 strain HFEM (Rösen et al. 1986). The distribution of the viral progeny in different organs of juvenile and adult tree shrews for the resulting recombinant virus HSV-R-Ml-C1, which is able to develop generalized herpesvirus infections in the animals, is shown in Table 5.

Table 5. Distribution of viral progeny in different organs of juvenile and adult tree shrews which developed generalized herpes-virus infections after intraperitoneal administration of recombinant virus HSV-R19 and HSV-R-Ml-C1

	Titer of HSV (PFU/g of tissue)[a] Inoculated virus			
	HSV-R19		HSV-R-M1-C1	
Tissue	Juvenile	Adult	Juvenile	Adult
Adrenal gland	6.2×10^6	2.1×10^5	1.2×10^6	7.8×10^3
Blood	3.0×10^4	3.0×10^4	9.1×10^2	2.7×10^3
Brain	5.5×10^2	3.0×10^4	1.0×10^4	2.6×10^2
Kidney	<1	1.7×10^4	1.0×10^5	9.5×10^3
Liver	1.7×10^5	1.1×10^7	2.5×10^7	2.7×10^6
Spleen	1.5×10^3	7.2×10^4	2.7×10^5	4.5×10^4

[a] The resulting titers (an average of two animals) of the virus obtained from the plaque assay were normalized to 1.0 g of the corresponding original tissue.

Thus, the ability of HSV-1 strain HFEM to induce generalized herpesvirus infections in tree shrews was restored by the replacement of the deleted DNA sequences in the BamHI DNA fragment B of the viral genome by the corresponding DNA sequences of the BamHI DNA fragment B of the virulent HSV-1 strain F. The conversion of the avirulent phenotype of HSV-1 strain HFEM to the virulent phenotype by manipulation of the BamHI DNA fragment B of the viral genome indicates that important genetic information is located within this particular region of the viral genome, which seems to be responsible for the virulent phenotype of HSV-1. From this point of view, the further characterization of the BamHI DNA fragment B was essential for the understanding of the molecular mechanisms participating in the complex process of the viral pathogenicity and latency.

Mapping of the Deletion within the DNA Sequences of the BamHI DNA Fragment B of Herpes Simplex Virus Type 1 HFEM

To determine the exact position of the deletion in the genome of HSV-1 strain HFEM, the DNA sequences of the BamHI DNA fragment B of the HSV-1 strains HFEM, F, and a variety of intratypic recombinant viruses were analyzed and compared with the known DNA sequences of the BamHI DNA fragment B of HSV-1 strain 17 (McGeoch et al. 1988) by Southern blot hybridizations (Rösen and Darai 1985a; Moyal et al. 1988) and DNA nucleotide sequencing (Koch et al. 1987). These studies revealed that the HpaI DNA fragment P of the intact viral genome of HSV-1 is not present in the genome of HSV-1 HFEM, indicating that the deletion in the genome of HSV-1 strain HFEM corresponds to this fragment, which spans the coordinates 0.762–0.790 of HSV-1 DNA.

The comparison of the DNA nucleotide sequences of this particular region of HSV-1 strain HFEM to the DNA nucleotide sequences of HSV-1 strain 17 revealed that the deletion in the genome of HSV-1 strain HFEM occurred between the nucleotide positions 117088 and 120641 of DNA sequences reported for HSV-1 strain 17 (McGeoch et al. 1988; Fig. 1). The 5′ terminus of the deletion is located 71 bp upstream of the UL/RL junction, and the 3′ terminus is located 36 bp upstream of the 3′ end of the IE110 mRNA (Koch et al. 1987; Rösen-Wolff et al. 1992).

Analysis of the RNA Transcripts Mapped within the DNA Sequences of the BamHI DNA Fragment B

In order to understand the functional role of the genetic information located within the BamHI DNA fragment B of HSV-1, which corresponds to the deleted genomic region of avirulent HSV-1 strain HFEM, the transcriptional activity of the whole BamHI DNA fragment B of HSV-1 strains HFEM, F, and their recombinant viruses R19, R26, and R-Ml-C1 was analyzed (Rösen-Wolff et al. 1988). Two novel RNA transcripts of 3.5 and 1.5 kb were detected which hybridize to the left terminus (nucleotide position 113322 to 114508) of the BamHI DNA

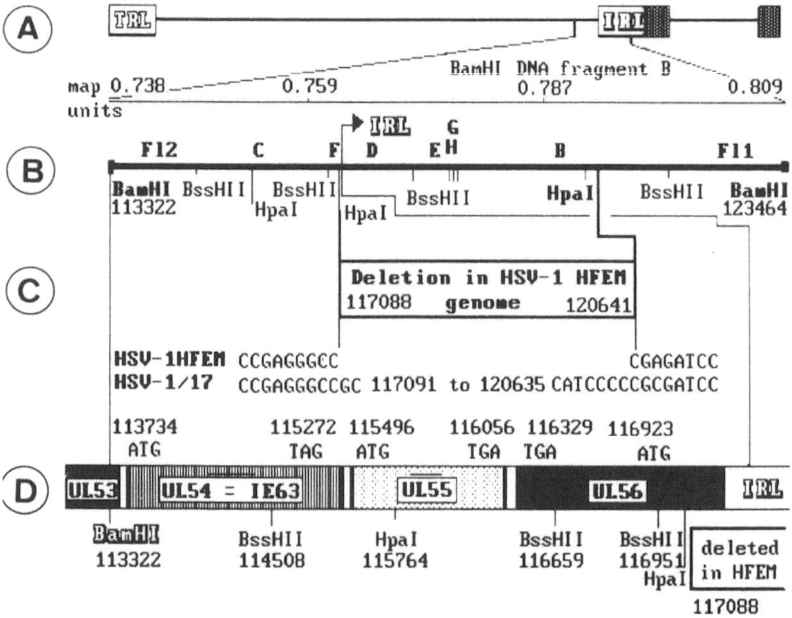

Fig. 1. A The HSV-1 genome. **B** Physical map of the BamHI DNA fragment B of HSV-1 (0.738–0.809 mu; nucleotide positions 113322–123464). The map presents the arrangement of the BssHII and HpaI DNA fragments within the BamHI DNA fragment B of HSV-1 F. The abbreviations *F11* and *F12* indicate both BssHII terminal fragments of the BamHI DNA fragment B. The position of the deletion in the genome of HSV-1 HFEM is indicated by an open *box* at the *bottom* of the map. The position of the beginning of *IRL* is indicated by an *arrow*. **C** Position of the deletion in the genome of HSV-1 HFEM compared with the DNA sequences of HSV-1 strain 17 (McGeoch et al. 1988). **D** Coding strategy of the unique region of the BamHI DNA fragment B

fragment B of the HSV-1 genome. The 1.5-kb RNA transcript was missing in cells infected with the avirulent HSV-1 strain HFEM. Hybridization with the BssHII DNA fragment F (nucleotide positions 116659–116951) led to the detection of a 3.5-kb RNA transcript by HSV-1 HFEM which was not transcribed by any of the other viruses tested. In contrast, a 1.5-kb RNA transcript was detectable in all the virus strains with the exception of HSV-1 HFEM (Rösen-Wolff et al. 1988). A more detailed analysis of the transcriptional profile of this region of the pathogenic prototype HSV-1 strain F by strand-specific hybridizations revealed that this 1.5-kb RNA species was transcribed in leftward orientation between 2 and 4 h after infection.

The corresponding open reading frame to this particular region in the HSV-1 genome (strain 17) had been predicted as the *UL56* gene (Fig. 1D). The fact that HSV-1 strain HFEM is not able to transcribe the 1.5-kb RNA is due to the fact that the promoter region of the *UL56* gene is located within those DNA sequences which are deleted in the HFEM genome (Fig. 2; Rösen-Wolff and Darai 1991; Rösen-Wolff et al. 1992).

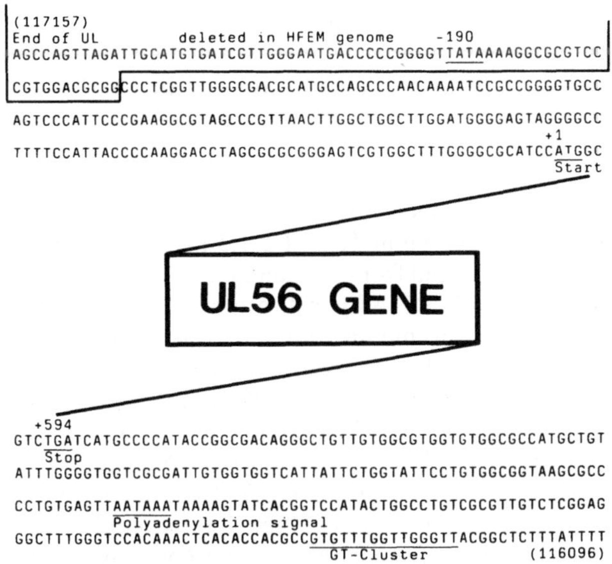

Fig. 2. DNA sequence of part of the DNA sequences of the BamHI DNA fragment B flanking the *UL56* gene. Although the *UL56* gene is transcribed in leftward orientation, the DNA sequence is shown in rightward orientation (inverted and complementary to the original sequence) to simplify the recognition of the transcriptional and translational signals. The TATA-box, the ATG start codon of *UL56*, the TGA stop codon of *UL56*, the polyadenylation signal AATAAA, and the GT-cluster are *underlined.* The *UL56* gene is indicated by a *closed box.* The 5′ terminus of the DNA sequences deleted within the genome of HSV-1 HFEM are indicated by an *open box* at the *top*. The position of this region within the HSV-1 genome is indicated by nucleotide positions given in *brackets*

The state of the *UL56* gene was investigated by determining the cDNA sequence of the *UL56* gene transcript of HSV-1 F using PCR products obtained after amplification of the cDNA with oligonucleotide primers (33mer) corresponding to the translational start and stop codons of the *UL56* gene (primer 1, 5'-GCGAATTCGTCGTGGCTTTGGGGCGCATCCATG-3′, and primer 2, 5′-CGGGATCCCTGTCGCCGGTATGGGGCATGATCA-3′). This analysis revealed that the DNA sequences of the *UL56* gene of HSV-1 F differed from those DNA sequence determined for the genomic DNA of HSV-1 strain 17. Between nucleotide positions 116343 and 116344, two nucleotides – AG – are inserted, which prolongs the open reading frame (ORF) of the *UL56* gene to 234 amino acids with a predicted relative molecular mass of 30 kDa (Rösen-Wolff et al. 1991).

Recently, McGeoch et al. (1991) corrected the sequence of the *UL 56* gene of HSV-1 strain 17 by adding two nucleotides (CG) at position 116343. Accordingly, the deduced amino acid sequence of this gene of HSV-1 strain 17 now comprises 234 amino acids, which is in agreement with the data described above.

The discovery of an open reading frame coding for a polypeptide of 21 kDa between nucleotide positions 116923 and 116332 of the HSV-1 genome was first predicted from the nucleotide sequence analysis of the HSV-1 genome and termed *UL56* (McGeoch et al. 1988). In our studies, we have shown the transcriptional activity of the *UL56* gene and demonstrated that this gene is indeed active in vivo (Rösen-Wolff et al. 1988, 1989, 1991). From this point of view, it was rational to investigate the involvement of this particular gene of HSV-1 in the process of the viral pathogenicity and latency. The transcriptional activity of the *UL56* gene of seven avirulent HSV-1 strains (HFEM, 1752/2,1752/3,1752/11,1469,1475, and 1618), two virulent HSV-1 strains (F and 17), and three virulent intratypic recombinant viruses (R19, R26, and R-Ml-C1) were screened. The virulent HSV-1 strains colonize the ganglia, but the avirulent virus strains are only able to persist in the spleen of infected animals. As shown in Table 6 a 1.5-kb RNA transcript was detectable in all virus strains recovered from the ganglia. The identity of this RNA transcript was confirmed by DNA–RNA hybridizations. In contrast, it was found that the 1.5-kb RNA transcript was missing or its size was changed in cells infected with those HSV-1 strains which were recovered from the spleens of latently infected animals. The state of the viral latency of three defined deletion variants of HSV-1 strain 17 (1704, 1705, and 1706; MacLean and Brown 1987) whose genomes harbor deletions (2.2–5.3 kbp) comprising the DNA sequences of the HSV-1 genome at the particular region between 0.760 and 0.762 mu was investigated. These studies revealed that all three deletion variants could only be recovered from the spleens of latently infected tree shrews (Rösen-Wolff et al. 1989). These results indicate that the *UL56* gene seems to be participating (directly or indirectly) in those mechanisms of virus–host interactions which are involved in the complex mechanisms of virulence and latency of HSV-1.

Table 6. Transcriptional activity of the UL56 gene of HSV-1 strains isolated from the spleen and ganglia of latently infected animals

Virus strain	Pathogenicity in tree shrew[b]	Site of recovery of latent virus	Size of RNA transcript hybridizing to the DNA sequences between nucleotide positions 116659 and 116951
HSV-1 F	+	Ganglia	1.5 kb
HSV-1 17	+	Ganglia	1.5 kb
HSV-1HFEM	–	Spleen	3.5 kb
HSV-1 1752/2			–
HSV-1 1752/3			–
HSV-1 1752/11	–	Spleen	–
HSV-1 1469			1.2 kb
HSV-1475			1.2 kb
HSV-1618			–
HSV-R19			1.5 kb
HSV-R26	+	Ganglia	1.5 kb
HSV-R-M1-C1			1.5 kb
HSV-1 1704[a]			–
HSV-1 1705[a]	–	Spleen	–
HSV-1 1706[a]			–

[a] The deletion variants were kindly provided by Dr. J. Subak-Sharpe and Dr. S.M. Brown, University of Glasgow. The mapping of the deletions in the genome has been published previously (MacLean and Brown 1987).
[b] Induction of generalized herpesvirus infection in tree shrews.

For further clarification of the role of the *UL56* gene in the processes determining viral pathogenicity and latency, a recombinant HSV-1 virus was constructed in which the DNA sequences between nucleotide positions 116030 and 121753 were replaced by the Escherichia coli β-galactosidase gene (*LacZ*; Rösen-Wolff et al. 1989). The recombinant virus HSV-1-M-LacZ generated under these conditions possesses a deletion within its genome corresponding to the nucleotide position 116030 (from the carboxy terminus of *UL55*) to the nucleotide position 121753 (to the second exon of IE110 mRNA). Due to this genomic manipulation (Rösen-Wolff et al. 1991), the DNA sequences of the *UL56* gene (nucleotide positions 116923–116329) were completely eliminated. As expected, it was found that HSV-1-M-LacZ lost its virulent phenotype and was not able to develop acute infection in tree shrews. However, the host range and growth kinetics in vitro of the recombinant virus HSV-1-M-LacZ were comparable to the parental strain HSV-1 F (Rösen-Wolff et al. 1991).

Insertion of the UL56 Gene of Virulent Herpes Simplex Virus Type 1 Strain 17 Together with its Promoter Region into the Genome of Avirulent Herpes Simplex Virus Type 1 Strain HFEM

In order to investigate whether the lack of the *UL56* gene expression is the only reason for the avirulent phenotype of HSV-1 strain HFEM, the following experimental strategy was applied: the recombinant plasmid pHSHF–B–B, which harbours the DNA sequences of the BamHI DNA fragment B of HSV-1 strain HFEM within the bacterial plasmid vector pAT153, was linearized with the restriction endonuclease *Asc*I. As shown in Fig. 3, this enzyme has a restriction site at nucleotide position 116659 of the HSV-1 genome which is placed within the DNA sequences corresponding to the *UL56* gene. A DNA linker containing DNA sequences which possess cleavage sites for the restriction enzymes *Asc*II–*Bfr*I–*Xba*I-*Asc*II was inserted into the corresponding site of the restriction endonuclease *Asc*II. This manipulation allowed the insertion of a *LacZ* cassette as a BfrI–XbaI DNA fragment (4.4 kbp; *E. coli* β-galactosidase gene flanked by a Rous Sarcoma Virus (RSV) promoter and a SV40 polyadenylation site). The resulting recombinant plasmid was linearized with the restriction enzyme *Xba*I, and a DNA linker containing cleavage sites for the restriction enzymes *Xba*I–*Sac*II–*Xba*I was inserted downstream of the *LacZ* cassette. This allowed the insertion of the *UL56* gene of HSV-1 strain 17 as a 1.4-kbp SacII DNA fragment (nucleotide positions 115967–117395). The SacII DNA fragment contains the corresponding promoter region as well as authentic transcriptional termination signals. The resulting recombinant plasmid was termed pHSHF–BB–LZ+UL56/17. The insert of the recombinant plasmid pHSHF–BB–LacZ+UL56/17 was used in cotransfection experiments with native DNA of HSV-1 strain HFEM with the goal of transfering the *UL56* gene of virulent HSV-1 strain 17 to the genome of HSV-1 strain HFEM. The resulting recombinant viruses were screened for their ability to express β-galactosidase from the inserted *LacZ* cassette. Recombinant viruses which were able to express β-galactosidase were analyzed for the transcriptional activity of the transferred *UL56* gene. Those recombinant viruses which showed *UL56* gene transcription were tested in tree shrews.

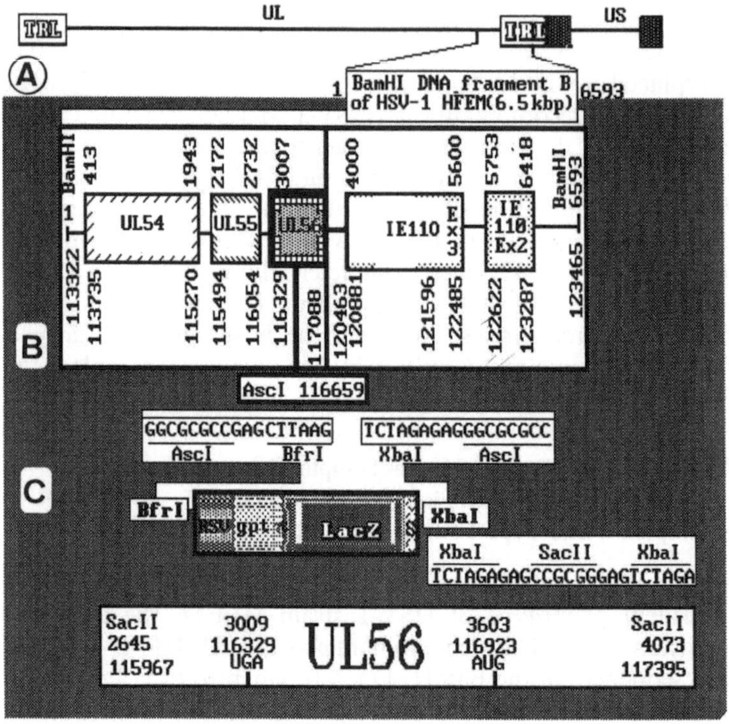

Fig. 3A–C. The construction of the recombinant plasmid pHSHF–BB–LacZ+UL56/17. **A** The position of the BamHI DNA fragment B within the genome of HSV-1. **B** The arrangement of the BamHI DNA fragment B of avirulent HSV-1 strain HFEM is indicated by a *box* in which the position of the HSV-1 genes is marked. The nucleotide positions (np) within the HSV-1 strain 17 genome (McGeoch et al. 1988) are indicated *below the line* and the nucleotide positions within the BamHI DNA fragment B of HSV-1 strain HFEM (Rösen-Wolff et al. 1992) *above the line*. **C** The position of the *AscI* site (np 116659) is magnified. Into this restriction site, a DNA linker containing restriction sites for *Asc*I–*Bfr*I–*Xba*I–*Asc*I was inserted and afterwards a LacZ-cassette was inserted as a BfrI–XbaI DNA fragment. The resulting recombinant plasmid was linearized with XbaI and a DNA linker containing restriction sites for *Xba*I–*Sac*II–*Xba*I was inserted. Into this newly generated *SacII* site, the *UL56* gene cassette (np 115967-117395) of HSV-1 strain 17 was introduced. The positions of the transcriptional start (np 116923) and stop (np 116329) signals are indicated

Juvenile tree shrews were infected intraperitoneally with an infectious dose of 1×10^6 PFU/animal. After 4 days, the animals developed generalized herpesvirus infections and died. The distribution of the recombinant virus within the organs of the animals was similar to that occurring after infections with wild-type virus. The highest virus titers were found in the livers (3.7×10^6 PFU/g tissue), adrenal glands (4.8×10^6 PFU/g tissue), and brain (8×10^3 PFU/g). These results indicate that the restauration of *UL56* gene expression by insertion of an intact *UL56* cassette of a pathogenic HSV-1 strain into the genome of HSV-1 strain HFEM can restore the pathogenic phenotype of this virus.

Biological Properties of the UL56 Gene

In order to investigate the functional properties of the viral *UL56* gene, it was necessary to express the resulting protein efficiently in vitro. The DNA sequences corresponding to the open reading frame of the *UL56* genes of HSV-1 strains F and 17 were amplified from genomic viral DNA by PCR. The primers used for amplification harbored recognition sites for the restriction endonucleases *Eco*RI and *Xba*I at their 5′ ends, respectively (primer 1, 5′-GCGAATTC*ATG*GCTTCGGAGGCGG-3′; primer 2, 5′-GCTCTAGACTCACAGGGGCGC*TTA*-3′), and correspond to the translational start (ATG) and termination (TAA) signals of the *UL56* ORF. The PCR product (705 bp) was inserted into the *Eco*RI–*Xba*I recognition sites of the bacterial expression vector pMal–c2. As shown in Fig. 4, this procedure allowed the expression of the viral *UL56* gene as a fusion protein with the malE gene encoding maltose-binding protein (MBP) of *E. coli* and subsequent cleavage of the fusion protein with the restriction protease factor Xa. The resulting recombinant plasmids were termed pMal–HSF–*UL56* and pMal–HS17–UL56, respectively. Expression of the fusion protein was carried out in *E. coli* TB-1 cells after induction with 1 mM isopropyl thiogalactopyranoside (IPTG) for 2 h. The induced fusion protein was purified by affinity chromatography using amylose columns and subsequently cleaved with factor Xa. As shown in Fig. 5, the apparent relative molecular masses of the fusion proteins were about 70 kDa. Factor Xa cleaves the fusion protein into two subfragments of 42 kDa (MBP) and 33 kDa (UL56).

In order to investigate the biological properties of the UL56 protein, rabbits were immunized with the purified fusion proteins of HSV-1 strains F and 17, and the resulting sera were screened in immunofluorescence and immunoblot assays. As shown in Fig. 6B, the antiserum raised against the UL56 protein of HSV-1 strain F does not interact with uninfected RC–37 cells. In contrast, a strong cytoplasmic fluorescence was observed (Fig. 6A) with cells which had been infected with HSV-1 strain F (6–16 h p.i.). These results are in agreement with the observation that *UL56* gene transcription takes place at 2–4 h p.i. (Rösen-Wolff and Darai 1991) and that the deduced amino acid sequences of UL56 proteins harbor specific sites for transport into the nucleus.

In order to investigate whether or not *UL56* gene expression is essential for development of the viral pathogenicity in vivo, juvenile tree shrews were immunized twice (intervals of 10–20 days) with 1mg recombinant UL56 protein of HSV-1 strain F. Two weeks after the last immunization, the animals were infected with a lethal dose of pathogenic HSV-1 strain F (3 × 10^6 PFU/animal p.i.). As shown in Table 7, the control animals which had not been immunized with the recombinant UL56 protein developed a lethal generalized infection and died (4–7 days p.i.). In contrast, those animals which had been immunized with recombinant UL56 protein did not develop any signs of disease and survived the HSV-1 infection. These results indicate that antibodies against the UL56 protein can prevent the development of lethal generalized herpes infections in tree shrews. These data underline the importance of the role of the viral UL56 protein in the mechanisms which influence the viral pathogenicity.

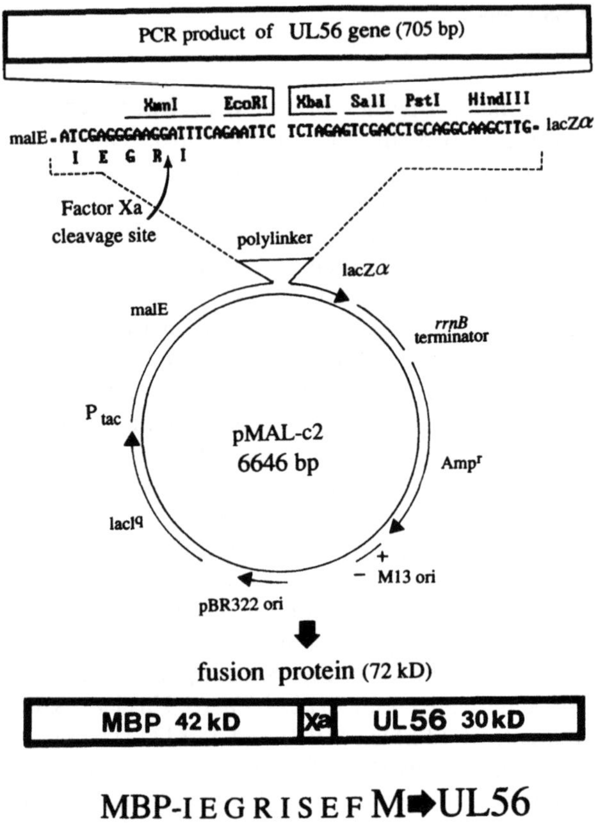

Fig. 4. The construction of the recombinant plasmid pMal–HSF–UL56. The Polymerase chain reaction (*PCR*) product (705 bp) of the UL56 gene was inserted into the *Eco*RI–*Xba*I recognition site of the bacterial expression vector pMal–c2. The position of the factor Xa cleavage site, which enables the separation of the UL56 protein from the maltose-binding protein *MBP* is indicated; the composition of the resulting fusion protein is indicated by *boxes* representing the MBP, the recognition sequence of protease factor *Xa*, and the UL56 protein. *Below the box*, the amino acid sequence of the joint region between MBP and UL56 within the fusion protein is shown

UL56 Gene Does Not Influence the Expression of *UL44* (Glycoprotein C) Gene

In order to answer the question as to whether the expression of *UL44* gene (glycoprotein C, gC) is involved in the mechanisms which influence viral pathogenicity, purified virions of virulent HSV-1 strains F and 17, avirulent HSV-1 strain HFEM, virulent recombinant viruses HSV-R-HFehx-C19, -C26, -CMlC1, and avirulent recombinant virus HSV-1-M-LacZ were analyzed for the presence of gC in western blots. These analyses revealed that all viruses tested were able to express viral gC without any significant differences (Fig. 7).

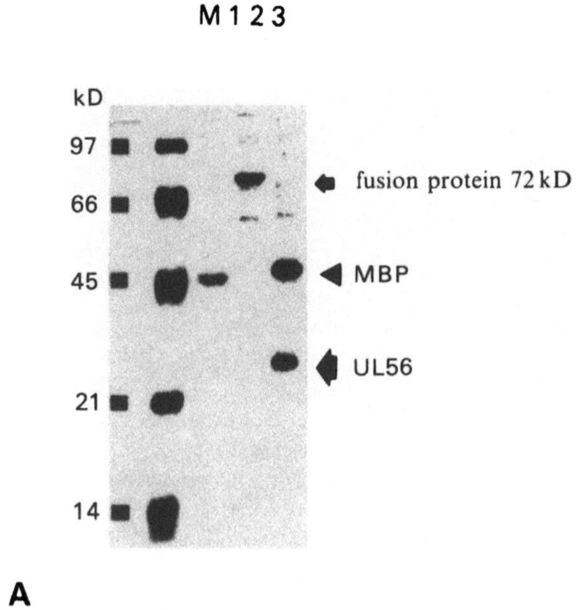

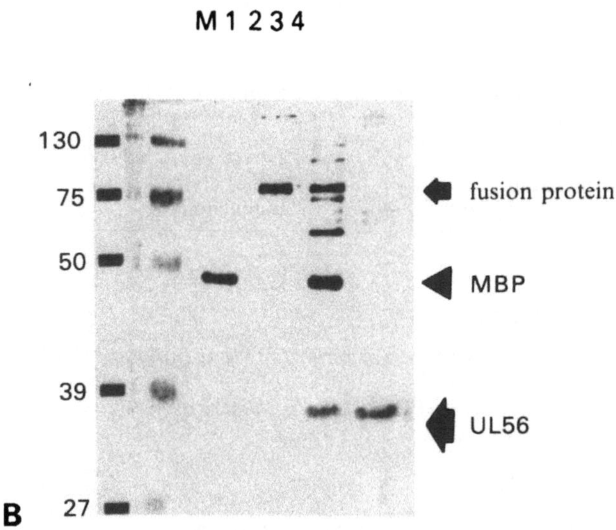

Fig. 5. A Coomassie blue-stained sodium dodecyl sulfate polyacrylamide gel electrophoresis of purified fusion protein expressed from recombinant plasmid pMal–HSF–UL56. *Lane 1*, 1 μg maltose-binding protein (*MBP*); *lane 2*, 5 μg purified *fusion protein* before cleavage with protease factor Xa; *lane 3*, 5 μg purified fusion protein after digestion with protease factor Xa for 20 h at 4°C. *Lane M*, molecular weight markers. The positions of the fusion protein, the MBP, and the UL56 protein are indicated by *arrows*. **B** The detection of UL56 antigen in HSV-1 virions by immunoblot analysis using rabbit antiserum (1:50 dilution) raised against purified recombinant fusion protein expressed from pMal–HSF–UL56. *Lane 1*, 5 μg maltose-binding protein (*MBP*); *lane 2*, 5 μg purified *fusion protein* expressed from pMal–HSF–UL56 before cleavage with protease factor Xa; *lane 3*, 5 μg purified fusion protein expressed from pMal–HSF–UL56 after cleavage with factor Xa for 20 h at 4°C; *lane 4*, 1 μg proteins of purified virions of HSV-1 strain F. The positions of the fusion protein, MBP, and UL56 protein are indicated by *arrows*. *Lane M* indicated the position of molecular weight markers

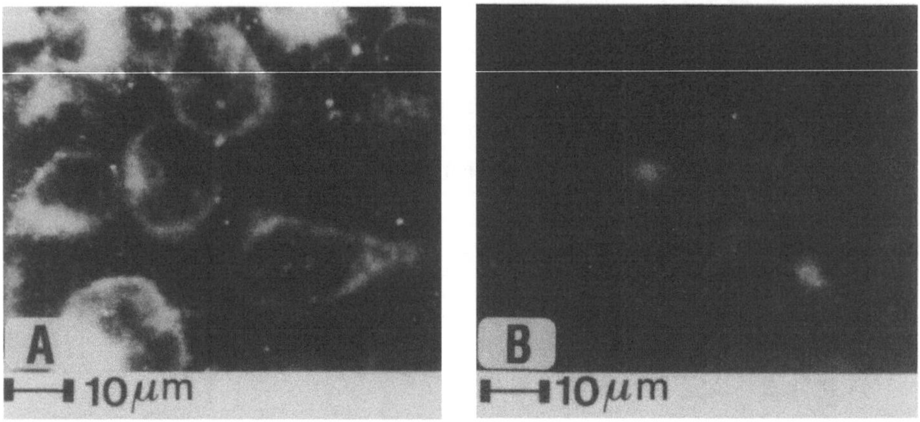

Fig. 6A,B. Photomicrograph of immunofluorescence staining of HSV-1 infected (**A**) and mock-infected (**B**) RC-37 cells 8 h post-infection. A specific cytoplasmic fluorescence of UL56 antigens in RC-37 cells stained by the indirect immunofluorescence antibody test. Cells were infected with HSV-1 strain F at an multiplicity of infection (M.O.I) of 50 PFU/cell and fixed in acetone. Incubation was carried out with 1:10 diluted rabbit antiserum raised against purified recombinant fusion protein expressed from pMal–HSF-UL56. After extensive washing, the slides were incubated with fluorescein isothiocyanate conjugated anti-rabbit immunoglobulin G. Microphotographs were taken under UV light. Magnification, × 1000

Table 7. Immunization of juvenile tree shrews with recombinant UL56 protein of HSV-1 strain F and challenge with virulent HSV-1 strain F

Recombinant UL56 protein[a]	Inoculated virus strain (1 × 10 PFU i.p.)[b]	No. of animals survived/ no. of animals inoculated
–	HSV-1 F	0/4
+	HSV-1 F	4/4

i.p., intraperitoneally
[a] The juvenile tree shrews were immunized twice with 1 mg recombinant UL56 protein of HSV-1 strain F subcutaneously.
[b] The immunized tree shrews were inoculated with HSV-1 strain F 2 weeks after the second administration of recombinant UL56 protein.

Identification of Genes of Herpes Simplex Virus Type 2 Which Influence Viral Pathogenicity

Intraperitoneal Virulence of Herpes Simplex Virus Type 2 is Influenced by Ribonucleotide Reductase Activity

In order to identify those genes of HSV-2 which are involved in the virulence phenotype of HSV-2, juvenile and adult tree shrews and Balb/c mice were infected intraperitoneally with 1×10^5 PFU wild-type strain HSV-2 3345 and 1×10^6 PFU

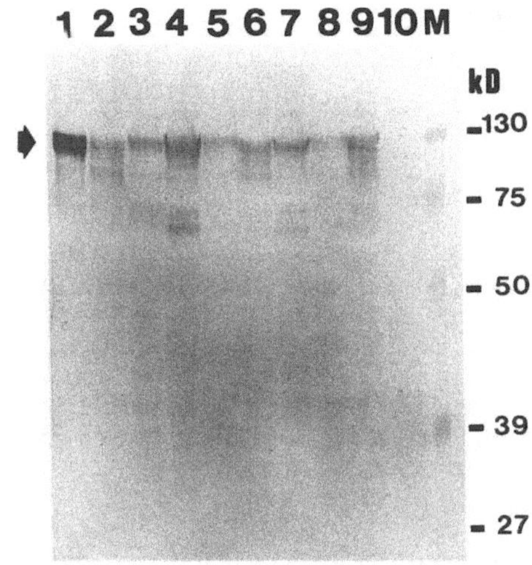

Fig. 7. Detection of *UL44* gene (glycopotein C) expression of a variety of pathogenic and apathogenic HSV-1 strains and recombinant viruses using immunoblot analysis. Proteins of purified virions of pathogenic HSV-1 strain ANG (*lane 1*); pathogenic HSV-1 strain F (*lane 2*); pathogenic HSV-1 strain 17 syn (*lane 3*); apathogenic HSV-1 strain HFEM (*lane 4*); pathogenic recombinant virus HSV-R-Ml-C1 (*lane 5*); apathogenic recombinant virus HSV-1-LacZ-C11-9 (*lane 6*); apathogenic recombinant virus HSV-R-HFehx-C15 (*lane 7*); pathogenic recombinant virus HSV–R–HFehx–C19 (*lane 8*); pathogenic recombinant virus HSV-R-HFehx-C26 (*lane 9*); and apathogenic HSV-1 strain MP (*lane 10*). The anti-HSV-1 UL44 monoclonal antibody gC-1-IV.8.2 (Kühn et al. 1990) was used to detect glycoprotein C expression

of several intertypic HSV-2 × HSV-1 recombinant viruses (Halliburton et al. 1980) and intratypic recombinant viruses, respectively (Rösen and Darai 1985b). The results of these experiments, which were performed as described previously (Rösen et al. 1985; Rösen and Darai 1985b), are summarized in Table 8.

Those animals which survived an acute infection of HSV-2 were killed 3 months after infection, and different organs were screened for the presence of latent virus. The analysis was carried out by cocultivation assays using RC-37 cell

Table 8. Biological behavior of HSV-2-R21 compared with the wild-type strain HSV-2 3345

	Strain of virus	
Properties	HSV-2 3345	HSV-2-R21
Pathogenicity[a]:		
Tree shrew	+	–
Mice	+	–
Latency[b]:		
Tree shrew	Ganglia	Ganglia
Mice	Ganglia	Ganglia

[a] Juvenile animals were infected with 1×10 PFU of each virus intraperitoneally. The animals died of generalized herpesvirus infections (3-7 days postinfection) or survived without any signs of disease, respectively.

[b] Latent virus could be reactivated from spinal cord ganglia.

Table 9. Determination of ribonucleotide reductase activity of HSV-2-R21 compared with the wild-type strain HSV-2 3345

Extract of cells infected	Protein/assay (μg)	RR activity (pmol H dCTP formed/h)
HSV-2-R21	250	9.5
HSV-2 3345	250	281.2
Mock-infected	250	8.3

dCTP, deoxycytidine triphosphate; RR, ribonucleotide reductase, M.O.I., multiplicity of infection

[a] To determine the RR activity, RC-37 cells were infected at an M.O.I. of 5 PFU for each virus per cell. Cells were harvested 24 h postinfection and resuspended in a minimal volume of 50 m*M* HEPES, pH 7.6, 2 m*M* dithiothreitol. RR activity was measured by the method of Darling et al. (1987). Assays were performed three times; and average value of all three assays is given.

cultures, as described elsewhere (Rösen and Darai 1985b). The reactivated viruses which were isolated from the ganglia of latently infected animals were screened for their virulence phenotype in tree shrews and mice. It was found that the reisolated virus HSV-2-R21 had completely lost its virulent phenotype in both animal model systems tested. The results are summarized in Table 9.

In the second step of the experiments, those animals which survived an infection with the reisolated virus HSV-2-R21 were killed 3 months after infection, and different organs including brain, spinal cord ganglia, and spleen were screened for the presence of the latent virus. As shown in Table 8, it was found that HSV-2-R21 latently persisted in the ganglia of infected animals. These results indicated that HSV-2-R21 was able to colonize the ganglia of latently infected animals, but unexpectedly lost the ability to develop acute infection by the intraperitoneal administration route. Therefore, it was of particular interest to characterize HSV-2-R21 genotypically in more detail, with the goal of identifying those genomic alterations which led to the development of the avirulent phenotype of this virus.

In order to characterize HSV-2-R21 genotypically, the DNA cleavage patterns (*Bam*HI, *Bgl*II, *Eco*RI, *Hin*dIII, *Hpa*I, *Cla*I, *Kpn*I, and *Xba*I) of the genome of HSV-2-R21 were analyzed and compared with the cleavage patterns of the virulent HSV-2 strain 3345. This analysis revealed that the DNA cleavage patterns of the avirulent HSV-2-R21 and the virulent HSV-2 strain 3345 were identical for all restriction enzymes tested with the exception of *Kpn*I. An example of this analysis is shown in Fig. 8.

The cleavage pattern of HSV-2-R21 DNA which resulted after digestion with the restriction endonuclease *Kpn*I showed one significant difference from that of HSV-2 3345, as shown in Fig. 8 (lanes 3 and 4, marked with an arrow and asterisks). It was found that the KpnI DNA fragment D (18.9 kbp; 0.48–0.62 map units of the HSV-2 genome; Cortini and Wilkie 1978) was not present in the cleavage pattern of HSV-2-R21 (Fig. 8, lane 3). In contrast, two novel DNA fragments of 10.2 and 8.7 kbp were detected (Fig. 8, lane 3, 4 marked with asterisks) which comigrate with the KpnI HSV-2 DNA fragments G (10.2 kbp) and H (8.7 kbp).

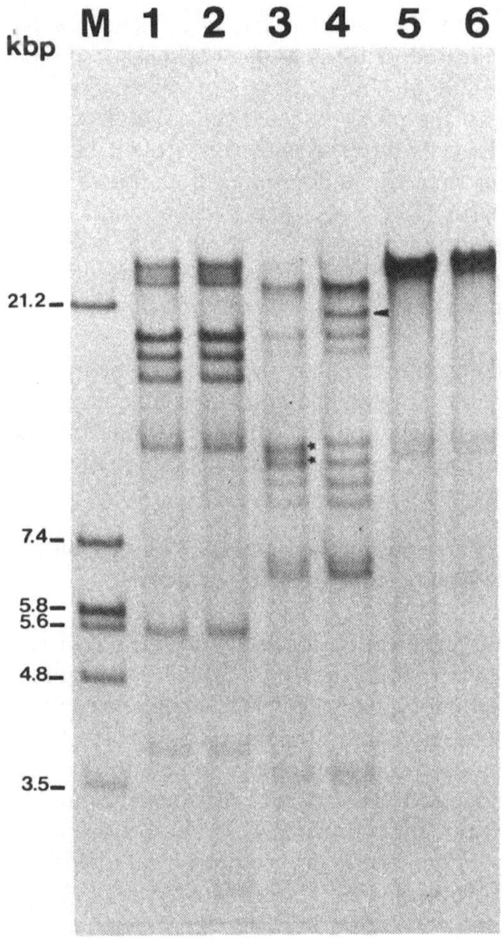

Fig. 8. Autoradiogram of ^{32}p-labeled DNAs of the reisolated virus HSV-2-R21 (*lanes 1,3, and 5*) and HSV-2 3345 (*lanes 2,4, and 6*) digested with *Cla*I (*lanes 1 and 2*), *Kpn*I (*lanes 3 and 4*), and *Hpa*I (*lanes 5 and 6*), respectively. Phage lambda-DNA digested with *Eco*RI (*M*) served as molecular weight marker. The position of the KpnI DNA fragment D of HSV-2 3345 is indicated by an *arrow* and the positions of the two subfragments of HSV-2-R21 are indicated by *asterisks*

These genomic variations indicated that the genetic information between the map coordinates 0.48 and 0.62 must have undergone some alterations. Due to these alterations, an additional KpnI recognition site must have been created which led to the division of the KpnI HSV-2 DNA fragment D (18.9 kbp) into two subfragments of 10.2 and 8.7 kbp. This alteration was the only detectable genotypic variation between the avirulent HSV-2-R21 and the virulent HSV-2 3345. Since no alteration was detected in the DNA cleavage patterns of all other restriction enzymes tested, it is possible that only a point mutation occurred in this area of the viral genome. This indicates that the avirulent phenotype of HSV-2-R21 could be due to the elimination of the gene functions located in this particular region of the virus.

Since the suspected genomic region (0.48–0.62 map units) of HSV-2-R21 corresponds to the DNA sequences located within the KpnI HSV-2 DNA fragment D coding for the viral ribonucleotide reductase (Dutia 1983), ribonucleotide reductase activity of the two virus strains was determined by the method of

Darling et al. (1987). In this study, RC-37 cell cultures (Rösen and Darai 1985b) were infected with an M.O.I. of 5 PFU/cell of each virus. The infected cells were harvested (24 h p.i.) and ribonucleotide reductase activity was measured. As shown in Table 9, the avirulent HSV–2–R21 showed levels of ribonucleotide reductase activity which were similar to the mock-infected cells. The levels of ribonucleotide reductase activity of the cells infected with the virulent HSV-2 strain 3345 were about 30 times higher than the background level. These data indicate that the avirulent reisolated virus HSV-2-R21 expressed no viral ribonucleotide reductase. These results are in agreement with results obtained for HSV-1, which showed that ribonucleotide reductase activity plays a significant role in the mechanisms influencing the HSV-1 virulence phenotype (Dutia et al. 1986).

In conclusion, the results presented here indicate that the avirulent phenotype of HSV-2-R21 which is able to colonize the ganglia of latently infected animals is due to the genomic alterations between the genome coordinates 0.48 and 0.62, leading to the abolishment of the expression of the viral ribonucleotide reductase.

Influence of the Genetic Information Located within the BamHI DNA Fragment V of the Herpes Simplex Virus Type 2 Genome on Viral Pathogenicity

Recently, it was shown that the genetic information located within the BamHI DNA fragment V (0–0.02 and 0.81–0.83 map units) of the HSV-2 HG52 genome was responsible for the neurovirulent phenotype of this virus (Taha et al. 1988; 1989a,b). Therefore, the DNA nucleotide sequence of the BamHI–Asp718 DNA fragment which harbors the DNA sequences between the genome coordinates 0.81 and 0.84 of the genome of the pathogenic HSV-2 strain 186 were determined. The analysis revealed that this DNA fragment comprises 4799 nucleotides with an extremely high GC content (77.1%). The genetic information located within this genomic region was suspected to harbor the HSV-2 correlate to the *ICP34.5* gene of HSV-1 (McGeoch and Barnett 1991). The *ICP34.5* gene seems to be involved in the mechanisms which influence the neurovirulent phenotype of HSV-1 (Dolan et al. 1992). In order to identify the presence of an HSV-2 *ICP34.5* correlate, a computer-assisted analysis of the DNA nucleotide sequence of the BamHI/Asp718 DNA fragment was performed. This analysis revealed that a possible ORF corresponding to the *ICP34.5* gene of HSV-1 strains F or 17syn+ could be located at this particular region of the genome of HSV-2. However, the sequences are interrupted by a cluster of 16 repetitions (ACCCCCTGACCCG) which terminate the ORF in all three possible frames. Similar genomic features had been reported for the genome of HSV-2 strain HG52 (McGeoch et al. 1991).

Discussion

The genetic information located between the genome coordinates 0.7 and 0.8 has already been documented to be involved in HSV-1 pathogenicity using the animal model systems rabbit (Centifanto-Fitzgerald et al. 1982), mice (Thompson et al.

1983), and tree shrews (Scholz et al. 1983). In our studies, we succeeded in demonstrating that the ability of avirulent HSV-1 strain HFEM to induce generalized herpesvirus infections in tree shrews was restored by the replacement of the deleted DNA sequences localized within the BamHI DNA fragment B of the viral genome.

We have shown that the avirulent deletion mutant HSV-1 HFEM, whose genome possesses a deletion corresponding to the nucleotide area of the genome of HSV-1 strain 17 between 117088 and 120641, could never be reactivated from nervous tissue of latently infected tree shrews. However, this strain shows a new phenotypic property reflected in the ability to persist as latent virus in the spleen cells of infected animals. Some recombinant viruses generated by replacing the deleted sequences within the genome of HSV-1 strain HFEM with the corresponding DNA sequences of the pathogenic HSV-1 strain F regained the ability to colonize ganglia.

In order to detect the genes involved in or regulating the viral pathogenicity and latency of HSV-1, a detailed analysis of the transcriptional activity of this particular region was carried out. A novel 1.5-kb RNA transcript was detected which seems to play a functional role in the process of the viral pathogenicity and latency of HSV-1. The identity of this 1.5-kb RNA to the *UL56* gene of HSV-1 has been demonstrated (Rösen-Wolff and Darai 1991). The absence of the *UL56* transcript in HSV-1 HFEM-infected cells is due to the fact that the promoter region of the *UL56* gene is located within those DNA sequences which are deleted in the HSV-1 HFEM genome.

The use of modern molecular biological tools and genetic engineering allowed a variety of laboratories to obtain interesting results concerning localization of those genes which seem to be essential for pathogenicity and latency of HSV-1. However, until now it was not possible to identify the gene and/or gene cascade which is indeed involved in the mechanisms determining and regulating the viral pathogenicity, organotropism, and latency of HSV-1. Therefore, more attempts and unconventional experimental strategies are required in future for clarifying the complex mechanism of pathogenicity and latency of HSV.

The genetic analysis of the BamHI DNA fragment B of HSV-1 strain HFEM revealed that in addition to the described alterations of the transcriptional activity of the *UL56* gene, other novel transcripts might exist which correspond to the deleted region of the HSV-1 HFEM genome (Rösen-Wolff and Darai 1991). The structural and functional analysis of these transcripts is one of the ways for further investigation of HSV-1 activity in vivo.

HSV-1 possesses several regions on its genome which have been shown to be involved in the mechanisms determining the virulence of this virus (Centifanto-Fitzgerald et al. 1982; Thompson et al. 1983; Rösen and Darai 1985a; Rösen et al. 1986; Chou and Roizman 1990). In contrast, our knowledge on the genes responsible for the virulence phenotype of HSV-2 is limited. Therefore, the identification of genes involved in the virulence and latency of HSV-2 and the elucidation of their functions was of special importance. As shown in this report, viral genes such as ribonucleotide reductase which are involved in the replication of the virus have been identified as influencing the virulence of HSV-2. The DNA sequence analysis of the genome of HSV-2 between genome coordinates 0.81 and

0.84 revealed that this particular region is able to encode an amino acid sequence homologous to *ICP34.5* gene product of HSV-1. However, the corresponding ORF was interrupted due to the existence of a cluster of repetitions. This phenomenon had been discussed by McGeoch et al. (1991) according to the data found by the sequence analysis of the genome of HSV-2 strain HG52. These investigators discuss the possibility that the detected repetitions might be introns which are located within the coding DNA sequences of this particular gene. However, the real existence of this putative HSV-2 protein has not been demonstrated up to now.

The eventual functional activity of the HSV-2 gene analogous to *UL56* gene of HSV-1 and its influence in the process of the viral pathogenicity of HSV-2 is another aspect which merits more attention in future. This investigation is of special importance for the understanding of the pathomechanisms and virus–host interactions of HSV-2.

Acknowledgement. This study was supported by the Deutsche Forschungsgemeinschaft, Schwerpunktprogramm "Persistierende Virusinfektionen: Molekulare Mechanismen und Pathogenese" (Project II B6–Da 142/1-1 to 1-5) and Schwerpunktprogramm "Virulenzfaktoren und Wirtstropismus animaler Viren" (project II B6–Da 142/6-1).

References

Centifanto–Fitzgerald ZM, Yamaguchi T, Kaufman HE, Togon M, Roizman BJ (1982) Ocular disease pattern induced by herpes simplex virus is genetically determined by a specific region of viral DNA. J Exp Med 155: 475–489

Chou J, Roizman B (1990) The herpes simplex virus 1 gene for ICP34.5, which maps in inverted repeats, is conserved in several limited-passage isolates but not in strain 17syn+. J Virol 64: 1014–1020

Cook ML, Stevens JG (1976) Latent herpetic infections following experimental viraemia. J Gen Virol 31: 75–80

Cortini R, Wilkie NM (1978) Physical maps of HSV type 2 with five restriction endonucleases. J Gen Virol 39: 259–280

Darai G, Scholz J (1984) Latency of herpes simplex virus in tree shrew; generation of intratypic recombinants of herpes simplex virus in vivo. Latent herpes virus infections in veterinary medicine. Nijhoff, Boston, pp 21–32

Darai G, Schwaier A, Komitowski D, Munk K (1978) Experimental infection of tupaia belangeri (tree shrews) with herpes simplex virus type 1 and 2. J Infect Dis 137: 221–226

Darai G, Matz B, Schröder CH, Flügel RM, Berger U, Munk K, Gelderblom H (1979) Characterization of a tree shrew herpesvirus isolated from a lymphosarcoma. J Gen Virol 43: 541–551

Darai G, Matz B, Flügel RM, Grafe A, Gelderblom H, Delius HJ (1980a) An adenovirus from tupaia (tree shrew), growth of the virus, characterization of viral DNA, and transformation ability. J Virol 104: 122–138

Darai G, Zöller L, Matz B, Schwaier A, Flügel RM, Munk K (1980b) Experimental infection and the state of viral latency of adult tupaias with herpes simplex virus type 1 and 2 and infection of juvenile tupaia with temperature-sensitive mutants of HSV type 2. Arch Virol 65: 311–318

Darai G, Scholz J, Kurz W, Koch HG (1982) Intertypic superinfection of herpes simplex virus in vivo leads to reisolation of latent viruses with newly acquired lymphotropic phenotype. Dev Biol Stand 52: 67–80

Darai G, Scholz J, Koch HG (1983a) Development of an in vivo model system (tree shrew) for investigation of organotropic function of herpes simplex virus. Viral and immunological diseases in primates. Liss, New York, pp 213–216

Darai G, Rösen A, Scholz J, Gelderblom H (1983b) Induction of generalised and lethal herpesvirus infection in the tree shrew by intrahepatic transfection of herpes simplex virus DNA. J Virol Methods 7: 305–314

Darling AJ, Dutia BM, Marsden HS (1987) Improved method for the measurement of ribonucleotide reductase activity. J Virol Methods 180: 281–290

Dolan A, McKie E, MacLean AR, McGeoch DJ (1992) Status of the ICP34.5 gene in herpes simplex virus type 1 strain 17. J Gen Virol 73: 971–973

Dutia BM (1983) Ribonucleotide reductase induced by herpes simplex virus has a virus-specified constituent. J Gen Virol 64: 513–521

Dutia BM, Frame MC, Subak-Sharpe JH, Clarke WN, Marsden HS (1986) Specific inhibition of herpes virus ribonucleotide reductase by synthetic peptides. Nature 321: 439–441

Fiedler W (1956) Übersicht über das System der Primaten. Primatology 1: 1–266

Goodpasture EW, Teague O (1923) Transmission of herpes febrilis along nerves in experimentally infected rabbits. J Med Res 44: 139–184

Halliburton IW, Morse LS, Roizman B, Quinn KE (1980) Mapping of the thymidine kinase genes of types 1 and 2 herpes simplex virus using intertypic recombinants. J Gen Virol 49: 235–253

Koch HG, Rösen A, Ernst F, Becker Y, Darai G (1987) Determination of the nucleotide sequence flanking the deletion (0.762–0.789 map units) in the genome of an intraperitoneally avirulent HSV-1 strain HFEM. Virus Res 7: 105–115

Kühn JE, Kramer MD, Willenbacher W, Wieland U, Lorentzen EU, Braun RW (1990) Identification of herpes simplex virus type 1 glycoproteins interacting with the cell surface. J Virol 64: 2491–2497

MacLean AR, Brown SM (1987) Deletion and duplication variants around the long repeats of herpes simplex virus type 1 strain 17. J Gen Virol 68: 3019–3031

McGeoch DJ, Barnett BC (1991) Neurovirulence factor. Nature 353: 609

McGeoch DJ, Dalrymple MA, Davison AJ, Dolan A, Frame MC, McNab D, Perry L, Scott JE, Taylor P (1988) The complete DNA sequence of the long unique region in the genome of herpes simplex virus type 1. J Gen Virol 69: 1531–1574

McGeoch DJ, Cunningham C, McIntyre G, Dolan A (1991) Comparative sequence analysis of the long unique and adjoining parts of the long unique regions in the genomes of herpes simplex virus types 1 and 2. J Gen Virol 72: 3057–3075

Moore WG, Goodman M (1968) A set theoretical approach to immunotaxonomy: analysis of species comparisons in modified Ouchterlony plates. Bull Math Biophys 30: 279–289

Moyal M, Raibstein I, Rösen A, Darai G, Becker Y (1988) Heterogeneity of BamHI DNA fragments B and E in several HSV-1 strains and recombinants. Virus Genes 1: 165–174

Plummer G, Hollingsworth DC, Phuangsab A, Bowling CP (1970) Chronic infections by herpes simplex viruses and by the horse and cat herpesviruses. Infect Immunol 1: 351–355

Rösen A, Darai G (1985a) Mapping of the deletion in the genome of HSV-1 strain HFEM responsible for its avirulent phenotype. Med. Microbiol Immunol 173, 329–343.

Rösen A, Darai G (1985b) Investigation of the virulence genes of herpes simplex virus 2 by experimental infections in vivo with defined intratypic recombinants of HSV-2 × an avirulent HSV-1. Med Microbiol Immunol 174: 237–248

Rösen A, Gelderblom H, Darai G (1985) Transduction of virulence in herpes simplex virus type 1 from a pathogenic to an apathogenic strain by a cloned viral DNA fragment. Med Microbiol Immunol 173: 257–278

Rösen A, Ernst F, Koch HG, Gelderblom H, Darai G, Hadar J, Tabor E, Ben-Hur T, Becker Y (1986) Replacement of the deletion in the genome (0.762–0.789 mu) of avirulent HSV-1 HFEM using cloned MluI DNA fragment (0.7615-0.796 mu) of virulent HSV-1 F leads to generation of virulent intratypic recombinant. Virus Res 5: 157–175

Rösen-Wolff A, Ben-Hur T, Becker Y, Darai G (1988) Comparative analysis of the transcripts mapped in the BamHI DNA fragment B of avirulent HSV-1 HFEM, virulent HSV-1 F, and their intratypic recombinant viruses. Virus Res 10: 315–324

Rösen-Wolff A, Darai G (1991) Identification and mapping of the UL56 gene transcript of herpes simplex virus type 1. Virus Res 19: 115–126

Rösen-Wolff A, Scholz J, Darai G (1989) Organotropism of latent herpes simplex virus type 1 is correlated to the presence of a 1.5 kb RNA transcript mapped within the BamHI DNA fragment B (0.738 to 0.809 map units). Virus Res 12: 43–52

Rösen-Wolff A, Lamade W, Berkowitz C, Becker Y, Darai G (1991) Elimination of UL56 gene by insertion of LacZ cassette between nucleotide position 116030 to 121753 of herpes simplex virus type 1 abrogates intraperitoneal pathogenicity in tree shrews and mice. Virus Res 20: 205–221

Rösen-Wolff A, Frank S, Raab K, Moyal M, Becker Y, Darai G (1992) Determination of the coding capacity of the BamHI DNA fragment B of apathogenic herpes simplex virus type 1 strain HFEM by DNA nucleotide sequence analysis. Virus Res 25: 189–199

Scholz J, Rösen A, Darai G (1983) Characterization of the genome of herpes simplex virus reisolates generated and isolated after intertypic superinfection of HSV-1 and 2 in vivo. Zentralbl Bakteriol Mikrobiol Hyg I [A] 254:183

Scriba M (1977) Extraneural localisation of herpes simplex virus in latently infected guinea pigs. Nature 267: 529–531

Stevens JG (1975) Latent herpes simplex virus and the nervous system. In: Arber W et al. (eds) Current topics in microbiology and immunology, vol 70. Springer, Berlin Heidelberg New York, pp 31–50.

Stevens JG, Cook ML (1971) Restriction of herpes simplex virus by macrophages. An analysis of the cell-virus interaction. Science 173: 843–845

Taha MY, Brown SM, Clements GC (1988) Neurovirulence of individual plaque stocks of herpes simplex virus type 2 strain HG 52. Arch Virol 103: 15–25

Taha MY, Clements GC, Brown SM (1989a) A variant of herpes simplex virus type 2 strain HG52 with a 1.5 kb deletion in RL between 0 to 0.02 and 0.81 to 0.83 map units is non-neurovirulent for mice. J Gen Virol 70: 705–716

Taha MY, Clements GC, Brown SM (1989b) The herpes simplex virus type 2 (HG52) variant JH2664 has a 1488 bp deletion which eliminates neurovirulence in mice. J Gen Virol 70: 3073–3078

Thompson RL, Wagner EK, Stevens JG (1983) Physical location of a herpes simplex virus type-1 gene function(s); specifically associated with a 10-million-fold increase in HSV virulence. Virology 131: 180–192

Chapter 11 Computer Analysis of the Protein Coded by Herpes Simplex Virus Type 1 *UL56* Gene

Mirta Grifman and Yechiel Becker

Summary

The *UL56* gene of herpes simplex virus (HSV) was studied with the aid of computer programs. These programs analyze the protein amino acid sequence and determine the hydrophilicity, surface probability, flexibility, and antigenicity index. We were able to recognize three different putative domains within this protein which may suggest a role for the *UL56* gene product in HSV-1 intraperitoneal infection in mice.

Introduction

The genetic information in herpes simplex virus type 1 (HSV-1) DNA coordinates 0.7 and 0.8 mu correlates with the viral pathogenicity by different routes of inoculation (Thompson et al. 1983). The HSV-1 strain HFEM, which harbors a deletion between nucleotide positions 117088 and 120641, has been shown to be apathogenic in mice (Becker et al. 1986) and tree shrews (Darai et al. 1983) when injected by the intraperitoneal route (Fig. 1). In addition, this strain lost the capacity to be reactivated from the nervous tissues of latently infected animals. The replacement of the deleted sequences within the genome of HSV-1 strain HFEM with the corresponding sequences of the pathogenic HSV-1 strain F resulted in the restoration of viral pathogenicity and the ability to colonize ganglia (Rösen et al. 1986).

It has recently been shown that *UL56* gene expression was missing in HFEM-infected cells. This was due to the deletion of the UL56 promoter in the HFEM genome (Rösen-Wolff and Darai 1991). The expression of the *UL56* gene has been shown to correlate with the pathogenicity and latency phenotype of several HSV-1 strains (Rösen-Wolff et al. 1988). In order to determine whether the lack of the *UL56* gene leads to a change in pathogenicity, the recombinant virus HSV-1-M-LacZ was constructed, in which the *UL56* gene of HSV-1 F was replaced by the *Escherichia coli LacZ* cassette. Although it has been proven to be apathogenic in mice and tree shrews by the intraperitoneal administration route, HSV-1-M-LacZ

Department of Molecular Virology, Faculty of Medicine, Hebrew University of Jerusalem, Jerusalem, Israel

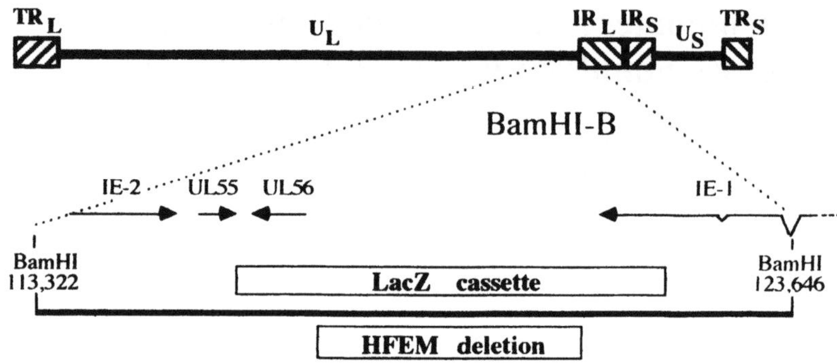

Fig. 1. HSV-1 genome; the BamHI fragment B is enlarged indicating the location of *IE-1*, *IE-2*, *UL55*, and *UL56* genes, the insertion of the *LacZ* cassette and the deletion in the HFEM genome

is able to replicate in cell cultures and attain similar titers to that of wild-type virus (Rösen-Wolff et al. 1991). Moreover, the recombinant HSV-1-M-LacZ is unable to replicate in the adrenal glands of intraperitoneally infected mice, which suggests that the *UL56* gene might be important for HSV-1 replication in the adrenal glands.

We present a computer analysis of the predicted amino acid sequence of the UL56 protein and suggest a function for the *UL56* gene product, according to its putative structure.

Computer Analysis

The complete nucleotide sequence of HSV-1 syn17 was determined and published (McGeoch et al. 1988). It was later found that an error existed in the published sequence, which fell within the *UL56* gene. Two nucleotides, CG, should be inserted after residue 116343 of the complete HSV-1 syn17 sequence, as numbered by McGeoch et al. (1991). It had already been reported by Rösen-Wolff et al. (1991) that HSV-1 strain F possesses two nucleotides, AG, at the same position, but this was interpreted as a mutation specific to HSV-1 F. The insertion of two nucleotides leads to a frameshift of the deduced amino acid sequence. The effect of this change on the amino acid sequence as predicted from the *UL56* gene is the removal of the last two amino acids and the addition of 40 amino acids. This additional region contains an uncharged and highly hydrophobic section of 18 amino acids.

The sequence of the *UL56* gene (nucleotide positions 116923-116264) was translated into the amino acid sequence using the program TRANSLATE of the UWGCG software package (Devereaux et al. 1984). The secondary structure of the UL56 protein was predicted by the program PEPTIDESTRUCTURE (Wolf et al. 1988). It was found (Fig. 2) that the UL56 protein contains three domains of interest. The first is a proline-rich domain; the second is a large and very positively charged α-helix domain; and the third is a hydrophobic region located at the C terminus of the protein.

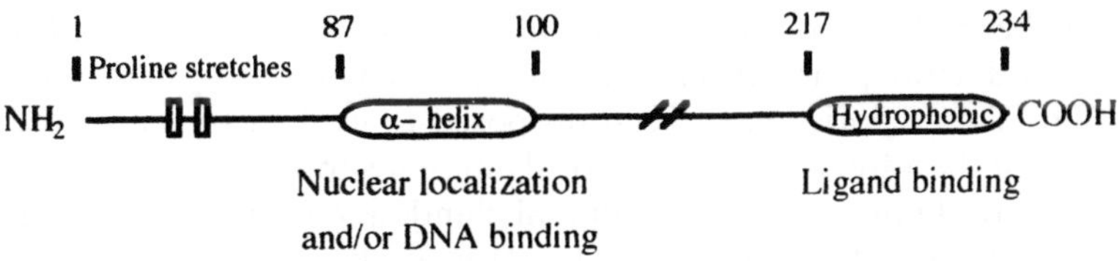

Fig. 2. Computer-predicted secondary structure of the UL56 protein

Table 1. Computer-predicted domains in HSV-1 UL56 protein

Domain	Amino acid position (number of amino acids)	Hydrophilicity per amino acid (average)	Surface probability per amino acid (average)	Secondary structure CF	GOR
1	22–25 (4)	1.48	3.01	T	T+C
2	45–51 (7)	1.53	2.92	T	T+C
3	87–100(14)	2.33	3.79	H	H+T
4	217–234(18)	–2.84	0.23	B	B+C

CF, Chou-Fasman; GOR, Garnier-Osguthorpe-Robson (Algorithms): T, β-turn; H, α-helix; B, β-sheet; C, coil.

The N-terminal region of the UL56 protein contains 23% proline (15 of 64 amino acids) and it is negatively charged as well. These features are common to the activator domains of many transcription factors (Bohman and Tjian 1989). Within this region, there are two repeats of four prolines (see Table 1 domains 1 and 2).

The basic region spans from amino acids 87 to 100. This α-helix has a very high surface probability, suggesting that this sequence is on the outside surface of the folded protein (see Table 1). Such basic "motifs" have been implicated in nuclear targeting (Dingwall and Laskey 1986; Kalderon et al. 1984). Many nuclear transport signals have been defined by mutational analysis and by their ability to direct nuclear localization of nonnuclear proteins (Yamasaki et al. 1989). These signals are characteristically short amino acid sequences with a high number of basic residues, which are often associated with a proline. They can reside in any exposed region of the protein and can occur more than once in a given protein (Wagner et al. 1990). Such a nuclear localization signal (NLS) has already been identified for the IE175 (ICP4) protein of HSV-1 (de Luca and Schaffer 1988), which is known to translocate into the nucleus and is involved in transcriptional activation.

The third region is an uncharged and highly hydrophobic section located at the C terminus of the protein. The hydrophobic region spans from amino acids 217 to 234. This structure could function as a transmembrane anchor domain, although the absence of a signal peptide sequence at the N terminus makes this improbable.

It has already been observed that the adrenal glands are the primary site for viral replication and viral penetration into the central nervous system after systemic infection. Two theories have been proposed to explain the enhanced tropism of HSV-1 for the adrenal glands: (1) local immunosuppression caused by the high levels of steroid hormones in the adrenal glands and (2) the ectodermal embryonic origin (the neural crest) of the adrenal medulla, which is an origin common to most of the tissues infected by HSV-1 (Peles et al. 1990). Another possible interpretation is that the hormone-rich environment is utilized by HSV-1 to activate its own gene expression through the nuclear receptor pathway.

The nuclear receptor gene family includes a large number of genes encoding receptors for various hydrophobic ligands such as steroids, vitamin D, retinoic acid, and thyroid hormones. Upon ligand binding, nuclear receptors bind to the DNA target sequence and enhance gene expression (Green and Chambon 1988). Nuclear receptors are composed of several domains important for transcriptional activation, DNA binding, ligand binding, and dimerization (Laudet et al. 1992).

The combination of a basic domain, proline stretches and a hydrophobic domain in the UL56 protein resembles the composition of hormone-dependent transcriptional activators.

In conclusion, we propose that the UL56 protein, essential for HSV-1 replication in the adrenal glands, might be a viral product that mimics the function of nuclear receptors, taking advantage of the high concentration of hormones in the adrenal glands to activate viral transcription.

References

Becker Y, Hadar J, Tabor E, Ben-Hur T, Raibstein I, Rösen A, Darai G (1986) A sequence in the *HpaI*-P fragment of herpes simplex virus type-1 DNA determines intraperitoneal virulence in mice. Virology 149: 255–259

Bohmann D, Tjian R (1989) Biochemical analysis of transcriptional activation by Jun: differential activity of c- and v-jun. Cell 59: 709–717

Darai G, Rösen A, Scholz J, Gelderblom H (1983) Induction of generalized and lethal herpesvirus infection in tree shrew by intrahepatic transfection of herpes simplex virus DNA. J Virol Methods 7: 3055

De Luca, NA, Schaffer PA (1988) Physical and functional domains of the herpes simplex virus transcriptional regulatory protein ICP4. J Virol 62: 732–741

Devereaux J, Haeberli P, Smithies O (1984) A comprehensive set of sequence analysis programs for the VAX. Nucleic Acids Res 12: 387–395

Dingwall C, Laskey RA (1986) Protein import into the cell nucleus. Ann Rev Cell Biol 2: 367–390

Green S, Chambon P (1988) Nuclear receptors enhance our understanding of transcription regulation. Trends Genet 4: 309–314

Kalderon D, Roberts BL, Richardson WD, Smith AE (1984) A short amino acid sequence able to specify nuclear localization. Cell 39: 499–509

Laudet V, Hanni C, Coll J, Catzeflis F, Stehelin D (1992) Evolution of the nuclear receptor gene superfamily. EMBO J 11: 1003–1013

McGeoch DJ, Dalrymble MA, Dolan DA, Frame MC, Mc Nab D, Perry LG, Scott JE, Taylor P (1988) The complete DNA sequence of the long unique region in the genome of herpes simplex virus type 1. J Gen Virol 69: 1531–1574

McGeoch DJ, Cunningham C, Mc Intyre G, Dolan A (1991) Comparative sequence analysis of the long repeat regions and adjoining parts of the long unique regions in the genomes of herpes simplex viruses types 1 and 2. J Gen Virol 72: 3057–3075

Peles E, Rosen H, Rösen-Wolff A, Becker Y (1990) Importance of the *HpaI*-P sequence for herpes simplex virus 1 replication in the adrenal glands. Arch Virol 113: 151–163

Rösen A, Ernst F, Koch H, Gelderblom H, Darai G, Hadar J, Tabor E, Ben-Hur T, Becker Y (1986) Replacement of the deletion in the genome (0.7615-0.796) of virulent HSV-1 F leads to generation of virulent intratypic recombinants. Virus Res 5: 157–175

Rösen-Wolff A, Ben-Hur, T, Becker Y, Darai, G (1988). Comparative analysis of the transcripts mapped in the *Bam*HI DNA fragment B of avirulent HSV-1 HFEM, virulent HSV-1 F and their intratypic recombinant viruses. Virus Res 10: 315–324

Rösen-Wolff A, Darai G (1991) Identification and mapping of the UL56 gene transcript of herpes simplex virus type 1. Virus Res 19: 115–126

Rösen-Wolff A, Lamade W, Berkowitz C, Becker Y, Darai G (1991) Elimination of UL56 gene of herpes simplex virus type 1 genome abrogates intraperitoneal pathogenicity in tree shrews and mice. Virus Res 20: 205–221

Thompson RL, Wagner EK, Stevens JG (1983) Physical location of a herpes simplex virus function(s) specifically associated with a ten million-fold increase in HSV neurovirulence. Virology 131: 180–192

Wagner P, Kunz J, Koller A, Hall MN (1990) Active transport of proteins into the nucleus. FEBS Lett 275: 1–5

Wolf H, Modrow S, Motz M, Janeson BA, Hermann G, Fortsch B (1988) An integrated family of amino acids programs. CABIOS 4: 187–191

Yamasaki L, Kanda P, Landford RE (1989) Identification of four nuclear transport signal binding proteins that interact with diverse transport signals. Mol Cell Biol 9: 3028–3036

[illegible]

IV. Genes and Latency and Intracerebral Pathogenicity

Chapter 12 The Herpes Simplex Type 1 Virus Latency Gene

Edward K. Wagner

Summary

The ability of herpesviruses to establish, maintain, and reactivate from latent infections is a hallmark of the group. Latent phase gene expression is under different control than is herpesvirus lytic phase infection. Recent experimental analysis of herpes simplex virus latency indicates that the latent phase is characterized by the expression of a single transcript unit (the latency-associated transcription unit, LAT). Transcripts expressed from this LAT are complexly processed to yield a stable intron and a number of, as yet, not fully characterized polyadenylated forms. One or more of these processed transcripts play an important role in the reactivation process, but the biological mechanism underlying this role is not understood.

Introduction

The ability of herpes simplex virus (HSV) to establish, maintain, and reactivate from the latent state has long been a subject of experimental inquiry in research laboratories throughout the world. A general picture of gene expression in the latent phase as compared to the lytic phase of HSV and other herpesvirus infections has recently been the subject of a general review, while specific aspects of α-herpesvirus latency has also just been reviewed in detail (Wagner 1991; Rice et al. 1991). In the present review, I have summarized the data generated from my own laboratory as well as those of two close collaborators: J. G. Stevens, who pioneered many of the animal studies on HSV latency, and L. T. Feldman, who has recently brought some powerful techniques of molecular genetics into the study of this subject. This should not be taken as a desire to denigrate the contributions of other active laboratories. In the latent state of infection of sensory nerve ganglia with HSV-1, only one single transcription unit, the latency-associated transcription unit (LAT), is expressed. This unit lies on the opposite DNA strand from that encoding both the α0 and α4 transcriptional activators

---

Department of Molecular Biology and Biochemistry and The Program in Animal Virology, University of California, Irvine, CA 92717, USA

(Stevens et al. 1987). With one exception (Sawtell and Thompson 1992), researchers agree that this limited gene expression in HSV-1 is not important for the establishment or maintenance of the latent phase of infection in laboratory animals (Izumi et al. 1989; Javier et al. 1988; Sedarati et al. 1989; Steiner et al. 1989), but it is strongly correlated with efficient induced HSV reactivation in the rabbit eye model (Hill et al. 1990) and in in vitro neuron explant virus recovery assays (Steiner et al. 1989; Leib et al. 1989; Block et al. 1990).

The primary HSV-1 LAT is a 8.5-kb transcript from which several poly (A) species are derived by splicing and accumulate in the nucleus as stable introns (Farrell et al. 1991; Devi-Rao et al. 1991). We have devoted a great deal of effort in assesing the role of such stable introns as well as one or more low-abundance processed polyadenylated derivatives of the primary transcript in reactivation. Present data indicate that the LAT introns are *not* involved in reactivation per se (M. Farrell, J. Hill, T. Margolis, J. Stevens, E. Wagner and L. Feldman, unpublished), but the fact that such introns are very common feature of latent phase transcription by α-herpesviruses suggests that they have some important role in the latency process.

Latent Infections by Herpes Simplex Virus

With HSV, latent infection can be viewed as having three separable phases: establishment, maintenance, and reactivation. The establishment of latent infections by HSV in experimental animals is the result of the failure of the lytic cascade in a neuronal cell. It is not clear whether a cell in which the lytic cycle has begun can be converted to a latently infected one, but it is quite clear that replication-deficient viruses and those lacking α4 (*ICP4*, Vmw175) gene products can establish latent infections with appreciable efficiency (Kosz-Vnenchak et al. 1990; Dobson et al. 1990; Gordon et al. 1990; Katz et al. 1990; Valyi-Nagy et al. 1991). The maintenance of the HSV genome in latently infected neurons appears to be a passive phenomenon, since the cells harboring the viral genomes do not replicate. However, HSV DNA is maintained as a circular molecule in latent infections, and some workers have suggested that low levels of genome replication could occur in the latent phase of infection (Sears and Roizman 1990). It is fairly clear that no expression of viral genes is required for maintenance of the genome for the reasons cited above and because of the fact that genomes of viruses unable to express LAT are efficiently maintained in the latent state.

Reactivation of HSV results in the appearance of infectious virus at the site of initial primary infection; this is the result of the appearance of complete viral genomes at the sensory nerve ending in the peripheral tissue, where lytic phase replication can proceed until the host immune response clears it. As noted, stress-induced reactivation of HSV in the rabbit eye model involves the expression of LAT (Hill et al. 1990); however, low-level spontaneous reactivation in rabbits does occur at frequencies comparable to wild-type virus. This has led us to speculate that there is more than one single pathway leading to expression of virus at the periphery of a sensory neuron. It is not yet fully established which portion of the latent phase transcription unit is involved in reactivation.

Latent Phase Transcription

Visualization of Latent Phase Transcription

The exquisite cellular resolution and sensitivity of in situ hybridization have been of great value in the detection and mapping of HSV-1 latent phase transcription products in neuronal tissue (Dobson et al. 1989; Stevens et al. 1988; Croen et al. 1987, 1988; Spivack and Fraser 1987; Gordon et al. 1988; Deatly et al. 1987,1988; Rock et al. 1987). Both neuronal and nonneuronal cells in acutely infected ganglia yield nuclear and cytoplasmic signals when hybridized with probes covering any region of the viral genome. In contrast, hybridization of latently infected tissues with defined probes covering 85% or more of the genome demonstrate no signal at all. Signals with intensities equivalent to those seen for acutely infected tissue are only seen with probes corresponding to DNA sequences within the stable LAT introns while a significantly weaker signal in fewer cells is seen with probes extending both 3′ and 5′ of the region exhibiting strong hybridization signals. Such a situation is fully consistent with latent phase transcription being controlled by a single latent phase promoter, with this difference in hybridization intensity corresponding to different abundances of RNA species processed from the primary transcription unit.

The Latency-Associated Transcript Promoter

All detectable transcription is abolished in latent infections by a recombinant virus (ΔPst) with a deletion of approximately 200 bases containing *pol*II promoter elements located at –1700 bases 3′ of the α0 (*ICP0* or Vmw110) transcript terminus (Dobson et al. 1989). This deletion lies between two PstI situated at bases 118,659 and 118,862 of the IRL and corresponding locations in the TRL in the syn17$^+$ strain of HSV-1 (McGeoch 1987). Transcription is also abolished with more extensive deletions encompassing this region (Izumi et al. 1989; Javier et al. 1988; Dobson et al. 1989; Mitchell et al. 1990). Confirming the location of the LAT promoter, RNAse protection experiments have shown the LAT primary transcript initiates within the 202 base region deleted in the recombinant just 3′ of a canonical TATA box (Devi-Rao et al. 1991).

Such data define the location of critical elements involved in the control of LAT expression; transient expression assays suggest, however, that DNA sequence elements 5′ of this region also have a role (Devi-Rao et al. 1991; Zwaagstra et al. 1989,1990; Batchelor and O'Hare 1990). The sequence of the LAT promoter contains a number of potential control elements which could have a role in the expression of LAT during latent infections. These include a canonical cyclic adenosine monophosphate (AMP) response element, CAT box, and Sp1-binding sites. Further, the sequence just at the cap site is quite similar to the strong α4 DNA-binding site which is involved in that regulatory protein's autoregulation (Muller 1987). Although the actual importance of such sequences in the control of LAT expression is unclear, LAT expression has been reported to be responsive to cyclic AMP (cAMP) levels in cell culture, and the α4-binding site has been reported to repress the ability of LAT expression to be induced by α4 in transient assays (Batchelor and O'Hare 1990; Leib et al. 1991).

In order to examine possible neuronal specificity of the LAT promoter, we placed the β-galactosidase gene under control of portions of this promoter in order to assess potential promoter activity (Devi-Rao et al. 1991). Our results can be summarized as follows: a construct containing 360 bases upstream of the LAT cap expressed somewhat more β-galactosidase activity than a lytic phase β-promoter in uninfected rabbit skin cells; however, levels of indicator enzyme activity only increased approximately threefold in infected cells, compared to a more than 50-fold increase with the lytic phase promoter; this latter result is presumably due to the presence of the α4 autoregulatory site at the cap site. Several constructs containing small deletions within the LAT promoter region showed slightly higher activities in uninfected rabbit skin cells and in cells after HSV infection; it is not clear how the removal of these sequences can partially overcome the putative repression of lytic phase LAT expression by α4. When we did transient expression assays in murine neuroblastoma, a significantly different pattern of expression was seen. Here, efficient expression of the LAT promoter required sequences upstream of the distal PstI site, since deletions of the 120 or so bases 5′ of this region significantly lowered LAT promoter activity in neuroblastoma cells. Interestingly, elements of the DNA sequence in question seem to reduce LAT promoter activity in rabbit skin cells, since increased activity was observed with the deletions in such cells.

Both the relatively high basal activity of the LAT promoter and its low level of activation by HSV superinfection compared with lytic phase promoters has been reported by workers in several other labs (Batchelor and O'Hare 1990; Zwaagstra et al. 1989). An enhancement of promoter activity in cells of neuronal origin has also been reported with the LAT promoter, and a region of the LAT promoter roughly corresponding to the region that we have implicated in high neuroblastoma activity has been shown to bind to a possible transcription factor found in neuronally derived IMR-32 cells (Batchelor and O'Hare 1990). Such data suggest a degree of neuronal specificity of the LAT promoter.

Latent Phase Transcripts

Poly(A–) Latency-Associated Transcripts are Stable Introns

Northern blots of RNA from latently infected ganglia hybridized with defined DNA and short oligonucleotide probes demonstrate the presence of two partially colinear, relatively abundant poly (A–) RNA species mapping within the limits of the strong in situ hybridization signal noted above. These RNA species are approximately 2.0 and 1.4 kb in size. The evidence for these poly (A–) RNA species being introns is now overwhelming. Precise mapping of the 5′ and 3′ ends of the poly (A–) LAT species from latently infected tissue has been carried out using both

S_1 nuclease and RNAse protection analysis. Both the 5′ and 3′ ends map to canonical splice signals. Further, poly (A–) LAT is uncapped (Devi-Rao et al. 1991), and most convincingly, Feldman and coworkers have expressed the LAT primary transcription unit in transfected cells and have demonstrated the processing of this unit to produce poly(A–) LAT (Farrell et al. 1991). Their data also demonstrate that this intron is stable in transfected cells and accumulates as it does in neurons, although we have found that the intron is not stable in all cells in which it is expressed – for example, mouse neuroblastoma cells (Devi-Rao et al. 1991).

The smaller species is related to the larger by the removal of another 600-base intron. This spliced, poly(A–) LAT species is seen only in latently infected tissue, and the relative amounts seen are dependent upon the strain of virus used to establish latent infection. The smaller intron is probably a result of some sort of secondary processing of the 2-kb stable intron, but we have not yet examined this process in any detail.

The Primary Latency-Associated Transcript

Northern blot analysis of poly (A+) RNA from lytically infected cells reveals the presence of a 9-kb transcript expressed from the LAT strand beginning just 3′ of the LAT promoter and extending to a polyadenylation signal in the short repeat region (Dobson et al. 1989). It should be noted that although other workers have reported detection of the primary transcript using northern blots of RNA from latently infected neural tissue (Zwaagstra et al. 1990), we have not. Despite this, the presence of such a primary transcript can be inferred from in situ hybridization data as well as detection of latent phase transcripts initiating at the expected cap site and extending to near the putative polyadenylation signal. Further, we have recently used polymerase chain reaction (PCR) amplification of oligo deoxythymidine (oligo-dT)-primed cDNA from latently infected tissue to detect the presence of polyadenylated LAT (G. B. Devi-Rao and E.K.Wagner, unpublished. RNAse protection assays demonstrate that the 5′ end of the primary poly (A+) transcript is located approximately 25 bases 3′ of the TATA box homology of the LAT promoter (Dobson et al. 1989), and we have also shown that poly (A–) RNA from latently infected murine sensory nerve ganglia protects that same fragment as seen with lytic phase poly (A+) RNA (Devi-Rao et al. 1991).

Although not quite as precise as the mapping of the 5′ end of the primary transcript, it is clear that the polyadenylation signal situated in the short repeat region just 3′ and on the opposite strand of DNA from that controlling the polyadenylation of α4 (base 127,140 and 127,190 of IRS and corresponding sites in the TRS, respectively) is utilized in lytic infection. RNAse protection probes from that region are protected by poly (A)-containing RNA isolated from lytically infected cells (Devi-Rao et al. 1991), and more recently we have isolated and sequenced a number of clones from an oligo-dT-primed cDNA library made against such RNA which correspond to transcripts terminating approximately 20 bases 3′ of this polyadenylation signal (G. B. Devi-Rao and E. K. Wagner, unpublished). Further, although the situation is not so clear with RNA from latently infected cells, we have been able to generate PCR signals using oligo-dT-primed cDNA from latently infected rabbit and murine sensory nerve ganglia with primers located upstream, but not downstream, of this polyadenylation signal. This strongly suggests that a major portion of latent phase transcription terminates here.

Processed Polyadenylated Forms of Latency-Associated Transcript

Since the 2 and 1.4-kb stable, poly(A–) LATs are introns, it is very clear that processed forms of poly(A+)LAT must occur. Oligo-nucleotide-probed northern blots of poly(A+)RNA isolated from lytically infected cells reveals low-abundance bands of RNA ranging in size from 2 to greater than 5 kb in addition to the primary LAT (Dobson et al. 1989). To date, these have not been extensively characterized, but we expect that they will utilize the splice donors already characterized and, perhaps, others. Feldman and coworkers were able to detect processed RNA from which the major LAT intron had been spliced out in their in vitro system, demonstrating the stability of the LAT intron (Farrell et al. 1991), but we have not been able to isolate such a poly(A+) containing RNA from lytically or latently infected tissue using RNAse protection probes (Devi-Rao et al. 1991). A 2-kb poly (A)-containing RNA extending from the LAT cap site to a position near the 3′ end of α0 mRNA (base 120,659 in the IRL and the corresponding location in the TRL) can be readily isolated from lytically infected cells (Devi-Rao et al. 1991). Other processed forms such as one utilizing the splice signals seen in the 1.4-kb LAT intron can also be posited, but no direct evidence for their occurring in either lytic or latent infections exits at this time. Our ability to map the biologically functional part of the latent phase transcription unit (see below) should direct attention to specific spliced species, which can then be confirmed using PCR.

Herpes Simplex Virus Reactivation and Latency-Associated Transcript Function

Reactivation

We are currently engaged in a very careful analysis of HSV gene expression during the reactivation process in rabbits and during explant cocultivation recovery of virus from latently murine sensory ganglia (E. K. Wagner, J. G. Stevens, D. G. Bloom, and G. B. Devi-Rao, in progress). Our approach has been to use defined PCR probes to investigate expression of lytic phase transcripts during the earliest times after induction of reactivation following either epinepherine induction or explantation. All data to date are consistent with reactivation involving induction of lytic phase gene expression within no more than 1–4 h after induction; this is a surprisingly short time. Indeed, the reactivation event is phenomenologically equivalent to viral gene expression in cells undergoing an extremely low multiplicity lytic infection. Such a finding indicates that there is no major physical block to lytic transcription of the viral genome in those latently infected cells destined to reactivate and that the reactivation process does not involve significant alterations in the earliest transcriptional events seen in lytic phase infection. These data provide important constraints on the generation of possible models for the investigation of the mechanism of LAT in the reactivation process.

The Biologically Important Part(s) of Latency-Associated Transcript in Reactivation

Since the genomic region from which latent phase transcripts are expressed by HSV-1 is large and could potentially encode processed, poly(A)-containing mRNA as well as stable poly(A–) introns, it is important to define the regions of the transcription unit which are involved in its biological function. Correlation of such regions with sequence and other data should provide critical insight into the mechanism of action of latent phase transcript. We (and other laboratories) have approached this problem via molecular genetic techniques with some success. Unfortunately, the use of more conventional biochemical approaches and attempts to use comparative sequence analysis between closely related α-herpesviruses known to express analogs of the HSV latent phase transcripts have not been particularly illuminating to date.

We have generated recombinant virus with polyadenylation–transcription termination signals inserted into regions of the primary latent phase transcript in order to generate a nest set of truncated transcripts. This approach is complicated by the presence of the α0 and the ICP34.5 transcription units. To date, termination signals have been inserted: (a) just 3′ of the α0 polyadenylation site (Devi-Rao et al. 1991); (b) into the the BamHI site in the major α0 intron (Farrell et al., unpublished); and (c) into an RsrII site approximately 900 bases 5′ of the LAT polyadenylation site (L. T. Feldman, J. G. Stevens, and E. K. Wagner, in progress; sites between bases 120,298 and 120,467, at base 123,460, and at base 126,317 in the IRL and corresponding locations in the TRL, respectively). Northern blot analysis of RNA expressed from the latent phase transcription unit following lytic infection with these recombinant viruses demonstrates that the first recombinant expresses an approximately 2-kb poly(A+) species, the second recombinant expresses an approximately 5-kb poly(A+) species, and the third recombinant expresses a nearly full-length primary transcript. RNA from latently infected sensory nerve ganglia demonstrate that the second and third recombinants express normal amounts of both poly(A–) LAT introns, while the first recombinant expresses none; thus, transcription can be shown to terminate at the appropriate location. Biological assays for the reactivation phenotype demonstrate that the second recombinant does not reactivate efficiently in the rabbit eye model; thus, information expressed 3′ of the BamHI site may be required. These data also indicate that the expression of stable LAT introns have no role in it. This latter conclusion is consistent with data from another laboratory where introduction of a phage λ-DNA sequence into the DNA encoding the LAT introns led to their destabilization with no loss of efficient recovery from in vitro explant assays (Block et al. 1990). Experiments in progress with the third recombinant tend to confirm the region between the BamHI site and the LAT polyadenylation site as being important for the reactivation phenotype.

Is Latency-Associated Transcript Function Mediated by a Protein?

Since the primary latent phase transcript is polyadenylated, processed species can be inferred, and as PCR suggests the presence of polyadenylated species in latently infected neurons, it is natural to expect that one or more proteins that encoded a LAT will have a major role in its biological activity. A number of open translation frames are contained within the extent of the primary transcript, but only a few contain more than 1000 bases. Of course, one or several of these could be utilized in an appropriately spliced transcript. Hopefully, the modifications of the LAT outlined above will lead to a specific region for further study.

We investigated expression of the open reading frame 2 (ORF-2) encompassing the 3′ portion of the LAT introns and overlapping the C-terminal region of the α0 translation frame which spans bases 120,419–121,241 in the IRL and corresponding bases in the IRL, before our molecular genetic studies indicated that this area is not involved in reactivation. Despite this, ORF-2 is a striking feature of the DNA sequence within this region of the viral genome and is conserved despite different multiples of a 16- base repeat within this ORF in three strains of HSV-1 (Rock et al. 1987; Wagner et al. 1988). As described in a recent review (Rice et al. 1991), we expressed this ORF in the pET system developed by Studier and colleagues (Rosenberg et al. 1987). An antibody generated against this protein reacted weakly, but specifically, in immunofluorescence assays with cells infected with wild-type virus, but not with cells infected with a virus mutant which lacks the LAT promoter. However, although the antibody reacted with western blots of the bacterial fusion protein, it did not react with HSV-infected cells extracts. Despite this, the data suggest that this portion of the latent phase transcription unit, at least, can be expressed as protein, but the role of such a protein in latent infections cannot be assesed. Other workers have used a different, independently derived antibody against protein expressed from this ORF to detect its expression in an in vitro latency system (Doerig et al. 1991). Again, such data suggest that this region of the latent phase transcription unit is expressible as protein via an as yet uncharacterized processed form of the primary transcript.

Other α-herpesviruses express transcripts analogous to the HSV-1 LAT (reviewed in Rice et al. 1991); unfortunately, comparative analysis of potential translational reading frames within the DNA sequence encoding such LAT does not indicate that any appreciably sized protein is conserved between these closely related viruses. Indeed, the extent of the primary latent phase transcripts is quite different, and the extent of complimentary overlap with immediate-early regulatory proteins is quite different.

The situation with potential similarities of proteins encoded by translational reading frames complementary to immediate-early transcriptional activators illustrates a problem with comparative analysis. The high degree of protein sequence conservation between regions of such regulatory proteins encoded by related α-herpesvirus can lead to artifactual homologies being scored on the opposite strand. A good case in point is seen with the ORF-2 translational reading frame of HSV-1, which is not involved in the REAC+ phenotype (see above). Despite its conservation in several strains of HSV-1, the only portion conserved between HSV-1 and two independent strains of HSV-2 (HG52 and MS) is in a short region complementary to the α0 protein (Mitchell et al. 1990; K. P. Anderson and E. K. Wagner, unpublished).

Future Directions for Studies on Latent Phase Transcription

Latent phase gene expression is readily observed with a number of, but not all, α-herpesviruses. In HSV-1 infections, latent phase gene expression clearly has a role in reactivation, but if this function is expressed via the action of a protein or proteins, any extended sequence of such as protein also does not appear to be conserved. Indeed, the most common feature of latent phase gene expression among those α-herpesviruses exhibiting it may well be the expression of a stable intron whose function is unknown at this time. Clearly, identification of LAT-encoded proteins will be an important accomplishment and efforts to this end continue to be made in many laboratories. A full description of the processed forms of LAT expressed both during the latent phase itself and during the earliest stages of reactivation will be very important in such studied as well as of interest in its own right.

Analysis of LAT promoter activity suggests a measure of neuronal specificity which may be of great interest in terms of tissue-specific transcriptional activity. There is, however, one rather obvious problem with interpretation of available data as evidence of promoter specificity alone. Since LAT promoter activity appears to be repressible by α4 protein and the LAT promoter has rather high basal levels of activity in transient assays, it would be predicted that LAT should act somewhat like an α-transcript during lytic infection. However, although we can detect low levels of LAT expression in certain cell lines treated with cyclohex-imide (conditions which lead to overexpression of α-transcripts), other labs have reported no LAT expression under similar conditions with other cells (Spivack and Fraser 1987; Wagner et al. 1988). Such results suggest that the actual state of the transcription template as well as the LAT promoter itself is important in LAT expression in latent infection.

If the information available cannot provide incontrovertible clues as to the mechanism of LAT function, they do readily suggest a scheme for the expression of LAT in the absence of any other viral gene products during the latent phase. This involves: (a) the failure of α-gene transcriptional activation in neurons, with the resulting lack of significant expression of any transcript requiring viral transcriptional activation and (b) the ability of the LAT promoter to be continu-ously expressed from the latent genome, despite its low basal activity. Processing of those LAT generated can yield both stable LAT introns and latent phase mRNAs which can function to allow viral recrudescence without gross cytopathology to the latently infected neurons.

Acknowledgements. Work described in this review was carried out using support from grant Ca11861 from the National Cancer Institute, grant AI06246 from the National Institutes of Allergy and Infectious Disease (co-pi with J. G. Stevens), and further support from the U. C. Irvine Program in Animal Virology. I thank M. Rice and G. Devi-Rao as well as all the other members of my laboratory for their efforts on this and related projects.

References

Batchelor AH, O'Hare P (1990) Regulation and cell-type-specific activity of a promoter located upstream of the latency-associated transcript of herpes simplex virus type 1. J Virol 64: 3269–3279

Block TM, Spivack JG, Steiner I, Deshmane S, McIntosh MT, Lirette RP, Fraser NW (1990) A herpes simplex virus type 1 latency-associated transcript mutant reactivates with normal kinetics from latent infection. J Virol 64: 3417–3426

Claoué CMP, Hodges TJ, Darville JM, Hill TJ, Blyth WA, Easty DL (1990) Possible latent infection with herpes simplex virus in the mouse eye. J Gen Virol 71: 2385–2390

Croen KD, Ostrove JM, Dragovic LJ, Smialek JE, Straus SE (1987) Latent herpes simplex virus in human trigeminal ganglia. Detection of an immediate early gene "anti-sense" transcript by in situ hybridization. N Engl J Med 317: 1427–1432

Croen KD, Ostrove JM, Dragovic LJ, Straus SE (1988) Patterns of gene expression and sites of latency in human nerve ganglia are different for varicella-zoster and herpes simplex viruses. Proc Natl Acad Sci USA 85: 9773–9777

Deatly AM, Spivack JG, Lavi E, Fraser NW (1987) RNA from an immediate early region of the type 1 herpes simplex virus genome is present in the trigeminal ganglia of latently infected mice. Proc Natl Acad Sci USA 84: 3204–3208

Deatly AM, Spivack JG, Lavi E, O'Boyle Dr, Fraser NW (1988) Latent herpes simplex virus type 1 transcripts in peripheral and central nervous system tissues of mice map to similar regions of the viral genome. J Virol 62: 749–756

Devi-Rao GB, Goodart SA, Hecht LB, Rochford R, Rice MK, Wagner EK (1991) The relationship between polyadenylated and non-polyadenylated herpes simplex virus type 1 latency associated transcripts. J Virol 65: 2179–2190

Dobson AT, Sedarati F, Devi-Rao GB, Flanagan WM, Farrell MJ, Stevens JG, Wagner EK, Feldman LT (1989) Identification of the latency-associated transcript promoter by expression of rabbit beta-globin mRNA in mouse sensory nerve ganglia latently infected with a recombinant herpes simplex virus. J Virol 63: 3844–3851

Dobson AT, Margolis TP, Sedarati F, Stevens JG, Feldman LT (1990) A latent, nonpathogenic HSV-1-derived vector stably expresses β-galactosidase in mouse neurons. Neuron 5: 353–360

Doerig C, Pizer LI, Wilcox CL (1991) An antigen encoded by the latency-associated transcript in neuronal cell cultures latently infected with herpes simplex virus type 1. J Virol 65: 2724–2727

Farrell MJ, Dobson AT, Feldman LT (1991) Herpes simplex virus latency-associated transcript is a stable intron. Proc Natl Acad Sci USA 88: 790–794

Gordon YJ, Johnson B, Romanowski E, Araullo-Cruz T (1988) RNA complementary to herpes simplex virus type 1 ICP0 gene demonstrated in neurons of human trigeminal ganglia. J Virol 62: 1832–1835

Gordon YJ, McKnight JLC, Ostrove JM, Romanowski E, Araullo-Cruz T (1990) Host species and strain differences affect the ability of an HSV-1 ICPO deletion mutant to establish latency and spontaneously reactive in vivo. Virology 178: 469–477

Hill JM, Sedarati F, Javier RT, Wagner EK, Stevens JG (1990) Herpes simplex virus latent phase transcription facilitates in vivo reactivation. Virology 174: 117–125

Izumi KM, McKelvey AM, Devi-Rao GB, Wagner EK, Stevens JG (1989) Molecular and biological characterization of a type 1 herpes simplex virus (HSV-1) specifically deleted for expression of the latency-associated transcript (LAT). Microb Pathog 7: 121–134

Javier RT, Stevens JG, Dissette VB, Wagner EK (1988) A herpes simplex virus transcript abundant in latently infected neurons is dispensable for establishment of the latent state. Virology 166: 254–257

Katz JP, Bodin ET, Coen DM (1990) Quantitative polymerase chain reaction analysis of herpes simplex virus DNA in ganglia of mice infected with replication-incompetent mutants. J Virol 64: 4288–4295

Kosz-Vnenchak M, Coen DM, Knipe DM (1990) Restricted expression of herpes simplex virus lytic genes during establishment of latent infection by thymidine kinase-negative mutant viruses. J Virol 64: 5396–5402

Leib DA, Coen DM, Bogard CL, Hicks KA, Yager DR, Knipe DM, Tyler KL, Schaffer PA (1989) Immediate-early regulatory gene mutants define different states in the establishment and reactivation of herpes simplex virus latency. J Virol 63: 759–768

Leib DA, Nadeau KC, Rundle SA, Schaffer PA (1991) The promoter of the latency-associated transcripts of herpes simplex virus type 1 contains a functional cAMP-response element: role of the latency-associated transcripts and cAMP in reactivation of viral latency. Proc Natl Acad Sci USA 88: 48–52

McGeoch DJ (1987) The genome of herpes simplex virus: structure, replication and evolution. J Cell Sci Suppl 7: 67–94

Mitchell WJ, Deshmane SL, Dolan A, McGeoch DJ, Fraser NW (1990) Characterization of herpes simplex virus type 2 transcription during latent infection of mouse trigeminal ganglia. J Virol 64: 5342–5348

Mitchell WJ, Steiner I, Brown SM, MacLean AR, Subak-Sharpe JH, Fraser NW (1990) A herpes simplex virus type 1 variant, deleted in the promoter region of the latency-associated transcripts, does not produce any detectable minor RNA species during latency in the mouse trigeminal ganglion. J Gen Virol 71: 953–957

Muller MT (1987) Binding of the herpes simplex virus immediate-early gene product ICP4 to its own transcription start site. J Virol 61: 858–865

Rice MK, Devi-Rao GB, Wagner EK (1991) Latent phase transcription by alphaherpesviruses. In: Adloph KW (ed) Techniques and applications of genome Research. Academic, Orlando, pp (in press)

Rock DL, Nesburn AB, Ghiasi H, Ong J, Lewis TL, Lokensgard JR, Wechsler SL (1987) Detection of latency-related viral RNAs in trigeminal ganglia of rabbits latently infected with herpes simplex virus type 1. J Virol 61: 3820–3826

Rosenberg AH, Lade BN, Chui D-s, Lin S-W, Dunn JJ, Studier FW (1987) Vectors for selective expression of cloned DNAs by T7 RNA polymerase. Gene 56: 125–135

Sawtell NM, Thompson RL (1992) Herpes simplex virus type 1 latency-associated transcription unit promotes anatomical site-dependent establishment and reactivation from latency. J Virol 66: 2157–2169

Sears AE, Roizman B (1990) Amplification by host cell factors of a sequence contained within the herpes simplex virus 1 genome. Proc Natl Acad Sci USA 87: 9441–9444

Sedarati F, Izumi KM, Wagner EK, Stevens JG (1989) Herpes simplex virus type 1 latency-associated transcription plays no role in establishment or maintenance of a latency infection in murine sensory neurons. J Virol 63: 4455–4458

Spivack JG, Fraser NW (1987) Detection of herpes simplex virus type 1 transcripts during latent infection in mice [published erratum appears in J Virol (1988) 62(2): 663]. J Virol 61: 3841–3847

Steiner I, Spivack JG, Lirette RP, Brown SM, MacLean AR, Subak-Sharpe JH, Fraser NW (1989) Herpes simplex virus type 1 latency-associated transcripts are evidently not essential for latent infection. EMBO J 8: 505–511

Stevens JG, Haarr L, Porter DD, Cook ML, Wagner EK (1988) Prominence of the herpes simplex virus latency-associated transcript in trigeminal ganglia from seropositive humans. J Infect Dis 158: 117–123

Stevens JG, Wagner EK, Devi-Rao GB, Cook ML, Feldman LT (1987) RNA complementary to a herpesvirus alpha gene mRNA is prominent in latently infected neurons. Science 235: 1056–1059

Valyi-Nagy T, Deshmane SL, Spivack JG, Steiner I, Ace CI, Preston CM, Fraser NW (1991) Investigation of herpes simplex virus type 1 (HSV-1) gene expression and DNA synthesis during the establishment of latent infection by an HSV-1 mutant, in1814, that does not replicate in mouse trigeminal ganglia. J Gen Virol 72: 641–649

Wagner EK (1991) Herpesvirus transcription-general aspects. In: Wagner EK (ed) Herpesvirus transcription and its regulation. CRC Press, Boca Raton, pp 1–15

Wagner EK, Devi-Rao GB, Feldman LT, Dobson AT, Zhang YF, Flanagan WM, Stevens JG (1988) Physical characterization of the herpes simplex virus latency-associated transcript in neurons. J Virol 62: 1194–1202

Wagner EK, Flanagan WM, Devi-Rao GB, Zhang YF, Hill JM, Anderson KP, Stevens JG (1988) The herpes simplex virus latency-associated transcript is spliced during the the latent phase of infection. J Virol 62: 4577–4585

Zwaagstra JC, Ghiasi H, Nesburn AB, Wechsler SL (1989) In vitro promoter activity associated with the latency-associated transcript gene of herpes simplex virus type 1. J Gen Virol 70: 2163–2169

Zwaagstra JC, Ghiasi H, Slanina SM, Nesburn AB, Wheatley SC, Lillycrop K, Wood J, Latchman DS, Patel K, Wechsler SL (1990) Activity of herpes simplex virus type 1 latency-associated transcript (LAT) promoter in neuron-derived cells: evidence for neuron specificity and for a large LAT transcript. J Virol 64: 5019–5028

Chapter 13 Neurovirulence of Herpes Simplex Virus Type 1 Accessory Gene Mutants

Myron Levine[1], David J. Fink[2], Ramesh Ramakrishnan[1], Prashant Desai[1,3], William F. Goins[3], and Joseph C. Glorioso[3]

Summary

A striking feature of the genome of herpes simplex virus (HSV) is that approximately half of the genes are not essential for replication of the virus in cell culture. It might be predicted that many of these accessory gene products are important for viral growth and spread in neurons and thus contribute to neurovirulence. Mutants altered in both essential and nonessential functions were studied for their ability to infect, replicate, spread, and establish latency in the central nervous system of rats. Some accessory gene mutants had little effect on viral pathogenesis, others showed reduced neurovirulence, but still caused neuropathogenesis, while a third class was attenuated and replicated in the rat brain to a limited degree, but failed to spread, and caused little or no neuropathology.

Introduction

Herpes simplex viruses types 1 and 2 (HSV-1 and HSV-2) are important human pathogens which infect a large fraction of the general population. Primary infection involves virus replication in mucocutaneous tissues causing relatively benign lesions of skin (reviewed in Fenner et al. 1974; Roizman and Batterson 1986) or potentially serious lesions of the cornea. Herpetic eye disease is the leading cause of blindness in industrialized countries (Whitley 1985). During primary infection, virus passes into nerve endings and is transported to the neuronal bodies in sensory ganglia, in which it establishes latency in these cells (reviewed in Stevens 1989; Roizman and Sears 1990). Viral gene expression is not required in the establishment of latency, and available evidence suggests that the virus immediately enters the latent state without the expression of replication functions. The viral genome circularizes and becomes bound with nucleosomes forming a chromatin-like structure (Mellerick and Fraser 1987). The latent state may persist for the life of the host, although periodically the virus may reactivate, resulting in

[1] Department of Human Genetics, University of Michigan, Ann Arbor, MI 48109, USA
[2] Department of Neurology and VA Medical Center, University of Michigan, Ann Arbor, MI 48109, USA
[3] Department of Molecular Genetics and Biochemistry, University of Pittsburgh School of Medicine, Pittsburgh, PA 15261, USA

production of infectious virus particles and the characteristic herpetic lesions. In rare instances, replication of virus in the central nervous system (CNS) can cause a transverse myelitis or encephalitis. HSV is the single most important cause of acute sporadic necrotizing encephalitis in the United States (Fenner et al. 1974).

The ability of HSV to cause disease in the nervous system depends on viral functions that determine the ability of the virus to (a) invade the nervous system (neuroinvasiveness), including spread from the peripheral nervous system (PNS) to the CNS and (b) to replicate and spread within the CNS. For animal models, some naturally occurring virus strains are more neuroinvasive than others when tested by inoculation in peripheral sites. However, all HSV strains are highly neurovirulent upon direct intracranial injection. Although little is known about the precise molecular mechanisms underlying neurovirulence, recent work using mutant viruses has allowed the determination of the importance of individual viral genes in neuropathologic processes and virus spread in brain.

HSV possesses a large genome consisting of 152,000 bp within which about 72 open reading frames (ORF) were initially identified on the basis of DNA sequence analysis (McGeoch et al. 1985, 1988). A large majority of the ORF now have assigned genes and gene products. The genes of HSV may be divided into two categories: those expressed during lytic infection and those expressed during latency. A striking feature of the genetic make-up of HSV is that approximately half of the lytic cycle genes can be deleted without substantially impairing the ability of the mutant virus to replicate in cell culture (Fig. 1). These genes include structural, enzymatic, and other nonstructural functions, and their conservation suggests that although they are nonessential for production of active virus in vitro, they are nevertheless required to carry out various aspects of the complex life cycle of the virus in the host. It is reasonable to predict that many of these products are important for viral growth and spread in nondividing cells such as neurons, thereby contributing to neurovirulence. They therefore represent potential sites at which functions can be disrupted to assess their role in neuropathogenesis. As described below, a series of mutants altered in both essential and nonessential functions were studied for their ability to infect, replicate, spread, and establish latency in the CNS (Fig. 2). These mutants were chosen on the basis of their potential ability to disrupt different steps in the replication cycle.

During latency, the only transcriptionally active region of the viral genome is a part of the inverted repeat sequences of the long unique (UL) region downstream of the *ICP0* (infected cell polypeptide 0) gene. The latency-associated transcripts (LAT) 1.8- to 2.3-kb transcripts, are transcribed from the strand opposite the *ICP0* gene and are partially complementary to *ICP0* mRNA (Deatly et al. 1987; Puga and Notkins 1987; Rock et al. 1987; Spivak and Fraser 1987; Stevens et al. 1987; Wagner et al. 1988). No gene product has yet been described for the LAT sequences. LAT deletion mutant viruses establish and maintain latency normally, but some mutants are impaired in reactivation (Ho and Mocarski 1989; Leib et al. 1989; Sedarati et al. 1989; Steiner et al. 1989). Despite the mystery which still surrounds the role of the LAT, expression of these transcripts is a hallmark of latency.

The genes coding for nonessential or accessory functions map throughout the HSV-1 genome, in UL and US (unique short segment of the HSV chromosome)

	Gene	Essential genes	Accessory genes
a	ICP34.5		Neurovirulence
b	ICPO		Regulatory protein
	LAT		Latency-assoc. transcript
	UL1	Fusion / Glycoprotein L	
	UL2		Uracil DNA glycosylase
	UL3		UL3
	UL4		UL4
	UL5	DNA dependent ATPase	
	UL6	Virion capsid protein	
	UL7		
	UL8	Helicase / Primase	
	UL9	ori-binding protein	
	UL10		Membrane protein
	UL11		Nucleocapsid envelopment/egress
	UL12		Alkaline nuclease
	UL13		Protein kinase
	UL14		
	UL15	Capsid assembly	
	UL16		UL16
	UL17		UL17
	UL18	Capsid protein	
	UL19	DNA packaging	
	UL20		Virion component
	UL21		UL21
	UL22	Glycoprotein H	
	UL23		Thymidine kinase
	UL24		UL24
	UL25	Capsid formation	
	UL26	DNA packaging/proteinase	
	UL27	Glycoprotein B	
	UL28	Capsid formation	
	UL29	Major ssDNA binding protein	
	UL30	DNA polymerase	
	UL31		
	UL32	Capsid formation	
	UL33	Packaging protein	
	UL34	Virion protein - DNA rep	
	UL35	Capsid protein	
	UL36	Tegument protein	
	UL37	Capsid assembly	
	UL38	DNA-binding capsid protein	
	UL39		Ribonucleotide reductase (large)
	UL40		Ribonucleotide reductase (small)
	UL41		Host-shut-off protein

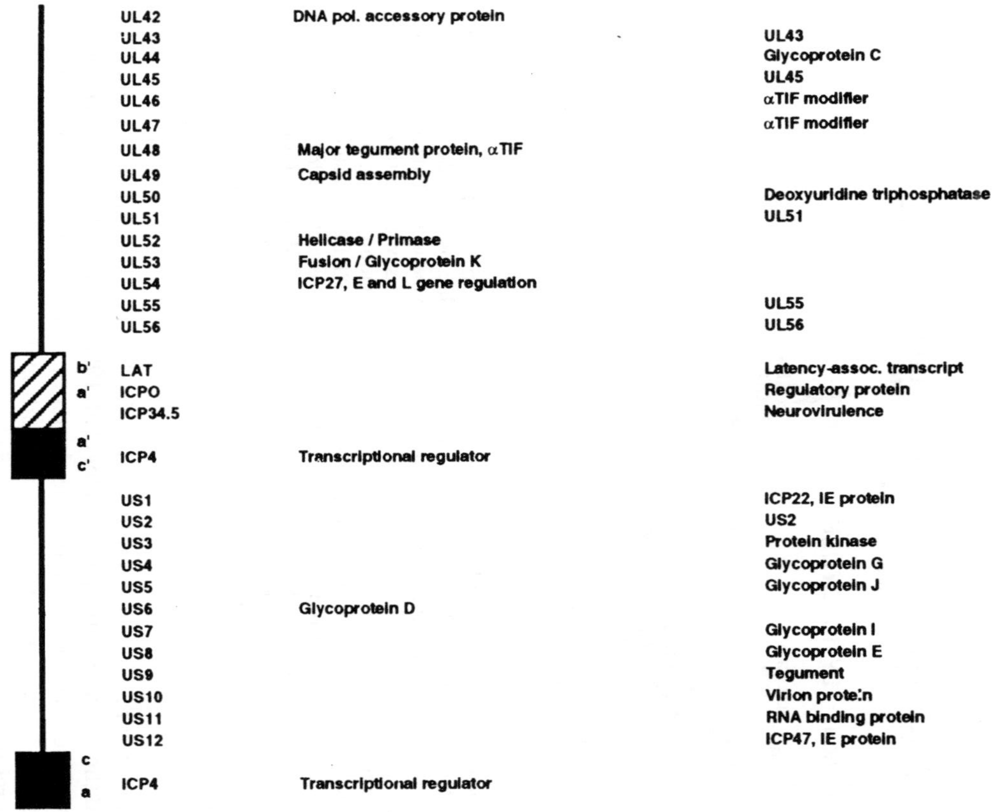

Fig. 1. Genetic map of herpes simplex virus (HSV). The unique long (*UL*) and unique short (*US*) components of the prototypic HSV-1 genome bracketed by the inverted repeat sequences are depicted. The map displays those HSV-1 genes for which a defined function has been assigned, listed adjacent to the numbered open reading frame. The genes have been divided into two subcategories based on whether they are *essential* or *accessory* to the growth of the virus in cell culture (*LAT*, latency-associated transcripts; *ICP*, infected cell polypeptide) (McGeoch et al. 1990; Roizman and Sears 1990; and references therein)

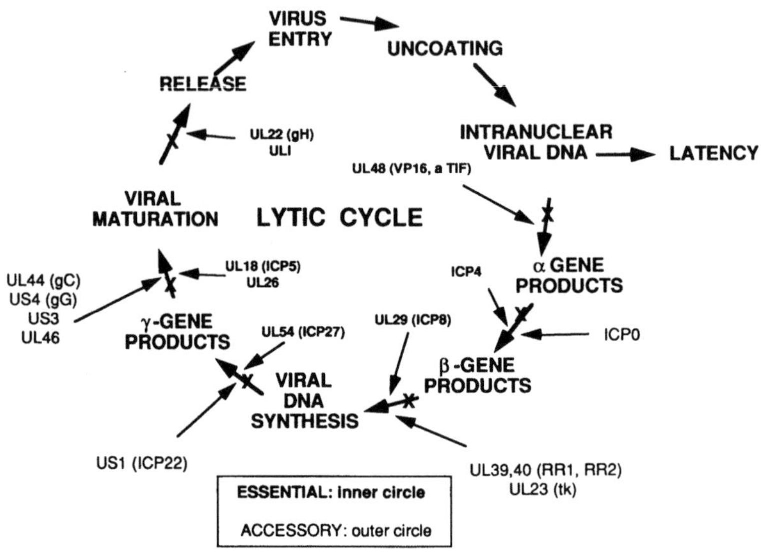

Fig. 2. Mutagenesis strategies used in the construction of herpes simplex virus type 1 (HSV-1) recombinant viruses. The various stages of the HSV-1 life cycle are depicted along with the viral gene products, which, when mutated, block virus replication at that specific stage in the life-cycle. The *inner circle* depicts essential viral gene products, whereas those found in the *outer circle* are accessory gene products (*UL*, unique long: *US*, unique short; *ICP*, infected cell polypeptide; *RR*, Ribonucleotide reductase) (Addison et al. 1990; Buckmaster et al. 1994; Campbell et al. 1984; Chalberg and Kelly 1989; Coen et al. 1989; Costa et al. 1984; DeLuca et al. 1985; Desai et al. 1988; Everett 1988; Fink et al. 1992; Glorioso et al. 1984; Gompels et al. 1991; Hutchinson et al. 1992; Jacobsen et al. 1989; Pellet et al. 1985; Purves et al. 1991; Rice and Knipe 1990; Schrag et al. 1989; Sears et al. 1985; Sekulovich et al. 1985; Weber et al. 1987; Zhang et al. 1991)

and in the inverted repeats surrounding UL (Fig. 1). Our laboratories (Weber et al. 1987) and the Roizman group (Meignier et al. 1988) showed that the US region of HSV-1 genome presents a remarkable cluster of accessory genes. Eleven of 12 genes encoded in US are dispensable for replication in tissue culture. Representatives of all three major temporally regulated classes of genes, α, β, and γ, are found among the accessory genes (Fig. 2). These include genes for immediate-early regulatory proteins (*US1* or *ICP22*; *US12* or *ICP47*), for proteins involved in viral DNA synthesis (ribonucleotide reductase, RR, large subunit, *ICP6* or *UL39*; RR small submit or *UL40*; thymidine kinase, TK or *UL23*, dUTPase or *UL50*, U glycosylase or *UL2*), for other enzyme activities such as protein kinase (*US3* and *UL39*), and genes for virion structural proteins, particularly glycoproteins (gC or *UL44*, gE or *US8*, gG or *US4*, gI or *US7*, and gJ or *US5*), which locate to the virion envelope. Indeed, the LAT gene itself is not essential for replication in cell culture.

The traditional method for assessing neurovirulence is to inject varying concentration of virus into brains of rodents to determine the number of plaque-forming units (PFU) needed to kill 50% of the animals (LD_{50}). For most wild-type strain of HSV, the LD_{50} ranges from 10^1 to 10^2 PFU. Mutants affected in neurovirulence show LD_{50} ranging from 10^4 to 10^8 PFU, reaching in effect complete avirulence by this test. Using this experimental approach, our studies (Sunstrum et al. 1988; Chrisp et al. 1989) demonstrated the capacity of wild-type HSV-1 strain KOS to replicate in neuronal cells of the limbic system with rapid development of encephalitis after intracranial inoculation of mice. KOS virus had an LD_{50} of 10^2. Clearly, viruses defective in genes that are essential for replication in vitro will be incapable of replication in vivo and therefore attenuated in neurovirulence (Larder et al. 1986; Field and Coen 1986). However, the in vivo dynamics which underlie reduced virulence of viruses defective in accessory genes are not well understood.

In order to extend the analysis of the pathophysiologic basis of attenuated neurovirulence in vivo, we have carried out a series of studies of viral gene expression and promoter activity following intracranial inoculation of virus into specific brain regions. In the studies discussed below, we injected 5 μl aliquots containing different viral recombinants directly into the hippocampus of 175–225 g Sprague-Dawley rats. The well-defined neuronal architecture of the hippocampus provides one easily defined measure of the pathogenic effect of the virus. We used in situ hybridization with riboprobes specific for individual viral mRNAs to determine the location and extent of viral gene expression, and we used semiquantitative polymerase chain reaction (PCR) with primer pairs of different regions of the viral genome to detect and quantitate and amount of viral DNA in specific brain regions at different times after inoculation, as a measure of the number of viral genomes present. We used reverse transcriptase PCR (RT-PCR) to detect and quantitate specific viral RNAs as a measure of expression of different viral gene classes. The expression of LAT RNA in the absence of lytic gene expression has been used as an indication of viral latency. In a number of mutants, we inserted a minigene cassette containing the reporter gene *lacZ* under the control of a variety of different viral and nonviral promoters. The detection of functional β-galactosidase activity by enzyme histochemical staining allows convenient detection of the extent of viral spread within the brain by these recombinants.

Here we describe the results of studies carried out with mutant viruses that typify the neurovirulent phenotypes encountered. Some mutations in nonessential genes (e.g., gC) had little affect on viral neuropathogenesis and spread, while others (e.g., TK) showed reduced neurovirulence, yet retained the ability to cause neuropathology at higher doses and were observed to spread widely in brain from the site of inoculation via neuronal pathways. A third category of nonessential gene mutants (e.g., *US3* and *RR*) replicated to some degree locally, but failed to spread to other brain regions. All mutants in essential genes failed to spread, and mutants defective in the major immediate-early gene *ICP4*, were highly attenuated. All mutants studied retained the ability to establish latency in CNS neurons.

Inactivation of the Glycoprotein C Gene

gC plays a role in the early stage of HSV-1 replication by contributing to virus attachment to host cells (Herold et al. 1991). This contribution is redundant with other glycoproteins, and thus deletion of the gC gene does not appreciably impair the ability of the mutant virus to replicate in vitro. Virus mutant in the accessory gC gene retained pathogenicity on intracranial injection in mice (Sunstrum et al. 1988; Chrisp et al. 1989) with an LD_{50} nearly identical to wild-type virus. Two thirds of the rats injected in the hippocampus with 5×10^3 PFU of a gC-defective virus died between 7 and 14 days post-inoculation. The brain of one rat which survived for 3 weeks showed severe encephalomalacia both at the site of injection and contralaterally. The gC gene thus appears to code for an accessory function not associated with neurovirulence, and this may be related to its role as a complement receptor in evading complement-mediated attack on HSV-infected cells in the host (Friedman et al. 1984; McNeary et al. 1987; WuDunn and Spear 1989).

Inactivation of the Thymidine Kinase Gene

Mutants in the HSV-1 *TK* gene were among the first to be shown to have an attenuated neurovirulence phenotype in rodents (Field and Wildy 1978). We disrupted the *TK* locus of HSV-1 by the insertion of a plasmid carrying a gene cassette consisting of the human cytomegalovirus (HCMV) immediate-early promoter driving the *lacZ*-coding sequence. Two days following inoculation into the hippocampus, β-Galactosidase activity was detected in cells in the ipsilateral hippocampus, including dentate gyrus pyramidal cells. *LacZ* staining could also be detected within cells of the contralateral hippocampus, ipsilateral pyriform cortex, and basal forebrain, regions containing neurons whose axons project to the inoculated hippocampus (Fig. 3). Thus, it appears that this viral recombinant replicated and spread within the CNS by retrograde axonal transport. The *TK* defective virus is an example of an accessory gene mutant which may still replicate, spread, and cause neuropathology.

Inactivation of the *US3* Accessory Genes

US3 one of the cluster of accessory genes encoded in the unique short region of the HSV-1 genome. The *US3* gene codes for a protein kinase (McGeoch et al. 1985; McGeoch and Davison 1986; Purves et al. 1987; Leader and Katan 1988) whose phosphorylation target has recently been identified as the HSV-1 UL34 phosphoprotein (Purves et al. 1991, 1992). The *US3* gene is dispensable for virus replication in cell culture and for the establishment of latency in mouse sensory ganglia (Longnecker and Roizman 1987; Meignier et al. 1988). To examine the virulence of *US3* mutant virus, a series of recombinants, in which the *lacZ* gene were placed under the control of the late promoter of the HSV gC gene or under the control of the HCMV immediate-early promoter, was introduced into the *US3* locus, interfering with the production of the normal *US3* gene product.

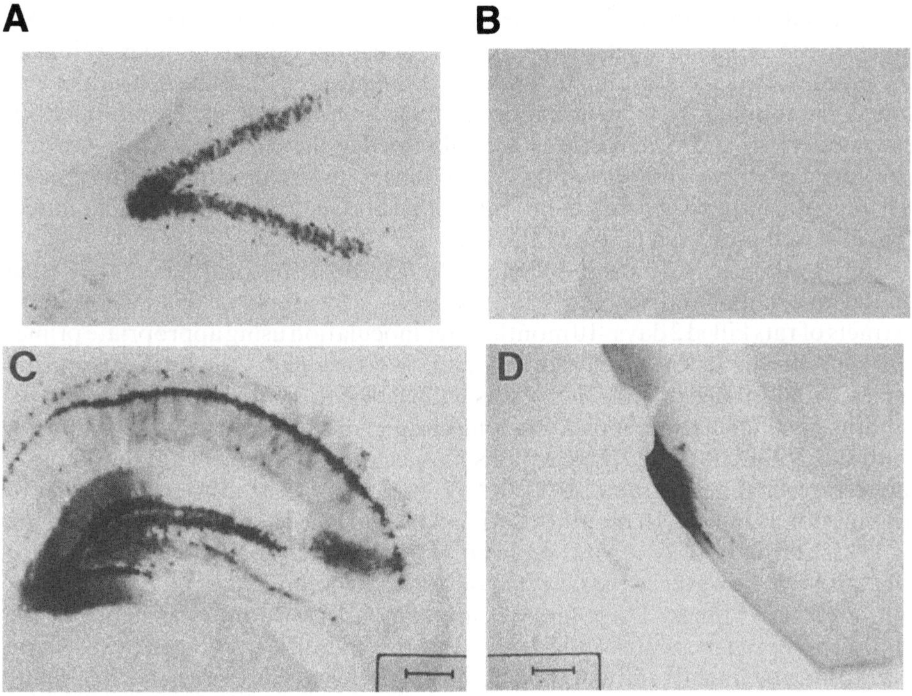

Fig. 3 A–D. In vivo analysis of mutations in accessory gene functions affecting the ability of the virus to replication and spread in the central nervous system (CNS). Rats were infected stereotactically into hippocampus with small volumes (1–5 μl) of viral mutants in the accessory genes *US3* (US3::HCMV-IEp-*lac*Z) or *tk* (tk::HCMV-IEp-*lap*Z), and the animals were killed at various times postinfection, the brains removed and sectioned, and stained with X-gal to identify β-gal-positive staining neurons in which the virus was present. **A** X-gal staining of β-galactosidase-producing neurons in the injected dentate gyrus of the hippocampus at 2 days postinoculation of US3::HCMV-IEp-*lac*Z. **B** No X-gal-positive neurons were detected in the contralateral hippocampal region following injection with *US3* mutant, suggesting that this mutant does not spread to other brain regions. **C** X-gal staining of the dentate gyrus and pyramidal cell layer at 2 days following inoculation of the tk::HCMV-IEp-*lac*Z recombinant. *LacZ* expression was detected in the pyriform cortex (**D**) and other regions of the brain which possess neural connections to the hippocampus. Thus, the tk^- recombinant is capable of replication and spread within the CNS by normal routes of axonal transport

The US3::pgc-lacZ virus replicated as well as the wild-type in cell culture, but showed an LD_{50} of $10^{4.6}$ PFU after intracranial inoculation of mice compared to the wild-type LD_{50} of 10^1 PFU (Fink et al. 1992), results similar to that obtained for another *US3*-defective mutant (Meignier et al. 1988). Following stereotactic infection into the right hippocampus, blue reaction product was seen in the injected hippocampus from 1 to 5 days postinoculation extending throughout the dorsal hippocampus, at least 3 mm in a rostral-to-caudal direction. Staining disappeared completely by 10 days. Unlike the TK^- recombinant, β-galactosidase-positive cells were confined to the dentate gyrus of the injected hippocampus. No β-galactosidase-positive cells in the contralateral hippocampus or other regions of the brain were detected (Fig. 3). The electron-dense X-gal reaction product was found predominantly in neurons on electron microscopic examination. Both enzyme activity and viral particles were observed in the same cells. However, few enveloped particles were seen and no infectious virus could be found after 7 days. All animals survived the treatment until killed and, in general neuronal architecture was well preserved (Fink et al. 1992).

PCR was used to determine whether the *US3*-defective viral genome persisted on the rat brains. Viral-specific DNA and RNA were amplified from whole brain extracts of rats killed 2 days–10 months after inoculation using appropriate primer pairs specific for *lacZ* (as a measure of the *US3* gene region), the LAT region, and the essential gB gene. Viral DNA was detected with each primer pair up to 10 months postinoculation, indicating long-term persistence of the viral genome in the brain. RT-PCR using RNA extracts was performed to detect viral specific gene expression in the rat brains. LAT RNA was detected in extracts of brains of animals for as long as 10 months postinoculation, consistent with the expression of LAT during both lytic viral replication and in latency. Messenger RNA for *lacZ* and *GB*, however, were detected only at 2 days and 7 days postinoculation, but not at later times. These findings are consistent with the persistence of the viral genome, expression of the LAT, and the shutoff of lytic gene expression during latency. Inactivation of the *US3* gene thus results in a virus which is highly attenuated in rodent brains, which undergoes limited replication, producing a self-limiting infection in a localized region of the brain, with the apparent establishment of latency.

Inactivation of the Accessory Genes Encoding the Subunits for Ribonucleotide Reductase

Similar results were obtained with virus mutant in the genes coding for the components of the HSV-1 KOS RR. RR catalyzes the reduction of ribonucleotides to deoxyribonucleotides (Thelander and Reichard 1979). The enzyme is made up of two subunits, RR1 and RR2 (Cohen 1972; Dutia 1983; Preston et al. 1984; Frame et al. 1985; Cohen et al. 1985; Bacchetti et al. 1986), as a complex of two molecules of each subunit (Ingemarson and Lamkinen 1987). Cameron et al. (1988) found that mutants in either subunit were 10^6-fold less virulent in mice than the parental strain 17 virus. In addition to its role in reduction of ribonucleotides, the RR1 molecules of both HSV-1 and HSV-2 have an intrinsic protein kinase domain located within the first 300 residues of their amino termini (Chung et al. 1989; Ali et al. 1990, 1992).

To examine the neurovirulence of RR mutants in rats, a virus was constructed in which the RR1 promoter drives chloramphenicol actyl transferase (CAT) and the RR2 promoter drives *lacZ* in this doubly deleted virus. Injection of 10^5 PFU of this recombinant virus into the hippocampus of rats showed expression of both

CAT and *lacZ* localized to the injected region at 2 days, with no evidence of spread to other areas of the brain. All animals survived with little pathology evident 8 weeks postinoculation. By 7 days, neither CAT no *lacZ* expression could be demonstrated. PCR analysis of these RR mutant-injected brains was undertaken for viral-specific DNA and RNA. An amplified signal from each of three genome regions, *gB*, *lacZ* and *LAT*, was detected from templates obtained from 10-μm tissue sections of the infected hippocampus of all the test animals from 2 day–8 weeks postinoculation. In contrast, no signals for any of these gene regions was detected from the uninfected contralateral hippocampus of the same sections. As a positive control, the host carbonic anhydrase II gene (Venta et al. 1991) was amplified from the extracts with the appropriate primers. These results imply that the viral genome persists in cells of the infected hippocampus and that the virus does not spread from the injected area. Semiquantitative PCR on the DNA extracts indicated a clear decrease in the amount of viral DNA between 2 and 7 days postinjection, but the amount of viral DNA in the rat brains appeared to hold constant between 7 days and 8 weeks.

Extracts from scrapings of 10-μm sections of the RR mutant-injected brains were also used for RNA PCR analysis. LAT and gB-specific RNA was detected following RT-PCR at day 2. By day 7, no signal was found using the gB primers, but LAT RNA was found at 8 weeks. RNAse pretreatment eliminated the amplification from the LAT primers. We conclude that the RR1–RR2-deleted virus is highly attenuated in the rat brain, fails to replicate and spread to any appreciable extent, and produces a self-limiting infection, leading to the establishment of latency in the region of the injection.

Inactivation of the Accessory *UL 34.5* Gene

The *ICP34.5* accessory gene has been studied by others (Chou et al. 1990; Chou and Roizman 1992; Robertson et al. 1992; Dolan et al. 1992). It maps to the inverted repeats flanking the long unique sequence of HSV-1 DNA and thus is present in two copies per molecule. Its regulatory region has no conventional TATA box, nor does it have *cis*-acting response elements associated with other TATA-less genes. The gene encodes a protein of 263 amino acids, containing the triplet Ala-Thr-Pro repeated ten times (Ackerman et al. 1986). Sixty-three amino acids near the carboxyl terminus of *ICP34.5* share 83% identity with a mouse protein, MyD116, present in a myeloid leukemic cell line which differentiates following exposure to interleukin 6 (Williams 1991; McGeoch and Barnett 1991; Lord et al. 1990). A gene with similar homologies has been identified in the African swine fever virus (Sussman et al. 1992). A mutant deleted for 1000 bp in the *ICP34.5* gene has an LD_{50} greater than 1.2×10^6 PFU, compared to the wild-type LD_{50} of 420 PFU. A second mutant carrying a stop codon in the gene is even less virulent, with an LD_{50} greater than 10^7 PFU, and is probably the most attenuated HSV-1 mutant so far studied (Chou and Roizman 1990). Little or no progeny virus was found in injected mouse brains. Infection of nonneuronal cells with either wild-type or mutant virus resulted in equivalent production of infection virions. Infection of a neuroblastoma cell line with *ICP34.5* mutants resulted in shutoff of host cell protein synthesis with the failure to produce active virus. These data suggest that the reason why *ICP34.5* mutants were avirulent in the mouse CNS is that no replication or spread of virus could occur in the absence of protein synthesis and that the function of *ICP34.5* is to sustain protein synthesis in neuronal cells (Chou and Roizman 1992).

The *ICP34.5* gene is most likely the neurovirulence factor detected as a consequence of intertypic recombination between HSV-1 and HSV-2 strains. Avirulent recombinant virus was isolated from the progeny of such crosses which were at least a million-fold less virulent than either parent (Thompson and Stevens 1983; Taha et al. 1989a, b; Thompson et al. 1989; Dolan et al. 1992). The avirulent recombinants replicated well in nonneuronal cells. Virulence was restored by replacing HSV-2 sequences with HSV-1 DNA from map units 0.82–0.832 (Thompson et al. 1989). The *ICP34.5* gene maps in this region of the HSV-1 genome.

Conclusions

Studies on HSV neurovirulence carried out in our laboratories and many others are beginning to shed light on HSV pathogenesis. These studies show that a virus must be capable of replicating and spreading in neuronal tissue to induce encephalitis and death. As discussed for *US3*, RR, and *ICP34.5* mutants, attenuation is accompanied by severe limitations on viral replication and spread in the CNS of mice and rats. These mutant viruses replicate relatively normally in cell culture, but not in vivo. Mutations in these accessory genes clearly affect the host range of cells in which the virus can replicate. For RR mutants, it has been shown that the endogenous cellular enzyme can complement for the viral deficiency in the production of deoxyribonucleotides at 37°C. The cellular enzyme is temperature sensitive and at higher temperatures in inactive. Under these conditions, the RR function becomes essential (Goldstein and Weller 1988). As the neurons of the CNS are nondividing, the cellular enzymatic machinery to complement RR defective virus may be absent. The protein kinase encoded by the *US3* gene may be unique to infected cells, since its target substrate is another protein encoded by the viral genome, the product of the *UL34* gene. The *UL34* gene product is associated with the infected cell plasma membrane. It is also found in the virion in small amounts. The function of the *UL34* gene product is unknown, but its unphosphorylated form appears to associate with four novel phosphoproteins which appear only in lysates of cells infected with *US3*-defective virus or with mutants containing a functional *US3* gene but with mutations in the *UL34* gene such that the *UL34* protein cannot be phosphorylated (Purves et al. 1992). The role of these phosphoproteins is also not understood. The kinase may not be available in the CNS in vivo. Chou and Roizman (1992) present evidence that the product of the *ICP34.5* gene allows sustained protein synthesis in the neuroblastoma cell.

They suggest it may act to prevent the cessation of protein synthesis rather than to produce a factor needed by the protein synthesizing machinery. The confluence of virological and neurobiological experimentation should resolve the questions posed by the neurovirulence of the herpes viruses.

Herpes Simplex Virus Type 1 as a Vector

HSV in the latent state has many features that would be ideal for a gene transfer vector. The viral genome remains extrachromosomal, so there is no risk of disrupting essential cellular genes or transactivating oncogenes. Viral genes expressed during lytic infection remain transcriptionally silent, and the latent state may persist for the life of the animal. As the experiments discussed above demonstrate, deletion of specific viral genes may attenuate neurovirulence, and foreign genes may be expressed in the brain from the context of the viral genome. Further work to produce a vector which is entirely nonpathogenic and which permits long-term expression of a biologically active gene product may allow the development of an HSV-based therapeutic gene-transfer vector which could be used to treat neurodegenerative conditions or brain tumors.

References

Ackerman M, Chou J, Sarmiento M, Lerner RA, Roizman B (1986) Identification by antibody to a synthetic peptide of a protein specified by a diploid gene located in the terminal repeats of the L component of herpes simplex virus genome. J Virol 58: 843–850

Addison C, Rixon FJ, Preston VG (1990) Herpes simplex virus type 1 UL28 gene product is important for the formation of mature capsids. J Gen Virol 71: 2377–2384

Ali MA, McWeeney D, Milosavljevic A, Jurka J, Jariwalla RJ (1991) Enhanced malignant transformation induced by expression of a distinct protein domain of ribonucleotide reductase large subunit from herpes simplex virus type 2. Proc Natl Acad Sci USA 88: 8257–8261

Ali MA, Prakash SS, Jariwalla RJ (1992) Localization of the antigenic sites and intrinsic protein kinase domain within a 300 amino acid segment of the ribonucleotide reductase large subunit from herpes simplex virus type 2. Virology 187: 360–367

Bacchetti S, Evelegh MJ, Muirhead B (1986) Identification and separation of the two subunits of the herpes simplex virus ribonucleotide reductase. J Virol 57: 1177–1181

Buckmaster EA, Gompels U, Minson AC (1984) Characterization and physical mapping of and HSV-1 glycoprotein of approximately 115×10^3 molecular weight. Virology 139: 408–413

Cameron JM, McDougall I, Marsden HS, Preston VG, Ryan DM, Subak-Sharpe JH (1988) Ribonucleotide reductase encoded by herpes simplex virus is a determinant of the pathogenicity of the virus in mice and a valid antiviral target. J Virol 69: 2607–2612

Campbell MEM, Palfreyman JW, Preston CM (1984) Identification of herpes simplex virus DNA sequences which encode a *trans*-acting polypeptide responsible for stimulation of immediate early transcription. J Mol Biol 180: 1–19

Chalberg MD, Kelly TJ (1989) Animal virus DNA replication. Annu Rev Biochem 58: 671–717

Chou J, Roizman B (1990) The herpes simplex virus 1 gene for ICP34.5, which maps in inverted repeats, is conserved in several limited-passage isolates but not in strain 17syn$^+$. J Virol 64: 1014–1020

Chou J, Roizman B (1992) The γ1 34.5 gene of herpes simplex virus 1 precludes neuroblastoma cells from triggering total shutoff of protein synthesis characteristic of programed cell death in neuronal cells. Proc Natl Acad Sci USA 89: 3266–3270

Chou J, Kern ER, Whitley RJ, Roizman B (1990) Mapping of herpes simplex virus-1 neurovirulence to γ1 34.5, gene nonessential for growth in culture. Science 250: 1262–1266

Chrisp CE, Averill DR, Sunstrum JC, Levine M, Glorioso JC (1989) Characterization of enchephalities in adult mice induced by intracerebral inoculation of herpes simplex virus type 1 (KOS) and comparison with mutants showing decreased virulence. Lab Invest 60: 822–830

Chung, TD, Wymer JP, Smith CC, Kulka M, Aurelian L (1989) Protein kinase activity associated with the large subunit of herpes simplex virus type 2 ribonucleotide reductase (ICP10). J Virol 63: 3389–3398

Coen DM, Kosz-Vnenchak M, Jacobson JG, Leib DA, Bogard CL, Schaffer PA, Tyler KL, Knipe DM (1989) Thymidine kinase-negative herpes simplex virus mutants establish latency in mouse trigeminal ganglion but do not reactivate. Proc Natl Acad Sci USA 86: 4736–4740

Cohen EA, Charron J, Perret J and Langelier Y (1985) Herpes simplex virus ribonucleotide reductase induced in infected BHK-1/C13 cells: biochemical evidence for the existence of two non-identical subunits, H1 and H2. J Gen Virol 66: 733–745

Cohen GH (1972) Ribonucleotide reductase activity of synchronized KB cells infected with herpes simplex virus. J Virol 9: 408–418

Costa RH, Cohen G, Eisenberg R, Long D, Wanger EK (1984) Direct demonstration that the abundent 6-kilobase herpes simplex virus type 1 mRNA mapping between 0.23 and 0.27 map units encodes the major capsid protein VP5. J Virol 49: 287–292

Deatly AM, Spivack JG, Lavi E, Fraser NW (1987) RNA from an immediate early region of the type 1 herpes simplex virus genome is present in the trigeminal ganglia of latently infected mice. Proc Natl Acad Sci USA 84: 3204–3208

DeLuca NA, McCarthy AM, Schaffer PA (1985) Isolation and characterization of deletion mutants of herpes simplex virus type 1 in the gene encoding immediate-early regulatory protein ICP4. J Virol 56: 558–570

Desai PD, Schaffer PA, Minson AC (1988) Excretion of non-infections virus particles lacking glycoprotein H by a temperature-sensitive mutant of herpes simplex virus type 1: evidence that gH is essential for virion infectivity. J Gen Virol 69: 1147–1156

Dolan A, McKie E, MacLean AR, McGeoch DJ (1992) Status of the ICP34.5 gene in herpes simplex virus type 1 strain 17. J Gen Virol 73: 971–973

Dutia BM (1983) Ribonucleotide reductase induced by herpes simplex virus has a virus-specified constituent. J Gen Virol 64: 513–521

Everett RD (1988) Analysis of the functional domains of herpes simplex virus type 1 immediate-early polypeptide Vmw110. J Mol Biol 202: 87–96

Fenner F, McAuslan BR, Mims CA, Sambrook J, White DO (1974) The biology of animals viruses. Academic, New York, p 393

Field HJ, Coen DM (1986) Pathogenicity of herpes simplex virus mutants containing drug resistance mutations in the viral DNA polymerase gene. J Virol 60: 286–289

Field HN, Wildy P (1978) The pathogenicity of thymidine kinase-deficient mutants of herpes simplex virus in mice. J Hyg 81: 267–277

Fink KJ, Sternberg LR, Weber PC, Mata M, Goins WF, Glorioso JC (1992) In vivo expression of β-galactosidase in hippocampal neurons by HSV-mediated gene transfer. Hum Gene Ther 3: 11–19

Frame MC, Marsden HS, Dutia BM (1985) The ribonucleotide reductase induced by herpes simplex virus type 1 involved minimally a complex of two polypeptides (136 K and 38 K). Gen Virol 66: 1581–1587

Friedman HM, Cohen GH, Eisenberg RJ, Seidel CA, Cines DB (1984) Glycoprotein C of herpes simplex virus 1 acts as a receptor for the C3b complement component on infected cells. Nature 309: 633–635

Glorioso JC, Schroder CH, Kumel G, Szczeisul M, Levine M (1984) Immunogenicity of herpes simplex glycoproteins gC and gB and their role in protective immunity. J Virol 50: 805–812

Goldstein DJ, Weller SK (1988) Factor(s) present in herpes simplex virus type-1-infected cells can compensate for the loss of the large subunit of the ribonucleotide reductase: characterization of an ICP6 deletion mutant. Virology 166: 41

Gompels NA, Carss AL, Saxby C, Hancock DC, Forrester A, Minson AC (1991) characterization and sequence analysis of antibody-selected antigenic variants of herpes simple virus show a conformationally complex epitope or glycoprotein H. J Gen Virol 65: 2393–2401

Herold BC, WuDunn D, Soltys N, Spear PG (1991) Glycoprotein C of herpes simplex virus type 1 plays a principal role in the adsorption of virus to cells and in infectivity. J Virol 65: 1090–1098

Ho DY, Mocarski EE (1989) Herpes simplex virus latent RNA (LAT) is not required for latent infection in the mouse. Proc Natl Acad Sci USA 86: 7596–7600

Hutchinson L, Browne H, Wagrant V, Davis-Poynter N, Primovac S, Goldsmith K, Minson AC, Johnson DC (1992) A novel herpes simplex virus glycoprotein, gL, forms a complex with glycoprotein H (gH) and affects normal folding and surface expression. J Virol 66: 2240–2250

Ingemarson R, Lankinen H (1987) The herpes simplex virus type 1 ribonucleotide reductase is a tight complex of the type composed of 40 K and 140 K proteins, of which the latter shows multiple forms due to proteolysis. J Virol 156: 417–422

Jacobsen JG, Leib DA, Goldstein DJ, Bogard CL, Schaffer PA, Weller SK, Coen DM (1989) A herpes simplex virus ribonucletide reductase deletion mutant is defective for productive acute and reactivatable latent infectious of mice and for replication in mouse cells. Virology 173: 276–283

Larder BA, Lisle JJ, Darby G (1986) Restoration of wild-type pathogenicity to an attenuated DNA polymerase mutant of herpes simplex virus type 1. J Gen Virol 67: 2501–2506

Leader DP, Katan M (1988) Viral aspects of protein phosphorylation. J Gen Virol 69: 1441–1464

Leib DA, Bogard CL, Kosz-Vnenchak, M, Hicks KA, Coen DM, Knipe DM, Schaffer PA (1989) A deletion mutant of the latency-associated transcript of herpes simplex virus type 1 reactivates from the latent state with reduced frequency. J Virol 63: 2893–2900

Longnecker R, Roizman B (1987) Clustering of gene dispensable for growth in culture in the S component of the HSV-1 genome. Science 236: 573–576

Lord K, Hoffman-Lieberman B, Lieberman D (1990) Sequence of MyD116 cDNA: a novel myeloid differentiation primary response gene induced by IL6. Nucleic Acids Res 18: 2823

McGeoch D, Barnett B (1991) Neurovirulence factor. Nature 353: 609

McGeoch DJ, Davison AJ (1986) Alphaherpesviruses possess a gene homologous to the protein kinase gene family of eukaryotes and retroviruses. Nucleic Acids Res 14: 1765–1777

McGeoch DJ, Dolan A, Donald S, Rixon FJ (1985) Sequence determination and genetic content of the short unique region in the genome of herpes simplex virus type 1. J Mol Biol 181: 1–13

McGeoch DJ, Dalrymple MA, Davison AJ, Dolan A, Frame MC, McNab D, Perry LJ, Scott JE, Taylor P (1988) the complete DNA sequence of the long unique region in the genome of herpes simplex virus type 1. J Gen Virol 69: 1531–1574

McGeoch DJ, Weller SK, Schaffer PA (1990) Herpes simplex virus. In: O'Brien SJ (ed) Genetic maps, locus maps of complex genomes. Cold Spring Harbor Laboratory Press, New York, pp 1115–1120

McNeary TA, Odell C, Holerts VM, Spear PG, Atkinson JP (1987) Herpes simplex virus glycoproteins gC-1 and gC-2 bind to the third component of complement and provide protection against complement-mediated neutralization of viral infectivity. J Exp Med 166: 1525–1535

Meignier B, Longnecker R, Mavromara-Nazos P, Sears AE, Roizman B (1988) Virulence of and establishment of latency genetically engineered deletion mutants of herpes simplex virus 1. Virology 162: 251–254

Mellerick DM, Fraser NW (1987) Physical state of the latent herpes simplex virus genome in a mouse model system: evidence suggesting an episomal state. Virology 158: 265–275

Pellet PE, McKnight JLC, Jenkins FJ, Roizman B (1985) Nucelotide sequence and predicted amino acid sequence of a protein encoded in small herpes simplex virus DNA fragment capable of trans-inducing α genes. Proc Natl Acad Sci USA 82: 5870–5874

Preston VG, Palfreyman JW, Dutia BM (1984) Identification of a herpes simplex virus type 1 polypeptide which is a component of the virus-induced ribonucleotide reductase. J Gen Virol 65: 1457–1466

Puga A, Notkins AL (1987) Continued expression of a poly (A)+ transcript of herpes simplex virus type 1 in trigeminal ganglia of latency infected mice. J Virol 61: 1700–1703

Purves FC, Longnecker RM, Leader DP, Roizman B (1987) Herpes simplex virus 1 protein kinase is encoded by open reading frame US3 which is not essential for virus growth in cell culture. J Virol 61: 2896–2901

Purves FC, Spector D, Roizman B (1991) The herpes simplex virus 1 protein kinase encoded by the US3 gene mediates posttranslational modification of the phosphoprotein encoded by the UL34 gene. J Virol 65: 5757–5764

Purves FC, Spector D, Roizman B (1992) UL34, the target of the herpes simplex virus US3 protein kinase, is a membrane protein which in its unphosphorylated state associates with novel phosphoproteins. J Virol 66: 4295–4303

Rice SA, Knipe DM (1990) Genetic evidence for two distinct transactivation functions of the herpes simplex virus 1 α protein ICP27. J Virol 64: 1707–1715

Robertson RM, MacLean AR, Brown SM (1992) Peripheral replication and latency reactivation kinetics of the non-neurovirulent herpes simplex variant 1716. J Virol 73: 967–970

Rock DL, Nesber AB, Chiashi H, Ong J, Lewis TL, Lokensgard JR, Wechsler S (1987) Detection of latency-related viral RNAs in trigeminal ganglia of rabbits latently infected with herpes simplex virus type 1. J Virol 61: 3820–3826

Roizman B, Batterson W (1986) Herpesviruses and their replication. In: Fields BN, Knipe DM (eds) Fundamental virology. Raven, New York, pp 607–636

Roizman B, Sears AE (1990) Herpes simplex viruses and their replication. In: Fields BN, Knipe DM, Chanock RM, Hirsch MS, Melnick JL, Monath TP, Roizman B (eds) Virology. Raven, New York, pp 1795–1841

Schrag JP, Peasad BVV, Rixon FJ, Chiu W (1989) Three dimensional structure of the herpes simplex virus type 1: evidence that gH is essential for virion infectivity. J Gen Virol 69: 1147–1156

Sears AE, Halliburton IW, Meignier B, Silver S, Rizman B (1985) Herpes simplex virus 1 mutant defective in the α 22 gene: growth and gene expression in permissive and restrictive cells and establishment of latency in mice. J Virol 55: 338–346

Sedarati F, Izumi KM, Wagner EK, Stevens JH (1989) Herpes simplex virus type 1 latency-associated transcription plays no role in establishment or maintenance of a latent infection in murine sensory neurons. J Virol 63: 4455–4458

Sekulovich RE, Leary K, Sandri-Goldin RM (1985) The herpes simplex virus type 1 α protein ICP27 can act as a *trans*-repressor or *trans*-activator in combination with ICP4 and ICP0. J Virol 62: 4510–4522

Spivack JG, Fraser NW (1987) Detection of herpes simplex virus type 1 transcripts during latent infection in mice. J Virol 61: 3841–3847

Steiner I, Spivack JG, Lirette RP, Brown SM, MacLean AR, Subak-Sharpe J, Fraser NW (1989) Herpes simplex virus type 1 latency-associated transcripts are evidently not essential for latent infection. EMBO J 8: 505–511

Stevens JG (1989) Human herpesviruses: a consideration of the latent state. Microbiol Rev 53: 318–332

Stevens JG, Wagner EK, Devi-Rao GB, Cook ML, Feldman LT (1987) RNA complementary to a herpesvirus gene mRNA is prominent in latently infected neurons. Science 235: 1056–1059

Sunstrum JC, Chrisp CE, Averill DR, Levine M, Glorioso JC (1988) Pathogenicity of glycoprotein C negative mutants of herpes simplex virus type 1 for the mouse central nervous system. Virus Res 11: 17–32

Sussman MD, Lu Z, Kutish G, Afonso CL, Roberts P, Rock DL (1992) Identification of an African Swine fever virus gene with similarity to a myeloid differentiation primary response gene and a neurovirulence-associated gene of herpes simplex virus. J Virol 66: 5586–5589

Taha MY, Clements GB, Brown SM (1989a) A variation of herpes simplex virus type 2 strain HG52 with a 1.5 kb deletion in RL between 0 to 0.02 and 0.82 to 0.83 map units is non-neurovirulent for mice. J Gen Virol 70: 705–716
Taha MY, Clements GF, Brown M (1989b) The herpes simplex virus type 2 (HG52) variant JH2607 has a 1488 bp deletion which eliminates neurovirulence in mice. J Gen Virol 70: 3073–3078
Thelander I, Reichard P (1979) Reduction of ribonucleotides. Annu Rev Biochem 48: 133–158
Thompson RL, Stevens JG (1983) Biological characterization of a herpes simplex virus intertypic recombinant which is completely and specifically non-neurovirulent. Virology 131: 171–179
Thompson RL, Rogers SK, Zerhusen MA (1989) Herpes simplex virus neurovirulence and productive infection of neural cells is associated with a function which maps between 0.82 and 0.832 map units on the HSV genome. Virology 172: 435–450
Venta PJ, Welty RJ, Johnson TM, Sly WS, Tashian RE (1991) Carbonic anhydrase II deficiency syndrome in a Belgian family is caused by a point mutation at an invariant histidine residue (107 His→Tyr): complete structure of the normal human CA II gene. Ann J Hum Genet 49: 1082–1090
Wagner E, Devi-Rao G, Feldman L, Dobson A, Zhang G, Glanagan M, Stevens J (1988) Physical characterization of the herpes simplex virus latency-associated transcript in neurons. J Virol 62: 1194–1202
Weber PC, Levine M, Glorioso JC (1987) Rapid identification of nonessential genes of herpes simplex virus type 1 by Tn5 mutagenesis. Science 236: 576–579
Whitley RJ (1985) Epidemiology of herpes simplex viruses. In: Roizman B (ed) The herpesviruses. Academic, New York, pp 1–44
Williams GT (1991) Programmed cell death: apoptosis and oncogenesis. Cell 65: 1097–1098
WuDunn D, Spear PG (1989) Initial interaction of herpes simplex virus with cells is binding to heparan sulfate. J Virol 63: 52–58
Zhang Y, Sirko DA, McKnight JLC (1991) Role of herpes simplex virus type 1 UL46 and UL47 in α-TIF-mediated transcriptional induction: characterization of three viral deletion mutants. J Virol 65: 829–841

Chapter 14 Herpes Simplex Virus Latency and Immediate Early Gene Repression by the Cellular Octamer-Binding Protein Oct-2

David S. Latchman

Summary

The repeated recurrence of epithelial lesions following initial infection with herpes simplex virus (HSV) is dependent upon the establishment of asymptomatic latent infections of sensory neurons from which the virus can repeatedly emerge to attack susceptible epithelial cells. The failure of the viral lytic cycle and consequent establishment of latent infections in sensory neurons is due to a lack of viral immediate-early (IE) gene expression following initial infection. By using neuronal cell lines, we have shown that this lack of IE gene expression is due to a neuronal cell-specific repressor factor which binds to the TAATGARAT motif in the IE promoters and prevents their transcription. This factor has been identified as a neuronal form of the octamer-binding transcription factor Oct-2, and this protein has been shown to be able to repress the IE promoter when Oct-2 is artifically expressed in normally permissive fibroblasts. Moreover, Oct-2 is expressed in primary sensory neurons both in intact ganglia in vivo and following culture in vitro, and its expression is increased by treatment with nerve growth factor, whose presence is necessary for the maintenance of HSV latency. It is likely therefore that Oct-2 plays a critical role in the establishment and maintenance of HSV latency in vivo.

Introduction

The mucocutaneous lesions produced by infection of humans with herpes simplex viruses (HSVs) types 1 and 2 are caused by lytic infection with the virus with consequent synthesis of viral proteins and viral DNA replication leading to the lysis and death of the infected cell (for review see Spear and Roizman 1980). These lytic infections are readily treatable with the drug acyclovir which is phosphorylated by the viral thymidine kinase enzyme and, when incorporated into DNA, inhibits viral DNA synthesis and prevents the lytic cycle from proceeding.

Department of Molecular Pathology, University College and Middlesex School of Medicine, The Windeyer Building, Cleveland Street, London, W1P 6DB, UK

Following the successful treatment of the initial infection, however, the lesions can recur repeatedly throughout life, resulting in considerable pain and discomfort. This recurrence is dependent upon the fact that, at the same time as the initial lytic infection of susceptible epithelial cells, the virus has moved up the nerve processes of sensory neurons innervating these cells and has established long lived latent infections of the sensory neuron cell bodies in the trigeminal or dorsal root ganglia (for general reviews of latency see Roizman and Sears 1987; Latchman 1990). These latent infections are entirely asymptomatic and do not contain replicating viral DNA, therefore rendering them insensitive to treatment with acyclovir. In response to a variety of stimuli, such as stress or fever, latent virus can be reactivated, and its subsequent migration to the periphery results in the recurrence of the infection in susceptible epithelial cells.

The successful treatment of herpetic disease will therefore require an understanding of the processes which occur in latent infection in order that the recurrent pattern of infection can be controlled. These latent infections also represent a challenging intellectual problem in understanding how a virulent virus which can kill most cell types in a matter of hours is able to establish life long stable asymptomatic infections in sensory neurons.

Although the involvement of the nervous system in HSV infection has been suspected since the early years of the twentieth century (Cushing 1905, for review of the earlier work see Stevens 1975), it is only in the last 10 years that the use of molecular biological approaches has led to some understanding of the processes occurring in latently infected cells. A major step forward was provided by the finding that, with the exception of one region encoding the latency associated transcript (Lat), the viral genome is completely silent in latently infected neuronal cells of humans (Croen et al. 1987) and mice (Stevens et al. 1987). In particular the viral genes encoding the immediate-early (IE) proteins are not transcribed in latently infected cells. Since the transcription of these genes is the first stage in the lytic cycle following infection of susceptible cells and their protein products play an essential role in the subsequent stages of viral gene expression (Preston 1979; Watson and Clements 1980), this explains why the lytic cycle does not occur in latently infected cells.

More recently, much data have emerged which suggest that the initial establishment of latency itself is dependent upon a failure of viral IE gene expression following the initial infection of neuronal cells. Thus mutants lacking the genes for viral proteins such as the virion transactivator Vmw65 (Valyi-Nagy et al. 1991) or the IE protein ICP4 (Katz et al. 1990) readily establish latency in neuronal cells despite the lack of significant viral gene expression following infection. Similarly, latent infection with wild-type HSV can be established in ganglionic neurons in which no prior viral IE gene expression or DNA replication has taken place (Speck and Simmons 1991; Margolis et al. 1992).

Hence both the establishment and maintenance of latent infections in neuronal cells are likely to be critically dependent upon a failure of viral IE gene expression, focussing attention on the manner in which IE gene expression is regulated. In lytic infections, the viral IE genes are *trans*-activated by a complex formed between the HSV virion protein Vmw65 and the cellular transcription factor Oct-1 (O'Hare and Goding 1988; Preston et al. 1988). The Oct-1 protein

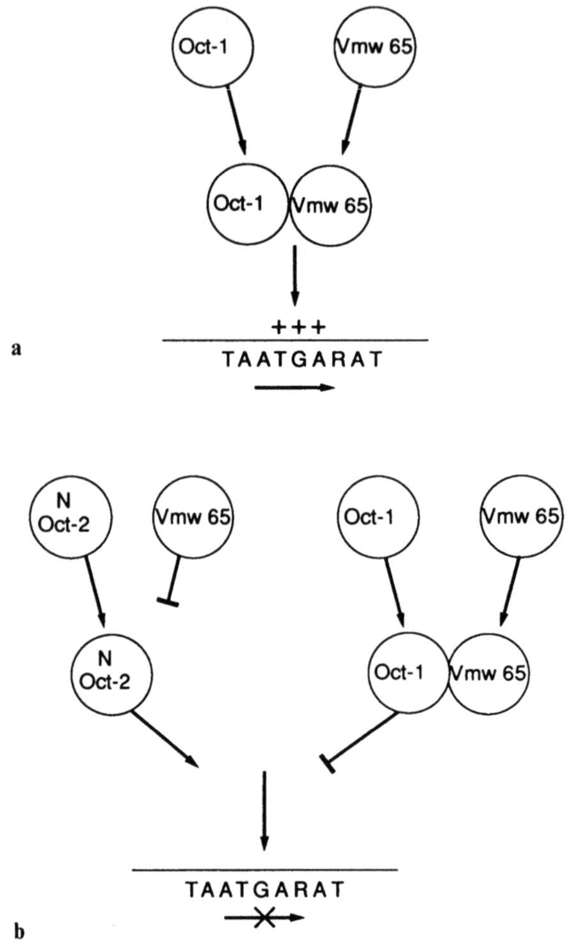

Fig. 1. a Transactivation of the viral IE promoters in permissive cells by binding to the TAATGARAT motif of a complex of the cellular transcription factor Oct-1 and the viral Vmw65 protein which possesses a strong activation domain and can therefore stimulate transcription. **b** Transactivation does not occur in neuronal cells due to the presence of a neuronal form of the Oct-2 protein which does not contain a strong activation domain and cannot complex with the Vmw65 protein. Binding of Oct-2 to the TAATGARAT motif does not therefore stimulate transcription and also prevents binding of the activating complex of Oct-1 and Vmw65

normally binds to the octamer motif present in a number of cellular genes such as those encoding histone H2B and the snRNAs (for review see Falkner et al. 1986). However, the complex of Vmw65 and Oct-1 binds with high affinity to the octamer-related TAATGARAT (R= purine) motif in the viral IE gene promoters and *trans*-activates their expression (Fig. 1a).

It has been suggested that the failure of IE gene expression in neuronal cells may arise from the loss of Vmw65 as the virus moves up the nerve process from the periphery resulting in a lack of *trans*-activation within the sensory neuron nucleus (Roizman and Sears 1987). This is rendered unlikely, however, by the observation that latency can be established in transgenic mice in which Vmw65 has been artificially expressed in all cell types including sensory neurons (Sears et al. 1991). It seems likely therefore that the block to IE gene expression in sensory neurons is due to differences in the complement of cellular octamer binding proteins in these cells compared to permissive cells. This may reflect the presence of insufficient levels of Oct-1 in these cells since this factor has been shown to be present at very low levels in most nondividing neurons (He et al. 1989). Alternatively one of the other octamer-binding proteins specifically present in neuronal cells (Scholer et al. 1989) might be able to repress HSV IE gene expression via the TAATGARAT motif preventing *trans*-activation by the Oct-1/Vmw65 complex.

An understanding of the process regulating latency will therefore require an analysis of the nature of the cellular octamer binding proteins present in sensory neurons and the manner in which they interact with the viral IE gene promoters. Unfortunately such a study is hampered by the small number of sensory neurons which can be obtained from the animal rendering the application of molecular techniques difficult. We have therefore chosen to study the interaction of HSV with a series of immortalized cell lines (ND cells) which were prepared by fusing primary sensory neurons with the C1300 mouse neuroblastoma cell line and selecting for a drug resistance marker present only in the primary cells (Wood et al. 1990). The resulting cell lines proliferate indefinitely in culture but possess many gene expression and electrophysiological characteristics of primary sensory neurons which are not found in the parental neuroblastoma (Wood et al. 1990; Suburo et al. 1992). As will be discussed insights obtained using the large amounts of material available from these cell lines can then be confirmed using the smaller amounts of material available from primary sensory neurons.

Materials and Methods

Cells and Viruses. The ND series of immortalized ganglionic neurons was prepared (Wood et al. 1990) by fusing the C1300 (N18TG2) azaguanine-resistant neuroblastoma (4×10^6 cells) with neonatal rat DRG neurons (5×10^5 cells) in serum-free L-15 medium (GIBCO) containing 50% polyethylene glycol 1500 (Kodak) at 37°C for 1.5 min. Cell lines were then selected in L-15 medium containing HAT, 10% fetal calf serum, and 10% bovine endothelial cell-conditioned medium. Routinely, ND cells were grown in RPMI medium supplemented with 10% fetal calf serum.

DNA Transfection. Transfection of plasmid DNA was carried out according to the method of Gorman (1985). The amounts of DNA in all samples in competition experiments were equalized by use of carrier DNA. Some 24 hrs after transfection, cells were harvested for chloramphenicol acetyl transferase (CAT) assays. All transfections included an internal control plasmid in which the MPSV promoter drives expression of β–galactosidase, to correct for differences in transfection efficiency and for any nonspecific effect of cotransfected plasmids on gene expression.

CAT Assays. Assays of CAT activity were carried out as described by Gorman (1985). Extracts were equalized for protein content as determined by the method of Bradford (1976).

DNA Mobility Shift Assays. Oligonucleotides for use in mobility shift assays were labelled following annealing by phosphorylation with [γ-^{32}P] ATP and T4 polynucleotide kinase. Nuclear extracts were made from about 5×10^7 cells as described by Dignam et al. (1983). For binding, 10 fmol of labelled probe was mixed with 1 µl of extract in 20 mM HEPES, 5mM $Mgcl_2$, 50 mM KCl, 0.5 mM DTT, 4% Ficoll, 2 µg of poly-(dLdC) in 20 µl. Competitor DNA was added at 100-fold molar excess as indicated. The binding reaction was incubated on ice for 40 min, and DNA-protein complexes were then separated by electrophoresis on a 4% polyacrylamide gel. Gels were pre-run for 2 h and then run for 2–3 h at 150 V and 4°C.

Polymerase Chain Reaction. PCR analysis of Oct-2 mRNA expression was carried out according to the method of Kawasaki (1990) using total RNA as a template for production of cDNA. Subsequent amplification of the Oct-2 mRNA was with specific oligonucleotide primers 250 bp apart in the POU domain.

Discussion

Neuronal Cell Lines Contain an Inhibitor of Viral IE Gene Transcription

The ND series of cell lines were derived by the fusion of rat primary sensory neurons (the natural site of HSV latency in vivo) with the immortal C1300 mouse neuroblastoma (Wood et al. 1990). Early studies showed that the C1300 cell line itself is nonpermissive for HSV infection (Vahlne and Lycke 1977, 1978) and this effect is due to a failure of viral IE protein synthesis following infection (Ash 1986) as occurs in latently infected neurons in vivo. In turn this failure of IE protein synthesis is caused by the lack of transcription of the IE genes in HSV infected C1300 cells (Kemp and Latchman 1989). Interestingly, however, the gene encoding

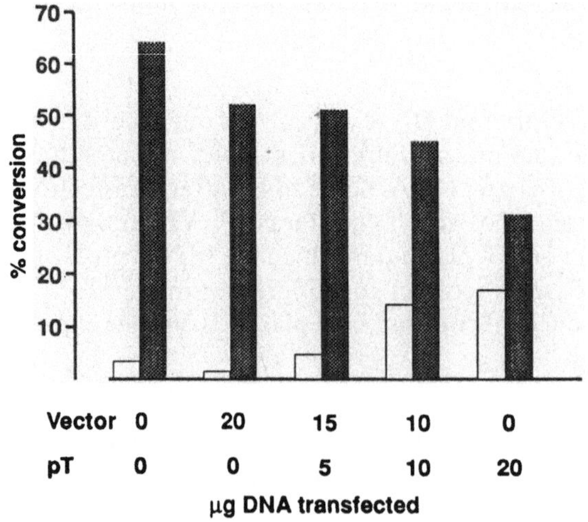

Fig. 2a

b

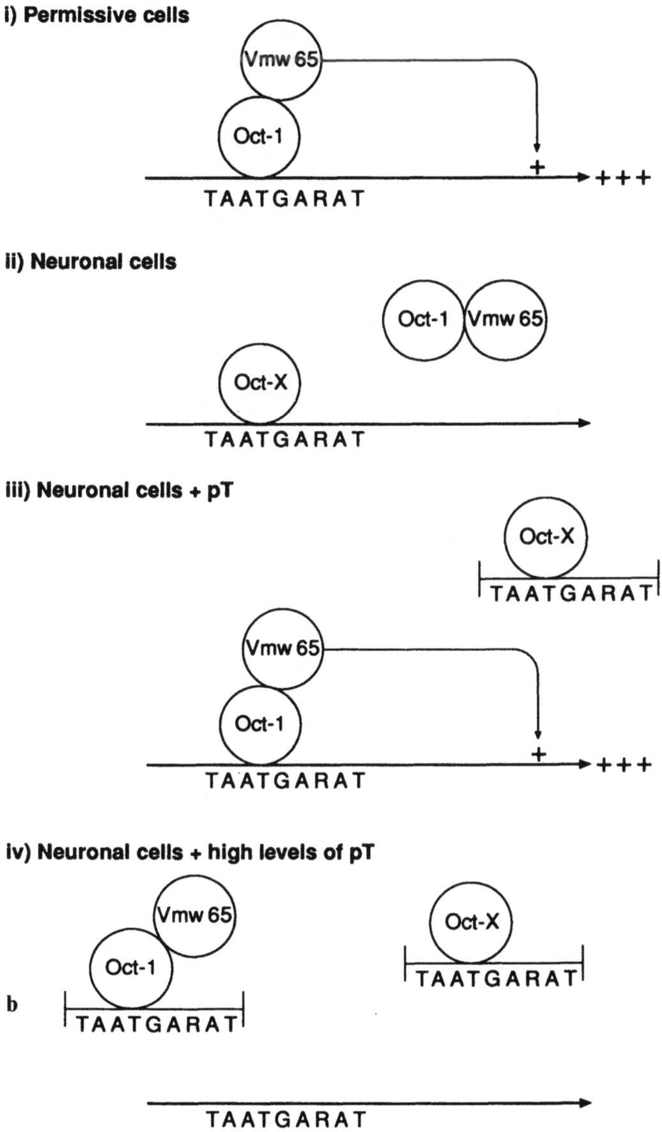

Fig. 2. a Percentage of chloramphenicol acetylated in ND cells (*open bars*) and permissive BHK cells (*closed bars*) following transfection with 5 μg of the IE-CAT plasmid (in which the HSV IE3 gene promoter drives expression of the chloramphenicol acetyltransferase gene) and the indicated amounts (in micrograms) of plasmid vector and the same vector containing a cloned TAATGARAT motif (*pT*). **b** Interaction of cellular octamer binding proteins with the TAATGARAT motif in permissive cells (*panel i*) and in neuronal cells in the absence (*panel ii*) or the presence (*panels iii, iv*) of increasing amounts of competitor plasmid (*pT*) containing a cloned TAATGARAT motif

the viral latency associated transcript (Lat) is not transcribed following infection of C1300 cells (Kemp and Latchman 1989) although this does occur in latently infected cells in vivo (Croen et al. 1987; Stevens et al. 1987), suggesting that C1300 cells may not entirely mimic the processes occurring following initial infection of neuronal cells in vivo.

We therefore studied the effect of HSV on a range of ND cell lines. In these experiments (Wheatley et al. 1990) the ND cell lines exhibited a range of permissiveness for the viral lytic cycle ranging from highly non-permissive to a level of permissiveness similar to that observed in fibroblast cells which readily support the viral lytic cycle in vitro. This range of permissiveness parallels that observed by others for populations of primary sensory neurons both in vitro and in vivo (Kennedy et al. 1983; Margolis et al. 1992). One cell line, ND3, was identified as highly nonpermissive in these experiments and, as in C1300 cells, this non-permissiveness was caused by the lack of IE gene transcription following infection. However, unlike C1300 cells, these cells did transcribe the Lat gene following infection (Wheatley et al. 1990) suggesting that they may represent a better model system. The ND3 cell line and another non-permissive ND cell line, ND7, were therefore used, together with C1300 cells, in subsequent experiments.

Transfection of these cells with constructs in which the HSV IE promoters drive expression of the readily assayable CAT gene revealed a very low level of IE promoter activity compared to that observed in permissive fibroblasts (Fig. 2a; Kemp et al. 1990a,b; Wheatley et al. 1991). Hence the failure of IE gene transcription following infection is dependent on the weak activity of the IE promoters in ND3, ND7 and C1300 cells.

This lack of IE promoter activity could be due to the absence in neuronal cells of a positively acting cellular transcription factor required for IE gene transcription or to the presence in neuronal cells of a negatively acting cellular factor which inhibits IE gene transcription. Our initial studies of both C1300 and ND cells showed that they contained high levels of the positively acting Oct-1 factor (Kemp et al. 1990a; Wheatley et al. 1991) and that this protein was capable of forming a functional complex with the viral Vmw65 protein (Estridge et al. 1990). We therefore considered the second possibility, namely, that neuronal cells might contain an inhibitory protein. If this possibility is correct it should be possible to increase the activity in neuronal cells of an IE promoter driving the CAT gene (IE-CAT) by cotransfecting with it plasmid DNA containing the binding site for the repressor. This will result in the removal of the repressor from the IE promoter in the IE-CAT plasmid resulting in activation of the promoter.

As shown in Fig. 2a this is indeed the case. Cotransfection of IE-CAT with increasing amounts of the pT plasmid containing a cloned TAATGARAT motif resulted in a steady increase in IE-CAT activity in neuronal cells and a similar increase could be observed using an octamer-containing plasmid but not with plasmid vector alone (Kemp et al. 1990a; Wheatley et al. 1991). Hence neuronal cells contain an inhibitory octamer binding protein (Oct-x in Fig. 2b) which following infection binds to the TAATGARAT motif in the IE promoters and prevents binding of the *trans*-activating complex of Oct-1/Vmw65 (Fig. 2b ii). Removal of this inhibitor by excess TAATGARAT sites allows activation of IE promoter by Oct-1/Vmw65 (Fig. 2b). Interestingly we did observe in some experiments that very high levels of the TAATGARAT plasmid could produce some decrease in IE promoter activity from the peak observed at an optimal level of pT. This presumably reflects the removal of Oct-1/Vmw65 from the IE promoter at very high levels of pT (Fig. 2b), and a similar effect is also likely to be responsible for the progressive decrease in IE-CAT activity produced by cotransfection of pT into permissive fibroblast cells containing only Oct-1 and lacking the octamer-binding repressor (Fig. 2a).

As well as showing effects of TAATGARAT and octamer-containing plasmids on IE promoter activity in cotransfection experiments we were also able to increase the permissivity of our neuronal cell lines for HSV infection itself by transfecting them with TAATGARAT or octamer-containing plasmids and subsequently infecting with HSV (Kemp et al. 1990a; Wheatley et al. 1991). Hence the inhibition of viral IE promoter activity by the octamer-binding repressor protein is directly responsible for the failure of the viral lytic cycle after initial infection of these cells.

The Neuronal Inhibitory Protein Is Oct-2

Having shown using functional assays that neuronal cells contain an inhibitory octamer-binding protein, we used DNA mobility shift assays in order to identify this protein. These experiments showed that whilst permissive cells contained only the ubiquitously expressed octamer binding protein Oct-1, our neuronal cells expressed both Oct-1 and another protein which bound specifically to the octamer motif forming a DNA-protein complex of higher mobility than that formed by Oct-1 (Kemp et al. 1990a; Wheatley et al. 1991).

A number of different octamer-binding proteins have been identified in neuronal cells by DNA mobility shift assays (Scholer et al. 1989) or hybridization analysis (He et al. 1989). In particular the Oct-2 protein, which was originally thought to be specific to B cells (Singh et al. 1986; Staudt et al. 1986), has been detected in neuronal cells in the brain on the basis of both these criteria (He et al. 1989; Scholer et al. 1989). In DNA mobility shift assays the additional octamer binding protein detectable in our neuronal cell lines showed a similar mobility to B cell Oct-2 and reacted with a polyclonal antibody to Oct-2 (Lillycrop et al. 1991). Moreover apart from the Oct-1 mRNA, the Oct-2 mRNA was the only mRNA encoding an octamer binding protein detectable in ND cells using PCR, and cDNA cloning of the PCR product revealed that the ND cell Oct-2 protein was identical to the B cell protein in the central POU domain which mediates DNA binding (Lillycrop et al. 1991; Latchman et al. 1992). Hence the octamer-binding repressor protein identified in our earlier experiments is a neuronally expressed form of the Oct-2 protein.

This identification of the repressor as Oct-2 raises a paradox: thus B cell Oct-2 was originally identified as a strong activator of immunoglobulin gene transcription acting via the octamer motif in the gene promoter or enhancer (Muller et al. 1988; Scheidereit et al. 1987) whilst in our experiments neuronal Oct-2 inhibited the IE promoter. Moreover, in additional experiments we showed that the same

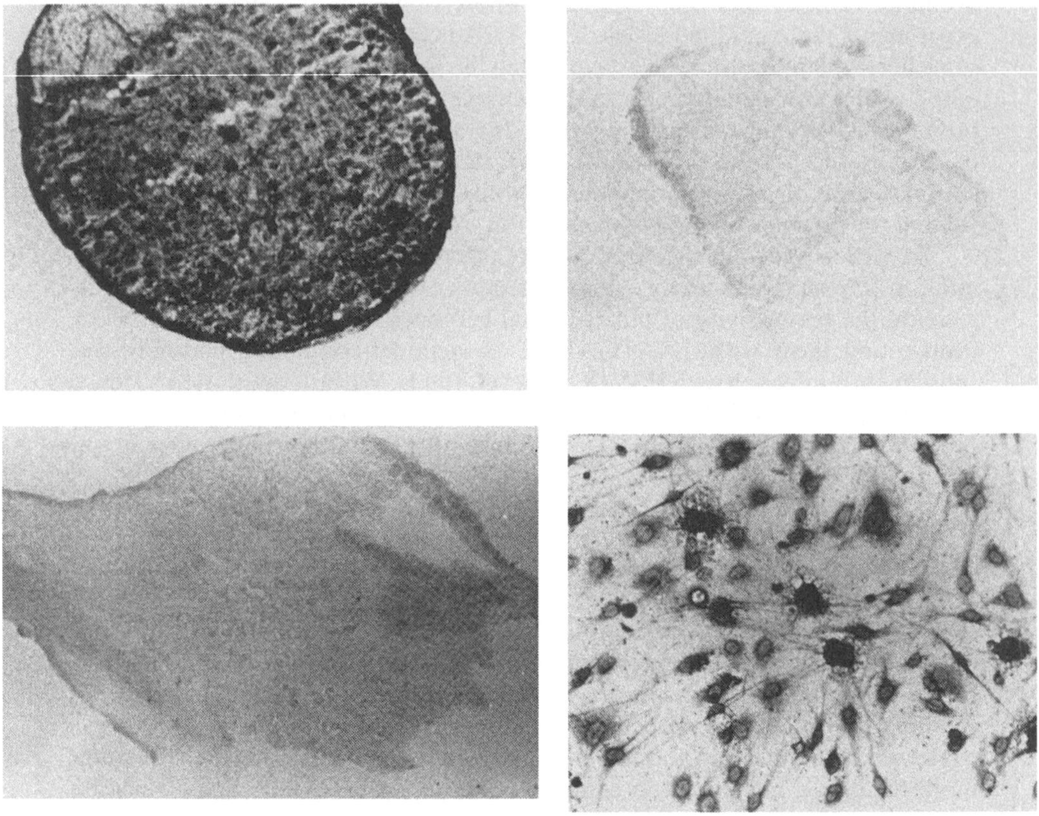

Fig. 3. In situ hybridization histochemical stain using an Oct-2 specific oligonucleotide probe coupled to alkaline phosphatase on adult rat dorsal root ganglion fresh frozen 10 μm sections (× 80 *top left*). Cell bodies of neurons are clearly visible, although not all neurons are stained. Pretreatment of the section with DNase-free ribonuclease A (× 80, *top right*) or addition of 50 n*M* complementary oligonucleotide (× 80, *bottom left*) abolished the signal. Analysis of primary cultures of adult sensory neurons (*bottom right*, × 200, bright-field) showed that staining was restricted to some (>50%), but not all, neuronal cell bodies, with considerable variation in the strength of the signal, and confirmed the fact that nonneuronal cells did not cross-hybridize with the Oct-2 probe

promoter could be stimulated by Oct-2 in B cells and repressed by Oct-2 in neuronal cells (Dent et al. 1991).

This paradox is explained by the finding that the Oct-2 mRNA exists in at least eight different alternatively spliced forms (Wirth et al. 1991; Stoykova et al. 1992). Two of these forms, Oct-2.4 and 2.5, which lack a gene activation domain at the C terminal of the molecule (Wirth et al. 1991) are overexpressed in neuronal cells whilst B cells contain predominantly the Oct-2.1 form, which contains a functional C terminal activation domain (Lillycrop and Latchman 1992). Moreover when the various alternatively spliced forms of Oct-2 are transfected into cells lacking any endogenous Oct-2, both Oct 2.4 and 2.5 can repress the IE promoter whereas Oct-2.1 stimulates it (Lillycrop and Latchman 1992). This experiment therefore directly demonstrates that Oct-2 can repress the IE promoter both in neuronal cells and when artificially expressed in other cell types and indicates that its ability to do so is dependent upon alternative splicing which produces predominantly inhibitory forms in B cells.

The ability of Oct 2.4 and 2.5 to inhibit the IE promoter is likely to be dependent upon both their lack of a strong C terminal activation domain and on their inability to form a complex with the viral Vmw65 protein. This inability of Oct-2 to interact with Vmw65 is mediated by differences between the POU domains of Oct-1 and Oct-2 (Gerster and Roeder 1988; Stern et al. 1989) and will therefore be observed for all the alternatively spliced forms of Oct-2 which are identical in the POU domain region (Lillycrop et al. 1991; Wirth et al. 1991). Hence Oct 2.4 and 2.5 cannot activate IE gene transcription following binding to the TAATGARAT motif since they lack their own activation domain and cannot recruit the strong activation domain of Vmw65 to the promoter (Fig. 1b). The binding of these proteins thus blocks activation of the promoter by the Oct-1/Vmw65 complex and therefore results in the inhibition of the IE promoter observed following infection of neuronal cell lines.

Oct-2 Is Expressed in Primary Sensory Neurons

The experiments described above show that the neuronal Oct-2 protein is responsible for the failure of IE gene expression and hence of the lytic cycle following infection of neuronal cell lines. Having demonstrated the role of Oct-2 using the large amounts of material available from the cell lines, it is necessary to confirm that Oct-2 is present in normal unimmortalized sensory neurons and hence could play a similar role in the failure of the HSV lytic cycle and consequent latent infection in sensory neurons in vivo. As illustrated in Fig. 3, this is indeed the case, the Oct-2 mRNA being readily detectable in neuronal cells in both intact dorsal root ganglia (DRG) dissected directly from the animal or in primary culture, using both PCR and in situ hybridization (Lillycrop et al. 1991). Moreover, the Oct-2 protein itself is detectable in DRG by DNA mobility shift assay (Wood et al. 1992).

Oct-2 and Latency

The demonstration of a role for Oct-2 in the inhibition of the viral lytic cycle, allowing the initial establishment of latency, raises the question of whether changes in the level of Oct-2 play any role in the reactivation of latent virus in response to specific stimuli. Thus a fall in Oct-2 levels would result in binding of Oct-1 to the TAATGARAT motif and the activation of IE gene expression resulting in reactivation of the virus (Fig. 4a).

This mechanism appears to operate in the case of reactivation of the virus by withdrawal of nerve growth factor (NGF). Thus in the original demonstration of

HSV latency by Cushing (1905) cutting of the trigeminal nerve in neuralgia patients resulted in the appearance of herpetic lesions consequent upon viral reactivation. This reactivation is now known to be due to the blockage of retrograde transport of NGF from the innervated tissue to the neuronal cell body, the presence of NGF being essential for the maintenance of viral latency (Wilcox and Johnson 1988). We have now shown that withdrawal of NGF from sensory neuron cultures results in a decrease in the Oct-2 mRNA and protein with no alteration in the level of Oct-1. Hence in the absence of NGF, Oct-2 will no longer be bound to the TAATGARAT motif in the promoter and Oct-1 can bind and activate gene expression (Fig. 4a).

It is likely, however, that other reactivation agents can act by mechanisms which do not involve Oct-2. Thus HSV can be reactivated by treatment of sensory neurons with cyclic AMP (Smith et al. 1992). In this case, however, IE gene

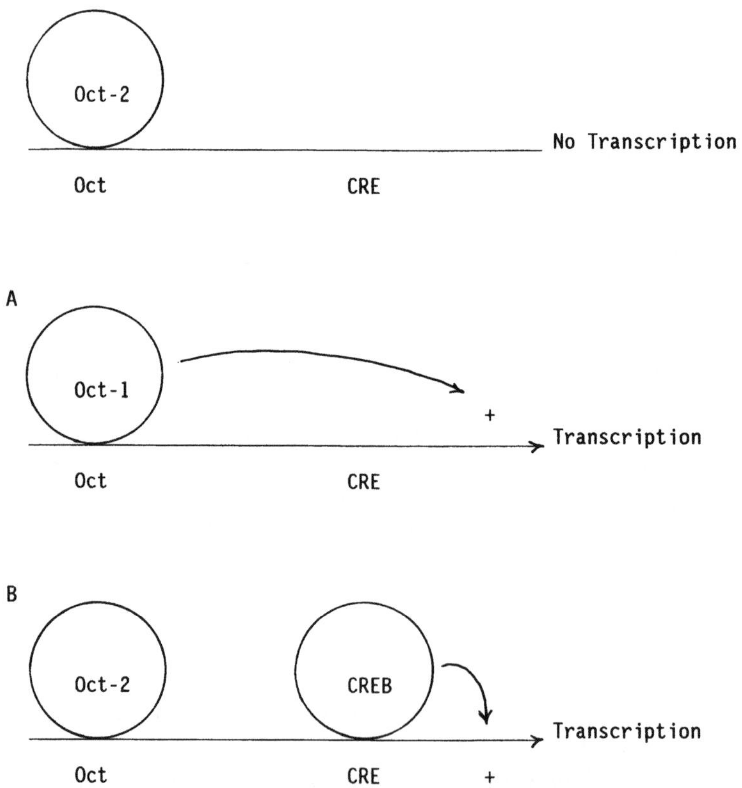

Fig. 4 A,B. Reactivation of latent HSV can occur either via stimuli such as nerve growth factor (NGF) removal which lower Oct-2 levels allowing Oct-1 to bind to the TAATGARAT motif (**A**) or via stimuli such as treatment with cyclic AMP that induce the activity of a positively acting transcription factor (CREB) which binds to a different region of the IE promoter (CRE) by-passing the need for *trans*-activation via the TAATGARAT motif (**B**)

transcription is directly stimulated via a cyclic AMP response element in the IE1 gene promoter (Wheatley et al. 1992). Binding of the CREB protein to this motif following stimulation by cyclic AMP will activate the IE promoter, bypassing the need for activation via the TAATGARAT motif which still has Oct-2 bound to it (Fig. 4b).

The Oct-2 protein is likely therefore to play a critical role in the maintenance of latency as well as its establishment, with some reactivation agents acting by lowering the level of Oct-2 whilst others act by stimulating the IE promoter via a TAATGARAT-independent mechanism.

Conclusion

The work described here has involved the use of cell lines to identify a neuronal form of the Oct-2 protein as being responsible for the failure of viral IE gene expression following initial infection of neuronal cells. The insights obtained using the large amounts of material available from these cells have subsequently been confirmed using the more limited amounts of material available from primary sensory neurons. Thus Oct-2 has been shown to be expressed in these cells both in vivo and in vitro and to be modulated by NGF which is known to play a role in the maintenance of HSV latency. Without the use of neuronal cell lines, it is unlikely that a factor first identified as an activator of immunoglobulin gene expression in B lymphocytes would have been shown to act as a repressor of HSV IE gene expression in neuronal cells and to thereby play a critical role in the initiation and maintenance of latent infection with HSV.

Acknowledgements. I would like to thank all the members of my laboratory and in particular Carolyn Dent, John Estridge, Keith Howard, Lynn Kemp, Karen Lillycrop and Susan Wheatley for their contribution to this work. I am also particularly indebted to John Wood who isolated the ND cell lines for his valuable advice and for an ongoing enjoyable and successful collaboration. Work from my laboratory described in this article is supported by Action Research, the Cancer Research Campaign, the Medical Research Council and the Sir Jules Thorn Charitable Trust.

References

Ash RJ (1986) Butyrate-induced reversal of herpes simplex virus restriction in neuroblastoma cells. Virology 155: 584–592

Bradford M (1976) A rapid and sensitive method for the quantitation of microgram quantitives of proteins utilizing the principle of protein-dye binding. Anal Biochem 72: 248–254

Croen KD, Ostrove JM, Dragovic LH, Smialek JE, Straus SE (1987) Latent herpes simplex virus in human trigeminal ganglia. Detection of an immediate-early gene anti-sense transcript by in situ hybridization. N Engl J Med 317: 1427–1432

Cushing H (1905) Surgical aspects of major neuralgia of trigeminal nerve: report of 20 cases of operation upon the Gasserian ganglion with anatomic and physiologic notes on the consequences of its removal. J Am Med Assoc 44: 1002–1008

Dent CL, Lillycrop KA, Estridge JK, Thomas NSB, Latchman DS (1991) The B cell and neuronal forms of the octamer binding protein Oct-2 differ in DNA binding specificity and functional activity. Mol Cell Biol 11: 3925–3930
Dignam JD, Lebovitz RM, Roeder RG (1983) Accurate transcription initiation by RNA polymerase II in a soluble extract from isolated mammalian nuclei. Nucleic Acids Res 11: 1475–1489
Estridge JK, Kemp LM, Latchman DS (1990) The herpes simplex virus protein Vmw65 can trans-activate both viral and cellular promoters in neuronal cells. Biochem J 271: 273–276
Falkner FG, Moickat R, Zachau HG (1986) Sequences closely related to an immunoglobulin promoter/enhancer element occur also upstream of other eukaryotic and of prokaryotic genes. Nucleic Acids Res 13: 7847–7863
Gerster T, Roeder RG (1988) A herpes virus trans-activating protein interacts with transcription factor OTF-1 and other cellular proteins. Proc Natl Acad Sci USA 85: 6347–6351
Gorman CM (1985) High efficiency gene transfer into mammalian cells. In: Glover DM (ed) DNA cloning, a practical approach, vol 2. IRL Press, pp 143–190
He X, Treacy MN, Simmons DM, Ingraham HA, Swanson LS, Rosenfeld MG (1989) Expression of a large family of POU-domain regulatory genes in mammalian brain development. Nature 340: 35–42
Katz J, Bodin T, Coen DM (1990) Quantitative polymerase chain reaction analysis of herpes simplex virus DNA in ganglia of mice infected with replication incompetent mutants. J Virol 64: 4288–4295
Kawasaki ES (1990) Amplification of RNA In: Innis MA, Gelford DH, Sminsky JJ, White TJ (eds) PCR protocols. Academic, San Diego, pp 21–27
Kemp LM, Latchman DS (1989) Regulated expression of herpes simplex virus immediate-early genes in neuroblastoma cells. Virology 171: 607–610
Kemp LM, Dent CL, Latchman DS (1990a) Transcriptional repression of herpes simplex virus immediate-early genes and of octamer containing cellular promoters in neuronal cells is mediated by the octamer DNA element. Neuron 4: 215–222
Kemp LM, Gelman IH, Silverstein SJ, Latchman DS (1990b). Regulation of herpes simplex virus immediate-early gene promoters in mouse neuroblastoma cells. Neurosci Lett 118: 185–188
Kennedy PGE, Clements GE, Brown SM (1983). Differential susceptibility of human neural cell types in culture to infection with herpes simplex virus. Brain 106: 101–119
Latchman DS (1990) Molecular Biology of herpes simplex virus latency. J Exp Pathol 71: 133–141
Latchman DS, Dent CL, Lillycrop KA, Wood JN (1992) POU family transcription factors in sensory neurons. Biochem Soc Trans 20: 627–631
Lillycrop KA, Latchman DS (1992) Alternative splicing of the Oct-2 transcription factor RNA is differentially regulated in neuronal cells and B cells and results in protein isoforms with opposite effects on the activity of octamer/TAATGARAT-containing promoters. J Biol Chem 267: 24960–24966
Lillycrop KA, Dent CL, Wheatley SC, Beech MN, Ninkina NN, Wood JN, Latchman DS (1991). The octamer-binding protein Oct-2 represses HSV immediate-early genes in cell lines derived from latently infectable sensory neurons. Neuron 7: 381–390
Margolis TP, Sedarati F, Dobson AT, Feldman LT, Stevens JG (1992). Pathways of viral gene expression during acute neuronal infection with HSV-1. Virology 189: 150–160
Muller MM, Ruppert S, Schaffner W, Matthias P (1988). A cloned octamer transcription factor stimulates transcription from lymphoid-specific promoters in non-B cells. Nature 336: 544–551
O'Hare P, Goding CR (1988) Herpes simplex virus regulatory elements and the immunoglobulin octamer domain bind a common factor and are both targets for virion transactivation. Cell 52: 435–445
Preston CM (1979) Control of herpes simplex virus type 1 mRNA synthesis in cells infected with wild-type virus or the temperature sensitive mutant tsK. J Virol 29: 275–284
Preston CM, Frame MC, Campbell MEM (1988) A complex formed between cell components and a herpes simplex virus structural polypeptide binds to a viral immediate-early gene regulatory DNA sequence. Cell 52: 425–434

Roizman B, Sears AE (1987) An inquiry into the mechanisms of herpes simplex virus latency. Annu Rev Microbiol. 41: 543–571

Scheidereit C, Heguy A, Roeder RG (1987) Identification and purification of a human lymphoid-specific octamer-binding protein (OTF-2) that activates transcription of an immunoglobulin promoter in vitro. Cell 51: 783–793

Scholer HR, Hatzopoulos AK, Balling R, Suzuki N, Gruss P (1989) A family of octamer-specific proteins present during mouse embryogenesis: evidence for germline-specific expression of an Oct factor. EMBO J 8: 2543–2550

Sears AE, Hukkanen V, Labow MA, Levine AJ, Roizman B (1991) Expression of the herpes simplex virus 1 α transinducing factor (VP16) does not induce reactivation of latent virus or prevent the establishment of latency in mice. J Virol 65: 2929–2935

Singh H, Sen R, Baltimore D, Sharp PA (1986) A nuclear factor that binds to a conserved sequence motif in transcriptional control elements of immunoglobulin genes. Nature 319: 154–158

Smith RL, Pizer LI, Johnson EM Jr, Wilcox, CL (1992) Activation of second-messenger pathways reactivates latent herpes simplex virus in neuronal cultures. Virology 188: 311–318

Spear PG, Roizman B (1980) Herpes simplex virus. In: Tooze J (ed) DNA tumour viruses, 2nd edition. Cold Spring Harbor Laboratory, Cold Spring Harbor, pp 615–746e

Speck PG, Simmons A (1991) Divergent molecular pathways of productive and latent infection with a virulent strain of herpes simplex virus type 1. J Virol 65: 4001–4005

Staudt LH, Singh H, Sen R, Wirth T, Sharp PA, Baltimore D (1986) A lymphoid-specific protein binding to the octamer motif of immunoglobulin genes. Nature 323: 640–643

Stern S, Tanaka M, Herr W (1989) The Oct-1 homoeodomain directs formation of a multiprotein complex with the HSV trans-activator VP16. Nature 341: 624–630

Stevens JG (1975) Latent herpes simplex virus and the nervous system. In: Arber W et al. (eds) Current topics in microbiology and immunology, vol 70: Springer, Berlin Heidelberg New York, pp 31–50

Stevens JG, Wagner EK, Devi-Rao GB, Cook ML, Feldman LT (1987) RNA complementary to a herpes virus alpha gene mRNA is prominent in latently infected neurons. Science, 235: 1056–1059

Stoykova AS, Steiner S, Erselius JR, Hatzopoulos AK, Gruss P (1992) Mini-Oct and Oct-2c:Two novel functionally diverse murine Oct-2 gene products are differentially expressed in the CNS. Neuron 8: 541–558

Suburo AM, Wheatley SC, Horn DA, Gibson SJ, Jahn R, Fischer-Colbrie R, Wood JN, Latchman DS, Polack JM (1992) Intracellular redistribution of neuropeptides and secretory proteins during differentiation neuronal cell lines. Neuroscience 46: 881–889

Vahlne A, Lycke E (1977) Herpes simplex virus infection of mouse neuroblastoma cells. Proc Soc Exp Biol Med 156: 82–87

Vahlne A, Lycke E (1978) Herpes simplex virus infection of in vitro cultured neuronal cells. J Gen Virol 39: 321–332

Valyi-Nagi T, Deshmane SL, Spivack JG, Steiner I, Ace CI, Preston CM, Fraser NW (1991). Investigation of herpes simplex virus type 1 (HSV-1) gene expression and DNA synthesis during the establishment of latent infection by an HSV-1 mutant in 1814 that does not replicate in mouse trigeminal ganglia. J Gen Virol 72: 641–649

Watson RJ, Clements JB (1980) A herpes simplex virus Type 1 function continuously required for early and late viral RNA synthesis. Nature 285: 329–330

Wheatley SC, Kemp LM, Wood JN, Latchman DS (1990) Cell lines derived from dorsal root ganglion neurons are non-permissive for HSV and express only the latency associated transcript following infection. Exp Cell Res 190: 243–246

Wheatley SC, Dent CL, Wood JN, Latchman DS (1991) A cellular factor binding to the TAATGARAT DNA sequence prevents the expression of the HSV immediate-early genes following infection of nonpermissive cell lines derived from dorsal root ganglion neurons. Exp Cell Res 194: 78–82

Wheatley SC, Dent CL, Wood JN, Latchman DS (1992) Elevation of cyclic AMP levels in cell lines derived from latently infectable sensory neurons increases their permissivity for herpes virus infection by activating the viral immediate-early 1 gene promoter, Mol Brain Res 12: 149–154

Wilcox C, Johnson EM (1988) Characterization of nerve growth factor-dependent herpes simplex virus latency in neurons in vitro. J Virol 62: 393–399

Wirth T, Priess A, Annweiler A, Zwilling S, Oeler B (1991) Multiple Oct-2 isoforms are generated by alternative splicing. Nucleic Acids Res 19: 43–51

Wood JN, Bevan SJ, Coote P, Darn PM, Hogan P, Latchman DS, Morrison C, Rougon G, Theveniau M, Wheatly SC (1990) Novel cell lines display the properties of nociceptive sensory neurons. Proc R Soc [B] 241: 187–194

Wood JN, Lillycrop KA, Dent CL, Ninkina NN, Beech MM, Willoughby JJ, Winter J, Latchman DS (1992) Regulation of expression of the neuronal POU protein Oct-2 by nerve growth factor. J Biol Chem 267: 17787–17791

Chapter 15 The Cell Fusion Protein Gene (*UL53*) of Herpes Simplex Virus Type 1 – A Pathogenicity Gene

Michal Moyal and Yechiel Becker

Summary

The pathogenicity of HSV-1 is determined by specific genes, designated pathogenicity genes. The *UL53* gene that codes for the cell fusion protein is considered a pathogenicity gene and determines the neurovirulence of the virus. HSV-1 R15, which has a mutation in the *UL53* gene, lost its neurovirulent phenotype.

The results presented here demonstrate that the defect in the *UL53* gene of HSV-1 R15 is not at the level of the expression of the gene. The mutations found in the amino acid sequence of the gene are thought to be responsible for its apathogenic phenotype. We suggest that the UL53 protein is translated from an internal ATG and is therefore shorter than expected. The sequence of the UL53 protein was analyzed by computer programs and its putative structure is described. The relevance of the cell fusion function of the protein to neurovirulence is discussed.

Introduction

The pathogenicity of HSV-1 for mice is usually determined by the ability of the virus to reach the central nervous system, to infect it, and to cause the death of the animal. The ability of the virus to kill the animal while present in the nervous system is referred to as its intracerebral (IC) pathogenicity. Many efforts have been made to identify the viral gene(s) responsible for IC pathogenicity. Although it is now clear that IC pathogenicity is a multigenic trait, most of the findings suggested the location of the gene to span coordinates 0.71–0.83 mu (Thompson and Stevens 1983; Choe and Roizman 1986; Choe et al. 1990). Thompson and Stevens (1983) created an intratypic HSV-1–HSV-2 recombinant (RE-6) which lost its neurovirulence. In further experiments, it was shown that this property could be rescued by a DNA fragment that contains the *BamHI* DNA fragment L (coordinates 107,948–113,322; Thompson et al. 1983; Thompson and Wagner 1989).

HSV-1 R15 is apathogenic to mice (Rösen et al. 1985), despite proliferation of the virus upon direct injection into the brain. A recombination experiment was carried out to reconstitute the IC pathogenicity of R15. It was shown that this

The Department of Molecular Virology, Faculty of Medicine, The Hebrew University of Jerusalem, Jerusalem, Israel

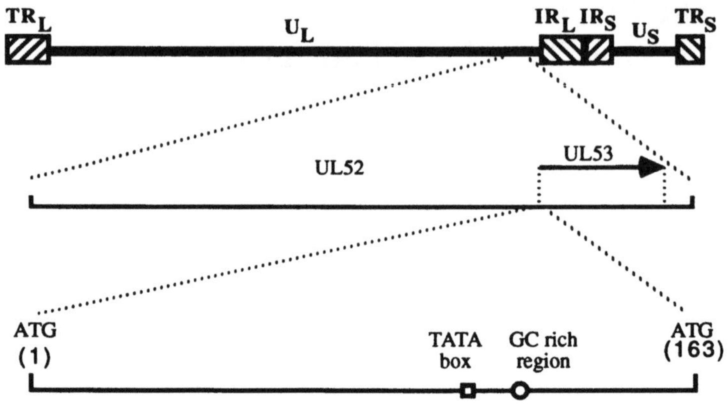

Fig. 1. The position of the *BamHI* DNA fragment L in the HSV-1 genome. The location of the *UL53* gene is indicated by an arrow and the putative promoter region is illustrated on the bottom line

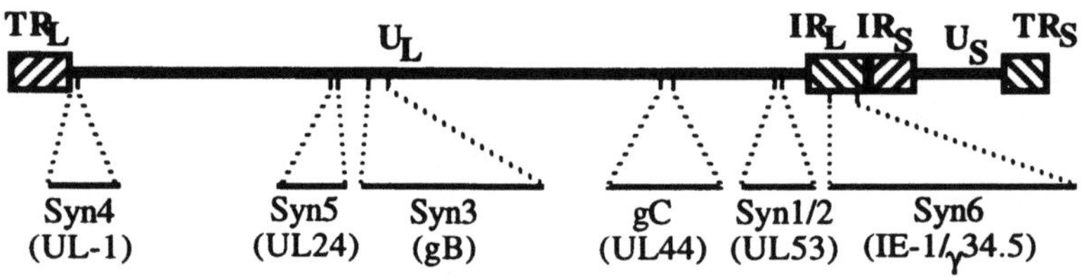

Fig. 2. The position of the syn loci in HSV-1 genome

property can be rescued by the *NruI–BamHI* DNA fragment (coordinates 111, 290–113,322) on the right-hand side of the *BamHI* fragment L (Ben-Hur et al. 1987). This DNA fragment carries the syn 1 locus that determines the ability of the virus to induce fusion of the infected cells and to form a syncytium (Debroy et al. 1985; Pogue-Guile et al. 1984, 1987). This property is encoded by the viral gene for the cell fusion protein designated UL53 (McGeoch et al. 1988; Fig. 1).

At least five additional viral genomic loci determine plaque morphology and size (designated syn loci). These sites code for viral components involved in adsorption and penetration of the virus into the cell (Bond and Person 1984; Machuca et al. 1986; Cai et al. 1988; Tognon et al. 1990; Romanelli et al. 1991; see Fig. 2). In this short review, we summarize our findings concerning the *UL53* gene and demonstrate the relevance of its expression to virus pathogenicity.

Results

UL53 is a Late Viral Gene

To preliminarily characterize the UL53 gene, an experiment was designed to determine the expression group to which it belongs. BSC-1 cells were infected with

HSV-1 in the presence of the inhibitors cycloheximide and phosphonoacetic acid, and the expression of *UL53* gene, as detected by the northern blot hybridization, was compared to that of the untreated cells. The expression of the UL53 gene was detected only in the untreated infected cells, suggesting its classification as a late (γ) virus gene.

HSV-1 R15 Expresses the *UL53* Gene in Infected Cell Cultures

As mentioned before, HSV-1 R15 lost the neurovirulent phenotype, but this could be restored by a DNA fragment containing the *UL53* gene of a neurovirulent strain (Ben-Hur et al. 1987). Therefore, the ability of R15 to express the *UL53* gene in cell culture was studied and compared to that of the pathogenic virus strains. We were able to demonstrate by northern blot experiments that R15, as well as the pathogenic strains, expressed *UL53*-specific mRNA to a similar extent (Moyal et al. 1992).

HSV-1 R15 Expresses the *UL53* Gene in Brains of Infected Mice

Since we assumed that the expression of the *UL53* gene in the brains of infected mice determines the neurovirulent phenotype of the virus, we studied the ability of HSV-1 R15 to express this gene in the infected brains. The use of the existing method to purify RNA from infected brains resulted in marked degradation of the RNA. Demonstration of the existence of viral-encoded mRNA was possible only by the use of a careful and rapid method for RNA extraction; the adsorption of the mRNA to oligo deoxythymidine (dT)-linked magnetic beads. The RNA purified by this method was then subjected to first-strand cDNA synthesis followed by polymerase chain reaction (PCR). (The experimental approach is presented in Fig. 3.) The combination of these techniques enabled us to detect the viral mRNA in brains of infected mice, and it was found that R15 was able to express the *UL53* gene in the infected brains, as was the pathogenic strains (Moyal et al. 1992).

The findings of the experiments described above did not reveal any defect in the control of the expression of the *UL53* gene; it was therefore possible to assume that R15 encodes for a defective *UL53* protein product which does not function properly and thus confers an apathogenic phenotype.

The Sequence of the *UL53* Gene of R15 is Mutated

The sequences of the UL53 gene of several HSV-1 strains have already been published by various research groups (Debroy et al. 1985; Pogue-Guile et al. 1984, 1987; McGeoch et al. 1988). We determined the sequence of the *UL53* gene of HSV-1 R15 and compared it with that of HSV-1 R19 (also determined by us; Moyal et al. 1992) and with the published sequences. Table 1 describes the mutations which were found in the amino acid sequence of the UL53 polypeptide of R15, as deduced from its nucleotide sequence. The effect of these mutations on

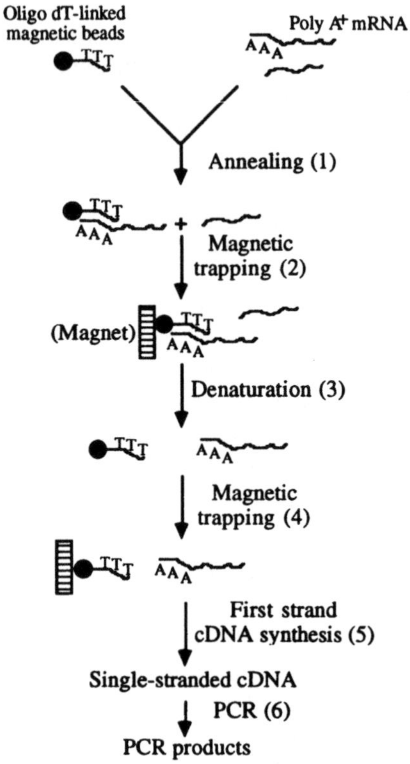

Fig. 3. The experimental approach for the isolation of mRNA from tissues and cells and the demonstration of its existence. Cell homogenate was mixed with oligo deoxythymidine (dT)-linked magnetic beads and RNA annealing was allowed (*1*). The poly A^+ mRNA (annealed to the beads) was trapped by a magnet (*2*) and separated from the beads by denaturation (*3*). Poly A^+ mRNA was purified (*4*) and subjected to first-strand cDNA synthesis (*5*) followed by polymerase chain reaction (*PCR*; *6*)

the nature of the mutated UL53 protein was studied by computer analysis using the program PEPTIDESTRUCTURE, which predicts the secondary structure of proteins (Devereux et al. 1984; Chou and Fasman 1987). The native UL53 protein has three hydrophobic regions that may serve as transmembrane domains. Two of the mutations of R15 are located in one of the hydrophobic regions and reduce its hydrophobicity. The changes might affect either the packaging of the protein in the infected cell membrane or its function. Therefore, we tend to believe that R15 expresses the *UL53*, but that the mutated product is not functional, resulting in an avirulent phenotype.

Table 1. Comparison of mutations in the *UL53* gene of HSV-1 R15 with HSV-1 R19, syn17, KOS and MP

Position of amino acids	R15	R19	Syn 17[a]	KOS[b]	MP[b]
82	C (TGC)	S (TCG)	C (TGC)	S (TCG)	S (TCG)
101	M (ATG)	M (ATG)	M (ATG)	T (ACG)	M (ATG)
223	M (ATG)	I (ATC)	M (ATG)	I (ATC)	I (ATC)
224	L (CTA)	V (GTA)	L (CTA)	V (GTA)	V (GTA)
309	V (GTG)	V (GTG)	V (GTG)	M (ATG)	M (ATG)

[a] McGeoch et al. (1988). [b] Debroy et al. (1985).

Translation of the *UL53* Gene Begins at an Internal ATG Codon

An RNase protection experiment was performed to map the 5′ terminus of the *UL53* gene. A DNA fragment, spanning the putative 5′ end of the *UL53* mRNA was cloned in a transcription vector and transcribed in vitro. The synthetic RNA transcripts were hybridized with mRNA extracted from cells infected with HSV-1 and subjected to single-stranded RNase digestion. Surprisingly, the protected fragment was shorter than expected, locating the 5′ end of the mRNA downstream of the ATG initiation codon. This result raised the possibility that the mRNA of the *UL53* gene does not include the complete open reading frame (ORF) and that the translation starts at an internal ATG codon. Analysis of the sequence revealed the presence of an additional ATG codon which is located 55 amino acids upstream of the first ATG codon (Moyal and Becker, submitted for publication).

To study the possibility that the UL53 protein is shorter than previously suggested (McGeoch et al. 1988), a hybrid selection experiment was done to isolate the *UL53* gene-specific mRNA, which was then translated in vitro. The translation product that was obtained in this experiment had a molecular mass of 32 kDa. This product was shorter than the expected size of the *UL53* original ORF (37.5 kDa) and corresponded to the size of the protein translated from the internal ATG (30.5 kDa). The computer analysis of the sequence also revealed the presence of promoter sequences downstream of the internal ATG which enable the transcription of the shorter mRNA. According to these findings, we also suggest a new location for the *UL53* ORF (coordinates 112,341–113,193).

Discussion

Special emphasis is given in our research to the study of the pathogenicity genes of HSV-1 and the roles of these genes in the pathogenic phenotype of the virus. In addition to several host factors, the genetic makeup of the virus is a major determinant in viral pathogenicity. In this short review, we describe our studies on the *UL53* gene and the relevance of its molecular properties to viral pathogenicity.

The *UL53* gene is a late viral gene and is expressed in the infected cells after the replication of the viral genome. The late proteins usually serve as structural components of the virion. The *UL53* gene is known to code for a cell fusion protein that determines plaque size and morphology (Debroy et al. 1985; Pogue-Guile et al. 1984, 1987). The protein might function in both the initial penetration of the virus into the infected cells and in the spread of the virus to neighboring cells after cycles of replication.

HSV-1 R15 lost intracerebral pathogenicity due to mutations in the *UL53* gene (Ben-Hur et al. 1987; Moyal et al. 1992). The virulent phenotype of R15 could be rescued by the replacement of a DNA fragment containing the *UL53* with an analogous fragment from a neurovirulent strain. No differences in the expression of the *UL53* gene of R15 and the pathogenic strains were observed in cell culture.

The expression of the *UL53* gene was also studied in the brains of HSV-1-infected mice. The demonstration of the presence of *UL53*-specific transcripts in the infected brains was made possible by the use of the combination of a rapid method to extract the RNA and a sensitive method to detect its existence. The RNA was purified from the brains of the infected mice with the aid of oligo dT-linked magnetic beads. This rapid method was found by us to be efficient for the purification of RNA from tissues rich in RNase. The short time of the purification of the RNA reduces the time of the exposure of the RNA to cellular RNases and therefore results in a high yield of intact mRNA. The existence of *UL53*-specific mRNA was demonstrated by the PCR method, preceded by first-strand cDNA synthesis. The combination of the two methods was applied by us to several additional systems in which it was found to be highly efficient.

The results of the PCR experiments revealed the expression of the *UL53* gene in the brains of infected mice in all of the tested strains, including R15. According to these results, it was possible to assume that the defect in the *UL53* gene of R15 is not at the transcriptional level, but might be either at the translational level or in the functioning of the protein.

The first step in studying these questions was the determination of the sequence of the *UL53* gene of HSV-1 strains R15 and R19. The comparison revealed the presence of mutations in the sequence of the *UL53* gene of R15. As was mentioned before, the *UL53* gene codes for a cell fusion protein, while the results of our experiments suggested the involvement of the *UL53* gene product in neurovirulence. These findings underlined the need for a theoretical model which offers a connection between the ability of the virus to induce cell fusion and its neurovirulent phenotype. Such a model was suggested by T. Ben-Hur (Chap. 16, this volume). According to this model, the neurovirulence of the virus is affected not only by its ability to proliferate in the brain, but also by the activity of the viral cell fusion protein. The model suggests that the cell fusion protein acts in the central nervous system as a neurotoxin that interferes with the normal functioning of the cells. The proposed mechanism of action of the UL53 protein as a neurotoxin involves interference in the conduction of nervous stimuli by the induction of electrical coupling (as a consequence of the fusion activity).

The size of the UL53 protein was found by us to be shorter than was deduced from the location of the ORF. According to our results, the size of the UL53 protein is 32 kDa and the location of the ORF spans coordinates 112,341–113,193.

The determination of the sequence of R15, the comparison with the pathogenic strains, and the identification of the internal ATG enabled us to analyze the structure of the UL53 gene and to suggest its organization in the membrane. The analysis of the amino acid sequence was done by the program PEPTIDESTRUCTURE from the UWGCG software package (Devereux et al. 1984; Wolf et al. 1984). This program predicts the secondary structure of the protein, the hydrophilicity of each amino acid, and its probability of appearing at the surface of the protein. Application of the PEPTIDESTRUCTURE program to the deduced amino acid sequence of the UL53 protein provided us with the following conclusions (see Table 2):

Table 2. Domains of interest in UL53 protein

Position[a]	Size	Property	Suggested function
55–102	48	Hydrophobic	Signal peptide
120–139	20	Hydrophobic	Membrane anchorage
205–236	32	Hydrophobic	Membrane anchorage
294–325	32	Hydrophobic	Membrane anchorage

[a] The position of the amino acids is indicated according to the coordinates of McGeoch et al. (1988). Position 55 of McGeoch is equivalent to position 1 according to our numbering system.

1. A highly hydrophobic sequence, 47 amino acids long, occurs in the amino terminus of the putative protein and can serve as a signal peptide. The sequence is followed by two successive arginine residues, which are often known as proteolytic cleavage sites.

2. Three hydrophobic sequences with the characteristics of membrane-anchorage domain occur in the sequence of the UL53 protein at positions 120–139, 205–236, and 294–325. All of these domains are longer than 19 amino acids, with an average hydrophobicity of 1.6 (these criteria are in agreement with those determined by Kyte and Doolittle (1982) to define a membrane-anchorage domain).

3. Two of the mutations in R15 at position 223 and 224 lie in the second hydrophobic domain and reduce its hydrophobicity.

According to these findings, we would like to suggest that the translation of the *UL53* gene begins at an internal ATG codon and spans coordinates 112,341–113,193. The UL53 protein is located in the cell membrane and its packaging in the membrane affects its function. The mutations in the UL53 protein of R15 reduce the hydrophobicity of one of the membrane-anchorage domains and probably change its organization. The mutated UL53 protein cannot function properly and therefore confers an avirulent phenotype to HSV-1 R15.

Acknowledgments: We would like to thank Prof. G. Darai and Dr. A. Rösen-Wolff for providing us with the HSV-1 R15 recombinant and for fruitful collaboration. This study was supported by the Foundation for Molecular Virology and Cell Biology, Phoenix, Arizona.

References

Ben-Hur T, Asher Y, Tabor E, Darai G, Becker Y (1987) HSV-1 virulence for mice by the intracerebral route is encoded by the BamHI-L DNA fragment containing the cell fusion gene. Arch Virol 96: 117–122

Bond VC, and Person S (1984) Fine structure physical map locations of alterations that affect cell fusion in herpes simplex type 1. Virology 132: 368–376

Cai W, Gu B, Person S (1988) Role of glycoprotein B of herpes simplex virus 1 in viral entry and cell fusion. J Virol 62: 2596–2604

Choe J, Roizman B (1986) The terminal 'a' sequence of herpes simplex virus genome contains the promoter of a gene located in the repeat sequence of the L component. J Virol 57: 628–637

Choe J, Kern ER, Whitley RJ, Roizman B (1990) Mapping of herpes simplex virus I neurovirulence to g 34.5 a gene nonessential for growth in culture. Science 250: 1262–1266

Chou PY, Fasman GD (1987) Prediction of the secondary structure of proteins from their amino acid sequence. Adv Enzymol 47: 45–147

Debroy C, Pederson N, Person S (1985) Nucleotide sequence of a herpes simplex virus type 1 gene that causes cell fusion. Virology 145: 36–48

Devereux J, Haeberli P, Smithies O (1984) A comprehensive set of sequence analysis programs for the VAX. Nucleic Acids Res 10: 305–321

Kyte J, Doolittle RF (1982) A simple method for displaying hydropathic character of a protein. J Mol Biol 157: 105–132

Machuca I, Jackumont B, Epstein A (1986). Multiple adjacent or overlapping loci affecting the level of gC and cell fusion mapped by intratypic recombinants of HSV-1. Virology 150: 117–125

McGeoch DJ, Dalrymple MA, Davison AJ, Dolan A, Frame MC, NcNab D, Perry LJ, Scott JE, Taylor P (1988). The complete DNA sequence of the long unique region in the genome of herpes simplex virus type 1. J Gen Virol 69: 1531–1574

Moyal M, Berkowitz C, Rösen-Wolff A, Darai G, Becker Y (1992). Mutations in the UL53 gene of HSV-1 abolish virus neurovirulence to mice by the intracerebral route of injection. Virus Res 26: 99–112

Pogue-Guile KL, Spear MP (1987). The single base pair substitution responsible for the syn phenotype of herpes simplex type 1 strain MP. Virology 157: 67–74

Pogue-Guile KL, Lee GTY, Shapira SK, Spear MP (1984). Fine mapping of mutations in the fusion-inducing MP strain of herpes simplex virus type 1. Virology 136: 100–109

Romanelli MG, Cattozzo EM, Faggiolio L, Tognon M (1991). Fine mapping and characterization of the syn 6 locus in herpes simplex virus 1 genome. J Gen Virol 72: 1991–1995

Rösen A, Gelderblom H, Darai G (1985) Transduction of virulence in herpes simplex virus type 1 from a pathogenic to an apathogenic strain by a cloned viral DNA fragment. Med Microbiol Immunol 173: 257–278

Thompson RL, Stevens JG (1983) Biological characterization of a herpes simplex virus intertypic recombinant which is completely and specifically non-neurovirulent. Virology 131: 171–179

Thompson RL, Wagner EK (1989) Partial rescue of herpes simplex virus neurovirulence with 3.2 kb cloned DNA fragment. Virus Genes 1: 261–273

Thompson RL, Wagner EK, Stevens JG (1983) Physical location of a herpes simplex virus type 1 gene function(s) specifically associated with a 10 million fold increase in HSV neurovirulence. Virology 131: 180–192

Tognon M, Guansalini R, Romanelli MG, Mansrvigi R, Trevisani B (1990). Phenotypic and genotypic characterization of locus syn 5 in herpes simplex virus 1. Virus Res 18: 135–150

Wolf H, Modrow S, Motz M, Janeson BA, Hernann G, Fortsch B (1988). An integrated family of amino acids programs. CABIOS 4: 187–191

V. Effect of Cellular Defenses on Virus Pathogenicity

Chapter 16 Role of Langerhans Cells and Other Dendritic Cells in the Pathogenesis of Herpes Simplex Virus Type 1 Infection

Eli Sprecher and Yechiel Becker

Summary

Langerhans cells are part of a vast system of potent antigen-presenting cells known under the name of dendritic cells. During the last decade, much has been learned about dendritic cell involvement in the immune response to viral diseases. This review briefly summarizes our current understanding of the role played by Langerhans cells and other dendritic cells in the pathogenesis of herpes simplex virus type 1 infection. These findings may foster the development of innovative approaches in the diagnosis, prevention, and treatment of herpetic infections.

Introduction

Herpes simplex virus type 1 (HSV-1) is responsible for the occurrence of primary and recurrent infections in mucocutaneous tissues. It is also the most commonly identified cause of sporadic viral encephalitis in the Western world (Wheeler 1988). Primary disease mostly affects young children and presents as a severe stomatogingivitis, although asymptomatic infection is frequent. Seroprevalence in adults is said to reach 90%. Primary cutaneous infection with HSV-1 initiates with viral replication in epithelial tissues and is followed by transfer of the virus to sensory ganglia, where latent infection is established (Roizman and Sears 1987; Wheeler 1988). In susceptible individuals, exposure to various, well-defined stimuli, including light trauma to the skin and solar irradiation (Roizman and Sears 1987; Wheeler 1988), is followed by the development of recurrent infection. Recurrent infection results from activation of latent virus and its transport back to the site of primary infection by axonal transport to the skin (Roizman and Sears 1987; Wheeler 1988).

Numerous studies have convincingly established the central role played by cell-mediated immunity in the response to both primary and recurrent HSV-1 infection (Bonneau and Jennings 1989; Larsen et al. 1984; Nash et al. 1987; Schmid

Department of Molecular Virology, Faculty of Medicine, The Hebrew University of Jerusalem, Jerusalem, Israel

1988). Cellular responses are dependent on initial processing and subsequent presentation of stimulating antigens to T lymphocytes by specialized antigen-presenting cells (Brodsky and Guagliardi 1991). Since Langerhans cells function as potent antigen-presenting cells in the epidermis (Cruz and Bergstresser 1990; Sprecher and Becker 1991a) and HSV-1 is known for its tropism to epithelial tissues (Roizman and Sears 1987; Wheeler 1988), the role of Langerhans cells in the pathogenesis of HSV-1 infection has become an important topic of investigation in the past years.

Langerhans cells are located in the suprabasal layer of the epidermis (Cruz and Bergstresser 1990; Sprecher and Backer 1991a). They display strong membranal adenosine triphosphatase (ATPase) activity and express major histocompatibility complex (MHC) class I and class II antigens, conditioned response 3 (CR3; CD11b/CD18) and Fc immunoglobulin G (Fc-IgG; FcRII) receptors, CD1a and CD4 antigens and as well as a series of adhesion proteins (Cruz and Bergstresser 1990; De Panfilis et al. 1991; Sprecher and Backer 1991a; Le Varlet et al. 1991; Simon et al. 1991). In addition, characteristic rod-shaped organelles, named Birbeck granules, are found in their cytoplasm (Bartosik 1992; Cruz and Bergstresser 1990; Sprecher and Becker 1991a).

Langerhans cells belong to the dendritic cell lineage (Metlay et al. 1989; Steinman 1991). Dendritic cells are unique among antigen-presenting cells in their ability to stimulate *primary* immune responses (Metlay et al. 1989; Steinman 1991). They also compare favorably with other antigen-presenting cells in their capacity to trigger primed lymphocyte *secondary* responses (Metlay et al. 1989; Steinman 1991).

Dendritic cells originate in the bone marrow, from where they migrate to peripheral tissues (Metlay et al. 1989; Steinman 1991). Dendritic cells are termed according to their location. In the epidermis, they are named Langerhans cells, while in the thymus and the paracortical area of the lymph nodes, they are known under the name of interdigitating cells (Cruz and Bergstresser 1990; Metlay et al. 1989; Sprecher and Becker 1991a; Steinman 1991). Langerhans cells differ from the dendritic cells found in lymphoid organs both phenotypically and functionally (Cruz and Bergstresser 1990; Metlay et al. 1989; Steinman 1991; Streilein et al. 1990). In vitro culture of Langerhans cells lead to their transformation into *mature* dendritic cells resembling those found in lymphoid organs (Cruz and Bergstresser 1990; Metlay et al. 1989; Steinman 1991; Streilein et al. 1990).

In contrast to Langerhans cells, mature dendritic cells do not display membranal ATPase activity and do not express Fc-IgG receptors (Cruz and Bergstresser 1990; Metlay et al. 1989; Sprecher and Becker 1991a; Steinman 1991). In addition, their cytoplasm does not contain Birbeck granules (Cruz and Bergstresser 1990; Metlay et al. 1989; Sprecher and Backer 1991a; Steinman 1991).

Langerhans cells function by taking up antigens in the epidermis and transferring them to the paracortical area of the draining lymph nodes, where they present them, as lymph node interdigitating cells, to T helper cells for a systemic immune response to develop (Cruz and Bergstresser 1990; Metlay et al. 1989; Sprecher and Becker 1991a; Steinman 1991). Langerhans cell migration to the lymph nodes is thought to be regulated by various cytokines, including tumor necrosis factor-α (TNF-α) (Cumberbatch and Kimber 1992). Epidermal

Langerhans cells process native antigens with great efficacy, but are unable to stimulate the proliferation of resting T cells (Cruz and Berstresser 1990; Steinman 1991; Streilein et al. 1990). During their migration to the draining lymph nodes. Langerhans cells progressively loose their processing capacity and acquire the phenotypic and functional characteristics of mature dendritic cells, including the ability to stimulate resting T cell proliferation (Aiba and Katz 1990; Cruz and Bergstresser 1990; Cumberbatch et al. 1991a,b, 1992; Steinman 1991; Streilein et al. 1990). In contrast to mature dendritic cells, other antigen-presenting cells such as macrophages and B cells, poorly stimulate resting T cells, although they efficiently stimulate primed T cell responses (Cruz and Bergstresser 1990; Cruz et al. 1991; Steinman 1991; Streilein et al. 1990).

To elucidate the role played by Langerhans cells in the responses to HSV-1 infection, several questions needed to be answered: (a) Are Langerhans cells able to present HSV-1 antigens to stimulate both naive and primed T lymphocytes in vitro? (b) Do Langerhans cells migrate to the site of HSV-1 infection in vivo to take up viral antigens in the epidermis? (c) Do Langerhans cells transfer HSV-1 antigens to the lymph nodes draining the site of infection to stimulate systemic immunity? (d) Do Langerhans cells play any regional role in the cutaneous immune response to the viral infection? (e) Do Langerhans cells play any role in the pathogenesis of recurrent HSV-1 infection? and (f) Do Langerhans cells represent a potential target for innovative antiviral strategies in herpetic diseases?

During the recent years, a number of studies have generated fascinating, but still fragmentary, answers to these and other questions. The present review briefly summarizes these recent advances.

Herpes Simplex Virus Type 1 Antigen Presentation by Dendritic Cells

Numerous reports have firmly demonstrated the outstanding ability of Langerhans cells and other dendritic cells, as compared with macrophages, to stimulate the in vitro proliferation of primed T cells in the presence of HSV-1 antigens (Braathen et al. 1980; Hayashi and Aurelian 1986; Otani and Mori 1987; Sprecher and Becker 1987; Vestey et al. 1990; Williams et al. 1991; Yasumoto et al. 1986, 1987). In addition, Langerhans cells were found to serve as efficient antigen-presenting cells in an in vitro antibody induction assay in which HSV-1 primed T cells, epidermal cells, and HSV-1 TNP, as an antigen, stimulated TNP-primed B cells (Howie et al. 1986a). The ability of Langerhans cells to present HSV-1 antigens to human, primed T lymphocytes was found to be dependent on the coordinate expression of DP, DQ and DR surface antigens (Vestey et al. 1990).

Williams et al. (1991) recently showed that although resting Langerhans cells are able to trigger secondary T helper cell responses, they are impaired in their capacity to stimulate primary responses to HSV-1 infection. In contrast, spleen antigen-presenting cells, most probably consisting in mature dendritic cells, efficiently induce the primary generation of HSV-1-specific T cells (Williams et al. 1991). Similarly, dendritic cells were shown to induce primary cytotoxic T cell responses to HSV-1 in vitro (Hengel et al. 1987).

Altogether, these data clearly indicate that dendritic cells are able to stimulate HSV-1-specific T cell responses in vitro although they do not establish the importance of dendritic cell involvement in the development of the immune response to HSV-1 infection in vivo.

Effect of Herpes Simplex Virus Type 1 Infection on Langerhans Cell Distribution in the Epidermis

To investigate the role played by Langerhans cells and other dendritic cells in the immune response to HSV-1 infection in vivo, several murine models were used. The local kinetics of Langerhans cell population during HSV-1 infection was initially assessed. Langerhans cells were found to accumulate at the site of primary infection with HSV-1 in mice inoculated through the skin (Sprecher and Becker 1986) or the cornea (Asbell and Kamenar 1987; Lewkowicz-Moss et al. 1987; Phillip and Göttinger 1991). A transient initial decrease in Langerhans cell density was observed following zosteriform spread of HSV-1 infection (Williams et al. 1992), possibly reflecting the migration of Langerhans cells to the draining lymph nodes, as previously reported in allergic reactions (Silberberg et al. 1976).

Increase in Langerhans cell density was found to be directly correlated with the intrinsic pathogenicity of the HSV strain used for infection (Sprecher and Becker 1986, 1987) and the age of the infected mice (Sprecher et al. 1991). Indeed, Langerhans cell migration to the site of HSV-1 inoculation was only observed when young adult mice were inoculated with a virulent strain of HSV-1 (Sprecher and Becker 1986; Sprecher et al. 1991). Whether these findings and the low density of epidermal Langerhans cells found in aged skin (Sprecher et al. 1990) are causally related to the higher susceptibility of old mice to HSV-1 infection (Sprecher et al. 1991) is not known.

Double staining of epidermal sheets for ATPase$^+$ Langerhans cells and HSV-1 glycoprotein D-expressing cells revealed the presence of viral antigens in Langerhans cells only 2 h after infection (Sprecher and Becker 1986). In conclusion, during cutaneous infection with HSV-1, Langerhans cells migrate towards the site of infection to take up viral antigens.

Transfer of Herpes Simplex Virus Type 1 Antigens from the Skin to the Lymph Nodes and Stimulation of Systemic Immunity

Following HSV-1 infection, Langerhans cells were observed to migrate to the lymph nodes draining the skin area of infection (Sprecher and Becker 1989). Infectious virus, albeit in small quantities, as well as viral antigens were detected in dendritic cells isolated from these lymph nodes (Knight and Macatonia 1988; Sprecher and Becker 1989). HSV-1 antigen-bearing dendritic cells were found to be able to stimulate the proliferative response of HSV-1-specific T cell lines (Knight and Macatonia 1988).

Little information regarding the status of the virus found in lymph node dendritic cells has been obtained. Dendritic cells infected in vitro with HSV-1 were found to be nonpermissive for virus replication (Sprecher and Becker 1991b), although it is not known whether dendritic cells activation prior to infection may create favorable conditions for virus replication, as previously demonstrated in macrophages (Ben-Hur et al. 1988).

Whether HSV-1 replicates or not in dendritic cells in vivo, it is clear that these cells are capable of processing the virus for efficient presentation to both naive and primed T cells (Braathen et al. 1980; Hayashi and Aurelian 1986; Otani and Mori 1987; Sprecher and Becker 1987; Vestey et al. 1990; Williams et al. 1991; Yasumoto et al. 1986, 1987). In addition, Langerhans cells are definitively needed for an efficient systemic response to the viral infection to develop. Indeed, lymph node cells obtained from mice infected through the skin with HSV-1 strongly proliferated in response to HSV-1 antigens in vitro (Sprecher and Becker 1989). In contrast, when HSV-1 was inoculated into a skin area previously depleted in Langerhans cells, cells isolated from the lymph nodes draining this area were unable to respond to HSV antigens, although they did proliferate in response to a standard mitogen (Sprecher and Becker 1989).

Thus, Langerhans cells play a central role in the development of systemic immunity to the viral infection (reviewed in Sprecher and Becker 1988). Langerhans cells were found to secrete several cytokines with strong antiviral activity, including interleukin-1 (IL-1), TNF-α, and interferon-γ (IFN-γ); (Knop et al. 1988; Larrick et al. 1989; Nickoloff et al. 1991; Sauder et al. 1984). These data provided the impetus to investigate Langerhans cells involvement in the regional immune response to cutaneous infection with HSV-1.

Langerhans Cell Regional Response to Cutaneous Infection with Herpes Simplex Virus Type 1

IL-1, TNF-α, and IL-6 are considered to be major mediators of inflammatory processes. These three cytokines display a very similar, but not identical, spectrum of activities, including differentiation and growth-regulating activities, inflammatory, hematopoietic, antitumoral, and antiviral activities (Akira et al. 1990; Van Damme and Opdenacker 1990; Van Snick 1990; Wong and Goeddel 1986).

A polymerase chain reaction (PCR) technique was used to analyze the expression of these three cytokine genes in epidermal cells during cutaneous infection with HSV-1 (Sprecher and Becker 1992). Epidermal cell suspensions obtained from mice infected with HSV-1 in the ear pinna were depleted or enriched in Langerhans cells by immunomagnetic fractionation. Langerhans cell-depleted epidermal cell suspensions consisted mainly of keratinocytes. HSV-1 infection in the skin was found to induce IL-1β, TNF-α , and IL-6 gene transcription in keratinocytes at 24 h postinfection (Sprecher and Becker 1992); gene transcription declined by 48 h postinfection. Specific induction of IL-1β and TNF-α , but not of IL-6 gene transcription was detected in Langerhans cells obtained from infected mice at 24 h postinfection (Sprecher and Becker 1992).

Keratinocytes may play an important role in the mechanisms of defense against HSV-1 infection in the skin. Keratinocytes were found to be able to present HSV-1 antigens to T cells and to function as a target for HSV-1 specific cytotoxic T cells (Cunningham and Noble 1989). Although cytokine release was not studied, it is likely that the three cytokine gene induction resulted in the actual release of cytokine proteins. Since HSV-1 is characterized by its epidermotropism and keratinocytes represent more than 95% of the epidermal cell population, IL-1, TNF-α, and IL-6 release may significantly contribute to the antiviral response to HSV-1 infection (Sprecher and Becker 1992). Consistent with this hypothesis is the fact that retinoic acid derivatives decrease the frequency of HSV-1 recurrency (Kanzler and Rasmussen 1988), while retinoic acid was also found to induce IL-1 production by keratinocytes (McGuire et al. 1988).

Langerhans cells were shown to synthesize IL-1β and TNF-α mRNA in response to HSV-1 infection in the skin (Sprecher and Becker 1992). Previous studies demonstrated IL-1 and TNF-α synthesis in epidermal Langerhans cells and their circulating precursors (Gothelf et al. 1989; Knop et al. 1988; Larrick et al. 1989; Nickoloff et al. 1991; Sauder et al. 1984). In contrast, mature dendritic cells found in lymphoid organs and in the peripheral blood do not synthesize IL-1 RNA or release IL-1 activity (Steinman 1988). Instead, IL-1 is known to amplify T-dependent immune responses by primarily acting on these dendritic cells rather than on T cells (Steinman 1988). Similarly, IL-1, in conjunction with granulocyte–macrophage colony-stimulating factor (GMCSF), promotes in vitro the transformation of resident Langerhans cells, solely capable of processing native proteins, into mature dendritic cells (Heufler et al. 1988; Metlay et al. 1989; Steinman 1991). During this maturation process, which takes place while they migrate from the epidermis to the lymph nodes, Langerhans cells become capable of inducing primary T cell response to viral and other antigens (Aiba and Katz 1990; Cruz and Bergstresser 1990; Cumberbatch et al. 1991a,b, 1992; Larsen et al. 1990; Macatonia et al. 1989b; Steinman 1991; Streilein et al. 1990). No IL-1 synthesis could be detected during HSV-1 infection in lymph node dendritic cells (Sprecher and Backer 1991b). Thus, during infection of the skin with HSV-1, Langerhans cell priming by keratinocyte- and Langerhans cell-derived IL-1 might occur in the epidermis, where high amounts of GMCSF are also known to be constitutively released (Heufler et al. 1988), and be followed by Langerhans cell maturation into functional dendritic cells and their migration to the draining lymph nodes (Sprecher and Becker 1989; Fig. 1).

Langerhans cells and keratinocytes and known to be part of the skin-associated lymphoid tissue (S.A.L.T.; Streilein 1985). Their ability to respond to HSV-1 infection by synthesizing antiviral cytokines (Sprecher and Becker 1992) emphasizes the importance of the regional immune response of the skin to the viral infection (Fig. 1) and suggests the possibility of developing new immunotherapeutic strategies based on stimulation of cytokine release by Langerhans cells and keratinocytes. Interestingly, the immunomodulator OK-432, a strong stimulator of Langerhans cell activity (Sprecher and Becker 1989; Tsujitani et al. 1988), has been shown to induce IL-1 production (Katsuta et al. 1989) and function as an efficient antiviral compound (Sprecher and Becker 1989).

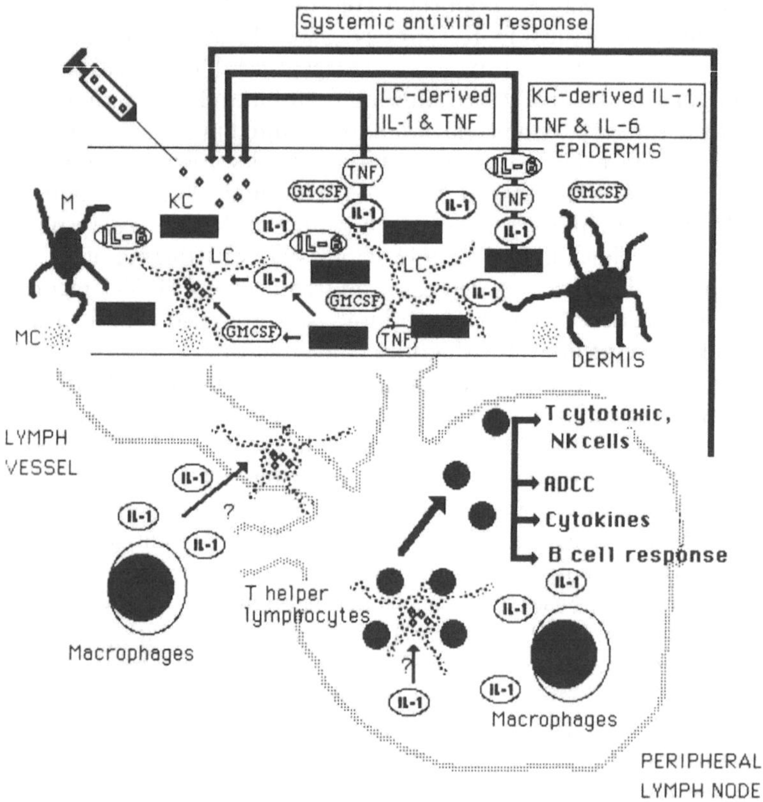

Fig. 1. Role of Langerhans cell in primary infection with herpes simplex virus type 1 (HSV-1). After infection with HSV-1, Langerhans cells take up the virus in the epidermis and migrate to the draining lymph nodes, where they present viral antigens to T helper lymphocytes and thereby stimulate systemic immunity. They acquire the capacity to stimulate T cells during their migration to the draining lymph nodes under the influence of interleukin-1 (*IL-1*) and granulocyte–macrophage colony-stimulating factor (*GMCSF*). In addition, both keratinocytes (*KC*) and Langerhans cells (*LC*) actively secrete cytokines displaying strong antiviral activities, including *IL-1*, *IL6* and tumor necrosis factor-α (*TNF*-α). *ADCC* antibody-dependent, cell-mediated cytotoxicity; *M*, melanocytes; *MC*, Merkel cells; *NK*, natural killer cells

Langerhans Cell Density and Activity in the Epidermis Determine Herpes Simplex Virus Type 1 Pathogenicity

To study the in vivo relevance of the above findings and to confirm the importance of the role played by Langerhans cells in the response to HSV-1 infection, the effect of changes in Langerhans cell density and functional state on the pathogenesis of the viral infection was studied.

Inoculation of HSV-1 into a skin area depleted in Langerhans cells was found to be associated with an increase in viral DNA synthesis (Sprecher and Becker

1991b) as well as in morbidity and mortality among the infected mice as compared with virus infection through a normal skin area (Sprecher and Becker 1986, 1987).

An inverse correlation could be established between Langerhans cell density at the time of virus inoculation in the infected skin area and the severity of infection with HSV-1 (Sprecher and Becker 1986). Similar results were reported in a murine model of vaccinia virus infection (Becker and Sprecher 1989).

In contrast, activation of Langerhans cells by local administration of the immunomodulator OK-432 prior to virus inoculation led to a marked decrease in HSV-1 pathogenicity (Sprecher and Becker 1989). These observation suggest the possibility of interfering with the course of HSV-1 infection by modulating Langerhans cell activity in the skin epidermis.

However, Langerhans cell antiviral activity may not always be beneficial to the infected host. Indeed, as indicated by several recent reports, Langerhans cells may be responsible for triggering the immune-mediated destruction of the cornea in HSV-1 induced stromal keratitis (Heindricks et al. 1992; Jagger et al. 1991).

Role of Langerhans Cells in the Pathogenesis of Corneal Infection with Herpes Simplex Virus Type 1

HSV-1-induced keratitis is a major cause of visual impairment and blindness worldwide. Corneal infection with HSV-1 can lead to a mild superficial keratitis or to a severe stromal disease (Heindricks et al. 1992). Langerhans cells are normally absent in the corneal epithelium (Asbell and Kamenar 1987; Lewkowicz-Moss et al. 1987). Following HSV-1 infection, Langerhans cells were observed to migrate from the periphery toward the center of the corneal surface (Asbell and Kamenar 1987; Lewkowicz-Moss et al. 1987; Pepose 1989; Phillip and Göttinger 1991). Unlike the observations made during cutaneous infection with HSV-1, no viral antigen could be detected in corneal Langerhans cells (Pepose 1989).

Corneal Langerhans cells are likely to play an important role in the development of an immune response aimed at clearing the viral infection. As described above (Sprecher and Becker 1988), they may be involved in the induction phase of systemic immunity as well as in the development of a regional and early immune response, possibly through cytokine secretion (Sprecher and Becker 1992). However, Langerhans cell antigen-presentation activity may also contribute to the pathogenesis of herpetic stromal keratitis (Heindricks et al. 1992; Jagger et al. 1991). Stromal keratitis is thought to result from an immune-mediated reaction, mostly dependent on $CD4^+$ T cell activation. Indeed, athymic mice as well as $CD4^+$ cell-depleted mice failed to develop stromal keratitis upon infection with HSV-1. In addition, stromal keratitis does not respond to conventional antiviral treatment, but can be brought under control by immunosuppressive regimens (Heindricks et al. 1992; Nevell et al. 1989).

Intraocular infection with HSV-1 was found to lead to stromal keratitis more frequently and to a more severe disease when the infected eye was being treated, prior to virus inoculation, in such a way as to induce Langerhans cell migration into the cornea (Jagger et al. 1991). In another murine model, the severity of stromal herpetic keratitis was shown to be dependent on the ability of the HSV-1 strain used for infection to induce Langerhans cell migration to the cornea (Heindricks et al. 1992). Thus, in the corneal model of HSV-1 infection, in contrast to the skin model of HSV-1 infection, Langerhans cell density was found to be directly correlated with the pathogenicity of HSV-1 (Heindricks et al. 1992; Jagger et al. 1991). Moreover, when mouse cornea was first exposed to UV irradiation before being inoculated with HSV-1, Langerhans cell migration to the cornea was partially inhibited and HSV-1 causes a less severe form of stromal keratitis (Heindricks et al. 1992).

Thus, the corneal model of HSV-1 infection illustrates a case where impairment in Langerhans cell function significantly alleviates the course of infection with HSV-1.

Role of Langerhans Cells in the Control of Herpes Simplex Virus Type 1 Recurrent Infections

One of the most distinguishing features of herpesviruses is their propensity for establishing latent infection and subsequently causing recurrent infections (Roizman and Sears 1987; Wheeler 1988). It has been demonstrated that HSV-1 establishes latent infection in the ganglia of the sensory nerves innervating the skin area where primary infection occurs (Roizman and Sears 1987; Wheeler 1988). Recurrent infection results from reactivation of latent virus and its transfer back to the site of primary infection by axonal transport (Roizman and Sears 1987; Stevens and Cook 1971; Wheeler 1988). It is associated with exposure to a number of stimuli such as light trauma to the skin, exposure to sun or UV light, and menstruation (Roizman and Sears 1987; Wheeler 1988). Latent virus is also detectable in peripheral epithelia (Clements and Subak-Sharpe 1988). However, reactivation of latent HSV-1 in the epidermis is thought not to contribute significantly to the overall incidence of recurrency, since neurotomy is followed by a striking decrease in the frequency of reactivation (Hill et al. 1983).

Two theories of HSV-1 recurrency have been proposed. According to the "ganglion trigger theory" (Stevens and Cook 1971), the reactivating stimuli directly affect the ganglion cells, causing them to release infectious virus, which is then transferred to the skin through sensory axons and, there, leads to the development of recurrent disease. However, it is not clear at all how stimuli such as exposure to external UV light sources affect the state of the virus in the ganglion (Hill and Blyth 1976). In addition, no correlation could be established between reactivation of virus in the ganglion and clinical recurrent episodes in the skin (Hill and Blyth 1976). The titer of antibodies to the virus remains stable for long periods, suggesting that silent reactivation continuously occurs (Hill and Blyth 1976). Finally, it is not possible to understand the actual kinetics of virus reactivation following exposure to reactivating stimuli if one considers those stimuli to act at the level of the ganglion only (Hill and Blyth 1976). In other words, the virus must be present near or in the epidermis at the time of exposure of the skin to reactivating stimuli. These considerations led Hill and Blyth to formulate in 1976 an alternative theory of HSV-1 recurrence. According to the "skin trigger theory" (Blyth et al. 1976; Hill and Blyth 1976), HSV-1 recurrence develops as a consequence of the coincident presence of virus in the skin and conditions favoring its proliferation, such as solar irradiation. Latent HSV-1 was shown to be spontaneously reactivated at various time intervals in the ganglion cells (Hill et al. 1980). Following reactivation, viral particles were shown to be released and transferred to the skin epidermis, where the formation of subclinical microfoci of infection, not associated with recrudescent blisters, was observed (Hill et al. 1980). Subclinical infections with recurrent HSV-1 are also common in humans (Hill and Blyth 1976). Thus, the skin trigger theory postulates that reactivation of latent virus in the sensory ganglia occurs with no need for prior exposure to reactivating stimuli. Accordingly to the skin trigger theory, those stimuli solely influence cutaneous regulatory elements which normally prevent the transformation of subclinical foci of infection into recrudescent lesions.

Hill et al. (Blyth et al. 1976; Hill and Blyth 1976) proposed that prostaglandin secretion in the epidermis might mediate the effect of the stimuli associated with recurrency, since prostaglandin E_2 is known to be produced in response to exposure of the skin to some HSV-1-reactivating agents. However, a number of these agents, such as steroids, are known to impair prostaglandin production (Williams 1981). Therefore, a novel interpretation of the skin trigger theory was proposed (Sprecher and Becker 1986; Fig. 2). Since Langerhans cells are known to play a pivotal role in the response to primary HSV-1 infection in the skin, they may also be able to prevent the development of infectious subclinical microfoci into recurrent overt lesions. Conversely, impairment in Langerhans cell function could promote the transformation of these microfoci of infection into recrudescent lesions (Fig. 2).

A number of findings support this hypothesis. Arachidonic acid, the precursor for prostaglandin E_2, was shown to induce a decrease in the numbers of Langerhans cells (Rheins and Nordlund 1986), a finding that might help integrating Hill's original hypothesis into this novel model. HSV recurrence follows exposure to UV light, steroids, hyperthermia, various chemicals, or tape stripping (Hill et al. 1978; Perna et al. 1987; Roizman and Sears 1987; Underwood and Weed 1974; Wheeler 1988; Williams and Hill 1991), and the same stimuli have independently been shown to induce Langerhans cell depletion in the skin, to decrease the capacity of Langerhans cells to present HSV antigens to sensitized T cells, and, as a consequence, to abrogate delayed-type hypersensitivity to the virus (Furue and Katz 1991; Howie et al. 1986b; Krueger and Mansoor 1984; Otani and Mori 1987; Simon et al. 1992; Sprecher and Becker 1987, 1988; Williams and Hill 1991; Yasumoto et al. 1987; Yoshioka et al. 1989). The effects of UV light on Langerhans cells and on the immune response to the virus have been related to the induction of suppressor T cells and the production of soluble immunosuppressive factors (Hayashi and Aurelian 1986; Yasumoto et al. 1987).

Reactivating stimuli may therefore promote the development of recurrent disease by damaging Langerhans cells, which may then be unable to restrict the transformation of subclinical microfoci of infection into recrudescent lesions

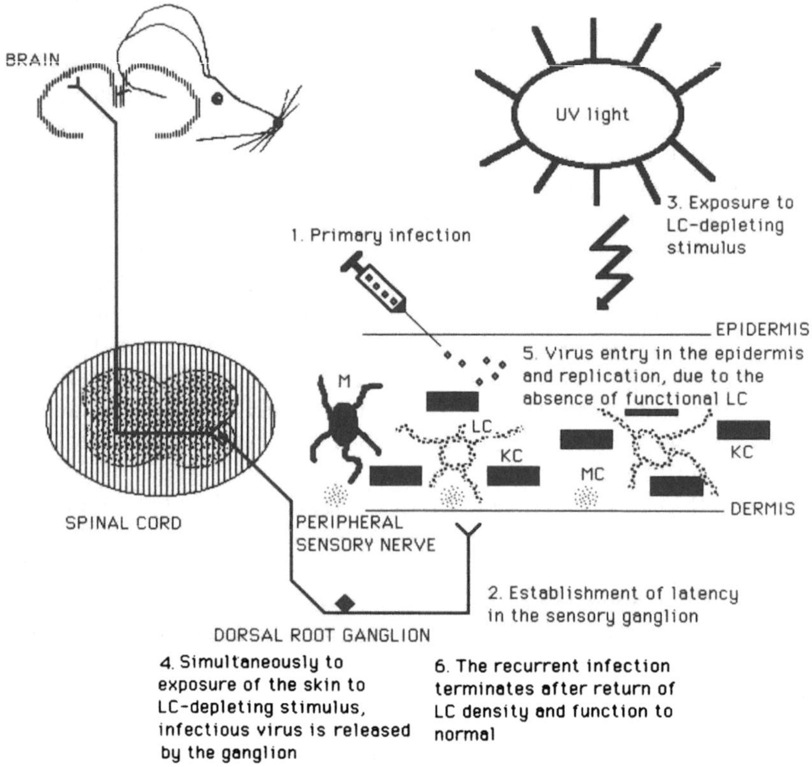

Fig. 2. Role of Langerhans cell (*LC*) in recurrent infection with herpes simplex virus type 1 (HSV-1). Following primary infection, the virus establishes latent infection in the sensory ganglion innervating the skin area where primary infection occurred. After establishment of latency, the ganglion spontaneously releases viral particles, which are transported by axonal flow to the skin epidermis. Those viral particles are thought to induce the formation of subclinical foci of infection which will develop into overt recrudescent lesions only in the absence of functional LC. Depletion of functional LC in the epidermis results from exposure of the skin to various stimuli such as solar irradiation. Reccurrent infection is cleared when LC function returns to normal. *KC*, keratinocytes; *M*, melanocytes; *MC*, Merkel cells

(Otani and Mori 1987; Sprecher and Becker 1986, 1988; Fig. 2). Therefore Langerhans cells may play a critical role not only in primary infections (Fig. 1), but also in the control of virus recurrency in the skin (Fig. 2). Prevention of HSV-1 recurrent episodes may thus be possible through administration of Langerhans cell-activating compounds (Gruner et al. 1990; Kaplan et al. 1992; Shiohara et al. 1987). Retinoids have been shown to enhance Langerhans cell function (Shiohara et al. 1987) and to be effective in decreasing the frequency of recurrency in HSV-1-infected patients (Kanzler and Rasmussen 1988).

Herpes Simplex Virus Type 1 Infection, Dendritic Cells, and the Acquired Immune Deficiency Syndrome

Dendritic cells have also been shown to be involved in the pathogenesis of human immunodeficiency virus type 1 (HIV-1) infection (Sprecher and Becker 1988; Stingl et al. 1990). HIV-1 was found to infect and replicate preferentially into these cells (Braathen et al. 1991; Langhoff et al. 1991; Naher et al. 1992; Patterson et al. 1991a,b; Rappersberger et al. 1988; Rodriguez et al. 1991; Tschachler et al. 1987; Zambruno et al. 1991), to damage dendritic cells, and to impair their function in the skin and elsewhere (Belsito et al. 1984; Dréno et al. 1988; Hughes et al. 1990; Knight and Macatonia 1991a; Knight et al. 1991; Macatonia et al. 1989a, 1990, 1991, 1992). Others claim that HIV-1 does not impair dendritic cell function and that dendritic cell-induced T cell activation is instrumental in the dissemination of the virus throughout the body (Cameron et al. 1992a,b). In addition, dendritic cells may represent a reservoir site for HIV-1 latent infection (Braathen et al. 1991) as well as an important target during primary infection (Lehner et al. 1991; Müller et al. 1987; Niedecken et al. 1987).

Unusually severe HSV infections are a hallmark of the acquired immune deficiency syndrome (AIDS; Cockerell 1991). Since dendritic cells function as critical elements in the mechanisms of defense against HSV-1 infection (Sprecher and Becker 1988), HIV-1-induced decreased dendritic cell function may partly explain the high susceptibility of AIDS patients to mucocutaneous infection with HSV-1. Moreover, HSV-1 has been shown to activate HIV-1 replication (Mosca et al. 1987). Thus, impaired dendritic cell function associated with HIV-1 infection may further amplify the deleterious effects of HSV-1 infection on the course of HIV-1 disease.

Although zidovudine therapy reverses some defects in dendritic cell function (Knight et al. 1992), it does not restore normal Langerhans cell density in HIV-1-infected individuals (French et al. 1990). Thus, correction of Langerhans cell dysfunction by administration of various immunomodulators (Gruner et al. 1990; Kaplan et al. 1992; Shiohara et al. 1987) could serve as a useful adjuvant therapy in the treatment of AIDS patients to prevent the occurrence of disseminated herpetic infections (Becker 1990; Stricker et al. 1991).

Conclusion

In conclusion, Langerhans cells were shown to play a pivotal role in the defense against primary HSV-1 infection and are probably involved in the control of recurrent infections as well (Sprecher and Becker 1988). As such, they may serve as unique targets for novel immunotherapeutic strategies based on the use of cytokines and various other agents known to modulate dendritic cell function (Gruner et al. 1990; Heufler et al. 1988; Kaplan et al. 1992; Shiohara et al. 1987). As mentioned above, Langerhans cell activation by the immunomodulator OK-432 was found to decrease the severity of primary HSV-1 infection (Sprecher and

Becker 1989), while oral administration of retinoids, which are known to stimulate dendritic cell function (Shiohara et al. 1987), was shown to reduce the frequency of recurrent HSV-1 infections (Kanzler and Rasmussen 1988). Langerhans cells may also serve in the near future as vectors in innovative vaccination strategies against HSV infections. Indeed, recent studies have unequivocally demonstrated that extracorporeal pulsing of dendritic cells with whole antigens and subsequently immunization with these antigen-loaded dendritic cells induce efficient cellular and humoral immune responses (Inaba et al. 1990; Sornasse et al. 1992). Dendritic cells are likely to process whole antigens into optimally immunogenic peptides, thereby circumventing the need for the design and construction of subunit vaccines (Inaba et al. 1990; Sornasse et al. 1992).

Altogether, the findings reviewed above may pave the way for novel approaches in the therapy of herpetic infections, based on judicious modulation of dendritic cell activity, which may prove to represent valuable alternatives to the currently available antiviral treatments.

Acknowledgements. This work was supported by the US-Israel Binational Science Foundation and by the Foundation for the Study of Molecular Virology and Cell Biology, Arizona, U.S.A. Eli Sprecher is a Fellow of the Foulkes Foundation, London, U.K. We are indebted to Dr. Laure Aurelian, Dr. Eytan Yefenof, Dr. Tamir Ben-Hur, Dr. Sidney Klaus, Dr. Claes Enk, Dr. Dan David, Yael Asher, Eynath Tabor, and Cheryl Berkowitz for stimulating discussions and collaboration.

References

Aiba S, Katz SI (1990) Phenotypic and functional characteristics of in vivo-activated Langerhans cells. J Immunol 145: 2791–2796

Akira S, Hirano T, Taga T, Kishimoto T (1990) Biology of multifunctional cytokines: IL-6 and related molecules (IL-1 and TNF) FASEB J 4: 2860–2867

Asbell PA, Kamenar T (1987) The response of Langerhans cells in the cornea to herpetic keratitis. Curr Eye Res 6: 179–182

Bartosik J (1992) Cytomembrane-derived Birbeck granuels transport horseradish peroxidase to the endosomal compartment in the human Langerhans cells. J Invest Dermatol 99: 53–58

Becker D, Neiß U, Neis S, Reske K, Knop J (1992) Contact allergens modulate the expression of MHC class I molecules on murine epidermal Langerhans cells by endocytic mechanisms. J Invest Dermatol 98: 700–705

Becker Y (1990) Rationale for anti-AIDS chemotherapy directed at HIV-1 infected dendritic cells in seropositive individuals prior to the appearance of ARC and AIDS. J Chemother 2: 152–155

Becker Y, Sprecher E (1989) Langerhans cells in vaccinia virus infection in mouse skin. Arch Virol 107: 307–313

Belsito DV, Sanchez MR, Baer RL, Valentine F, Thorbecke GJ (1984) Reduced Langerhans cell la antigen and ATPase activity in patients with the acquired immunodeficiency syndrome. N Engl J Med 310: 1279–1282

Ben-Hur T, Rösen-Wolff A, Lamade W, Darai G, Becker Y (1988) HSV-1 DNA sequence determining intraperitoneal pathogenicity in mice is required for transcription of viral immediate early genes in macrophages. Virology 163: 397–404

Blyth WA, Hill TJ, Field HJ, DA (1976) Reactivation of herpes simplex virus by ultraviolet light and possible involvement of prostaglandins. J Gen Virol 33: 547–550

Bonneau RH, Jennings SR (1989) Modulation of acute and latent herpes simplex virus infection in C57BL/6 mice by adoptive transfer of immune lymphocytes with cytolytic activity. J Virol 63: 1480–1486

Braathen LR, Berle E, Mobeck-Hansen U, Thorsby E (1980) Activation of human T lymphocytes to herpes simplex virus. Acta DermVenerol (Stockh) 60: 381–387

Braathen LR, Ramirez G, Kunze RO, Mork C, Strand O (1991) HIV infection of skin Langerhans cells In: Becker Y (ed) Skin Langerhans (dendritic) cells in virus infections and AIDS. Kluwer Academic, Boston, pp 131–140

Brodsky FM, Guagliardi LE (1991) The cell biology of antigen processing and presentation. Annu Rev Immunol 9: 707–744

Cameron PU, Forsum U, Teppler H, Granelli-Piperno A, Steinman RM (1992a) During HIV-1 infection, most blood dendritic cells are not productively infected and can induce allogeneic $CD4^+$ T cell clonal expansion. Clin Exp Immunol 88: 226–236

Cameron PU, Freudenthal PS, Barker JM, Gezelter S, Inaba K, Steinman RM (1992b) Dendritic cells exposed to human immunodeficiency virus type-1 transmit a vigorous cytopathic infection to $CD4^+$ T cells. Science 257: 383–387

Clements GB, Subak-Sharpe JH (1988) Herpes simplex virus establishes latency in mouse footpad. J Gen Virol 69: 375–383

Cockerell CJ (1991) Human immunodeficiency virus infection and the skin. A crucial interface. Arch Intern Med 151: 1295–1303

Cruz PD, Bergstresser PR (1990) Antigen processing and presentation by epidermal Langerhans cells. Dermatol Clin 8: 633–647

Cumberbatch M, Kimber I (1992) Dermal tumor necrosis factor-α induces dendritic cell migration to draining lymph nodes and possibly provides one stimulus for Langerhans cell migration. Immunology 75: 257–263

Cumberbatck M, Illingworth I, Kimber I (1991a) Antigen-bearing dendritic cells in the draining lymph nodes of contact sensitized mice: cluster formation with lymphocytes. Immunology 74: 139–145

Cumberbatch M, Gould SJ, Peters SW, Kimber I (1991b) MHC class II expression by Langerhans cells and lymph node dendritic cells: possible evidence for maturation of Langerhans cells following sensitization. Immunology 74: 414–419

Cumberbatch M, Peters SW, Gould SJ, Kimber I (1992) Intercellular adhesion molecule-1 (ICAM-1) expression by lymph node dendritic cells compared with epidermal Langerhans cells. Immunol Let 32: 105–110

Cunningham AL, Noble JR (1989) Role of keratinocytes in human recurrent herpetic lesions. Ability to present herpes simplex virus antigen and act as targets for T lymphocytes cytotoxicity in vitro. J Clin Invest 83: 490–496

De Panfilis G, Manara GC, Ferrari C, Torresani C (1991) Adhesion molecules on the plasma membrane of epidermal Langerhans cells III. Keratinocytes and Langerhans cells constitutively express the lymphocyte function-associated antigen 3. J Invest Dermatol 96: 512–517

Dréno B, Milpied B, Bignon JD, Stalder JF, Litoux P (1988) Prognostic value of Langerhans cells in the epidermis of HIV patients. Br J Dermatol 118: 481–486

French MAH, Cameron PU, Smyth LA, Dawkins RL (1990) Correction of human immunodeficiency virus-associated depression of delayed-type hypersensitivity (DTH) after zidovudine therapy; DTH, $CD4^+$ T-cell numbers and epidermal Langerhans cell density are independent variables. Clin Immunol Immunopathol 55: 86–96

Furue M, Katz SI (1991) Direct effects of glucocorticoids on epidermal Langerhans cells. J Invest Dermatol 146: 1717–1721

Gothelf Y, Dinarello CA, Yamin M, Sharon N, Milner Y, Gazit E (1989) IL-1 production by T6 (CD1a) positive cord blood mononuclear cells (Langerhans cell precursors?). Lymphokine Res 8: 373–382

Gruner S, Diezel W, Strunk D, Zwirner A, Sönnichen N, Anhalt GJ (1990) Stimulation of the recruitment of epidermal Langerhans cells by splenopentin. Arch Dermatol Res 281: 526–529

Hayashi Y, Aurelian L (1986) Immunity to herpes simplex type 2: viral antigen presenting capacity of epidermal cells and its impairment by ultraviolet irradiation. J Immunol 136: 1087–1092

Heindricks RL, Janowicz M, Tumpey TM (1992) Critical role of corneal Langerhans cells in the CD4- but not CD8-mediated immunopathology in herpes simplex virus-1- infected mouse corneas. J Immunol 148: 2522–2529

Hengel H, Lindner M, Wagner H, Heeg K (1987) Frequency of herpes simplex virus-specific murine cytotoxic T lymphocyte precursors in mitogen- and antigen-driven primary in vitro responses. J Immunol 139: 4196–4202

Heufler C, Koch F, Schuler G (1988) Granulocyte/macrophage colony-stimulating factor and interleukin 1 mediate the maturation of murine epidermal Langerhans cells into potent immunostimulatory dendritic cells. J Exp Med 167: 700–705

Hill TJ, Blyth WA (1976) An alternative theory of herpes simplex recurrence and a possible role for prostaglandins. Lancet i: 397–399

Hill TJ, Blyth WA, Harbour DA (1978) Trauma to the skin causes recurrence of herpesvirus in the mouse. J Gen Virol 39: 21–28

Hill TJ, Harbour DA, Blyth WA (1980) Recovery of herpes simplex virus from the skin of clinically normal mice during latent infection. J Gen Virol 47: 205–207

Hill TJ, Blyth WA, Harbour DA (1983) Recurrence of herpes simplex in the mouse requires an intact nerve supply to the skin. J Gen Virol 64: 2763–2766

Howie S, Norval M, Maingay JP (1986a) Alterations in epidermal handling of HSV-1 antigens in vitro induced by in vivo exposure to UV-B light. Immunobiology 57: 225–230

Howie S, Norval M, Maingay JP (1986b) Exposure to low-dose ultraviolet radiation suppresses delayed-type hypersensitivity to herpes simplex virus in mice. J Invest Dermatol 86: 125–128

Hughes RA, Macatonia SE, Rowe IF, Keat AC, Knight SC (1990) Detection of human immunodeficiency virus DNA in dendritic cells from the joints of patients with aseptic arthritis. Br J Rheumatol 29: 166–170

Inaba K, Metlay JP, Crowley MT, Steinman RM (1990) Dendritic cells pulsed with protein antigens in vitro can prime antigen-specific, MHC-restricted T cells in situ. J Exp Med 172: 631–640

Jagger MJ, Atherton S, Bradley D, Streilein JW (1991) Herpetic stromal keratitis: less reversibility in the presence of Langerhans cells in the central cornea. Curr Eye Res 10 [Suppl]: 69–75

Kanzler ZH, Rasmussen JE (1988) Isotretinoin therapy for recurrent herpes simplex lesions. Arch Dermatol 124: 323–325

Kaplan G, Walsh G, Laarni SG, Meyn P, Burkhardt ERA, Abalos RM, Barker J, Frindt PA, Tranquilino TF, Celona R, Cohn ZA (1992) Novel responses of human skin to intradermal recombinant granulocyte/macrophage-colony stimulating factor: Langerhans cell recruitment, keratinocyte growth and enhanced wound healing. J Exp Med 175: 1717–1728

Katsuta S, Kaneko Y, Tsutsui K, Tamaki S, Hibata H, Chen DR, Suzuki S, Natsuume-Sakai S, Yamamoto A (1989) Interleukin-1 inducing activity of a streptococcal preparation OK-432 and its fractions by human monocytes. J Clin Lab Immunol 28: 129–136

Knight SC, Macatonia SE (1988) Dendritic cells and viruses. Immunol Lett 19: 177–182

Knight SC, Macatonia SE (1991) Effect of HIV on antigen-presentation by dendritic cells and macrophages. Res Virol 142: 123–128

Knight SC, Patterson S, Macatonia SE (1991) Stimulatory and immunosuppressive effects of infection of dendritic cells with HIV-1. Immunol Lett 30: 212–218

Knight SC, Macatonia SE, Gompels M, Pinching AJ, Patterson S (1992) Zidovudine reverses the defect in dendritic cells in HIV infection. AIDS 6: 231–232

Knop J, Enk A, Heremans H (1988) Release of interferon-γ by epidermal Langerhans cells. J Invest Dermatol 90: 246

Krueger GG, Mansoor E (1984) Biology of Langerhans cells: analysis by experiments to deplete Langerhans cells from human skin. J Invest Dermatol 82: 613–618

Langhoff E, Terwilliger EF, Bos JH, Kalland KH, Poznansky MC, Bacon OML, Haseltine WA (1991) Replication of human immunodeficiency virus type 1 primary dendritic cell cultures. Proc Natl Acad Sci USA 88: 7998–8002

Larrick JW, Morhenn V, Chiang YL, Shi T (1989) Activate Langerhans cells release tumor necrosis factor. J Leukoc Biol 45: 429–433

Larsen HS, Feng M, Horohov DW, Moore RM, Rouse BT (1984) Role of T lymphocyte subsets in recovery from herpes simplex virus infection. J Virol 50: 56–60

Larsen CP, Steinman RM, Witmer-Pack M, Hankins DF, Morris PJ, Austyn JM (1990) Migration and maturation of Langerhans cells in skin transplants and explants. J Exp Med 172: 1483–1493

Lehner T, Hussain L, Wilson J, Chapman M (1991) Mucosal transmission of HIV. Nature 353: 709

Le Varlet B, Dezutter-Dambuyant C, Staquet MJ, Delorme P, Schmitt D (1991) Human epidermal Langerhans cells express integrins of the β1 subfamily. J Invest Dermatol 96: 518–522

Lewkowicz-Moss SJ, Shimeld C, Lipworth K, Hill TJ, Blyth WA, Easty DL (1987) Quantitative studies on Langerhans cells in mouse corneal epithelium following infection with herpes simplex virus. Exp Eye Res 45: 127–140

Macatonia SE, Patterson S, Knight SC (1989a) Suppression of immune responses by dendritic cells infected with HIV. Immunology 67: 285–289

Macatonia SE, Taylor PM, Knight SC, Askonase BA (1989b) Primary stimulation by dendritic cells induces antiviral proliferative and cytotoxic T cell responses in vitro. J Exp Med 169: 1255–1264

Macatonia SE, Lau R, Patterson S, Pinching AJ, Knight SC (1990) Dendritic cell infection, depletion and dysfunction in HIV-infected individuals. Immunology 71: 38–45

Macatonia SE, Patterson S, Knight SC (1991) Primary proliferative and cytotoxic T-cell response to HIV induced in vitro by human dendritic cells. Immunology 74: 399–406

Macatonia SE, Gompels M, Pinching AJ, Patterson S, Knight SC (1992) Antigen-presentation by macrophages but not by dendritic cells in human immunodeficiency virus (HIV) infection. Immunology 75: 576–581

McGuire J, Langdon R, Birchall N, Kupper T (1988) Interleukin-1 mRNA induced by cycloheximide, PMA and retinoic acid is reduced by dexamethasone in PAM-212 keratinocytes. Ann NY Acad Sci 548: 283–290

Metlay JP, Puré E, Steinman RM (1989) Control of the immune response at the level of antigen-presenting cells: a comparison of the function of dendritic cells and B cells. Adv Immunol 47: 45–116

Mosca JD, Bednarik DP, Raj NBK, Rosen CA, Sodroski JG, Haseltine WA, Pitha PM (1987) Herpes simplex virus type 1 can reactivate transcription of latent human immunodeficiency virus. Nature 325: 67–70

Müller H, Falk S, Stutte HJ (1987) Accessory cells as primary target of human deficiency virus HIV infection. J Clin Pathol 39: 1161

Naher H, Schule T, Petzoldt D (1992) Evidence for genetic variants from detection of HIV DNA. Lancet ii: 519–520

Nash AA, Jayasuriya A, Phelan J, Cobbold SP, Waldmann H, Prospero T (1987) Different roles for L3T4+ and Lyt 2+ T cell subsets in the control of an acute herpes simplex virus infection of the skin and nervous system. J Gen Virol 68: 825–834

Nevell CK, Martin S, Sendele D, Mercadal CM, Rouse BT (1989) Herpes simplex virus-induced stromal keratitis: role of T lymphocyte subsets in immunopathology. J Virol 63: 769–775

Nickoloff BJ, Karabin GD, Barker JN, Griffiths CE, Sarma V, Mitra RS, Elder JT, Kunkel SL, Dixit VM (1991) Cellular localization of interleukin-8 and its inducer, tumor necrosis factor-α in psoriasis. Am J Pathol 138: 129–140

Niedecken H, Lutz G, Bauer R, Kreysel HW (1987) Langerhans cells as primary targets and vehicle for transmission of HIV. Lancet ii: 519–520

Otani T, Mori R (1987) The effects of ultraviolet irradiation of the skin on herpes simplex virus infection: alteration in immune function mediated by epidermal cells and in the course of infection. Arch Virol 96: 1–15

Patterson S, Gross J, Bedford P, Knight SC (1991a) Morphology and phenotype of dendritic cells from peripheral blood and their productive and non-productive infection with human immunodeficiency virus type 1. Immunology 72: 361–367

Patterson S, Macatonia SE, Gross J, Bedford P, Knight SC (1991b) HIV infection of peripheral blood dendritic cells. In: Becker Y (ed) Skin Langerhans (dendritic) cells in virus infections and AIDS. Kluwer Academic, Boston, pp 141–156
Pepose JS (1989) The relationship of corneal Langerhans cells to herpes simplex antigens during dendritic keratitis. Curr Eye Res 8: 851–858
Perna JJ, Mannix ML, Rooney JF, Notkins AL, Strauss SE (1987) Reactivation of latent herpes simplex virus infection by ultraviolet light: a human model. J Am Acad Dermatol 17: 473–476
Phillip W, Göttinger W (1991) T6-Positive Langerhans cells in diseased corneas. Invest Ophtalmol Vis Sci 32: 2492–2497
Rappersberger K, Gartner S, Schenk P, Stingl G, Groh V, Tschaler E, Mann DL, Wolff K, Konrad K, Popovic M (1988) Langerhans cells are an actual site of HIV-1 replication. Intervirology 29: 185–194
Rheins LA, Nordlund JJ (1986) Modulation of the population density of identifiable epidermal Langerhans cells associated with enhancement or suppression of cutaneous immune reactivity. J Immunol 136: 867–876
Rodriguez ER, Nasim S, Hsia J, Sandin RL, Ferreira A, Hilliard BA, Ross AM, Garret CT (1991) Cardiac myocytes and dendritic cells harbor human immunodeficiency virus in infected patients with and without cardiac dysfunction: detection by the polymerase chain reaction in individually microdissected cells from right ventricular endomyocardial biopsy tissue. Am J Cardiol 68: 1511–1520
Roizman B, Sears AE (1987) An inquiry into the mechanisms of herpes simplex virus latency. Annu Rev Microbiol 41: 543–571
Sauder DN, Dinarello CA, Morhen VB (1984) Langerhans cell production of interleukin-1. J Invest Dermatol 82: 605–608
Schmid DS (1988) The human MHC-restricted cellular response to herpes simplex virus type 1 mediated by CD4+, CD8- T cells and is restricted to the DR region of the MHC complex. J Immunol 140: 3610–3616
Shiohara T, Koyobashi M, Narimatsu H, Nagashima M (1987) Effect of orally administered aromatic retinoid on murine Langerhans cells. Arch Dermatol Res 279: 198–202
Silberberg I, Baer RL, Rosenthal SA (1976) The role of Langerhans cells in allergic contact hypersensitivity. A review of findings in man and guinea pigs. J Invest Dermatol 66: 210–217
Simon JC, Cruz PD, Tigelaar RE, Sontheimer RD, Bergstresser PR (1991) Adhesion molecules CD11a, CD18 and ICAM-1 on human epidermal Langerhans cells serve a functional role in the activation of alloreactive T cells. J Invest Dermatol 96: 146–151
Simon JC, Krutman J, Elmets CA, Bergstresser PR, Cruz PD (1992) Ultraviolet B irradiated antigen-presenting cells display altered accessory signaling for T-cell activation. Relevance to immune responses initiated in the skin. J Invest Dermatol 98: 66s-69s
Sornasse T, Flamand V, De Becker G, Bazin H, Tielemans F, Thielemans K, Urbain J, Leo O, Moser M (1992) Antigen-pulsed dendritic cells can efficiently induce an antibody response in vivo. J Exp Med 175: 15–21
Sprecher E, Becker Y (1986) Skin Langerhans cells play an essential role in the defense against HSV-1 infection. Arch Virol 91: 341–349
Sprecher E, Becker Y (1987) Herpes simplex virus type 1 pathogenicity in footpad and ear skin of mice depends on Langerhans cell density, mouse genetics and virus strain. J Virol 61: 2515–2522
Sprecher E, Becker Y (1988) Role of epidermal Langerhans in viral infections. Arch Virol 103: 1–14
Sprecher E, Becker Y (1989) Langerhans cell number and activity in the skin and lymph nodes determine herpes simplex virus type 1 pathogenicity. Arch Virol 107: 191–205
Sprecher E, Becker Y, Kraal G, Hall E, Harrison D, Shultz LD (1990) Effect of aging on epidermal dendritic cell populations in C57BL/6J mice. J Invest Dermatol 94: 247–253
Sprecher E, Becker Y (1991a) Dendritic cells in the epidermis and the lymph nodes – a review. In: Becker Y (ed) Skin Langerhans (dendritic) cells in virus infections and AIDS. Kluwer Academic, Boston, pp 3–23

Sprecher E, Becker Y (1991b) Involvement of epidermal and lymph nodes dendritic cells in skin HSV-1 infection. In: Becker Y (ed) Skin Langerhans (dendritic) cells in virus infections and AIDS. Kluwer Academic, Boston, pp 241–260

Sprecher E, Shultz LD, Becker Y (1991) Epidermal dendritic cells in aged C57BL/6 mice. In: Becker Y (ed) Skin Langerhans (dendritic) cells in virus infections and AIDS. Kluwer Academic Boston, pp 99–118

Sprecher E, Becker Y (1992) Detection of IL-1β, TNF-α and IL-6 gene transcription by the polymerase chain reaction in keratinocytes, Langerhans cells and peritoneal exudate cells during infection with herpes simplex virus-1. Arch Virol 126: 253–269

Steinman RM (1988) Cytokines amplify the function of accessory cells. Immunol Lett 17: 197–202

Steinman RM (1991) The dendritic cell system and its role in immunogenicity. Annu Rev Immunol 9: 271–296

Stevens JG, Cook ML (1971) Latent herpes simplex virus in spinal ganglia of mice. Science 173: 843–846

Stingl G, Rappersberger K, Tschachler E, Gartner S, Groh V, Mann DL, Wolff K, Popovic M (1990) Langerhans cells in HIV-1 infection. J Am Acad Dermatol 22: 1210–1217

Streilein JW (1985) Circuits and signals of the skin-associated lymphoid tissue (S.A.L.T.). J Invest Dermatol 85: 10s–15s

Streilein JW, Grammer SF, Yoshikawa T, Demidem A, Vermeer M (1990) Functional dichotomy between Langerhans cells that present antigen to naive and to memory/effector T lymphocytes. Immunol Rev 117: 159–183

Stricker RB, Elswood BF, Abrams D (1991) Dendritic cells and dinitrochlorobenzene (DNCB): a new treatment approach to AIDS. Immunol Lett 29: 191–196

Tschachler E, Groh V, Popovic M, Mann DL, Konrad K, Safai B, Eron L, di Marzo Veronese F, Wolff K, Stingl G (1987) Epidermal Langerhans cells – a target for HTLV-III/LAV infection. J Invest Dermatol 88: 233–237

Tsujitani S, Okamura T, Baba H, Korenaga D, Haraguchi M, Sugimachi K (1988) Endoscopic intratumoral injection of OK-432 and Langerhans cells in patients with gastric carcinoma. Cancer 61: 1749–1753

Underwood GE, Weed SD (1974) Recurrent cutaneous HSV in hairless mice. Infect Immun 10: 471–474

Yan Damme J, Opdenakker G (1990) Interaction of interferons with skin reactive cytokines: From interleukin 1 to interleukin 8. J Invest Dermatol 95: 90s–93s

Van Snick J (1990) Interleukin-6. An overview. Annu Rev Immunol 8: 253–278

Vestey JP, Norval M, Howie SEM, Maingay JP, Neill W (1990) Antigen presentation in patients with recrudescent orofacial herpes simplex virus infections. Br J Dermatol 122: 33–42

Wheeler CE (1988) The herpes simplex problem. J Am Acad Dermatol 18: 163–168

Williams NA, Hill TJ (1991) Effects on murine epidermal Langerhans cells of drugs known to cause recrudescent herpes simplex virus infection in a mouse model. J Invest Dermatol 97: 933–937

Williams NA, Hill TJ, Hooper DC (1991) Murine epidermal antigen-presenting cells in primary and secondary T-cell proliferative responses to herpes simplex virus in vitro. Immunology 72: 34–39

Williams NA, Shimeld C, Hill TJ (1992) The distribution of viral antigens and Langerhans cells during zosteriform spread of herpes simplex virus to the skin of the mouse. Arch Virol 122: 349–358

Williams RH (ed) (1981) Textbook of endocrinology. Saunders, Philadelphia

Wong GHW, Goeddel DV (1986) Tumor necrosis factor α and β inhibit virus replication and synergize with interferons. Nature 323: 819–822

Yasumoto S, Okabe N, Mori R (1986) Role of epidermal Langerhans cells in resistance to HSV infection. Arch Virol 90: 261–271

Yasumoto S, Hayashi Y, Aurelian L (1987) Immunity to herpes simplex virus type 2: suppression of virus-induced immune responses in ultraviolet B-irradiated mice. J Immunol 139: 2788–2793

Yoshioka A, Miyashi Y, Imamura S, Hiraoka M, Jo S, Abe M (1989) Suppression of contact hypersensitivity by local hyperthermia treatment due to reduced Langerhans cell population in mice. Br J Dermatol 120: 493–501
Zambruno G, Mori L, Marconi A, Mongiardo N, De Rienzo B, Bertazonni U, Gianetti A (1991) Detection of HIV-1 in epidermal Langerhans cells of HIV-infected patients using the polymerase chain reaction. J Invest Dermatol 96: 979–982

B. Epstein Barr Virus

Chapter 17 Gene Expression in Burkitt's Lymphoma Cells

Marc Jacquemin, Alison J. Sinclair, and Paul J. Farrell

Summary

Development of Burkitt's lymphoma involves translocation of the c-*myc* proto-oncogene to one of the immunoglobulin loci and other genetic changes which may include changes to the *pim*-1, c-*fps*/c-*fes* and p53 genes. Epstein-Barr virus is linked to endemic Burkitt's lymphoma but Epstein-Barr virus gene expression in Burkitt's lymphoma differs from that in B lymphocytes immortalised by Epstein-Barr virus. Only the EBNA-1 and EBER RNA genes of Epstein-Barr virus are expressed in Burkitt's lymphoma cells. This restriction of virus gene expression may be a mechanism of evasion of immune surveillance and points to a role for the EBNA-1 gene in growth of Burkitt's lymphoma cells.

Introduction

Burkitt's lymphoma (BL) is a high-grade malignant B cell lymphoma, which was first recognised through its high incidence in children in central Africa (Burkitt 1983). Three different forms are now recognised (reviewed in Magrath 1990). The endemic form is found in Africa and New Guinea in areas where infection by *Plasmodium falciparum*, the parasite causing malaria, is frequent. This tumour is the most frequent cancer in African children and the malignant cells of the great majority (95%) of BL-cases occurring in high incidence areas carry Epstein-Barr virus (EBV) DNA.

The second, sporadic, form of BL occurs with a 20- to 100-fold lower incidence worldwide and is not associated with malarial infection. It represents less than 3% of childhood cancer. Viral markers are present in the tumour cells in less than 15% of cases. Thus the few EBV-negative tumours (5%) in areas of endemic BL probably represent the underlying incidence of sporadic EBV-negative tumours worldwide. The third form of BL is found in patients suffering from AIDS; the EBV genome is present in the tumours in 30%–45% of these cases of BL. These variations in EBV content occur even though the great majority of people worldwide carry EBV.

Ludwig Institute for Cancer Research, St. Mary's Hospital Medical School, Norfolk Place, London W2 1PG, UK

Independent of geographical distribution or association with EBV, all BLs present the same cell phenotype and display a characteristic translocation bringing the c-*myc* proto-oncogene on chromosome 8 into the proximity of one of the immunoglobulin genes leading to the deregulation of c-*myc*.

Typing of Burkitt's Lymphoma Biopsies and Cell Lines

BL cells were shown many years ago to synthesize immunoglobulins and therefore to be of B cell origin (Klein et al. 1967). The vast majority of tumours express surface IgM but a few percent of tumours lack IgM and express other heavy chain classes (Gunven et al. 1980). Some tumours have a pre-B phenotype in which μ chains are present in the cytoplasm but are not expressed on the surface (Cohen et al. 1987). All BL cells express the surface antigens CD19 and CD20 (Rowe et al. 1985). The expression of these membrane glycoproteins is restricted to cells in the B lymphoid lineage, from early pre-B to mature B cells and correlates with the presence of immunoglobulin (Ig) gene rearrangements characteristic of B cells.

BL cells consistently express two markers, CD10 (Rowe et al. 1985) and CD77 (Nudelman et al. 1983), which are characteristic of B cells at specific stages of maturation. Normal B cells expressing CD10 and CD77 are mostly found in germinal centres and have the morphology of centroblasts (Gregory et al. 1987). Centroblasts might therefore represent the normal correlate of the BL cell. As germinal centres are sites of Ig isotype switching and somatic mutation (Mac Lennan and Gray 1986), associated errors may give rise to the chromosome translocations characteristic of BL. However, unlike BL cells, centroblasts do not express surface Ig. Therefore the best physiological correlate of BL cells might be an intermediate stage between follicular B blasts (activated by antigen and T lymphocytes outside the germinal centre) and centroblasts in the germinal centre (Ling et al. 1989).

HLA class I antigens are usually found on BL cells but defective expression of alleles within the HLA-A and -C loci is a common feature (Masucci et al. 1987). BL cells also show a dramatic down-regulation of cell adhesion molecules, LFA-1, ICAM-1 and LFA-1 (Gregory et al. 1988).

Many cell lines have been established from BL tumours. Whilst EBV-negative cell lines have a homogeneous phenotype mostly similar to the original tumour cells (Rowe et al. 1986), EBV-positive cell lines display an important heterogeneity with regard to surface antigens and growth properties. This is due to a phenotypic drift during the early passage of the cell lines in vitro (Rowe et al. 1987; Rowe and Gregory 1989; Klein 1989). They have been divided into several groups according to their cell and viral gene expression.

Group I BL lines display the CD10, CD77 surface phenotype characteristic of freshly isolated BL cells (Rooney et al. 1986) and are the best cell culture equivalent of BL cells available (Rowe et al. 1987). Group III BL lines have acquired surface antigens normally found on activated B cells (CD23, CD30, CD39, CD70), lost the CD10 and CD77 markers characteristic of BL biopsy cells and also show up-regulation of surface adhesion molecules (LFA-1, ICAM-1 and

LFA-3; Gregory et al. 1988). Group II cells have an intermediate phenotype. These phenotypic changes are associated with modification of growth properties. Group III lines grow in multicellular clumps, whereas group I lines grow as a carpet of single cells. Many commonly studied EBV-positive BL cell lines fall into the group II or III category and are therefore atypical of authentic BL cells.

Cell Growth Regulatory Genes Altered in Burkitt's Lymphoma

c-*myc*

The c-*myc* locus on one of the two copies of chromosome 8 is always reciprocally translocated in BL to one of the Ig loci (reviewed in Klein 1983; Lenoir and Bornkamm 1987; Magrath 1990). Although 8:14 translocations to the heavy chain gene are the most common, 8:2 and 8:22 to the κ and λ light chain genes are also found. The translocations do not simply result in the immunoglobulin promoter driving c-*myc* expression but cause several complex and subtle changes to c-*myc* expression. The locations of the breakpoints in many cases of sporadic and endemic BL have been mapped and can occur 5′ or 3′ of the c-*myc* gene (Magrath 1990). There is a tendency for breakpoints in EBV-negative tumours to be very close to the c-*myc* gene and to be clustered near the 5′ end (Shiramizu et al. 1991). In EBV-positive tumours the breakpoints are more widely dispersed around the c-*myc* gene and can be hundreds of kilobases (kb) away.

Several different mRNAs are normally transcribed through the c-*myc* gene and the expression of the proto-oncogene is tightly regulated (reviewed in Spencer and Groudine 1991). Transcription of the gene can start from one of four promoters (P0 to P3; Fig. 1), resulting in at least four different mRNAs. The major coding sequence, which initiates with an AUG start codon and encodes a 64 kDa protein, is located within exons 2 and 3 which are common to all four mRNAs. Exon 1 also contains a CUG initiation codon and further coding sequence which extends the N terminus of the protein giving a 67 kDa product. The expression of c-*myc* is also controlled by premature termination of transcription at the 3′ end of exon 1 (Bentley and Groudine 1986, 1988; Eick and Bornkamm 1986). This control appears to be specific for transcripts which initiate at promoter P2, the most abundant in normal B cells.

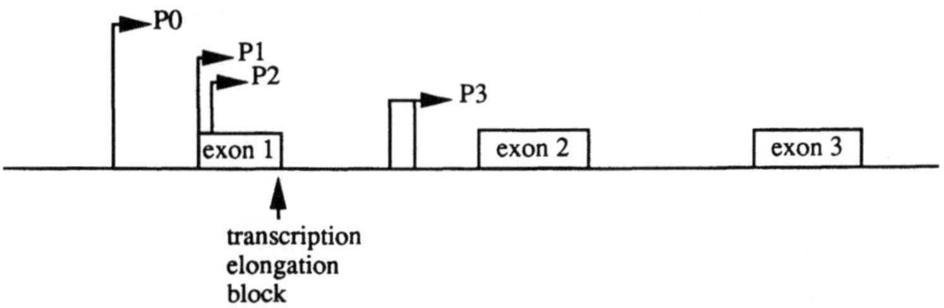

Fig. 1. The c-*myc* gene. Exons 1–3 and positions of transcription initiation points from promoter P0 to P3 are indicated. The position of premature transcription termination is shown by a *vertical arrow*.

In BL cells transcription from the normal allele is reduced to a very low level (Spencer and Groudine 1991) and is mainly from the P3 promoter (Eick et al. 1990). On the translocated allele a switch in promoter usage is observed and the P1 promoter is now used preferentially. This results in avoidance of the transcription termination controls specific for P2 initiated transcripts (Spencer et al. 1990). RNAs initiated at P0 are also enhanced from the translocated allele. In a few cell lines mutations in the first exon have also been found (Rabbitts et al. 1983) but there is no evidence that these affect c-*myc* functionally. The translocation results in the Ig enhancer being brought into the vicinity of the *myc* gene and at first it seemed obvious that this would cause the altered deregulation. But direct testing of this hypothesis using transfection of c-*myc* constructs including the enhancer did not result in the characteristic promoter usage shift (Polack et al. 1991). It is possible that further powerful enhancer sequences in the Ig locus (Petterson et al. 1990) more distant from the gene will prove to be the important element. These were not included in the constructs of Polack et al. (1991). Loss in tumour cells of a negative feedback regulation operating on c-*myc* may also be important in tumour development (Lombardi et al. 1990) and a stabilisation of c-*myc* RNA in response to EBV in some cell lines (Lacy et al. 1989) may also be relevant.

The next known gene along chromosome 8 is called PVT (Shtivelman et al. 1989) and a few translocation breakpoints are so far from the c-*myc* gene that they disrupt PVT. This, however, is uncommon and there is no evidence that it contributes to transformation. The importance of the c-*myc* deregulation to B lymphoma development has been demonstrated by constructing transgenic mice which contain a c-*myc* gene linked to the Eμ immunoglobulin enhancer (Adams et al. 1985; Harris et al. 1988; Schmidt et al. 1988). These mice develop plasmacytomas (a murine equivalent of BL) at a high rate, strongly implicating c-*myc* as the key gene on chromosome 8 in BL. The development of lymphomas in the Eμ-*myc* transgenic mice is strongly enhanced relative to normal mice but does not occur in every B lymphocyte, clearly indicating that other changes must accumulate in the lymphocyte for lymphoma development. There are many candidates for these changes but none has been fully defined yet.

pim-1

One approach to identifying other genes that might be important in BL has used retrovirus mediated lymphomagenesis in mice containing the Eμ-*myc* transgene (Haupt et al. 1991; van Lohuizen et al. 1991). Retroviral infection causes rapid tumour formation, apparently as a result of chance integration of the retrovirus near key growth control genes. Analysis of the proviral integration sites in the tumours revealed frequent integration near the *pim*-1 gene and identified a new gene called *bmi*-1 and three new loci called *pal*-1, *bla*-1 and *emi*-1, implicating these in tumour growth. Elevated levels of *pim*-1 RNA have been reported in a BL cell line (Meeker et al. 1990). A survey of group I BL cell lines showed several lines which overexpress *pim*-1 RNA (Fig. 2). The contributions of the *bmi*-1 gene and the *pa*l-1, *bla*-1 and *emi*-1 loci to BL have not yet been determined.

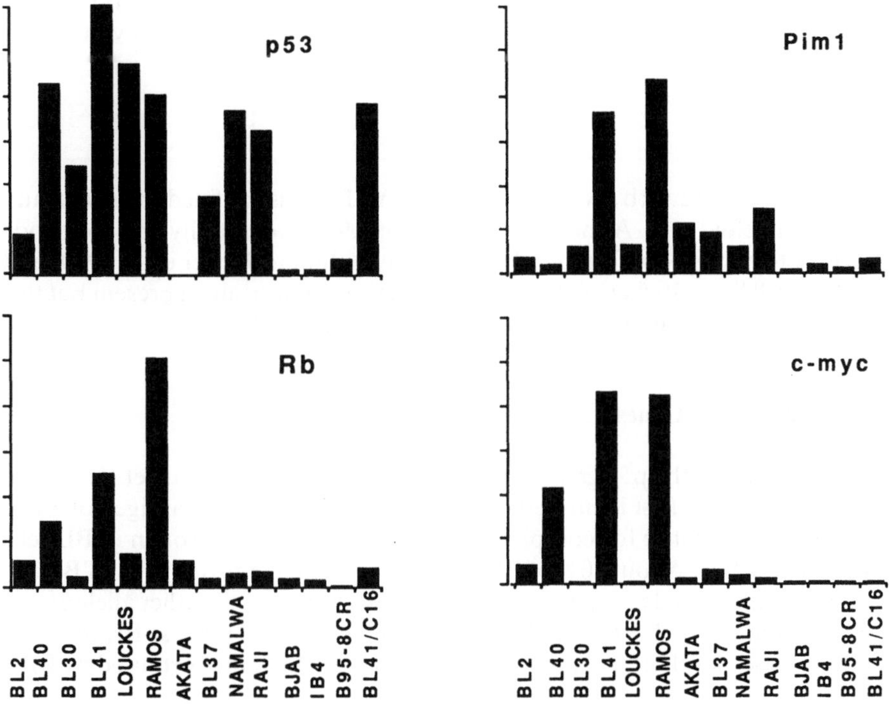

Fig. 2. Levels of *pim*-1, c-*myc*, p53 and Rb mRNAs in a series of lymphoid cell lines. BL2, BL40, BL30, BL41, Louckes and Ramos are all EBV-negative BL lines; Akata, BL37, Namalwa and Raji are EBV-positive BL lines; Akata and BL37 retaining the gpI phenotype. BJAB is a high grade B lymphoma that lacks the c-*myc* translocation characteristic of BL. BL41/C16 in cells were made by infecting BL41 cells with the P3HR1 (clone 16) strain of EBV. Poly A+ RNA was analysed by northern blotting; the filters were subsequently probed for actin and glyceraldehyde-3-phosphate dehydrogenase (GAPDH) RNA and autoradiographs scanned densito-matically to quantitate the RNA bands. Each mRNA type was then normalised relative to the actin or GAPDH and plotted in arbitrary units (the GAPDH results are shown but the histograms normalised to actin are indistinguishable). The p53 data have been reported in detail elsewhere (Farrell et al. 1991) and references on the cell lines may be found there.

bcl-2

Translocation of *bcl*-2 and consequent overexpression is a regular feature of the low grade, follicular lymphomas but this translocation is not found in BL (Klein 1991). Transgenic mice containing a *bc*1-2-IgH construct structually similar to the translocation in follicular lymphomas eventually develop malignant lymphomas, of which many contain a rearranged c-*myc* gene (McDonnell and Korsmeyer 1991). Expression of *bc*1-2 prevents apoptosis in BL cells (Gregory et al. 1991) and the EBV LMP protein causes induction of *bcl*-2 (Henderson et al. 1991). So if the EBV LMP gene were expressed in an EBV infected cell in an early stage of BL development, this could promote cell survival within the lymph node (where many B cells normally die through apoptosis). This could in turn expand the B cell pool and prolong the life of precursors to BL cells. Inappropriate expression of c-*myc* when cell growth is arrested can lead to apoptosis, at least in fibroblasts (Evan et al. 1992), and *bcl*-2 expression could have a role in maintaining cell viability in lymphocytes in those circumstances.

c-*fps*/c-*fes*

The expression of a 3.0 kb *fps*/*fes* proto-oncogene RNA is confined to normal and leukaemic myeloid cells. A shorter transcript of 0.9 kb, potentially able to encode a short protein, has been found in Hodgkin's lymphoma (Jucker et al. 1990) and 50% of BL cell lines (Jucker et al. 1992). Its function is uncertain at present but this change in expression makes it potentially important in BL.

Tumour Suppressor Genes p53 and Rb

Loss of function of the p53 gene is implicated in many types of cancer (Hollstein et al. 1991). p53 was first identified by its ability to bind to the T antigen of SV40 but there is no evidence for complex formation with an EBV protein in BL cells (Farrell et al. 1991). About 30% of BL biopsies and a higher proportion of BL cell lines contain a mutated p53 gene and frequently show loss of the other allele of p53 (Gaidano et al. 1991; Farrell et al. 1991; Wiman et al. 1991). Most BL cell lines also show raised levels of p53 mRNA relative to B cell lines immortalised by EBV (Fig. 2).

Mutation of the p53 gene can be detected relatively simply by use of the single strand conformation polymorphism (SSCP) technique (Fig. 3) to reveal the part of the gene which may be altered. Nucleotide sequencing can then be applied to determine the precise change. In several cases these mutations of p53 have been

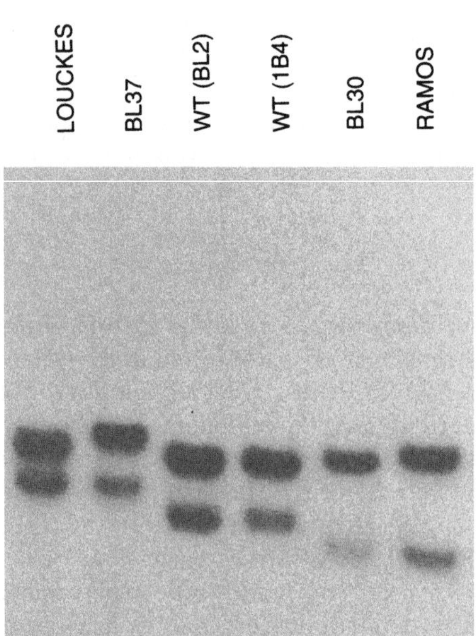

Fig. 3. Single strand conformation polymorphism (SSCP) analysis of part of the p53 mRNA for mutations. PCR products were radioactively labelled, boiled and electrophoresed on a nondenaturing gel. Banks corresponding to the two strands can be seen on the autoradiograph and the p53 mutants (Louckes, BL37, BL30 and Ramos) are readily distinguished from the wild-type genes

shown to alter p53 function, reducing its normal ability to suppress DNA synthesis in BL cell (Vousden et al. 1993). So it is likely that mutation of p53 is an important step in tumour development in some BL cases.

The Rb protein seems normally to bind and inactivate several cell transcription factors at certain stages in the cell cycle (Rustgi et al. 1991; Shirodkar et al. 1992). It can also be found in an inactive complex with the adenovirus Ela or HPV16 E7 transforming proteins but there is no direct evidence yet published for it binding to an EBV protein in BL cells (Farrell et al. 1991). It is noteworthy, however, that Rb and the EBNA-LP protein colocalise within the nucleus (Jiang et al. 1991). BL cells do not have frequent gross deletions of Rb since the protein is present and the mRNA is not obviously changed in size but the gene is very large and careful studies of possible point mutations have not yet been attempted. The levels of RNA for the Rb, p53, *pim*-1 and c-*myc* RNAs have all been compared by northern blotting in a panel of EBV-positive and -negative B cell lines (Fig. 2). No simple correlations emerge but it is perhaps noteworthy that the level of all the mRNAs generally seems to be higher in BL cell lines than in B lymphocytes immortalised by EBV (LCLs). In this series of lines the highest levels of mRNA from Rb, c-*myc* and *pim*-1 are all in EBV-negative BL cell lines. The same results were obtained using either glyceraldehyde-3-phosphate dehydrogenase (GAPDH) or actin RNA to normalise RNA levels so these are likely to reflect genuine differences between the cell lines.

Epstein-Barr Virus Gene Expression in Burkitt's Lymphoma

For many years it was assumed that EBV gene expression in BL cells was the same as in LCLs but careful analysis of group I BL cell lines that faithfully retain the BL cell phenotype showed that this is not the case.

An LCL immortalised by EBV normally expresses up to 11 EBV genes (Kieff and Liebowitz 1990) of which six encode nuclear proteins (EBNA1, 2, 3A, 3B, 3C, LP), three encode cytoplasmic membrane proteins LMP (LMP1) TP1, (LMP2A) and TP2 (LMP2B) and two are small RNAs (EBER1, EBER2). For a summary of the alternative nomenclatures used for EBV latent genes by different groups see Table 1. Of the genes expressed when EBV immortalises a B cell, EBNA-2 (Cohen et al. 1989; Hammerschmidt and Sugden 1989) and EBNA-LP (Mannick et al. 1991) have been shown by mutagenesis to be required for efficient immortalisation but these proteins are not expressed in group I BL cell lines. In BL biopsies and in group I BL cell lines EBNA-1 and the EBER RNAs are expressed are EBNA-3A, 3B, 3C, EBNA-LP and LMP are absent (Rowe et al. 1987). There is still some uncertainty about the expression of TP1 and TP2 in group I BL cells lines. TP1 (LMP2A) protein was not detected in a panel of group I BL lines (Rowe et al. 1990; Frech et al. 1990) but in some lines smaller proteins that react with one TP1 antiserum were found (Rowe et al. 1990). TP1 transcripts have been found in one group I line (Laux et al. 1988) but no TP2 transcripts were observed; lack of antisera reacting with TP2 has prevented analysis of its expression at the protein level. So the general conclusion is that the TP1 and TP2 (LMP2A, B) are not expressed in group I BL cell lines but there might be a few exceptions to this. BL

Table 1. Epstein-Barr virus gene nomenclature[a]

EBNA1		
EBNA2		
EBNA 3A	EBNA3	
EBNA 3B	EBNA4	
EBNA 3C	EBNA6	
EBNA-LP	EBNA5	EBNA 4
LMP1		LMP
LMP 2A		TP1
LMP 2B		TP2

[a] Names on the same horizontal line refer to the same gene.

cell lines classified as group III have become like LCLs in their surface markers and EBV gene expression. There may be a selective advantage in cell culture for cells expressing the full set of EBV immortalisation genes and this may be the reason for the spontaneous outgrowth of the aberrant group III lines.

If there is a continued role for EBV in the growth of mature EBV-positive BL cells, EBNA-1 seems to be the obvious candidate since it is the only EBV protein known to be expressed. The EBER RNAs are also expressed in group I BL cells (Rowe et al. 1987) but they play no obvious role in EBV immortalisation (Swaminathan et al. 1991) so it is more difficult to envisage a function for them in BL cell growth. The necessary maintenance replication function of EBNA-1 has so far precluded testing whether it has a further direct role in the immortalisation process. Unfortunately there are no data on EBV gene expression in the cells in the lymph node germinal centre which may be the precursors to BL cells. In BL cell lines and tumour biopsies we are only observing the end stage of presumably a multistep carcinogenic process. This is of great relevance in considering possible roles for EBV in BL.

The down-regulation of most of the EBNAs and LMPs in BL cells may be an important part of the mechanism by which BL cells evade cytotoxic T lymphocyte (CTL) mediated immune surveillance. Some of the EBNAs (excluding EBNA-1) and LMPs have been shown to encode peptide epitopes which mediate class I restricted CTL killing of cells infected with EBV (Murray et al. 1992). The reduction in surface LFAs and ICAM-1 (and sometimes class I HLA) doubtless also plays a role in this.

Organisation of Epstein-Barr Virus Gene Expression in Burkitt's Lymphoma and EBV-Immortalised B Lymphocytes

Although their coding sequences are distributed over a large part of the EBV genome, in LCLs the EBNA genes are all part of the same transcription unit. This extends over about 100 kb from the promoter (Cp) to the 3′ end of the EBNA-1 gene (Fig. 4). In BL cells the down-regulation of all the EBNAs except EBNA-1 is achieved by Cp being off and another promoter (Fp) becoming active (Sample et al. 1991; Schaefer et al. 1991; Smith and Griffin 1992). RNA from this promoter

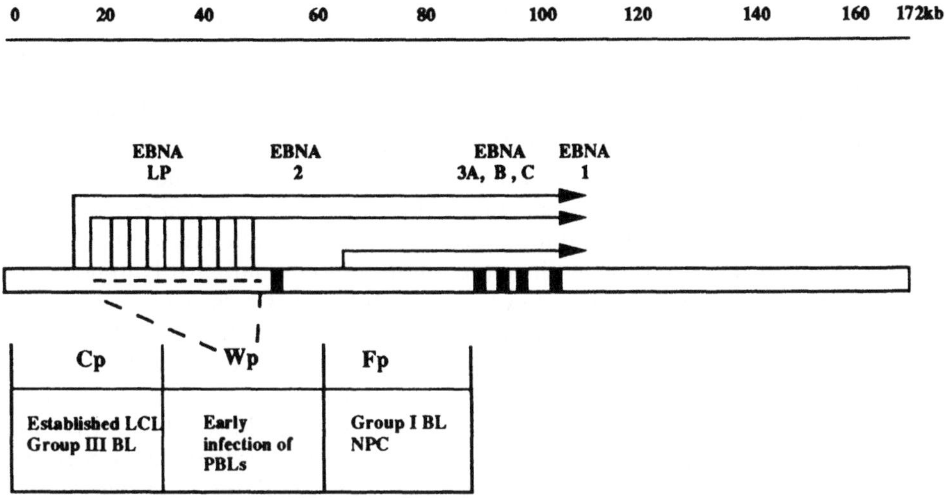

Fig. 4. Transcription of EBNA genes in and BL cells and cells immortalised by EBV

appears to splice only to EBNA-1 and not to the EBNA3 family. The Wp promoter, which mediates EBNA expression immediately after infection and becomes relatively down-regulated as immortalisation is established (Woisetschlaeger et al. 1990), is also off in BL cells. In group III BL cell lines where artefactual reactivation of the EBV genome has occurred, the Cp promoter is again active, as in LCLs.

The mechanism of EBNA promoter regulation in BL cells has two components, the inactivation of Cp and Wp and the activation of Fp. Little has yet been reported about the regulation of Fp but some studies have been made of Cp and Wp. DNA methylation may have a role, particularly in keeping Wp switched off (Jannson et al. 1992). Recent studies have suggested a mechanism for down-regulation of Cp in BL cells which involves a functional inactivation of the glucocorticoid receptor, a transcription factor upon which Cp is dependent (Sinclair et al. 1994). The promoters expressing LMP1 and 2A and B are also inactive in group I BL cells. For LMP1 and LMP2B this may partly be the result of lack of EBNA-2 expression since EBNA-2 activates these promoters (reviewed in Sinclair and Farrell 1992). Little is known about the regulation of LMP2A.

Role of Epstein-Barr Virus in Burkitt's Lymphoma

Arguments in favour of a role for EBV in BL remain circumstantial rather than directly mechanistic and are most readily applied to the endemic disease, in which the incidence of EBV-positive BL is highest. In EBV-positive cases of BL, every tumour cell retains the virus, perhaps indicating a continued role for EBV in growth of the mature BL cell. Clonality of EBV terminal repeat number (Raab-Traub and Flynn 1986) in BL cells has been used as an argument in favour of the virus being present from a very early stage (perhaps the beginning) of tumour development (Neri et al. 1991; Brown et al. 1988), again consistent with an aetiological role for EBV. It is clear that the factor that correlates with the much higher incidence of BL in Africa and New Guinea is hyperendemic malaria (reviewed in de Thé 1982). The vast majority of people worldwide are infected with EBV but the geometric mean titre of EBV antibodies is significantly raised in African children who go on to develop the disease (de Thé et al. 1978). If one accepts the arguments about monoclonality of BL with respect to the cell (all cells in a tumour show the same Ig rearrangement) and virus (terminal repeat copy number) this indicates that the tumour arose from a single cell that was probably infected with EBV at the time. Since the number of circulating B lymphocytes containing EBV is normally very low (less than 0.1% of B cells; Yao et al. 1985) but the proportion of EBV-positive tumours is high (90% in Africa) it follows that there was a much higher chance of malignant conversion occurring in a B cell infected with EBV than in an uninfected B cell. The weakness of this numerical argument is that we do not know the proportion of B cells infected with EBV in the lymph nodes of children in Africa with malaria. It is known that the proportion of circulating B cells infected with EBV is raised during malaria (Lam et al. 1991) but it is still a relatively small proportion of the B cells.

The role of EBV has been presented as enhancement of the size of the precursor B lymphocyte pool where Ig rearrangements are happening, this increasing the possibility of an abberant rearrangement occurring in the *myc* locus (Klein 1987). If EBV immortalising genes such as LMP are expressed early in BL development, prevention of apoptosis within the lymph node might be one mechanism that leads to this. Malaria seems more likely to be the factor that causes the B cell pool size to be enhanced (Lenoir and Bornkamm 1987) and it simultaneously causes a reduction in T cell mediated immune surveillance (altered T4/T8 ratios; Whittle et al. 1984). These may be important factors in the role of malaria. Arguments based on the immortalising capacity of EBV genes have become confused since we now know that most of the genes involved in B cell immortalisation are switched off in the mature tumour cell. It is still possible that the simple and obvious solution, that EBNA-1 is the important gene, will be correct but no role for EBNA-1 in cell growth control has yet been discovered. An intriguing possibility (Shiramizu et al. 1991) is that the location of the breakpoint relative to c-*myc* sufficient to cause a tumour is much less restricted in EBV-positive than EBV-negative BL; this might point to a role of EBV. If a higher proportion of erroneous translocations became tumorigenic in the presence of EBV this could cause the enhanced tumour incidence. What is greatly needed is a clear molecular insight into the mechanism which could then be tested properly. A solution where EBV substitutes for an otherwise obligatory oncogenic change in EBV-negative BL would be perhaps the most attractive and consistent with current general views on the role of viruses in multistep carcinogenesis.

Clinical Applications

Understanding the molecular mechanisms of tumour development in BL may allow the development of treatments specific for the tumour. One general problem in treatment of lymphomas is measurement of minimal residual disease and knowing when to stop the chemotherapy. In principle the presence of a chromosome translocation in BL should allow very sensitive polymerase chain reaction (PCR) methods to detect DNA fragments spanning the rearrangement which will act as a tumour cell marker. The wide distribution of breakpoints in different BL cases unfortunately means that the different PCR primers might be required for different cases. As the oncogenic changes important in BL are identified these gene products will represent specific targets for new classes of chemotherapeutic agents. Design of such drugs will require a detailed understanding of the complex network of growth control circuitry. Even though c-*myc* is an obvious target for such an approach we still have little idea of the important targets for c-*myc* action and its detailed mechanism of action.

The recent progress in analysing the genes expressed in BL could also allow treatment of the tumour by using cytotoxic agents coupled to monoclonal antibodies directed against markers present on the surface of the cells. Antigens like CD10 and CD77 that are expressed only on a limited subpopulation of normal cells could provide suitable targets. As most BL express surface Ig, monoclonal antibodies directed against idiotopes (i.e. antigenic determinants present in the variable part of Ig) could also be useful. However, treatment with anti-idiotypic monoclonal antibodies is difficult because production of suitable monoclonal antibodies takes a long time and the tumour cells are likely to produce mutated Igs having lost the relevant idiotype.

The characterisation of antigens exclusively expressed by BL cells also offers the possibility to develop treatments aiming to improve the recognition of tumours and their subsequent destruction by the immune system. This is of particular interest in endemic EBV-positive BLs. Patients presenting these tumours, like all people infected with EBV, develop a large number of CTLs recognizing the viral proteins. However, the immune defenses against BL cells are impaired by the fact that adhesion molecules critical for interactions with CTLs are down-regulated in the tumour cells. The restriction of EBV latent gene expression also helps to protect the tumour cells as EBNA1, the unique EBV protein expressed in BL, is poorly (if at all) recognized by CTLs (Murray et al. 1992). If it were possible to reactivate the expression of a complete set of viral latent proteins in the tumour cells, this would render the phenotype of BL similar to the phenotype of group III BL cell lines. These cells, expressing all the adhesion molecules required for interaction with T cells, would then be destroyed by CTLs recognizing products of viral latent genes. A prerequisite to this interesting approach is a comprehensive understanding of the mechanisms governing the switch between promoters controlling EBV latent gene expression.

Acknowledgements. We would like to thank Roger Watson and Fritz Propst for reviewing the manuscript, Anton Berns for the *pim*-1 probe and Frances Shanahan for excellent technical assistance.

References

Adams JM, Harris AW, Pinkert CA, Corcoran LM, Alexander WS, Cory S, Palmiter RD, Brinster RL (1985) The c-myc oncogene driven by immunoglobulin enhancers induce lymphoid malignancy in trangenic mice. Nature 318: 533–538

Bentley DL, Groudine M (1986) A block to elongation is largely responsible for decreased transcription of c-*myc* in differentiated HL60 cells. Nature 312: 702–706

Bentley DL, Groudine M (1988) Sequence requirements for premature termination of transcription in the human c-myc gene. Cell 53: 245–256

Brown NA, Liu CR, Wang YF, Garcia CR (1988) B-cell lymphoproliferation and lymphomagenesis are associated with clonotypic intracellular terminal regions of the Epstein-Barr virus. J Virol 62: 962

Burkitt D (1983) The discovery of Burkitt's lymphoma. Cancer 51: 1777–1786

Cohen JH, Revillard JP, Magaud JP, Lenoir G, Vuillaume M, Mane AM, Vincent C, Bryon PA (1987) B-cell mutation stage of Burkitt's lymphoma cell lines according to Epstein-Barr virus status and type of chromosome translocations. J Natl Cancer Inst 78: 235–242

Cohen JI, Wang F, Mannick J, Kieff E (1989) Epstein-Barr virus nuclear protein 2 is a key determinant of lymphocyte transformation. Proc Natl Acad Sci USA 86: 9558–9562

de-Thè G (1982) Epidemiological of Epstein-Barr virus and associated diseases in man. In: Roizman B (ed) The herpesviruses vol 1. Plenum, New York, pp 25–104

de-Thé G, Geser A, Day NE, Tukei PM, Williams EH, Beri DP, Smith PG, Dean AG, Bornkamm GW, Feorino P, Henle W (1978) Epidemiological evidence for a causal relationship between Epstein-Barr virus and Burkitt's lymphoma: results of the Ugandan prospective study. Nature 274: 756–761

Eick D, Bornkamm GW (1986) Transcriptional arrest within the first exon is a fast control mechanism in c-myc gene expression. Nucleic Acids Res 14: 8331–8346

Eick D, Polack A, Kofler E, Lenoir GM, Rickinson AB, Bornkamm GW (1990) Expression of P0- and P3-RNA from the normal and translocated c-myc allele in Burkitt's lymphoma cells. Oncogene 5: 1397–1402

Evan GI, Wyllie AH, Gilbert CS, Littlewood TD, Land H, Brooks M, Waters CM, Penn LZ, Hanock DC (1992) Induction of Apoptosis in Fibroblasts by c-myc Protein. Cell 69: 119–128

Farrell PJ, Allan GJ, Shanahan F, Vousden KH, Crook T (1991) p53 is frequently mutated in Burkitt's lymphoma cell lines. EMBO J 10: 2879–2887

Frech G, Zimber-Strobl U, Sventzenich KO, Palish O, Lenoir GM, Bornkamm GW, Mueller-Lantzsch N (1990) Identification of Epstein-Barr virus terminal protein (TP1) in extracts of four lymphoid cell lines, expression in insect cells and detection of antibodies in human sera. J Virol 64: 2759–2767

Gaidano G, Ballerini P, Gong JZ, Inghirami G, Meri A, Newcomb EW, Macgrath IT, Knowles DM, Dalla-Favera R (1991) p53 mutations in human lymphoid malignancies: association with Burkitt lymphoma and chronic lymphocytic leukemia. Proc Natl Acad Sci USA 88: 5413–5417

Gregory CD, Tursz T, Edwards CF, Tetaud C, Talbot M, Caillou B, Rickinson AB, Lipinski M (1987) Identification of a subset of normal B cells with a Burkitt's lymphoma (BL)-like phenotype. J Immunol 139: 313–318

Gregory CD, Muray RJ, Edwards CF, Rickinson AB (1988) Down-regulation of cell adhesion molecules LFA-3 and ICAM-1 in Epstein-Barr virus-positive Burkitt's lymphoma underlies tumour cell escape from virus-specific T cell surveillance. J Exp Med 167: 1811–1824

Gregory CD, Dive C, Henderson S, Smith CA, Williams GT, Gordon J, Rickinson AB (1991) Activation of Epstein-Barr virus latent genes protects human B cells from death by apoptosis. Nature 349: 612–614

Gunven P, Klein G, Klein E, Narin T, Singh S (1980) Surface immunoglobulins on Burkitt's lymphoma biopsy cells from 91 patients. Int J Cancer 25: 711–719

Hammerschmidt W, Sugden B (1989) Genetic analysis of immortalizing functions of Epstein-Barr virus in human B lymphocytes. Nature 340: 393–397

Harris AW, Pinkert CA, Crawford M, Langdon WY, Brinster RL, Adams JM (1988) The Eμ-myc transgenic mouse: a model for high-incidence spontaneous lymphoma and leukemia of early B cells. J Exp Med 167: 353–371

Haupt Y, Alexander WS, Barri G, Klinken SP, Adams JM (1991) Novel zinc finger gene implicated as myc collaborator by retrovirally accelerated lymphomagenesis in Eμ-*myc* trangenic mice. Cell 65: 753–763

Henderson S, Rowe M, Gregory C, Croom-Carter D, Wang F, Longnecker R, Kieff E,

Rickinson A (1991) Induction of bcl-2 expression by Epstein-Barr virus latent membrane protein 1 protects infected B cells from programmed cell death. Cell 65: 1107–1115
Hollstein M, Sidransky D, Vogelstein B, Harris CC (1991) p53 mutations in human cancers. Science 253: 49–53
Jannson A, Masucci M, Rymo L (1992) Methylation of discrete sites within the enhancer region regulates the activity of the Epstein-Barr virus BamHI W promoter in Burkitt lymphoma lines. J Virol 66: 62–69
Jiang W-Q, Szekely L, Wendel-Hausen V, Ringertz N, Klein G, Rosen A (1991) Colocalisation of the retinoblastoma protein and the Epstein-Barr virus encoded nuclear antigen EBNA-5. Exp Cell Res 197: 314–318
Jucker M, Schaadt M, Diehl V, Poppena S, Jones D, Tesch H (1990) Heterogeneous expression of proto-oncogenes in Hodgkin's disease derived cell lines. Hemat Oncol 8: 191–204
Jucker M, Roebroek AJM, Mautner J, Koch K, Eick D, Diehl V, Van de Ven WJM, Tesch H (1992) Expression of truncated transcripts of the proto-oncogene c-*fps*/*fes* in human lymphoma and lymphoid leukemia cell lines. Oncogene 7: 943–954
Keiff E, Liebowitz D (1990) Epstein-Barr virus and its replication. In: Fields, Knipe (eds) Virology, 2nd edn. pp 1889–1920
Klein E, Klein G, Nadkarni JS, Nadkarni JJ, Wigzell H, Clifford P (1967) Surface IgM specificity on cells derived from a Burkitt's lymphoma. Lancet ii; 1068–1070
Klein G (1983) Specific chromosomal translocations and the genesis of B cell lymphoma. Cell 32: 311–315
Klein G (1987) In defence of the "old" Burkitt lymphoma scenario. In: Klein G (ed) Advances in viral oncology, vol 7. Raven, New York, pp 207–211
Klein G (1989) Virus latency and transformation: the strategy of Epstein-Barr virus. Cell 58: 5–8
Klein G (1991) Comparative action of myc and bcl-2 in B cell malignancy. Cancer Cells 3: 141–143
Lacy J, Summers WP, Summers WC (1989) Post-transcriptional mechanisms of deregulation of MYC following conversion of a human B cell line by Epstein-Barr virus. EMBO J 8: 1973–1980
Lam KMC, Syed N, Whittle H, Crawford DH (1991) Circulating Epstein-Barr virus carrying B-cells in acute malaria. JAMA 265: 876–879
Laux G, Perricaudet M, Farrell PJ (1988) A spliced Epstein-Barr virus gene expressed in immortalized lymphocytes is created by circularization of the linear viral genome. EMBO J 7: 769–774
Lenoir GM, Bornkamm GW (1987) Burkitt's Lymphoma, a human cancer model for the multistep development of cancer: proposal for a new scenario. In: Klein G (ed) Advances in viral oncology. Raven, New York, pp 173–206
Ling NR, Hardie D, Lowe J, Johnson GD, Khan M, MacLennan ICM (1989) A Phenotypic study of cells from Burkitt lymphoma and EBV-B-lymphoblastoid lines and their relationship to cells in normal lymphoid tissues. Int J Cancer 43: 112–118
Lombardi L, Grinani F, Sternas L, Cechova K, Inghiraci G, Dalla-Favera R (1990) Mechanism of negative feedback regulation of c-myc gene expression in B-cells and its inactivation in tumour cells. In: Potter M, Melchers F (eds) Mechanisms in B-cell neoplasia 1990-springer, Berlin Heidelberg New York, pp 253–301 (Current topics in microbiology and immunology, vol 166)
Maclennan ICM, Gray D (1986) Antigen-driven selection of virgin and memory cells. Immunol Rev 91: 61–85
Magrath I (1990) The pathogensis of Burkitt's lymphoma. Adv Cancer Res 53: 132–270
Mannick JB, Cohen JI, Birkenback M, Marchini A, Kieff E (1991) The Epstein-Barr virus nuclear protein encoded by the leader of the EBNA RNAs is important in B-lymphocyte transformation. J Virol 65: 6826–6837
Masucci MG, Torsteinsdottir S, Colombani BJ, Brautbar C, Klein E, Klein G (1987) Down regulation of class 1 HLA antigens and of the Epstein-Barr virus-encoded latent membrane protein in Burkitt lymphoma lines. Proc Natl Acad Sci USA 84: 4567–4571
McDonnell TJ, Korsmeyer SJ (1991) Progression from lymphoid hyperplasia to high grade malignant lymphoma in mice transgenic for the t (14:18) Nature 349: 254–256

Meeker TC, Loeb J, Ayres M, Sellers W (1990) The human Pim-1 gene is selectively transcribed in different hemato-lymphoid cell lines in spite of a G+C-rich housekeeping promoter. Mol Cell Biol 10: 1680–1688

Murray RJ, Kurilla MG, Brooks JM, Thomas WA, Rowe M, Kieff E, Rickinson AB (1992) Identification of target antigens for the human cytotoxic T cell response of Epstein-Barr virus (EBV): Implications for the immune control of EBV-positive malignancies. J Exp Med (in press)

Neri A, Barriga F, Inghirami G, Knowles DM, Neequaye J, Magrath IT, Dalla-Favera R (1991) Epstein-Barr virus infection precedes clonal expansion in Burkitt's and acquired immunodeficiency syndrome-associated lymphoma. Blood 77: 1092–1095

Nudelman E, Kannagi R, Hakomori S, Parsons M, Lipinski M, WielsJ, Fellous M, Tursz T (1983) A glycolipid antigen associated with Burkitt lymphoma defined by a mono-clonal antibody. Science 220: 509–511

Pettersson S, Cook GP, Bruggemann M, Williams GT, Neuberger MS (1990) A second B cell specific enhancer 3′ of the immunoglobulin heavy-chain locus. Nature 344: 165–168

Polack A, Strobl L, Feederle R, Schweizer M, Koch E, Eick D, Wiegal J, Bornkamm GW (1991) The intron enhancer of the immunoglobulin kappa gene activates c-myc but does not induce the Burkitt-specific promoter shift. Oncogene 6: 2033–2040

Raab-Traub N, Flynn K (1986) The Structure of the termini of the Epstein-Barr virus as a marker of clonal cellular proliferation Cell 47: 883–889

Rabbitts TH, Hamlyn PH, Baer R (1983) Altered nucleotide sequences of a translocated c-myc gene in Burkitt's lymphoma. Nature 306: 760–765

Rooney CM, Gregory CD, Rowe M, Finerty S, Edwards C, Rupani H, Rickinson AB (1986) Endemic Burkitt's lymphoma: Phenotypic analysis of tumour biopsy cells and of the derived tumour cells lines. JNCI 77: 681–687

Rowe M, Gregory C (1989) Epstein-Barr Virus and Burkitt's Lymphoma. Adv Vir Oncol 8: 237–259

Rowe DT, Joab I, Laux G (1990) Identification of the Epstein-Barr virus terminal protein gene products in latently infected lymphocytes. J Virol 64: 2866–2875

Rowe M, Rooney CM, Rickinson AB, Lenoir GM, Rupani H, Moss DJ, Stein H, Epstein MA (1985) Distinctions between endemic and sporadic forms of Epstein-Barr virus-positive Burkitt's lymphoma. Int J Cancer 35: 435–441

Rowe M, Rooney CM, Edwards CF, Lenoir GM, Rickinson AB (1986) Epstein-Barr virus status and tumour cell phenotype in sporadic Burkitt's lymphoma. Int J Cancer 37: 3647–373

Rowe M, Rowe DT, Gregory CD, Young LS, Farrell PJ, Rupani H, Rickinson AB (1987) Differences in B cell growth phenotype reflect novel patterns in Epstein-Barr virus latent gene expression in Burkitt's lymphoma cells. EMBO J 6: 2743–2751

Rustgi AK, Dyson N, Bernards R (1991) Amino-terminal domains of c-*myc* and N-*myc* proteins mediate binding to the retinoblastoma gene product. Nature 352: 541–544

Sample J, Brooks L, Sample C, Young L, Rowe M, Gregory C, Rickinson A, Kieff E (1991) Restricted Epstein-Barr virus protein expression in Burkitt lymphoma is due to a different Epstein-Barr nuclear antigen 1 transcriptional initiation site. Proc Natl Acad Sci USA 88: 6343–6347

Schaefer BC, Woisetschlaeger M, Strominger JL, Speck SH (1991) Exclusive expression of Epstein-Barr virus nuclear antigen 1 in Burkitt lymphoma arises from a third promoter, distinct from the promoters used in latently infected lymphocytes. Proc Natl Acad Sci USA 88: 6550–6554

Schmidt EV, Pattengale PK, Weir L, Leder P (1988) Transgenic mice bearing the human c-myc gene activated by an immunoglobulin enhancer: a pre-B-cell lymphoma model. Proc Natl Acad Sci USA 85: 6047–6051

Shiramizu B, Barriga F, Neequaye J, Jafri A, Dalla-Favera R, Neri A, Guttierez M, Levine P, Magrath I (1991) Patterns of chromosomal breakpoint locations in Burkitt's lymphoma: relevance to geography and Epstein-Barr virus association. Blood 77: 1516–1526

Shirodkar S, Ewen M, DeCaprio, Morgan J, Livingston DM, Chittenden T (1992) The transcription factor E2F interacts with the retinoblastoma product and a p107-cyclin a complex in a cell cycle-regulated manner. Cell 68: 157–166

Shtivelman E, Henglein B, Groitl P, Lipp M, Bishop JM (1989) Identification of a human transcription unit affected by the variant chromosomal translocations 2;8 and 8;22 of Burkitt lymphoma. Proc Natl Acad Sci USA 86: 3257–3260

Sinclair AJ, Farrell PJ (1992) Epstein-Barr virus transcription factors. Cell Growth Diff 3: 557–563

Sinclair AJ, Jacquemin M, Brooks L, Shanahan F, Brimmell M, Rowe M, Farrell PJ (1994) Reduced signal transduction through glucocorticoid receptor in Burkitt's lymphoma cell lines. Virology (in press)

Smith PR, Griffin BE (1992) Transcription of the Epstein-Barr virus gene EBNA-1 from different promoters in nasopharyngeal carcinoma and B-lymphoblastoid cells. J Virol 66: 706–714

Spencer CA, Groudine M (1991) Control of c-myc Regulation in normal and neoplastic cells. Adv Cancer 56: 1–48

Spencer CA, LeStrange RC, Novak U, Hayward WS, Groudine M (1990) The block to transcription elongation is promoter dependent in normal and Burkitt's lymphoma c-*myc* alleles. Genes Dev 4: 75–88

Swaminathan S, Tomkinson B, Kieff E (1991). Recombinant Epstein-Barr virus with small RNA (EBER) genes deleted transforms lymphocytes and replicates in vitro. Proc Natl Acad Sci USA 88: 1546–1550

van Lohuizen M, Verbeek S, Scheijen B, Wientjens E, van der Gulden H, Berns A (1991) Identification of cooperating oncogenes in Eμ-*myc* transgenic mice by provirus tagging. Cell 65: 737–752

Vousden KV, Crook TR, Farrell PJ (1993) Biological activities of p53 mutants in Burkitt's lymphoma cells. J Gen Virol 74: 803–810

Whittle HC, Brown J, Marsh K, Greenwood BM, Seidelin P, Tighe H, Wedderburn L (1984) T-cell control of Epstein-Barr virus-infected B cells is lost during P. falciparum malaria. Nature 312: 449–450

Wiman WG, Magnusson KP, Ramqvist T, Klein G (1991) Mutant p53 detected in a majority of Burkitt lymphoma cell lines by monoclonal antibody PAb240. Oncogene 6: 1633–1638

Woisetschlaeger M, Yandava CN, Furmanski LA, Strominger JL, Speck SH (1990) Promoter switching in Epstein-Barr virus during the initial stages of infection of B lymphocytes. Proc Natl Acad Sci USA 87: 1725–1729

Yao QY, Rickinson AB, Epstein MA (1985) A re-examination of the Epstein-Barr virus carrier state in healthy seropositive individuals. Int J Cancer 35: 35–42

Chapter 18 Molecular Mechanisms of the Development of EBV-Related B Lymphomas: Functional Cooperation of EBV with IL-6 and HIV-1

Giuseppe Scala[1], Ileana Quinto[2], Maria R. Ruocco[2], Massimo Mallardo[2], Concetta Ambrosino[2], Francesca Baldassarre[2], Vincenzo Giordano[2], and Salvatore Venuta[2]

Summary

The biological role of interleukin 6 (IL-6) molecules in human B cell tumorigenesis was studied by using an episomal expression vector, pHEBoSV-IL6, to introduce stably the human IL-6 gene into human Epstein-Barr virus (EBV)-transformed B lymphoblasts. The constitutive expression of the IL-6 gene led to an altered pattern of growth and to a malignant phenotype, as shown by clonogenicity in soft agar cultures and tumorigenicity in nude mice. These data suggest that the combined action of EBV, which exerts an immortalizing function, and of the growth-promoting activity of IL-6 molecules can give rise to fully transformed B cell tumors in immunodeficient subjects.

Among immunodeficient patients, human immunodeficiency virus 1 (HIV-1) infected subjects show a high incidence of (EBV) infection. This suggests that EBV may function as a cofactor that affects HIV-1 activation and modulates the progression of AIDS. To test this hypothesis, we generated two EBV-negative human B cell lines stably expressing the EBNA2 gene of EBV. These EBNA2-positive cell lines were transiently transfected with plasmids that carry either the wild-type or deletion mutants of HIV-1 LTR fused to the chloramphenicol-acetyltransferase (CAT) gene. There was a consistently higher HIV-1 LTR activation in the EBNA2-expressing cells than in the control cells, which suggested that EBNA2 proteins could activate the HIV-1 promoter, possibly by inducing nuclear factors binding to HIV-1 *cis*-regulatory sequences. To test this possibility, we used CAT-based plasmids, carrying deletions of NF-kB (pNFA-CAT), or Sp1 (pSpA-CAT), regions of the HIV-1 LTR, and retardation assays, in which nuclear proteins from EBNA2-expressing cells were challenged with oligonucleotides encompassing the NF-kB or Sp1 regions of the HIV-1 LTR. We found that both the NF-kB and the Sp1 sites of the HIV-1 LTR are necessary for EBNA2 transactivation and that increased expression resulted from the induction of NF-kB-like factors. Transfection experiments with plasmids expressing the EBNA1,

[1] Dipartimento di Biochimica e Biotecnologie Mediche, Università Federico II, 80131 Naples, Italy

[2] Dipartimento di Medicina Sperimentale Clinica, Università di Reggio Calabria, 88100 Catanzaro, Italy

EBNA3 and EBNALP genes did not cause significant HIV-1 LTR activation. Thus, it appears that among the tested EBV latent genes, EBNA2 was the only EBV gene active on the HIV-1 LTR.

Introduction

EBV is a herpesvirus that preferentially infects B lymphocytes and induces proliferation, immunoglobulin (Ig) secretion, and immortalization, giving rise to long-term lymphoblastoid B cell lines (LCLs) (Kieff et al. 1985). EBV infection appears to play a pathogenetic role in the development of endemic Burkitt's lymphoma (eBL) and in the recently identified AIDS-associated Burkitt's lymphoma (AIDS-BL) (Epstein and Achong 1983; Birx et al. 1986; Pelicci et al. 1986). The nontumorigenic capabilities of LCLs suggest that the neoplastic EBV-infected clones present in eBL and AIDS-BL may have acquired a malignant phenotype through additional pathogenetic events. As shown for a variety of neoplastic cell types, the continuous proliferation of EBV-infected B cells appears to depend on autocrine secretion of cytokines, including IL-1 and B cell growth factor-like molecules (Sporn and Todaro 1980; Gordon et al. 1984; Scala et al. 1984, 1987). IL-6 is a glycoprotein secreted by a variety of cell types, e.g., monocytes, fibroblasts, certain T cell lines, and myeloma cells (Kishimoto 1989). Its biological action includes the growth stimulation of plasmacytomas and some EBV lymphoblastoid B cell lines (Scala et al. 1984, 1987; Van Damme et al. 1987; Tosato et al. 1988). To investigate whether EBV infection, which exerts a typical immortalizing function, cooperates with the B cell growth-promoting activity of IL-6 in inducing B cell tumorigenesis, we transfected ROHA-9 MC3 cells, an EBV B cell line (Scala et al. 1984, 1987), with an expression vector that allows constitutive expression of the human IL-6 cDNA. Our results indicate that hyper-expression of an exogenous IL-6 gene in EBV-immortalized B cells results in a dramatic decrease in the requirement for exogenous growth factors with the development of a fully tumorigenic phenotype.

Results and Discussion

To obtain constitutive expression of human IL-6 cDNA in EBV-infected lymphoblasts, an appropriate plasmid was developed using the vector pHEBoSV, which carries the hygromycin B resistance gene together with the EBV origin of replication (oriP) (Sugden et al. 1985; Lombardi et al. 1987). The IL-6 cDNA coding sequences were inserted downstream to the SV40 promoter element of pHEBoSV plasmid (Fig. 1). The resulting pHEBoSV-IL6 and pHEBoSV control vectors were transfected by electroporation into the MC3 EBV-B cell line. Following selection of the transfected cells in medium supplemented with hygromycin B (300 μg/ml), bulk cultures of IL-6-transfected MC3 cells (MC3-

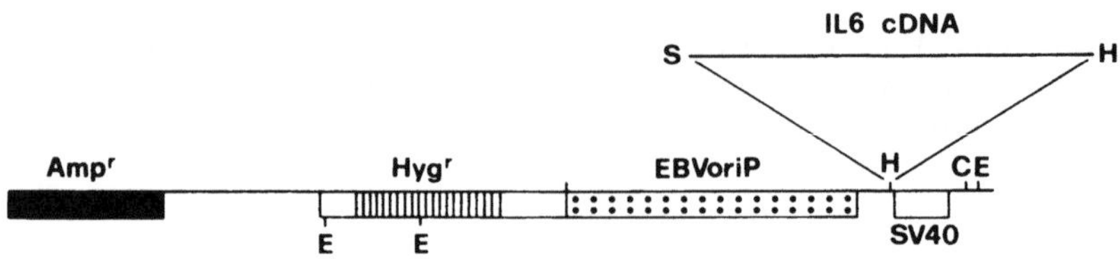

Fig. 1. The pHEBoSV-IL6 vector. Relevant regions of the pHEBoSV-IL6 plasmid are indicated by *boxes* marked as follows: SV40, SV40 enhancer/promoter region; IL-6 *cDNA*, full-length human IL-6 cDNA; *EBVorip*, EBV origin of replication; *Hyg*ʳ, hygromycin B resistance gene flanked by the thymidine kinase promoter and polyadenylation regions; *Amp*ʳ ampicillin resistance gene. Restriction sites: *H*, *Hind*III, *S*, *Sal*I; *E*, *Eco*RI

IL6) or MC3-pHEBoSV were obtained. The MC3-IL6 cells were then subjected to limiting dilution cultures (0.2 cells/well) to obtain cell clones. The MC3-IL6-9, MC-3IL6-11 bulk cultures and the MC3-IL6-2C4 and MC3-IL6-2A2 clones were randomly selected for the present study; MC3-pHEBoSV bulk cultures served as controls.

The presence of intact episomal vectors in transfected cells was confirmed by Southern analysis of low molecule weight DNA (Scala et al. 1990). The vectors were present as extrachromosomal plasmids with a copy number of approximately 5–10 per cell. No integrated copies of the plasmids were revealed by Southern analysis of genomic DNA. The expression of the transfected IL-6 cDNA was examined by northern blots of total RNA from the MC3-IL6 cells by using an IL-6-specific probe. All the transfected MC3-IL6 clones expressed high amounts of IL-6 mRNAs, whereas the MC3-pHEBoSV control cells showed low levels of specific IL-6 mRNA transcription (not shown). The secretion of IL-6 by the transfected cells was tested by the using the IL-6-dependent 7TD1 hybridoma cell line (Van Snick et al. 1986). On a per cell basis, the transcription and secretion of IL-6 was higher in MC3-9, MC3-11, MC3-2C4, and MC3-2A2 cells than in lipopolysaccharide (LPS)-activated monocytes, while the MC3-pHEBoSV control cells secreted very low amounts of IL-6 (Table 1).

To test whether IL-6 expression may lead to an altered requirement for serum growth factors, the MC3-IL6 cells were cultured in serum-free medium at different cell concentrations, as previously reported (Scala et al. 1987). As shown in Fig. 2, the MC3-IL6-9 cells were able to proliferate even when cultured at very low cell concentrations. Under the same culture conditions, the MC3-pHEBoSV control cells failed to proliferate significantly unless the cultures were supplemented with conditioned medium from MC3-IL6-9 cells. Similar results were obtained using the other IL-6-transfected cells (not shown). This indicated that the requirement for exogenous growth factors was decreased in the IL-6-expressing cells, suggesting that the IL-6 molecules could function as autocrine factors for the MC3-IL6 cells. In fact, IL-6 molecules were secreted by the cells and present in the culture medium. This possibility was confirmed by experiments in which control and IL-6-transfected cells were cultured at 5×10^4/ml cell in HB-101 serum-free medium and supplemented with optimal concentrations of IL-6. While the control cells retained the capacity to proliferate in response to IL-6, as previously reported for other EBV B cell lines (Tosato et al. 1988), the IL-6-transfected cells showed a

Table 1. IL-6 secretion, in vitro cloning efficiency and in vivo tumorigenicity of IL-6-transfected MC3 cells

Cell line	IL-6 secretion[a]	Cloning efficiency[b]	Tumorigenicity in vivo[c]	
			Number of mice with tumors/ number of inoculated mice	Average of tumors per mouse
MC3-pHEBoSV	6	0	0/6	0
MC3-IL 6-9	750	120	4/4	3.5
MC3-IL6-11	710	80	4/4	2.5
MC3-IL6-2C4	650	78	3/4	3.0
MC3-IL6-2A2	560	148	5/5	4.0
MØ	420	–	–	–

[a] IL-6 activity on the IL-6-dependent 7TD1 cells is expressed as pg/ml of IL-6 secreted by 5×10^5 cells over a 48 h culture time.
[b] Growth in soft agar was assayed as previously reported (Scala et al. 1990). Data are expressed as number of colonies per 10^4 cells and are the arithmetic mean of triplicate cultures. SEM were <10% of the mean values.
[c] Tumorigenicity was assayed by a single subcutaneous injection of 5×10^6 cells in 0.2 ml of PBS. Typically, tumors >5 mm in diameter developed after 2–3 weeks of latency at the challenge site and in the regional lymph nodes. Data are the number of tumors recorded through 6 weeks of observation.

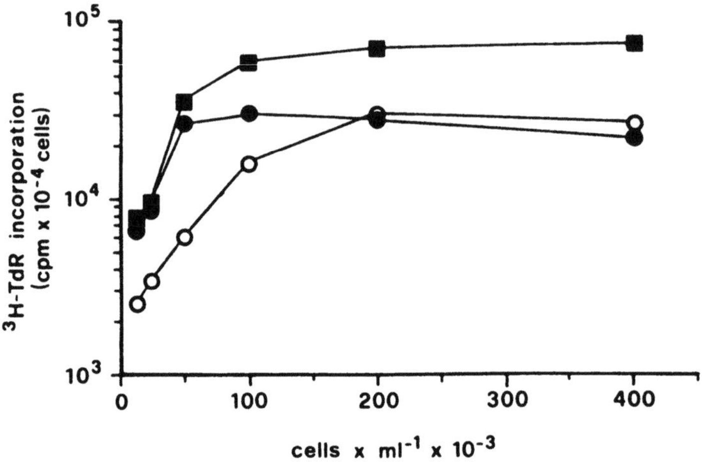

Fig. 2. Constitutive expression of the IL-6 gene alters the growth of MC3-IL6 cells. MC3-pHEBoSV (○) and MC3-IL6-9 (■) cells were cultured for 48 h at the indicated cell concentrations in HB-101 serum-defined medium as previously described (Scala et al. 1987). In parallel cultures (●), control MC3-pHEBoSV cells were cultured in the presence of dialyzed conditioned medium obtained from MC3-IL6-9 cells cultured in HB-101 medium and tested at 25% final dilution

baseline increase in [^{3}H]thymidine uptake and were unresponsive to the exogenous IL-6 molecules (Scala et al. 1990). This may depend on a saturation of IL-6 surface receptors by the constitutively secreted IL-6 molecules. The addition of a goat polyclonal antibody to IL-6 molecules resulted in a partial decreased in the constitutive proliferative of the MC3-IL6 cells, while the IL-6-induced proliferation of MC3-pHEBoSV cells was significantly inhibited (not shown).

The altered growth properties of the MC3-IL6 cells were also studied by examining their ability to form colonies in soft agar cultures, a property common to many naturally occurring B cell neoplasias (Magrath 1986). As shown in Table 1, the MC3-IL6 cells exhibited a significant increase in their clonogenic capabilities, while the control MC3-pHEBoSV cells did not give rise to colonies.

The clonogenic potential of transformed cells is often associated with a tumorigenic phenotype (Magrath 1986). To test whether the MC3-IL6 cells could have acquired a malignant phenotype, we subcutaneously injected various B cell populations into nude mice. As shown in Table 1, MC3-IL6 B cells gave rise to tumors in nude mice, whereas the control MC3-pHEBoSV cells were not tumorigenic. The tumors appeared subcutaneously 10–15 days after the injections, and showed a histological picture of homogeneous infiltrating B cell populations. These cells were characterized by eccentric nucleolated nuclei and large cytoplasms with peripheral condensation. These immunoblastic-like cells were shown to infiltrate periglandular lymph nodes (not shown). The secretion of IL-6 by MC3-IL6 cells could have induced a polyclonal proliferation of mouse B cells that could account for the tumoral masses. To test this possibility, tumor cells from mice injected with MC3-IL-6 cells were examined for the expression of CD20, CD23, MHC class I and class II molecules, and for mouse and human surface Igs (μ, δ, γ, α, specificities). In these experiments, both the original and tumor MC3-IL6 cells showed a similar flow cytometry profile (data not shown). Moreover, analysis of the low molecular weight DNA of tumor cells showed the presence of pHEBoSV-IL6 plasmids (not shown).

The nontumorigenic properties of in vitro-infected EBV B cells suggest that additional pathogenic events have occurred in spontaneous EBV-related B lymphomas, including eBL and the recently identified AIDS-BL (Epstein and Achong 1983; Birx et al. 1986; Pelicci et al. 1986). The growth of neoplastic lymphoid B cells is often dependent upon interleukin-like growth factors that share most of the biochemical and biological properties of the factors secreted and used by normal lymphoid cells (Scala et al. 1987). Among these growth factors, IL-6 has been recently identified as a growth and differentiation factor for both normal and transformed B cells (Kishimoto 1989). In this study, we attempted to drive an EBV B cell line, MC3, to a malignant phenotype by inserting a constitutively expressed IL-6 gene. Present in an episomal replicating plasmid, the gene was stably present in transfected cells with a high copy number and it was highly expressed. The resulting MC3-IL6 cells showed a clonogenic capacity and were tumorigenic in nude mice. These data are consistent with and support the recent observations that transgenic mice carrying a human IL-6 gene develop plasmacytomas (Suematsu et al. 1989), that myeloma cells secrete and use IL-6 as an autocrine growth factor (Kawano et al. 1988), and that IL-6 molecules play a role in the abnormal growth of neoplastic B cells.

The in vivo tumorigenicity of the human IL-6-secreting MC3 cells could be expressed in immunodeficient hosts such as nude mice. In this regard, it is noteworthy that variable levels of immunodeficiency are a common feature of EBV-related malignacies, including eBL and AIDS-BL (Epstein and Achong 1983; Birx et al. 1986; Pelicci et al. 1986). Our data suggest that the concomitant presence of EBV, which acts as an immortalizing factor, and deregulated secretion of a cytokine, acting as a progression growth factor, can give rise to fully transformed B cell tumors in T cell-deficient subjects.

Among immunodeficient subjects, AIDS patients have a high incidence of EBV infection, with up to 96% showing serologic evidence of EBV (Fauci et al. 1984). Moreover, a consistent percentage of AIDS associated lymphomas harbor EBV genome (Freter 1990) and HIV-1 can infect B cells (Dahl et al. 1987; de Rossi et al. 1990; Pomerantz et al. 1988), suggesting that the two viruses can coexist in HIV-infected patients. Furthermore, HIV-1 and EBV coinfection has been reported in clones of EBV B cells and in the cervical epithelium of AIDS patients (Dahl et al. 1987; de Rossi et al. 1990; Pomerantz et al. 1988). Thus, EBV may act as a cofactor that stimulates the expression of HIV-1, possibly through *trans*-activation of the HIV-1 LTR, so promoting a high level of HIV-1 replication and eventually reducing the clinical latency period of AIDS. The EBV genome persists in infected cells either as an episome or as integrated viral DNA, with only a limited number of EBV genes being expressed in latently infected cells. These include genes coding for EBNA1 EBNA2, EBNA3, EBNALP, and LMP (Kieff and Liebowits 1989). The blastic-like phenotype, together with the proliferation and Ig secretion induced by EBV, is similar to the phenotype induced by short-term antigen stimulation of normal B cells. This suggests that EBV may leads to B cell immortalization by activating a set of host cellular genes transiently expressed during physiological B cell activation. The EBNA2 gene product has been reported to activate the transcription of CD21, CD23 and c-fgr genes in human B cells (Knutson 1990; Wang et al. 1990). Moreover, EBNA2 protein expression could be crucial for EBV-mediated B-cell immortalization. In fact, the P3HR-1 strain of EBV, which does not express EBNA2, is unable to immortalize B cells (Bornkamm et al. 1982). These findings suggest that EBV latent gene products may activate HIV-1 transcription.

This possibility was addressed by using several eukaryotic expression plasmids carrying the EBNA1, EBNA2, EBNA3, and EBNALP coding sequences in cell transfection experiments to test whether the EBV genes modulate the expression of a reporter CAT gene linked to the LTR sequences of HIV-1 (Scala et al. 1993).

In the first experiments, expression of the HIV1-LTR-CAT plasmid in different B cell lines was examined by transiently transfecting pWT-CAT plasmid into the EW36 or De Fews EBV-negative NHL cells and into the EBV-positive MC3, or CB33 LCLs. These experiments indicated that the expression of the HIV-1 LTR in MC3 or CB33 cells was consistently higher than that observed in EBV-negative B cell lines (not shown), suggesting that latent EBV-infection may result in efficient of HIV-1, possibly mediated by the expression of the EBV latent genes. Among these genes, EBNA2 induces the expression of CD23 and CD21 B cell surface proteins (Wang et al. 1987) and transcriptionally activates the LMP and CD23 genes by modulating 5′ upstream regulatory regions (Wang et al. 1990, 1991).

To ascertain whether EBNA2 might induce a cell phenotype prone to HIV-1 expression, we transfected EW36 and De Few, two EBV-negative B cell lines, with the pZipSVneo-EBNA2 plasmid (Fig. 3) to obtain constitutive synthesis of EBNA2 proteins. Following electroporation with the pZipSVneo-EBNA2 or pZipSVneo(x) plasmids, stable transfectants of EW36 and De Few pZipSVneo-EBNA2 (hereafter referred to as EW36- and De Few-EBNA2) cells or EW36 and De Few-pZipSVneo(x) control cells were obtained after selection with G418 (geneticin, 3 mg/ml). The protein product of the EBNA2 gene in the transfected cells was identified in immunoblots with the anti-EBNA2 PE2 monoclonal antibody (MoAb) (Young et al. 1989). The EW36 and De Few EBNA2 cells showed a consistent level of EBNA2 protein. The correct subcellular localization of the expressed EBNA2 protein was tested by immunocytochemical analysis of the transfected cells using the PE2 MoAb. In these experiments, the EBNA2 protein localized in the nuclei of the transfected cells in a pattern very similar to that obtained with MC3 LCLs. These data indicated that the transfected EBNA2 gene efficiently expressed EBNA2 proteins of the expected molecular weight and with the correct unclear localization and thus it was possibly able to function as a *trans*-activator gene. We next transiently transfected EW36- and De Few-EBNA2 cells and the relative pZipSVneo(x) control cells with the pWT-CAT plasmid, which carries the wild-type HIV-1 LTR sequences linked to the bacterial CAT gene. The level of HIV-1 LTR expression was assayed after 48 h by evaluating CAT activity in cellular extracts. As shown in Fig. 4, both the EBNA2-transfected EW36 and De Few-EBNA2 cells efficiently expressed the transfected HIV-1 LTR-CAT plasmid to a greater extent than the pZipSVneo(x)-control cells.

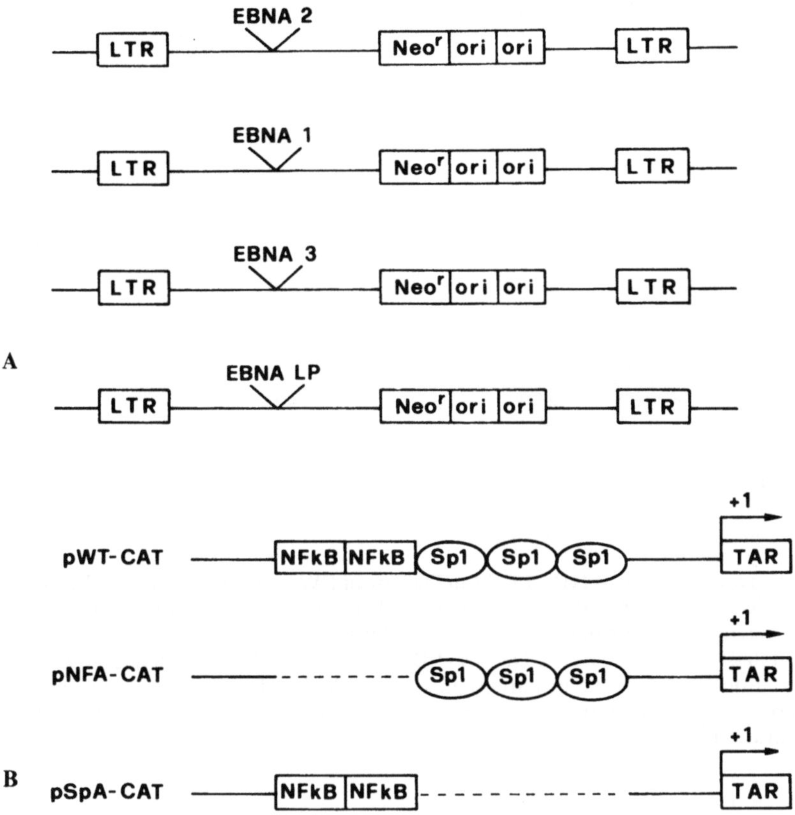

Fig. 3. A EBV eukaryotic expression plasmids. **B** HIV-1 LTR-CAT plasmids. Region of HIV-1 from –108 to +80 is represented

This indicated that the EBNA2 protein could act as *trans*-activator of the HIV-1 promoter.

Similar results were obtained when cell clones isolated from the pZipSVneo(x) or pZipSVneo-EBNA2 transfected bulk cultures were examined. AT least 30 subclones of EBNA2-expressing or De Few and EW36 control cells were individually examined for the constitutive expression of the transiently transfected pWT-CAT plasmid. There were no significant differences in the activation of the HIV-LTR among individual clones, and the levels of CAT expression were consistently higher in EBNA2-transfected cell clones than in the pZipSVneo(x)-control clones (not shown). These experiments indicated that stable expression of the EBNA2 gene in two EBV-negative B cell lines resulted in a higher constitutive activation of the HIV-1-LTR promoter-enhancer regions, thus mimicking the high activation of the HIV-1 LTR observed in a variety of EBV-infected B cells either isolated by in vitro infection or from naturally EBV-infected AIDS patients (data not shown).

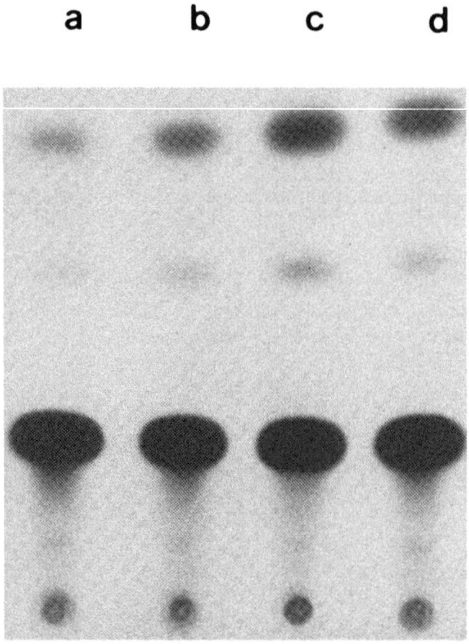

Fig. 4. Constitutive expression of HIV-1 LTR in De Few of EW36 transfected cells. Cells were transfected by electroporation with 10 µg of pWT-CAT plasmid. CAT activity was assayed in a cell lysate (50 µg) prepared 48 h posttransfection. *a*, *b*, De Few-pZipSVneo(×) (0.2) and EW36-pZipSVneo(×) (0.4) control cells; *c*, *d*, De Few-EBNA2 (5.8) and EW36-EBNA2 (4.9) cells. Percent acetylation, noted above in parentheses, was calculated by directly counting the acetylated spots, and is representative of several independent experiments using different plasmid preparations

The regulation of HIV-1 gene expression requires a complex interaction of cellular factors binding to transcriptional regulatory sequences in HIV-1 LTR. This includes a TATA box, three Sp1 binding sites, and a core enhancer region containing two κB sites (Franza et al. 1987; Jones et al. 1986). To identify the *cis*-regulatory sequences of the HIV-1-LTR target for EBNA2 activation, we performed cell transfection experiments in which the wild-type or mutant HIV-1 LTR plasmids carrying deletions of the κB enhancer region or of Sp1 sites were transiently cotransfected with the pZipSVneo(×) or pZipSVneoEBNA2 vectors in De Few or EW36 cells. After 48 h, cell aliquots from each transfection were analyzed for the expression of EBNA2 protein by immunoblots (Scala et al. 1993), and cell extracts were tested for CAT activity. As shown in Table 2, the EBNA2-induced HIV-1 LTR activation was drastically diminished when plasmid pNFA-CAT, deleted of the κB sites of HIV-1 LTR, was used. In the same experiments, deletions of the Sp1 sites also significantly affected the transcriptional activation of HIV-1 LTR by EBNA2. Similar results were obtained when EW36- and De Few-EBNA2 cells were transiently transfected with pWT-CAT and the relative deletion mutants (not shown).

The results shown in Table 2 suggested that EBNA2 may activate the HIV-1 LTR by inducing NF-κB- or Sp1-like binding factors. To address this issue, nuclear extracts from De Few- or EW36-EBNA2, and from De Few- or EW36-pZipSVneo(x) control cells were tested for the presence of specific NF-κB or Sp1 binding activity in an electrophoretic mobility shift assay (EMSA). These

Table 2. Responsiveness of wild-type or mutant HIV-1-LTR-CAT plasmids to the EBNA2 gene

Target plasmid[a]	pZip EBNA2[a]	Specific acetylation[b] (%)	Fold induction[b]
pwt-CAT	–	1.1	5.8
	+	6.4	
pNFA-CAT	–	0.6	0.8
	+	0.5	
pSpA-CAT	–	0.8	0.8
	+	0.7	

[a] De Few cells were transfected with 10 μg of the indicated plasmids alone or together with 30 μg of pZipSvneo-EBNA2 vector.
[b] CAT activity was determined at 48 h posttransfection by using 50 μg of cell extract. Fold induction is expressed as the ratio of percentages acetylated in the presence or absence of pZipSVneo-EBNA plasmids.

experiments showed that the expression of the EBNA2 gene in stable transfectants resulted in an increase in the NF-κB-binding factors. In the same experiments, no significant differences in Sp1 binding activity were observed between the EBNA2-expressing cells and the control cells (not shown). Similar results were obtained when De Few or EW36 cells were transiently transfected with pZipSV-EBNA2 or pZipSVneo(x) plasmid (data not shown). These findings indicate that EBNA2 may function by inducing active NF-κB-like transcription factors. These nuclear proteins share DNA-binding specificity with other cellular proteins, such as NF-κB and HIVEN 86A (Franza et al. 1987; Nabel and Baltimore 1987), which bind to κB consensus sequences present in a set of cellular genes which includes the α subunit of the IL-2 receptor and IL-6 (Franza et al. 1987; Nabel and Baltimore 1987; Lieberman and Baltimore 1990). Work is in progress to determine whether the kB-binding factors induced by EBNA2 are identical to NF-κB or HIVEN 86A factors or whether they are members of a closely related family of DNA-binding cell factors. The deletion of the three Sp1 sites in the pSpA-CAT plasmid prevents the response to EBNA2 and is consistent with the findings that Sp1-binding *cis*-regulatory sequences are essential for the activation of HIV-1 LTR (Jones et al. 1986). In fact, deletion or inactivation by point mutations of the Sp1 sites results in a marked reduction of HIV-1 LTR promoter activity and inducibility and greatly limits HIV-1 replication (Leonard et al. 1990).

In addition to EBNA2, EBV expresses a set of genes, including EBNA1, EBNA3, and EBNALP, during the latent phase of chronic infection (Kieff and Liebowits 1989). To test whether the expression of these EBV genes might result in the activation of the HIV-1 LTR sequences, we cotransfected De Few cells with the pWT-CAT plasmid and with the pZipSV-EBNA1, pZipSV-EBNA3 and pZipSV-EBNALP or pZipSV-EBNA2 plasmids. CAT activity in cell extracts after 48 h was considered a measure of possibly activation of the HIV-1 LTR. As reported in Table 3, no activation of the HIV-1 LTR-CAT construct was observed as a consequence of EBNA1, EBNA3 or EBNALP gene transfection, while EBNA2 gene transfection resulted in increased activation of HIV-1 LTR. Therefore, among the EBV latent genes tested, EBNA2 was the only one functional in transducing an activation signal to the HIV-1 promoter.

Table 3. Transactivating of HIV-1 LTR by EBV latent genes

Transactivating plasmid[a]	pWT-CAT	Acetylation[b](%)	Fold induction
pZipSVneo(×)	+	0.36	–
pZipSVneo-EBNA 1	+	0.30	0.8
pZipSVneo-EBNA 2	+	5.8	16.1
pZipSVneo-EBNA 3	+	0.32	0.8
pZipSVneo-EBNA LP	+	0.48	1.3

[a] EBV-negative De Few cells were transfected by electroporation with 30 μg of the indicated transactivating plasmids and with 10 μg of pWT-CAT plasmid.
[b] CAT activity was determind at 48 h posttransfection by using 50 μg of cell extract. Fold induction is expressed as the ratio of percentages acetylated in the presence or absence of pZipSVneo-EBNA plasmids.

Evidence is accumulating that EBV infection may play a pathogenetic role in the development of AIDS-associated B cell neoplasms (Freter 1990). The cell tropism of HIV-1 and EBV viruses is only partially known. Recent data suggest that HIV-1 can infect EBV-transformed lymphoblastoid B cells (Dahl et al. 1987; De Rossi et al. 1990, Pomerantz et al. 1988). This strongly suggests that EBV can coexist with HIV-1 at the cellular level and that it could modulate the clinical latency of HIV-1 infection, possibly by activating its cellular expression through induction of *trans*-activating factors that interact with HIV-1 *cis*-regulatory sequences. To test this possibility, we examined the capacity of EBNA1, EBNA2, EBNA3 and EBNALP latent genes of EBV to transactivate the HIV-1 LTR regulatory sequences. Among the genes tested, EBNA2 was able to transactivate the HIV-1 LTR sequences in B (De Few and EW 36) cells.

The molecular mechanisms of EBNA2 activation of HIV-1 expression are unknown. The EBNA2 gene product is a major trans-activating factor of EBV. EBNA2 is constitutively expressed at high levels in circulating latently infected EBV-B cells and may act through the induction of host cellular genes. Indeed, EBNA2 expression is required for immortalization of B cells, which results in continuous proliferation of lymphoblastoid-like cells (Bornkamm et al. 1982). These EBV-LCLs produce cytokines such as IL-1 and IL-6 (Scala et al. 1984, 1987; Tosato et al. 1989), which in turn could activate HIV-1 expression (Osborn et al. 1990; Poli et al. 1990). The ultimate mechanisms whereby EBNA2 exerts its effects on target DNA sequences are obscure. EBNA2 proteins do not bind directly to DNA sequences, however, they can associate in different forms of oligomerization and possibly may interact with host cellular proteins (Grasser et al. 1992). We found that the functional deletion of κB sites abolished EBNA2 *trans*-activation, suggesting that EBNA2 may activate the HIV-1 promoter by inducing active NF-κB factors. This possibility was confirmed by in vitro binding of HIV-1 κB sequences to nuclear extracts of EBNA2-transfected cells. The levels of EBNA2-induced *trans*-activation of HIV1 LTR is probably underevaluated with respect to naturally occurring EBV infection. In fact, the EBNA2 gene product induces expression of the LMP gene of EBV and cooperates with LMP proteins in promoting full induction of host cellular genes such as CD23 (Wang et al. 1990) and in protecting B cells from apoptosis (Gregory et al. 1991). Indeed, we have recently obtained a synergistic higher HIV1 LTR-driven CAT activity by transfecting De Few-or EW36 EBNA2 cells with the LPM-expressing pgpt-LMP plasmid (G. Scala et al., unpublished observation).

Acknowledgements. This work was supported by grants from A.I.R.C., C.N.R., and A.I.D.S. grants from the Istituto Superiore di Sanità. F.B. and V.G were supported by fellowships from Associazione Italiana per la Ricerca sul Cancro (A.I.R.C.) M.M. was supported by a fellowship from the Consiglio Nazionale delle Ricerche (C.N.R.). M.R.R. was supported from a fellowhip from the Istituto Superiore di Sanità, progetto AIDS.

References

Birx DL, Redfield RR, Tosato G (1986) Defective regulation of Epstein-Barr virus infection in patients with acquired immunodeficiency syndrome (AIDS) or AIDS related disorders. N Engl J Med 314: 874–878

Bornkamm G, Hudewantz J, Freese UK, Zimber U (1982) Deletion of the nontransforming Epstein-Barr virus stain P3HR-1 causes fusion of the large internal repeat to the DS_L region. J Virol 43: 952–961

Dahl K, Martin K, Miller G (1987) Differences among human immunodeficiency virus strains in their capacities to induce cytolysis or persistent infection of a lymphoblastoid cell line immortalized by Epstein-Barr virus. J Virol 61: 1602–1608

De Rossi A, Roncella S, Calabro ML, D'Andrea E, Pasti M, Panozzo M, Mammano F, Ferrarini M, Chieco-Bianchi L (1990) Infection of Epstein-Barr virus-transformed lymphoblastoid B cells by the human immunodeficiency virus: evidence for a persistent and productive infection leading to B cell phenotypic changes. Eur J Immunol 20: 2041–2049

Epstein MA, Achong BG (1983) The relationship of the EBV virus to Burkitt's lymphoma. In: Epstein MA, Achong BG (eds) The Epstein-Barr-virus. Springer, Berlin Heidelberg New York, pp 322–327

Fauci AS, Macher AM, Longo DL, Lane HC, Rook AH, Masur H, Gelman EP (1984) Acquired immunodeficiency syndrome: epidemiologic, clinical, immunologic, and therapeutic considerations. Ann Intern Med 100: 92–106

Franza BR, Josephs SF, Gilman MZ, Ryan W, Clarkson B (1987) Characterization of cellular proteins recognizing the HIV enhancer using a microsale DNA affinity precipitation assay. Nature 330: 391–395

Freter CE (1990) Acquired immunodeficiency syndrome: epidemiologic, clinical, immunologic, and therapeutic considerations. Ann Intern Med 100: 92–106

Gordon J, Ley SC, Melamed MA, Aman P, Hughes-Jones NC (1984) Soluble factor requirements for the auto-stimulatory growth of B lymphoblasts immortalized by Epstein-Barr virus. J Exp Med 159: 1554–1558

Grasser FA, Haiss P, Gottel S, Mueller-Lantzsch N (1992) Biochemical characterization of Epstein-Barr virus nuclear antigen 2A. J Virol 65: 3779–3788

Gregory CD, Dive C, Henderson S, Smith C, Williams GT, Gordon J, Rickinson J (1991) Activation of Epstein-Barr virus latent genes protects human B cells from death by apoptosis. Nature 349: 612–614

Jones KA, Kadonaga JT, Luciw PA, Tijan R (1986) Activation of the AIDS retrovirus promoter by the cellular transcription factor, Sp1. Sience 232: 755–759

Kawano M, Hirano T, Matsuda T, Taga T, Horii Y, Iwato K, Asoaku H, Tang B, Tanabe O, Tanaka H, Kuramoto A, Kishimoto T (1988) Autocrine generation and requirements of BSF-2/IL6 for human multiple myelomas. Nature 332: 83–86

Kieff E, Liebowits D (1989) The Epstein-Barr virus. In: Fields B, Knipe D (eds) Virology. Raven, New York, pp 1889–1920

Kieff E, Hennessy T, Dambaugh T, Marsuo T, Fennewald S, Heller M, Petti L, Hummel M (1985) Persistence and expression of the Epstein-Barr virus genome in latent infection and growth transformation of lymphocytes. In: Levine PH, Ablashi DV, Pearson GR, Kottaridis SD (eds) Epstein-Barr virus and associated diseases. Nijhoff, Boston, pp 221–247

Kishimoto T (1989) The biology of interleukin-6. Blood 74: 1–9

Knutson JC (1990) The level of c-fgr RNA is increased by EBNA-2, an Epstein-Barr virus gene required for B-cell immortalization. J Virol 64: 2530–2536

Leonard J, Parrott C, Buckler-White AJ, TurneR W, Ross EK, Martin MA, Rabson AB (1990) The NF-κB binding sites in the human immunodeficiency virus type 1 long terminal repeat are not required for virus infectivity. J Virol 63: 4919–4924

Liebermann TA, Baltimore D (1990) Activation of interleukin 6 gene expression through the NF-κB transcription factor. Mol Cell Biol 10: 2327–2339

Lombardi L, Mewconb EW, Dalla-Favera R (1987) Pathogenesis of Burkitt lymphoma: expression of an activated c-myc oncogene causes the tumorigenic conversion of EBV-infected human B lymphoblasts. Cell 49: 161–168

Magrath I (1986) Clinical and pathobiological features of Burkitt's lymphoma and their relevance to treatment. In: Levine PH, Ablashi, Pearson GR, Kittaridis SD (eds) Epstein-Barr virus and associated diseases. Nijhoff Boston, pp 631–643

Nabel G, Baltimore D (1987) An inducible transcription factor activates expression of human immunodeficiency virus in T cells. Nature 326: 711–713

Osborn L, Kunkel S, Nabel GJ (1989) Tumor necrosis factor alpha and interleukin-1 stimulate the human immunodeficiency virus in T cell. Proc Natl Acad Sci USA 86: 2336–2340

Pelicci PG, Knowles DM, Arlin ZA, Wieczorek R, Luciw P, Dina D, Basilico C, Dalla-Favera R (1986) Multiple monoclonal B-cell expansion and c-myc oncogene rearrangements in acquired immune deficiency syndrome related lymphoproliferative disorders; implication for lymphomagenesis. J Exp Med 164: 2049–2061

Poli G, Bressler P, Kinter A, Duh A, Timmer WC, Rabson A, Justement HS, Stanley S, Fauci AS (1990) Interleukine 6 induces human immunodeficiency virus expression in infected monocytic cells alone and in synergy with tumor necrosis factor α by transcriptional and post-transcriptional mechanisms. J Exp Med 172: 151–158

Pomerantz RJ, de la Monte SM, Donegan SP, Rota TR, Vogt MW, Craven DE, Hirsch MS (1988) Human immunodeficiency virus (HIV) infection of the uterine cervix. Ann Intern Med 108: 321–328

Scala G, Kuang YD, Hall RE, Muchmore AV, Oppenheim JJ (1984) Accessory cell functions of human B cells. I. Production of both interleukin-1 like activity and an interleukin-1 inhibitory factor by an EBV-transformed human B cell line. J Exp Med 159: 1637–1652

Scala G, Morrone G, Tamburrini M, Alfinito F, Pastore CI, D'Alessio G, Venuta S (1987) Autocrine growth function of human interleukin 1 molecules on ROHA-9, and EBV-transformed human B cell line. J Immunol 138: 2527–2533

Scala G, Quinto I, Ruocco MR, Arcucci A, Mallardo M, Caretto P, Forni G, Venuta S (1990) Expression of an exogenous interleukin 6 gene in human Epstein Barr virus B cells confers growth advantage and in vivo tumorigenicity. J Exp Med 172: 61–68

Scala G, Quinto I, Ruocco MR, Mallardo M, Ambrosino C, Squitieri B, Tassone P, Venuta S (1993) The Epstein-Barr virus nuclear antigen 2 trans-activates the long terminal repeat of human immunodeficiency virus type 1. J Virol (in press)

Sporn MB, Todaro T (1980) Autocrine secretion and malignant transformation of cells. N Engl J Med 303: 878–890

Suematsu S, Matsuda T, Aozasa K, Akira S, Nakano N, Ohno S, Miyazaki J, Yamamura K, Hirano T, Kishimoto T (1989) IgG1 plasmacytosis in interleukin 6 transgenic mice. Proc Natl Acad Sci USA 86: 7547–7551

Sugden B, Marsh, Yates J (1985) A vector that replicates as a plasmid and can be efficiently selected in B-lymphoblasts transformed by Espstein-Barr virus. Mol Cell Biol 5: 410–418

Tosato G, Seamon, KB, Goldman ND, Sehgal PB, Map LT, Washington GC, Jones KD, Pike SE (1988) Identification of a monocyte-derived human B cell growth factor as interferon-2 (BSF-2; IL-6). Science 239: 502–505

Tosato G, Tanner J, Jones KD, Revel M, Pike SE (1990) Identification of interleukin-6 as an autocrine growth factor for Epstein-Barr virus-immortalized B cells. J Virol 64: 3033–3042
Van Damme J, Opdenakker G, Simpson RJ, Rubira MR, Cayhas S, Vink A, Billiau A, Van Snick J (1987) Identification of the human 26-kD protein, interferon β2 (IFN-β 2), as a B cell hybridoma/plasmacytoma growth factor induced by interleukin 1 and tumor necrosis factor. J Exp Med 165: 914–919
Van Snick J, Cayphas S, Vink A, Uyttenhove, Coulie PG, Rubira MR, Simpson (1986) Purification and NH2-terminal amino acid sequence of a T-cell-derived lymphokine with growth factor activity for B-cell hybridomas. Proc Natl Acad Sci USA 83: 9679–9682
Wang F, Gregory CD, Rowe M, Rickinson AB, Wang D, Birkenbach M, Kilutani H, Kishimoto T, Kieff E (1987) Epstein-Barr virus nuclear antigen 2 specifically induces expression of the B-cell activation antigen CD23. Proc Natl Acad Sci USA 8: 3452–3456
Wang F, Tsang S, Kurilla MG, Cohen JI, Kieff E (1990) Epstein-Barr virus nuclear antigen 2 transactivates latent membrane protein LMP1. J Virol 64: 3407–3416
Wang F, Kikutani H, Tsang S, Kishimoto R, Kieff E (1991) Epstein-Barr virus nuclear protein 2 transactivates a cis-acting CD23 DNA element. J Virol 65: 4104–4106
Young L, Alfieri C, Hennessy K, Evans H, O'Hara C, Anderson K, Ritz J, Shapiro RS, Rickinson A, Kieff E, Cohen JI (1989) Expression of Epstein-Barr virus transformation associated genes in tissues from patients with EBV lymphoproliferative disease. N Engl J Med 321: 1080–1085

[illegible]

C. Human Cytomegalovirus

Chapter 19 Murine Cytomegalovirus Genes Influencing Virus Growth and Tropism for Salivary Gland

Jessica M. Boname, Laurel A. Lagenaur, and Edward S. Mocarski

Summary

The salivary gland is an important tissue in cytomegalovirus (CMV) pathogenesis. In the mouse, the salivary gland is the primary site from which virus is transmitted from animal to animal and is also a site of viral persistence and latency. Several murine CMV mutants that exhibit reduced salivary gland tropism have been described in the literature. Many temperature-sensitive mutants grow poorly in the salivary gland; however, the mutations responsible for this poor growth phenotype have not been mapped to particular genes. One variant of the Smith strain of murine CMV, the Vancouver strain, carries a deletion that was rescued by the *Hin*dIII E fragment from wild-type murine CMV. The ability to grow in the salivary gland correlated with the expression of a 42-kDa early protein. The salivary gland growth gene-1 (*sgg*1) of murine CMV has been the most thoroughly characterized determinant of salivary gland tropism. This gene encodes two overlapping transcripts (1.5 and 1.8 kb) which are initiated from different promoters but utilize the same polyadenylation signal. Sequence analysis revealed an open reading frame encoding a putative type 1 integral membrane protein. A 37-kDa protein absent in an *sgg*1 mutant could be detected in both rescued and wild type virus. The 37 and 42 kDa proteins represent distinct murine CMV proteins that correlate with salivary gland tropism in the mouse. It remains to determine how many other genes are involved in the salivary gland tropism of murine CMV and the mechanism by which these genes act during murine CMV pathogenesis.

Introduction

Human cytomegalovirus (CMV; human herpesvirus 5) causes disease following congenital infection and after primary or reactivated infection of immunocompromised individuals. Human CMV is the prototypic representative of the β-herpesviruses, a subgroup with members that are widely distributed in many different animal species (Weller 1970; Plummer 1967, 1973; Wildy 1973; Ho 1991). The CMVs were initially distinguished from other herpesviruses by characteristic

---

Department of Microbiology and Immunology, Stanford University School of Medicine, Stanford, CA 94305-5402, USA tropism, cytopathology, species specificity, and slow growth in cultured cells (Roizman et al. 1981; Mocarski 1993). More recently, genome organization and sequence characteristics have been used to compare human CMV with related members of the β-herpesvirus subgroup. Thus, the β-herpesviruses now include the human T lymphotropic herpesviruses, human herpesvirus (HHV) -6 and -7 (Efstathiou et al. 1988; Lawrence et al. 1990; Littler et al. 1990; Martin et al. 1991).

The success of herpesviruses may be attributed to their efficient transmission between hosts due to lifelong persistence with recurrent shedding. Spread of CMV occurs through direct exposure to infected body secretions, including saliva, urine and milk. The salivary glands have long been recognized as significant sites of replication for CMV, and, within the salivary gland, epithelial cells are important targets for replication. Human and murine CMVs were initially observed in and isolated from salivary gland (Cowdry and Scott 1935; Smith 1954, 1956) and are referred to by the alternative designation, salivary gland viruses. This site is also important for persistence and dissemination of other herpesviruses, including Epstein-Barr virus (EBV) (Wolf et al. 1984; Deacon et al. 1991) and HHV-6 and -7 (Frenkel et al. 1990), which all replicate in salivary glands and disseminate via saliva.

The lack of a laboratory animal in which to directly study human CMV has stimulated the use of animal CMVs, particularly mouse, rat and guinea pig CMVs, as models of the human virus (Plummer 1973; Osborn 1982). Many parallels in the biology of human and murine CMV have been described (Hudson 1979; Griffiths and Grundy 1987). Following infection, both viruses disseminate widely in the host via a leukocyte-associated viremia and infect many different tissues. During the course of acute infection, the salivary glands are a major site of virus replication where virus titers reach very high levels and virus is shed in saliva. For murine CMV, serous acinar cells (Henson and Strano 1972; Mercer and Spector 1986; Jonjic et al. 1989), and for human CMV, ductal cells are the most prominent sites of virus replication during acute infection (Ho 1991), and both of these epithelial cell types allow for efficient release of virus into saliva. Epithelial cells are important sites of viral replication in other organs, including the kidney and mammary gland, where human CMV can also be shed for extended periods of time. A recently described experimental SCID-hu model for studying human CMV replication found that thymic medullary epithelial cells can be sites of persistent viral replication (Mocarski et al. 1993). Virus shed in saliva, milk and urine is believed to contribute to transmission by direct contact between individuals (Spector 1983; Pass et al. 1987; Ho 1990; Adler 1991; Alford 1991).

In experimentally infected mice, intraperitoneal inoculation of sublethal doses of murine CMV results in the spread of virus to the salivary glands between 3 and 6 days post-inoculation. Natural killer cells help restrict viral replication in some organs during this early phase (Quinnan and Manischewitz 1979; Bancroft et al. 1981; Shellam et al. 1981; Bukowski et al. 1984; Welsh et al. 1991; Scalzo et al. 1992). After the onset of the specific immune response at 6–9 days postinoculation, virus is cleared efficiently from most organs but continues to actively replicate in the salivary glands for at least 4 weeks regardless of the strain of mouse infected (Allan and Shellam 1984; Manning et al. 1992). In addition to the role for natural immunity (Bukowski et al. 1984; Shanley 1990), passive administration of neutralizing antiserum directed at murine CMV decreases the amount of virus present in the salivary glands during acute infection (Shanley et al. 1981) and helper ($CD4^+$), rather than cytotoxic ($CD8^+$) T cells are important in clearance (Lucin et al. 1992). Following resolution of the acute infection, murine CMV enters latency in salivary glands. Latent infection of the salivary glands has been demonstrated both by explant reactivation and immunosuppression (Jordan et al. 1977; Mayo et al. 1977, 1978; Sandford and Burns 1988; Schmader et al. 1991). The viral genome persists in this organ and can be detected using the polymerase chain reaction (PCR) (Klotman et al. 1990; Schmader et al. 1991). As yet, the cell type in which the virus remains latent has not been identified.

Viral Pathogenesis in the Salivary Glands

Following parenteral inoculation, dissemination of murine CMV to the salivary glands is via a leukocyte-associated viremia (Bale and O'Neil 1989; Manning et al. 1992). Virus is delayed in reaching peak titers in this organ and occurs well after the immune response has cleared virus from other organs. This difference in clearance can be evaluated directly in animals that have been inoculated with murine CMV carrying the *E. coli lacZ* gene capable of expressing β-galactosidase in infected cells and tissues. Replication in the salivary gland is limited to a small proportion of total acinar cells concentrated in serous acini of the submaxillary gland. Although the appearance of virus in the salivary gland is delayed relative to other organs, once present, virus replicates to much higher titers and persists much longer. The number or size of foci do not appear to increase over the period of weeks that virus continues to replicate at this site. Importantly, there is little change in the pattern or extent of infection even though specific immunity has cleared virus from other tissues. Although only a small percentage of the total cells are infected, an individual infected acinar cell may be estimated to yield $10^4 - 10^5$ pfu based on peak titers of 10^9 pfu/g of salivary gland tissue reached at 14–21 days postinoculation of a BALB/c mouse (Manning et al. 1992). This level is orders of magnitude higher than the levels reached in mouse fibroblasts in culture, and has long been believed to be due to cellular or viral factors only active in this cell type (Osborn and Walker 1971).

When the immune system is debilitated by sublethal γ-irradiation, murine CMV spreads into connective tissue fibroblasts in mouse salivary glands, thus implicating the immune response in restricting viral replication in the normal host. Both humoral and cellular components of the immune system have been implicated in the control of viral persistence in salivary gland. When infected mice are depleted of $CD4^+$ T cells, replication continues in acini without the eventual clearance that normally occurs within 6 weeks (Jonjic et al. 1989), suggesting that helper T cells are important for the control of virus replication in salivary gland. In contrast, depletion of $CD8^+$ T cells has little effect on the growth patterns in salivary gland and virus is cleared normally (Jonjic et al. 1990). Interferon-γ, a cytokine product of helper T cells, may be one means by which $CD4^+$ T cells mediate clearance of murine CMV from the salivary gland (Lucin et al. 1992). $CD8^+$ T cells are certainly required to restrict murine CMV replication in a wide variety of cell types in the infected animal (Koszinowski et al. 1990) and may play a role in preventing virus from spreading to cell types outside of the acini within the salivary gland. The reason that $CD8^+$ cells are ineffective at clearance of infected acinar cells remains unknown, however MHC class I expression has not been detected on this cell type (Savage et al. 1987; Hirozane et al. 1992). The role of $CD4^+$ cells in the eventual clearance may result from production of cytokines, by more direct cytotoxicity or via generation of antibodies (Jonjic et al. 1989).

Virus that is released by acinar cells is at least two orders of magnitude more infectious than virus produced in other organs or in cultured fibroblasts (Osborn and Walker 1971), however, cultured epithelial cells do not maintain this capacity (Jordan and Takagi 1983). The phenomenon of rapid attenuation of virulent murine CMV by passage in vitro and subsequent restoration of virulence by a single passage of virus in vivo does not depend upon whether susceptible or resistant strains of mice are used (Selgrade et al. 1981). Because a single passage is able to confer either virulence or attenuation, this process probably results from some epigenetic change in the virus during maturation from acinar cells. Glycosylation patterns or processing of envelope proteins, acquisition of some cellular protein and a role for nonneutralizing antibody have been proposed contributors to this phenomenon, but the underlying mechanism remains uncharacterized (Osborn and Walker 1971; Selgrade et al. 1981; Jordan and Takagi 1983; Inada et al. 1985; Ravindranath and Graves 1990).

Viral Mutants That Alter Growth in Salivary Glands

Restriction enzyme cleavage maps published 10 years ago (Ebeling et al. 1983; Mercer et al. 1983) showed that two strains of murine CMV are in common use in laboratories, and that they have both been referred to as "Smith." The Smith strain, originally obtained from Margaret Smith (1954) and still provided by the ATCC as mouse salivary gland virus, was maintained by passage through mice. A more virulent derivative, K181, a virus also widely called Smith, was derived by June Osborn (Misra and Hudson 1980; Osborn 1982; Hudson et al. 1988; J. E. Osborn, personal communication). The distinction between the genomes of these two viruses is readily established by either *Hin*dIII or *Eco*RI digestion. Many manuscripts have appeared without clarifying the variety of Smith strain being used. Here, Smith and K181 will refer to the different virus strains when it is possible to discern, otherwise the term "murine CMV" will be employed.

Temperature-Sensitive Mutants

Several studies have shown that temperature-sensitive mutants of murine CMV grow poorly in the salivary glands of inoculated mice (Tonari and Minamishima 1983; Sandford and Burns 1988; Sammons and Sweet 1989; Kumura et al. 1990), a phenomenon that may be attributed to the relatively high body temperature of the mouse (38.5°C). In one study, however, interpretation of this work has been complicated because rescue of the temperature-sensitive lesion did not alter the growth phenotype. In this situation, the viral mutants apparently carried multiple independent mutations influencing growth in cell culture and in the infected mouse (Kumura et al. 1990). No further investigations of the specificity of the block(s) or the nature of the mutation(s) responsible for the poor growth phenotype have been reported.

The Vancouver Strain of Murine Cytomegalovirus

A derivative of the Smith strain of murine CMV was isolated based on its more efficient growth in cell culture and referred to as the "Vancouver" strain (Boname and Chantler 1992), and this virus was probably utilized in earlier strain comparisons to K181under the name "Smith" (Hudson et al. 1988). The Vancouver strain was severely limited in its ability to grow in salivary glands or in other organs following intraperitoneal inoculation and the genome was found to carry a 9.4 kbp deletion within *Hin*dIII D and a 0.9 kbp insertion within *Hin*dIII E as compared to the Smith strain (Fig. 1). In addition, the Vancouver strain lacked expression of a 42 kDa protein normally expressed at early and late times postinfection in cells infected with the Smith strain (Fig. 2; Boname and Chantler 1992).

Three recombinant viruses (R1, R2, R3) that have regained the ability to replicate in salivary gland have been recovered following cotransfection of the *Hin*dIII E fragment from the K181 strain genome along with Vancouver strain DNA. The progeny from this cotransfection were inoculated into mice and viruses that acquired the capacity to replicate in salivary glands were identified. All three recombinants had a normal growth phenotype in the mouse and were found to have replaced the 9.4 kbp deletion within *Hin*dIII E with a full-size version of *Hin*dIII E and to have restored the expression of the 42 kDa protein (Fig. 2; J. M. Boname and J. K. Chantler, unpublished). Thus, the viral gene(s) responsible for the growth defect were localized to the region at the extreme right hand end (0.960 to 0.995 map units) of the viral genome.

The *sgg*1 Gene

A recombinant murine CMV, RM427, was initially isolated as a spontaneous mutation adjacent to a *lac*Z insertion into the *ie*2 gene of the K181 strain of murine CMV (Fig. 3). RM427 was found to be severely deficient in its ability to replicate in the salivary glands after intraperitoneal, intranasal or intraglandular inoculation (Manning et al. 1992). RQ401, a derivative of RM427 that restored *ie*2 but retained the second mutation, exhibited a greatly reduced ability to grow in the salivary glands of mice (Manning et al. 1992). RQ401 grew to titers similar to wild-type virus in the spleen and liver of experimentally infected mice following intraperitoneal inoculation. However, viral titers in submaxillary salivary gland were three to five orders of magnitude lower than titers of either parental wild-type virus or rescued virus RQ427 (Fig. 4).

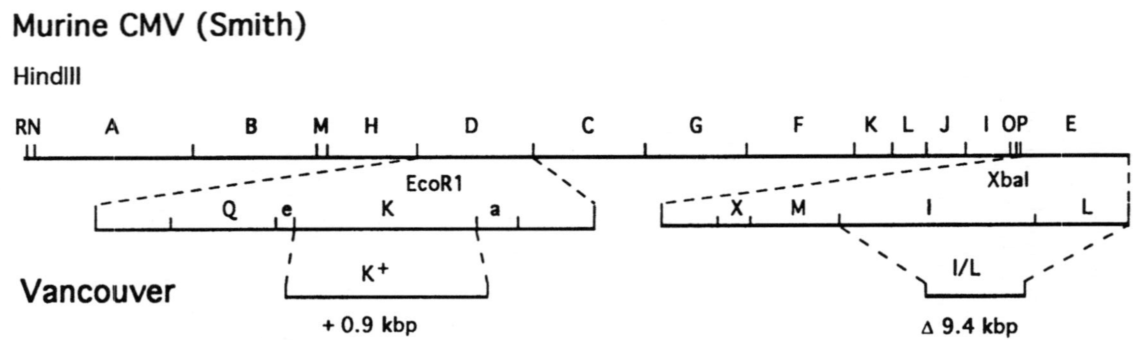

Fig.1. *Hin*dIII endonuclease map of the murine cytomegalovirus (CMV) (Smith strain) genome including *Eco*RI sites in an expanded *Hin*dIII D region and *Xba*l sites in an expanded *Hin*dIII E region (Boname and Chantler 1992). The mutations in the murine CMV (Vancouver strain) genome are shown in expanded regions below. They include a 0.9 kbp insertion (+0.9 kbp) into the *Eco*RI K fragment (K^{+}) and a 9.4 kbp deletion (*Δ9.4 kbp*) which spans the *Xba*l I/L junction of murine CMV (Smith)

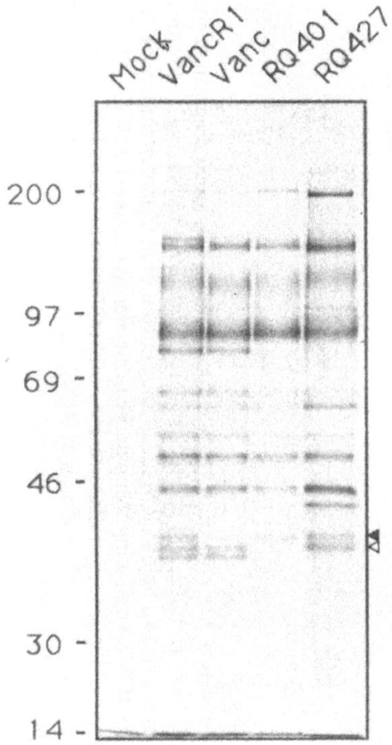

Fig. 2. Radioimmune precipitation of early proteins from mock-, Vancouver R1- (*VancR1*), Vancouver- (*Vanc*), RQ401- and RQ427-infected cells. NIH3T3 cells were infected with an MOI of 10 pfu/cell. Early proteins were labeled with 45 μCi/ml of [^{35}S] methionine from 6 to 8 h post-infection in the presence of 200 μg/ml of phosphonoacetic acid. Proteins were precipitated with anti-murine cytomegalovirus (CMV) (Smith strain) mouse serum and subjected to 8% polyacrylamide gel electrophoresis under reducing conditions prior to autoradiography. *Filled arrow*, 42 kDa protein missing in the Vancouver strain; *Open arrow*, 37 kDa protein missing in *sgg*1 mutant RQ401. Protein size standards (kDa) are indicated to the *left* of the figure

Characterization of the mutation in RQ401 revealed a 323 bp deletion within the *Hin*dIII J fragment that encompassed the polyadenylation signal of a β gene encoding a major 1.5 kb transcript and a minor 1.8 kb transcript (Fig. 5). These transcripts were absent in RNA isolated at 8 hpi from cells infected with RM427 or RQ401 but were readily detected in parental or rescued wild-type viruses (Manning et al. 1992). RQ401 failed to produce a 37 kDa protein that was rescued in RQ427 (Fig. 2). Sequence and structural analysis of the two transcripts revealed a 312 amino acid open reading frame (ORF) capable of expressing this protein, and the ORF has characteristics of a type 1 integral membrane protein (Fig. 6). This 37 kDa protein is distinct from the 42 kDa protein that is absent in the Vancouver strain (Fig. 2).

Murine CMV disseminates in a leukocyte-associated viremia (Bale and O'Neil 1989). The *sgg*1 mutant RQ401 and the rescued RQ427 replicated to similar levels in peripheral blood leukocytes during acute infection and exhibited equivalent LD_{50} doses (Manning et al. 1992), suggesting that viral dissemination was not altered. The ability of *sgg*1-deficient viruses to disseminate is also supported by the fact that *lac*Z-tagged *sgg*1 mutants can be demonstrated in salivary gland acini by staining for β-galactosidase at 7 or 9 days postinoculation. Thus, the *sgg*1 mutation does not decrease virus dissemination in general, but rather reduces the ability of virus to replicate to high titers in the serous acinar epithelial cells in the salivary glands.

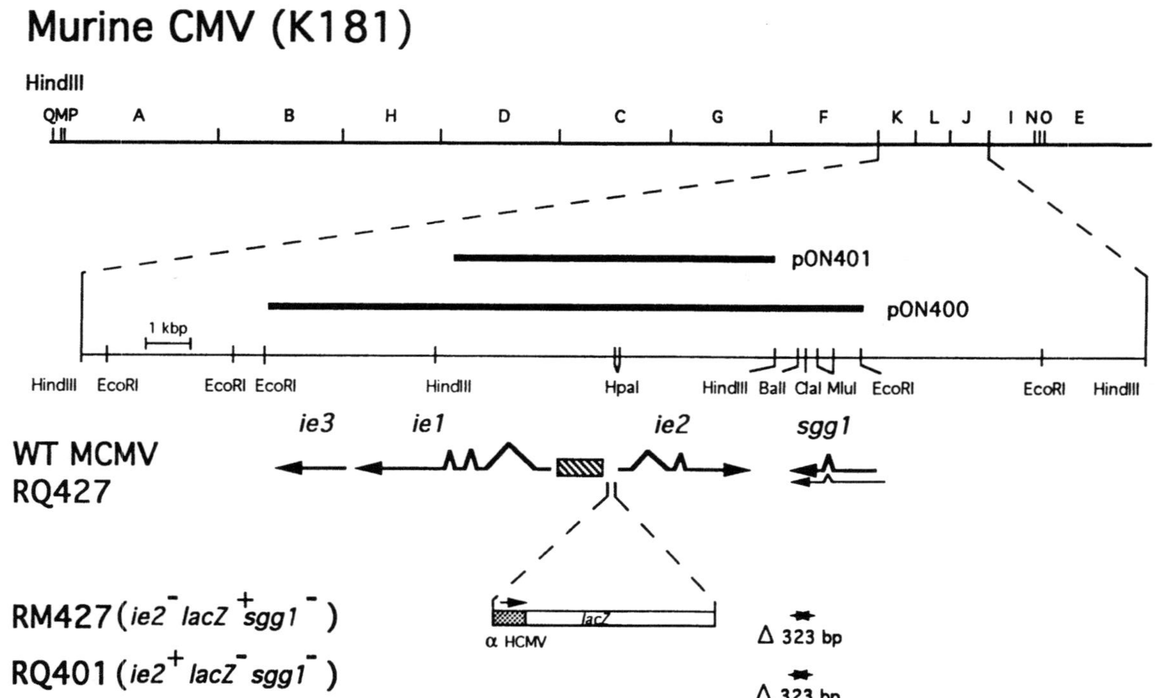

Fig. 3. *Hin*dIII map of the murine cytomegalovirus (CMV) (strain K181) genome with an expanded diagram of the *Hin*dIII K, L, and J regions to show the α gene locus (*ie*1, *ie*2, *ie*3) and the salivary gland growth gene (*sgg*1; Manning et al. 1992). The recombinant virus constructs are shown below. RM427 is an insertion of *lac*Z under control of the *ie*1/*ie*2 promoter–enhancer of human CMV (αHCMV) replacing 79 bp in the promoter of murine CMV *ie*2. Shown to the *right* is the 323 bp deletion in the adjacent early gene *sgg*1 also present in RM427. RQ401, which contains the 323 bp deletion in *sgg*1, was constructed by rescuing the *lac*Z insertion in RQ427 with pON401(*Hin*dIII L). RQ247 is a rescue to wild type using the *Eco*RI fragment pON400

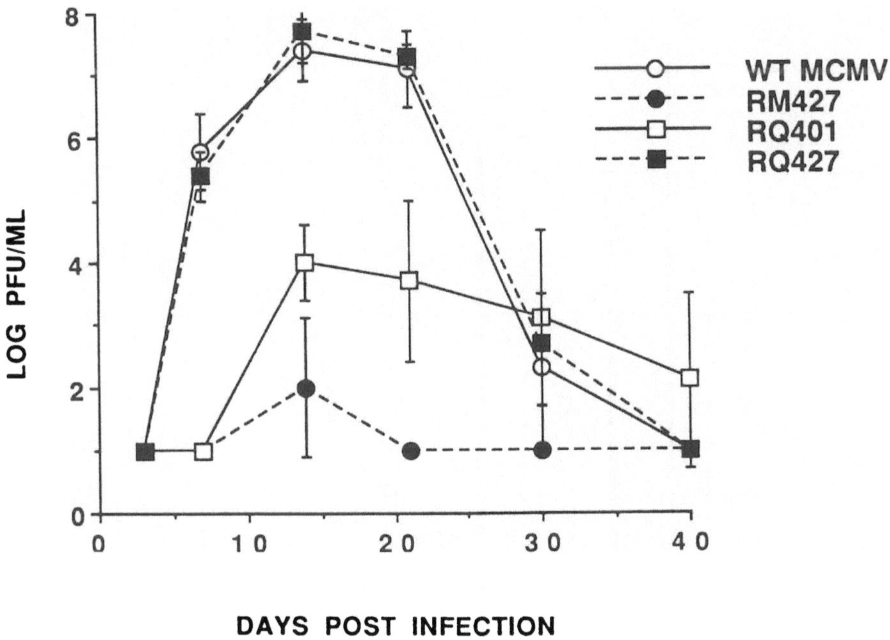

Fig. 4. Growth of wild-type (K181⁺), rescued (RQ427), and *sgg*1 mutant (RM427 and RQ401) viruses in the salivary gland following i.p. inoculation. Three-week-old BALB/c. ByJ mice were infected with 10^6 pfu of virus. Salivary gland virus was harvested at 3, 7, 14, 21, 30, and 40 days postinoculation. The virus titer was determined by plaque assay on NIH3T3 cells and results are presented as $\log_{10}$ pfu/ml of sonicated salivary gland tissue (10% suspension)

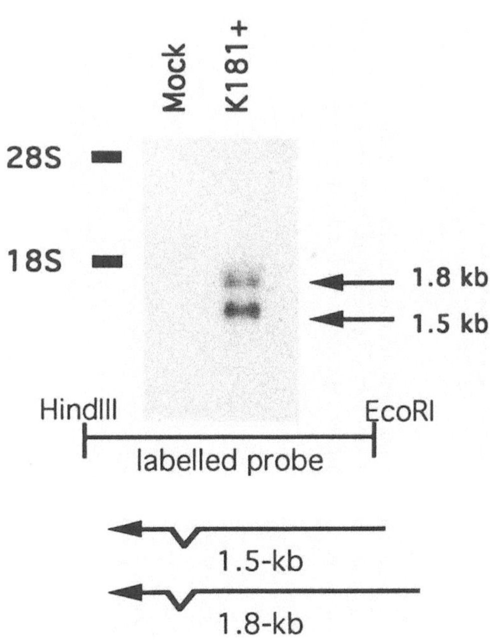

Fig. 5. Early RNA from mock-infected or wild-type (K181⁺) virus-infected NIH3T3 cells. The major 1.5 kb and minor 1.8 kb transcripts are detected with an *Eco*RI-*Hin*dIII fragment spanning the *sgg*1 region

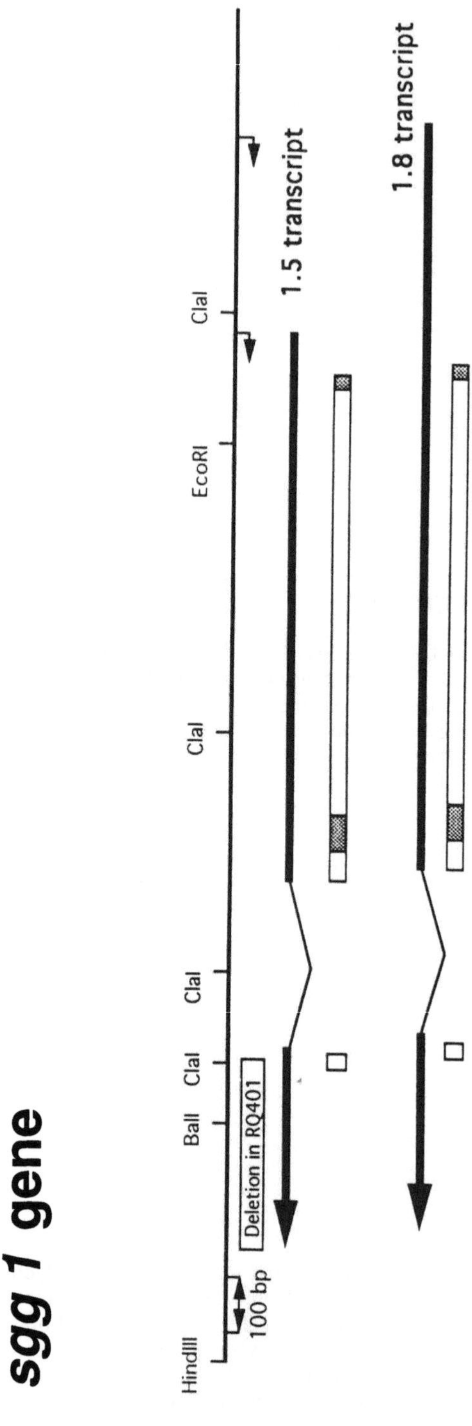

Fig. 6. The *sgg*1 gene region contains two early transcripts (a major 1.5 kb and minor 1.8 kb). The two overlapping transcripts are expressed from separate promoters. The 323 bp deletion in RQ401 (and RM427) encompasses the polyadenylation signal shared by these transcripts. The *open box* indicates an open reading frame of 312 amino acids with two hydrophobic domains (*hatched regions*)

Comparison of Different Murine Cytomegalovirus Mutants

These studies demonstrate that many murine CMV genes can dramatically influence viral replication in a variety of organs without altering growth in cultured cells. The limited results so far suggest that a significant number of the over 200 predicted murine CMV genes encode such accessory determinants. Interestingly, it appears that growth in the salivary gland is the most easily disrupted by such mutations and that many viral functions may ultimately be found to play a role in allowing efficient replication in this organ. The role that *sgg*1 plays as a determinant of growth in the salivary glands of mice is understood in greatest detail; however, the mechanism through which this putative membrane glycoprotein can influence the levels of growth in specific target cells in not known. Although an animal model has been established for human CMV (Mocarski et al. 1993), viral genes that carry out tissue tropism functions will still be more difficult to evaluate than those shown here to exist in murine CMV. Thus studies on murine CMV should provide insights into the function of genes related to those in human CMV.

Conclusions

Further studies are needed to show whether or not the mutations responsible for the phenotype in the mutants described by Sammons and Sweet (1989) and Kumura et al. (1990) map to the *Hin*dIII J or E region of the genome, or whether there are other regions of the genome involved in the interaction of murine CMV with the salivary glands. It will also be important to evaluate the sequence of the right hand end of the murine CMV genome for genes that are functional homologues of genes in the U_s region of the human CMV genome. It is possible that genes required for growth in the salivary glands encode viral attachment protein(s), regulatory functions, receptors for host growth factors, proteins that enable the survival of the host cell or functions that are required for specific viral replication functions in fully differentiated epithelial cells of the salivary glands. As both the 37 kDa Sgg1 in RQ401 and the 42 kDa protein in the Vancouver variant of the Smith strain are expressed as β proteins, it is more likely that they play roles in early replication events rather than in late maturation or virus structure.

Characterization of the genes responsible for the replication of CMV in specific organs could be useful for designing a live attenuated vaccine. The large deletion within *Hin*dIII E of the Vancouver variant resulted in a virus that not only failed to replicate in the salivary glands of infected animals, but also showed decreased replication in all other organs examined. Even though this virus was debilitated, an immune response very similar to fully virulent virus was induced following inoculation. Immunoprecipitation experiments using the anti-Vancouver serum demonstrated that a response to all major classes of viral proteins was elicited. Elucidation of the mechanisms behind the genes responsible for tissue tropism will provide insight into CMV pathogenesis and viral tropism.

References

Adler SP (1991) Molecular epidemiology of cytomegalovirus: a study of factors affecting transmission among children at three day-care centers. Pediatr infect Dis J 10: 584–590

Alford C (1991) Breast milk transmission of cytomegalovirus (CMV) infection. Adv Exp Med Biol 310: 293–299

Allan JE, Shellam GR (1984) Genetic control of murine cytomegalovirus infection: virus titres in resistant and susceptible strains of mice. Arch Virol 81: 139–150

Bale JF Jr, O'Neil ME (1989) Detection of murine cytomegalovirus DNA in circulating leukocytes harvested during acute infection of mice. J Virol 63: 2667–2673

Bancroft GJ, Shellam GR, Chalmer JE (1981) Genetic influences on the augmentation of natural killer (NK) cells during murine cytomegalovirus infection: correlation with patterns of resistance. J Immunol 126: 988–994

Boname JM, Chantler JK (1992) Characterization of a strain of murine cytomegalovirus which fails to grow in the salivary glands of mice. J Gen Virol 73: 2021–2029

Bukowski JF, Woda BA, Welsh RM (1984) Pathogenesis of murine cytomegalovirus infection in natural killer cell-depleted mice. J Virol 52: 119–128

Cowdry EV, Scott GH (1935) Nuclear inclusions suggestive of virus action in salivary glands of the monkey Cebus fatuellus. Proc Soc Exp Biol Med 32: 709–711

Deacon EM, Matthews JB, Potts AJ, Hamburger J, Bevan IS, Young LS (1991) Detection of Epstein-Barr virus antigens and DNA in major and minor salivary glands using immunocytochemistry and polymerase chain reaction: possible relationship with Sjogren's syndrome. J Pathol 163: 351–360

Ebeling A, Keil GM, Knust E, Koszinowski UH (1983) Molecular cloning and physical mapping of murine cytomegalovirus DNA. J Virol 47: 421–433

Efstathiou S, Gompels UA, Craxton MA, Honess RW, Ward K (1988) DNA homology between a novel human herpesvirus (HHV-6) and human cytomegalovirus. Lancet 1: 63–64

Frenkel N, Schirmer EC, Wyatt LS, Katsafanas G, Roffman E, Danovich RM, June CH (1990) Isolation of a new herpesvirus from human CD4$^+$ T cells. Proc Natl Acad Sci USA 87: 748–752

Griffiths PD, Grundy JE (1987) Molecular biology and immunology of cytomegalovirus. Biochem J 241: 313–324

Henson D, Strano AJ (1972) Mouse cytomegalovirus. Clin J Pathol 68: 183–195

Hirozane A, Fujikura Y, Sawada T, Shinozaki F, Fukumoto T (1992) Analysis of major histocompatibility complex (MHC) class I antigen in the rat salivary gland. Arch Oral Biol 37: 147–151

Ho M (1990) Epidemiology of cytomegalovirus infections. Rev Infect Dis 12 [Suppl 7]: 701–710

Ho M (1991) Cytomegalovirus: Biology and infection, 2nd edn. Plenum, New York

Hudson JB (1979) The murine cytomegalovirus as a model for the study of viral pathogenesis and persistent infections. Arch Virol 62: 1–29

Hudson JB, Walker DG, Altamirano M (1988) Analysis in vitro of two biologically distinct strains of murine cytomegalovirus. Arch Virol 102: 289–295

Inada T, Chong KT, Mims CA (1985) Enhancing antibodies, macrophages and virulence in mouse cytomegalovirus infection. J Gen Virol 66: 871–878

Jonjic S, Mutter W, Weiland F, Reddehase MJ, Koszinowski UH (1989) Site-restricted persistent cytomegalovirus infection after selective long-term depletion of CD4$^+$ T lymphocytes. J Exp Med 169: 1199–1212

Jonjic S, Pavic I, Lucin P, Rukavina D, Koszinowski UH (1990) Efficacious control of cytomegalovirus infection after long-term depletion of CD8$^+$ T lymphocytes. J Virol 64: 5457–5464

Jordan MC, Takagi JL (1983) Virulence characteristics of murine cytomegalovirus in cell and organ cultures. Infect Immun 41: 965–970

Jordan MC, Shanley JD, Stevens JG (1977) Immunosuppression reactivities and disseminates latent murine cytomegalovirus. J Gen Virol 37: 419–423

Klotman ME, Henry SC, Greene RC, Brazy PC, Klotman PE, Hamilton JD (1990) Detection of mouse cytomegalovirus nucleic acid in latently infected mice by in vitro enzymatic amplification. J Infect Dis 161: 220–225

Koszinowski UH, Del Val M, Reddehase MJ (1990) Cellular and molecular basis of the protective immune response to cytomegalovirus infection. In: McDougall JK (ed) Cytomegaloviruses. Springer, Berlin Heidelberg New York, pp 189–220 (Current topics in microbiology and immunology, vol 154)

Kumura K, Ibusuki K, Minamishima Y (1990) Independent existence of mutations responsible for temperature sensitivity and attenuation in a mutant of murine cytomegalovirus. Virology 175: 572–574

Lawrence GL, Chee M, Craxton MA, Gompels UA, Honess RW, Barrell BG (1990) Human herpesvirus 6 is closely related to human cytomegalovirus. J Virol 64: 287–299

Littler E, Lawrence G, Liu MY, Barrell BG, Arrand JR (1990) Identification, cloning, and expression of the major capsid protein gene of human herpesvirus 6. J Virol 64: 714–722

Lucin P, Pavic I, Polic B, Jonjic S, Koszinowski UH (1992) Gamma interferon-dependent clearance of cytomegalovirus infection in salivary glands. J Virol 66: 1977–1984

Manning WC, Stoddart CA, Lagenaur LA, Abenes GB, Mocarski ES (1992) Cytomegalovirus determinant of replication in salivary glands. J Virol 66: 3794–3802

Martin ME. Thomson BJ, Honess RW, Craxton MA, Gompels UA, Liu MY, Littler E, Arrand JR, Teo I, Jones MD (1991) The genome of human herpesvirus 6: maps of unit-length and concatemeric genomes for nine restriction endonucleases. J Gen Virol 72: 157–168

Mayo D, Armstrong JA, Ho M (1978) Activation of latent murine cytomegalovirus infection: cocultivation, cell transfer, and the effect of immunosuppression. J Infect Dis 138: 890–896

Mayo DR, Armstrong JA, Ho M (1977) Reactivation of murine cytomegalovirus by cyclophosphamide. Nature 267: 721–723

Mercer JA, Spector DH (1986) Pathogenesis of acute murine cytomegalovirus infection in resistant and susceptible strains of mice. J Virol 57: 497–504

Mercer JA, Marks JR, Spector DH (1983) Molecular cloning and restriction endonuclease mapping of the murine cytomegalovirus genome (Smith Strain). Virology 129: 94–106

Misra V, Hudson JB (1980) Minor base sequence differences between the genomes of two strains of murine cytomegalovirus differing in virulence. Arch Virol 64: 1–8

Mocarski ES (1993) Cytomegalovirus Biology and Replication. In: Roizman B, Whitley RJ, Lopez, C (eds) The human herpesviruses. Raven, New York, 173–226

Mocarski ES, Bonyhadi M, Salimi S, McCune JM, Kaneshima H (1993) Human cytomegalovirus replication in the SCID-hu mouse: thymic epithelial cells are prominent targets of viral replication. Proc Natl Acad Sci USA 90: 104–108

Osborn JE (1982) Cytomegaloviruses and other herpesviruses. In: Foster HL, Small JD, Fox JG (eds) The mouse in biomedical research. Academic, New York, 267–292

Osborn JE, Walker DL (1971) Virulence and attenuation of murine cytomegalovirus. Infect Immun 3: 228–236

Pass RF, Little EA, Stagno S, Britt WJ, Alford CA (1987) Young children as a probable source of maternal and congenital cytomegalovirus infection. N Engl J Med 316: 1366–1370

Plummer G (1967) Comparative virology of the herpes group. Prog Med Virol 9: 302–340

Plummer G (1973) Cytomegaloviruses of man and animals. Prog Med Virol 15: 92–125

Quinnan GV, Manischewitz JE (1979) The role of natural killer cells and antibody-dependent cell-mediated cytotoxicity during murine cytomegalovirus infection. J Exp Med 150: 1549–1554

Ravindranath RM, Graves MC (1990) Attenuated murine cytomegalovirus binds to N-acetylglucosamine, and shift to virulence may involve recognition of sialic acids. J Virol 64: 5430–5440

Roizman B, Carmichael LE, Deinhardt F, de The G, Nahmias W, Plowright F, Rapp F, Sheldrick P, Takahashi M, Wolf K (1981) Herpesviridae. Definition, provisional nomenclature, and taxonomy. The Herpesvirus Study Group, the International Committee on Taxonomy of Viruses. Intervirology 16: 201–217

Sammons CC, Sweet C (1989) Isolation and preliminary characterization of temperature-sentive mutants of mouse cytomegalovirus of differing virulence for 1-week-old mice. J Gen Virol 70: 2373–2381
Sandford GR, Burns WH (1988) Use of temperature-sensitive mutants of mouse cytomegalovirus as vaccines. J Infect Dis 158: 596–601
Savage NW, Walsh LJ, Seymour GJ (1987) Expression of class I and class II major histocompatibility complex antigens on oral mucosal epithelium. J Oral Pathol 16: 153–157
Scalzo AA, Fitzgerald NA, Wallace CR, Gibbons AE, Smart YC, Burton RC, Shellam GR (1992) The effect of the Cmv-1 resistance gene, which is linked to the natural killer cell gene complex, is mediated by natural killer cells. J Immunol 149: 581–589
Schmader KE, Rahija RJ, Porter KR, Hamilton JD (1991) Murine cytomegalovirus gene amplification and culture after submaxillary salivary gland biopsy. Lab Anim Sci 41: 396–400
Selgrade MK, Nedrud JG, Collier AM, Gardner DE (1981) Effects of cell source, mouse strain, and immunosuppressive treatment on production of virulent and attenuated murine cytomegalovirus. Infect immun 33: 743–749
Shanley JD (1990) In vivo administration of monoclonal antibody to the NK 1.1 antigen of natural killer cells: effect on acute murine cytomegalovirus infection. J Med Virol 30: 58–60
Shanley JD, Jordan MC, Stevens JG (1981) Modification by adoptive humoral immunity of murine cytomegalovirus infection. J Infect Dis 143: 231–237
Shellam GR, Allan JE, Papadimitriou JM, Bancroft GJ (1981) Increased susceptibility to cytomegalovirus infection in beige mutant mice. Proc Natl Acad Sci USA 78: 5104–5108
Smith MG (1954) Propagation of salivary gland virus of the mouse in tissue cultures. Proc Soc Exp Biol Med 86: 435–440
Smith MG (1956) Propagation in tissue cultures of a cytopathogenic virus from human salivary gland (SGV) disease. Proc Soc Exp Biol Med 92: 424–430
Spector SA (1983) Transmission of cytomegalovirus among infants in hospital documented by restriction-endonuclease-digestion analyses. Lancet 1: 378–381
Tonari Y, Minamishima Y (1983) Pathogenicity and immunogenicity of temperature-sensitive mutants of murine cytomegalovirus. J Gen Virol 64: 1983–1990
Weller TH (1970) Cytomegalovirus: the difficult years. J Infect Dis 122: 532–539
Welsh RM, Brubaker JO, Vargas-Cortes M, O'Donnell CL (1991) Natural killer (NK) cell response to virus infections in mice with severe combined immunodeficiency. The stimulation of NK cells and the NK cell-dependent control of virus infections occur independently of T and B cell function. J. Exp Med 173: 1053–1063
Wildy P (1973) Herpes: History and classification. In: Kaplan AS (ed) The herpesviruses. Academic, New York, pp 1–22
Wolf H, Haus M, Wilmes E (1984) Persistence of Epstein-Barr virus in the parotid gland. J Virol 51: 795–798

D. Human Herpesvirus-6

Chapter 20 Pathogenicity of Human Herpesvirus-6

Abdur Razzaque[1], Koichi Yamanishi[2], and Donald R. Carrigan[3]

Summary

Human herpesvirus-6 (HHV-6) is a recently discovered lymphotropic virus. Serologic evidence indicates a high incidence of HHV-6 antibody in almost all areas of the world, and most children are infected by 2 years of age. HHV-6 can productively infect CD4$^+$ lymphocytes, but CD4 does not appear to be used by the virus as its primary receptor. The virus can also infect monocytes, macro-phages, epithelial cells, and cells of other lineages. Virus transmission probably occurs horizontally, as the virus is secreted in the saliva of healthy individuals, and is unlikely to be transmitted through the placenta. HHV-6 has been identified as the etiologic agent of exanthem subitum. It can cause interstitial pneumonitis in bone marrow transplant patients, in whom it is associated with suppression of bone marrow function. HHV-6 can also cause infectious mononucleosis, lymphadenitis, liver dysfunction, and is associated with various lymphoprolifera-tive disorders and the chronic fatigue syndrome. It has been proposed that the virus plays a role in the pathogenesis of AIDS. HHV-6 is suggested to be an oncogenic virus because of its lymphoproliferative linkage and because HHV-6 DNA can transform murine fibroblast and human epidermal keratinocyte cell lines. These transformed cells are tumorigenic in nude mice. Primary HHV-6 infection may establish latent infection in monocytes or macrophages, and the virus can be reactivated from a latent state in immunocompromised patients and in individuals with various malignant and nonmalignant diseases.

Introduction

Human herpesvirus-6 (HHV-6), previously termed human B-lymphotropic virus, is a recently discovered member of the human herpesvirus family. This chapter will first provide brief general background information on HHV-6 and then will focus in more detail on its role in some human diseases including experimental data.

[1] Division of Viral Products, CBER, FDA, Bethesda, MD 20892, USA
[2] Department of Virology, Research Institute for Microbial Diseases, Osaka University, Osaka, Japan
[3] Department of Pathology and the Bone Marrow Transplant Program, Medical College of Wisconsin, Milwaukee, WI 53226, USA

HHV-6 was first isolated in 1986 from the peripheral blood lymphocytes of six patients with lymphoproliferative disorders, two of whom had AIDS-related lymphoma (Salahuddin et al. 1986). Since then, other isolates have been recovered from patients with AIDS (Downing et al. 1987; Lopez et al. 1988; Tedder et al. 1987), lymphoma (Becker et al. 1989), leukemia (Becker et al. 1988), exanthem subitum (ES) (Yamanishi et al. 1988) fatal hepatitis (Asano et al. 1990), bone marrow transplants (Carrigan et al. 1990), and renal transplants (Okuno et al. 1990). The virus has also been isolated from the saliva of healthy adults (Levy et al. 1990a; Pietroboni et al. 1988a). By DNA sequence analysis, HHV-6 has homology with human cytomegalovirus (Lawrence et al. 1990; Josephs et al. 1991; Neipel et al. 1991), but the virus is antigenically and genetically distinct from the other human herpesviruses. HHV-6 isolates form two closely related but distinct groups, U1102- and GS- like viruses, typified as "HHV-6 variant A" and Z29-like viruses, as "HHV-6 variant B." These viral types can be differentiated from one another with respect to both molecular and biologic properties (Schirmer et al. 1991; Ablashi et al. 1991).

Seroepidemiology of HHV-6

More than 80% of the normal human population have antibodies to the virus, and primary infection generally occurs by 2 years of age (Okuno et al. 1989; Saxinger et al. 1988). The prevalence of HHV-6 IgG antibody is generally high in most parts of the world, but a lower seroprevalence is found in some far eastern countries. Several studies show a decrease in HHV-6 seropositivity with age; however, other studies have shown increased HHV-6 specific antibody titers in elderly individuals (Briggs et al. 1988; Brown et al. 1988; Yanagi et al. 1990).

Biology and Cellular Tropism of HHV-6

HHV-6 is a lymphotropic virus that can productively infect $CD4^+$ T lymphocytes both in vitro (Lusso et al. 1988) and in vivo (Takahashi et al. 1989). Mitogen stimulation is necessary for HHV-6 replication in cord blood mononuclear cells (Ablashi et al. 1991). HHV-6 infected cells from human peripheral blood, cord blood, and thymus have been identified as being $CD2^+$, $CD5^+$, $CD7^+$, and $CD4^+$ and leukocyte antigen positive (Lusso et al. 1988). Cells with $CD8^+$ antigen can occasionally be infected with type A strains of HHV-6. Infection of $CD4^+$ cells cannot be blocked by CD4 specific monoclonal antibodies, indicating that CD4 is not the primary receptor for HHV-6. Studies with human tissues suggest that HHV-6 can infect salivary gland epithelial cells (Fox et al. 1990; Krueger et al. 1990), renal tubular epithelial cells (Okuno et al. 1990), and macrophages (Carrigan et al. 1991). Type A variants of HHV-6 can also infect B lymphocytes and cells of other lineages, such as megakaryocytes, glioblastomas (Ablashi et al.

1988), and fibroblasts (Luka et al. 1990). In studies of in vivo cellular tropism of HHV-6 in patients with ES, virus was consistently isolated from $CD4^+CD8^-$ and $CD3^+CD4^+$ cells, but never from $CD4^-CD8^+$,$CD4^-CD8^-$, or $CD3^-$ or B cells (Takahashi et al. 1989). This CD4 tropism of HHV-6 suggests that the virus may be involved in the pathogenesis of T cell related disorders and possibly in acquired immunodeficiencies. HHV-6 has been suggested as a cofactor in AIDS since both HIV and HHV-6 can infect $CD4^+$ lymphocytes and HHV-6 has been frequently isolated from AIDS patients. However, it remains unclear whether HHV-6 enhances (Lusso et al. 1989) or suppresses the replication of HIV (Pietroboni et al. 1988b; Carrigan et al. 1990; Levy et al. 1990b). Recent evidence suggests that HHV-6 may establish latent infection in monocytes or macrophages (Kondo et al. 1991).

HHV-6 in Human Diseases

Primary HHV-6 infection in infants causes ES (Yamanishi et al. 1988). In adults, HHV-6 has not yet been clearly associated with any specific disease. However, the virus has been implicated as the causative agent of 12% of heterophile antibody negative infectious mononucleosis cases (Steeper et al. 1990). HHV-6 is thought to be involved in lymphoproliferative disorders (Salahuddin et al. 1986; Torelli et al. 1991; Krueger et al. 1989), lymphoma and sarcoidosis (Biberfeld et al. 1988), lymphadenopathy (Eizuru et al. 1989), pneumonitis (Carrigan et al. 1991), chronic infections (Krueger et al. 1987) and hepatitis (Dubedat and Kappaguta 1989). In spite of these studies, pathogenic potential of HHV-6 has not yet been clearly defined. The remainder of this chapter will focus upon studies of HHV-6 as a pathogen in ES, in bone marrow transplant patients, and on its oncogenic potential using cell culture and animal models.

HHV-6 Causes Exanthem Subitum

ES is a common disease of childhood characterized by high fever and skin rash. Yamanishi et al. (1988) first reported HHV-6 as the causative agent of ES. The virus was isolated from peripheral blood mononuclear cells (PBMC) of ES patients by cultivating the cells with interleukin-2 and phytohemaglutinin (PHA). After about 10 days of culture, cytopathic effects (CPE) were observed consisting of balloon-like cells which gave positive immunofluorescent staining with sera from convalescent phase and not from acute phase ES patients. By electron microscopy, typical herpesvirus particles were seen in the nucleus (diameter 90–110 nm) and cytoplasm (diameter 150–170 nm) of infected lymphocytes (Fig. 1). Sera specific for the five other human herpesviruses did not stain these cells. Cells infected with the Z29 strain of HHV-6 stained with sera from convalescent phase ES patients but not with sera from individuals in the acute phase. These data indicated that the virus isolated from ES patients was HHV-6.

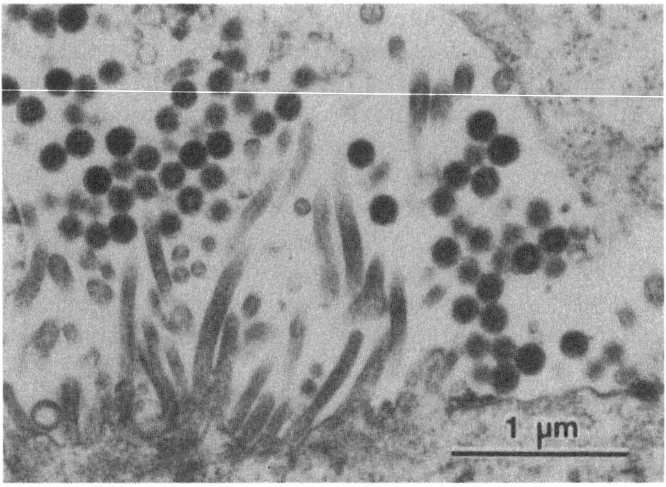

Fig. 1. Herpesvirus particles in the infected lymphocytes (kindly provided by Dr. S. Nii, Okayama University)

ES is usually a mild and self-limited illness, and often primary HHV-6 infections occur without rash, fever, or any overt clinical symptoms (Asano et al. 1989; Yoshikawa et al. 1991; Pruksananonda et al. 1992). However, serious complications such as convulsions (Segondy et al. 1992), severe hepatitis (Asano et al. 1990), and hemophagocytic syndrome (Huang et al. 1990) have been reported. The frequencies of such complications are unknown due to the relatively low incidence of the pathognomonic rash in primary HHV-6 infections (Pruksananonda et al. 1992).

Regarding the mode of transmission of HHV-6, there are two possibilities: vertical (transplacental) and horizontal transmission. Since IgM antibody was not found in peripheral blood of newborns (Farr et al. 1990), and HHV-6 DNA was not detected by the polymerase chain reaction (PCR) in umbilical cord blood cells (Yamanishi et al., unpublished), it is unlikely that HHV-6 infection is routinely transmitted through the placenta. The data of outbreak of ES in a nursery in Japan (Okuno et al. 1991), the detection of HHV-6 DNA in saliva and throat swab by PCR (Kido et al. 1990), and the isolation of virus from saliva suggest that virus transmits horizontally, probably from mother to infant.

HHV-6 and Bone Marrow Transplantation

HHV-6 can cause severe interstitial pneumonitis in bone marrow transplant (BMT) patients (Carrigan et al. 1991), but it is unknown what portion of idiopathic interstitial pneumonitis in those patients can be accounted for by HHV-6 infections. A retrospective study of tissues from BMT patients dying with pneumonitis demonstrated that lung infections with HHV-6 are common (Carrigan

1991). HHV-6 infected cells were identified in the lungs of three of the nine patients examined. In all three the infected cells were lymphocytes and intra-alveolar macrophages scattered throughout the tissue. Two of the three HHV-6 positive cases had coexistent cytomegalovirus (CMV) infections. In both of these at least as many HHV-6 infected cells were present as cells infected with CMV, raising the question of which virus was actually the cause of the pneumonitis. These coinfections suggest that HHV-6 may be able to synergize with other pathogens in causing lung disease. One possible mechanism for such a synergism in HHV-6 mediated destruction or disabling of monocytes and macrophages in the infected tissues, thus inducing a state of local immunodeficiency that may allow other pathogens to replicate unchecked.

Another important clinical manifestation of HHV-6 infections in BMT patients is suppression of grafted marrow (Carrigan et al. 1991; Carrigan 1991). In a prospective investigation (Drobyski et al. 1993), 29 adult patients who underwent allogeneic bone marrow transplantation were studied for the first 100 days posttransplant to define the incidence and severity of infection due to HHV-6. Of the 29 patients who developed unexplained febrile illness, 16 were assessed for the presence of HHV-6 in their peripheral blood and bone marrows. HHV-6 was present in the blood of six of these patients. The clinical courses of these six patients were reviewed to determine whether isolation of the virus was associated with clinical features other than fever. Otherwise unexplained marrow graft suppression contemporaneous with isolation of HHV-6 from the peripheral blood was observed in four of the six patients. This was clinically manifested as either a delay in engraftment or the development of postengraftment myelosuppression as defined by a decline in the absolute neutrophil count to $\leq 500/mm^3$ for at least 2 consecutive days. Marrow suppression in each of these four patients occurred in the absence of other known bacterial, fungal, or viral pathogens. All had persistent donor chimerism as determined by restriction fragment length polymorphism (RFLP) analysis without evidence of graft rejection. None of these patients were receiving myelosuppressive drug therapy. HHV-6 was also isolated from the bone marrow of all four patients at the time of marrow graft suppression.

In analyzing the clinical courses of the remaining ten patients in whom HHV-6 infection was not documented, two of them had marrow graft suppression. One had graft rejection with evidence of recipient cells by RFLP analysis, while the second had idiopathic marrow graft failure. In this patient, HHV-6 could not be cultured from the peripheral blood at the time of impaired marrow function despite multiple isolation attempts. The incidence of idiopathic marrow graft impairment was significantly ($p<0.05$ by Fisher's exact test) greater in the HHV-6 viremia positive (4/6) than in negative (1/10) patient groups. Thus, it appears that HHV-6 may be responsible for a large portion of idiopathic marrow suppression.

The ability of HHV-6 infection to suppress bone marrow function has been confirmed in two separate in vitro experimental systems. First, a methylcellulose based colony formation assay was used to evaluate the impact of HHV-6_{KF} infection on marrow cell differentiation and proliferation (Knox and Carrigan

1992). In four separate experiments utilizing marrows from different normal individuals, the mean number of colonies seen in the infected vs mock infected cultures were 30 vs 120 (75% suppression) for the erythroid burst-forming units (BFU-E), 13.5 vs 24 (44% suppression) for the colony-forming units of granulocyte and macrophage lineages (CFU-GM), and 1.75 vs 6.5 (73% suppression) for the CFU of granulocyte, erythrocyte, macrophage, and megakaryocyte lineages (CFU-GEMM). The predominant mechanism of this HHV-6_{KF} associated suppression appeared to be the induction of interferon-α in the infected marrow cells. However, other data indicate that HHV-6_{GS} exerts an even stronger suppressive effect on marrow stem cell proliferation and differentiation by a mechanism that is unrelated to interferon production (K. K. Knox and D. R. Carrigan, unpublished).

The second in vitro system that has been exploited to confirm the ability of HHV-6 to suppress bone marrow function involved measurement of the ability of marrow monocytic precursors to respond to two growth factors, interleukin-3 (IL-3) and granulocyte/macrophage-colony stimulating factor (GM-CSF) (Burd et al. 1993). In a quantitative assay for the effects of these growth factors on normal human marrow, samples of purified marrow mononuclear cells were exposed to the factors in plastic culture vessels. After incubation for 7 days, nonadherent cells were removed and the remaining adherent cells were stained for nonspecific esterase. Dramatic outgrowth of plastic adherent, nonspecific esterase positive macrophages occurred in response to both IL-3 and GM-CSF with the effect of GM-CSF being noticeably stronger in most experiments. When marrow cells were treated with culture medium from HHV-6 infected cultures immediately prior to their exposure to the growth factors, a dramatic suppression of the macrophage outgrowth was observed. This suppression was statistically significant for both the GM-CSF ($p = 0.0001$) and IL-3 ($p = 0.0001$) treated cultures and occurred with all seven of the HHV-6 strains tested, including HHV-6_{GS}, HHV-6_{Z29}, and HHV-6_{KF}. The observed effect was not due to a general cytopathic or toxic effect of the viral material since similar numbers of viable nonadherent cells were present in HHV-6 infected and uninfected cultures, as determined by trypan blue exclusion.

As with the marrow CFU suppression by HHV-6_{GS} exposure, inhibition of the marrow cells' responses to growth factors by the virus was largely independent of the presence of infectious virus in the viral preparation. This conclusion was supported by four lines of evidence. First, only low levels of infected cells (1/10 000–1/100 000) were observed by immunofluorescent staining for a viral early antigen. Second, incubation of HHV-6_{Z29} exposed marrow in viral inhibitory concentrations of foscarnet did not abrogate the virus associated suppression of macrophage outgrowth in response to GM-CSF. Third, a heat inactivated preparation of HHV-6_{Z29} suppressed GM-CSF induced macrophage outgrowth by 60% compared to 90% inhibition by the same viral preparation containing infectious virus. Fourth, macrophage outgrowth in response to GM-CSF treatment was suppressed by 64% and 95% for HHV-6_{Z29} and HHV-6_{GS}, respectively, using infected culture materials from which infectious virus had been removed by ultracentrifugation.

Oncogenicity of HHV-6

Since HHV-6 DNA sequences have been detected in some lymphoma cases (Torelli et al. 1991; Josephs et al. 1988; Jarrett et al. 1988) and high antibody titers have been detected in these patients, the question of HHV-6 as an oncogenic virus was of interest. Active HHV-6 infection has been suggested to cause atypical polyclonal lymphoproliferation and thus may increase the risk of malignancy. To address this question, several cloned portions of the HHV-6 genome (GS) were first tested for transforming activity in mouse 3T3 fibroblasts and then in human cells to establish their relationship to human malignancies. The assay system consisted of : (1) transfection of cells by calcium phosphate precipitation, (2) selection of cells by focus formation or selection on the neomycin analogue G418, (3) anchorage independent growth of selected cells in agarose, and (4) tumorigenicity in nude mice.

Using the above assay system, it was reported that HHV-6 genome (170 kb, GS isolate from AIDS patient), and one subgenomic clone pZVH14 (8.7 kb *Hin*dIII insert in pBluescript) exhibited similar transforming (focus forming) activity in NIH 3T3 cells and this activity was more than 100-fold higher than the background level (Razzaque 1990). Another clone, pZVB70 (21 kb *Bam*HI insert in pBluescript), showed four fold lower transforming activity than pZVH14. Focal cell lines established from these DNA-induced individual foci showed morphological alteration, grew in agarose as clumps, and produced tumors in nude mice (Fig. 2). Similar results were also obtained with cells selected on G418 using pSV2neo DNA cotransfection. By Southern blot analysis, more than single copy HHV-6 DNA segment could be detected in G418 selected transformants but not in focus derived transformants. The tumors were identified to be fibrosarcomas. The tumor cell lines derived from these tumors could metastasize to the lungs of immunocompetent NIH Swiss mice when injected through the tail vein (Razzaque and Puri 1992). Two other strains of HHV-6, Z29 and GD (from chronic fatigue syndrome), also showed transforming activity (A. Razzaque, unpublished).

In experiments with human cell lines, the transforming activity of HHV-6 DNA was tested in a human epidermal keratinocyte cell line, RHEK-1. This cell line is established by Ad12-SV40 hybrid virus infection of primary foreskin epidermal keratinocytes and is flat in morphology, nontumorigenic, contains SV40 tumor antigens, but no detectable transcripts from the early region 1 of Ad12 (Rhim et al. 1985). The two clones pZVH14 and pZVB70 that could transform 3T3 cells, could also cause tumorigenic transformation of RHEK-1 cells (Fig. 3). In this study, the DNAs were introduced into the cells by the polybrene method of transfection (Rhim et al. 1989) and the G418 selected cells could produce tumors in nude mice (Razzaque et al. 1993). Vector DNA or other HHV-6 plasmids were used as controls and were found to be inactive in transformation. By Southern blot analysis, the pZVH14 and pZVB70 insert DNAs could be detected in the respective transformants. Chromosome analysis of pZVH14 DNA induced tumors indicated the tumors to be of human origin, but contained some chromosomal aberrations with respect to the control untransformed cells. The tumors were diagnosed as poorly differentiated carcinomas by histological analysis.

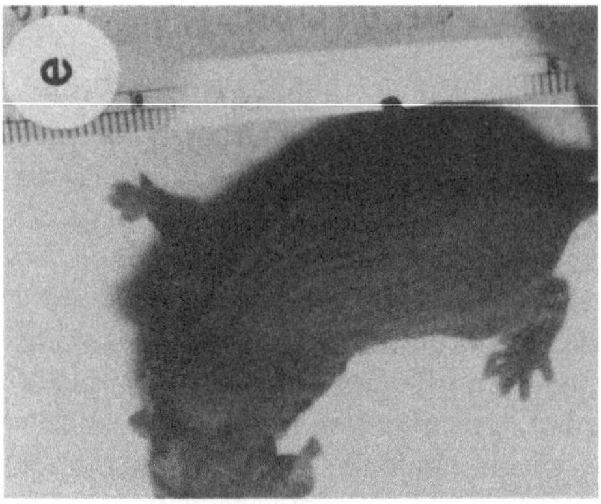

Fig. 2a–e. Morphology, growth in agarose, and tumorigenicity of HHV-6 pZVH14 DNA transfected NIH3T3 cells. Untransfected cells (**a**, **b**); pZVH14 DNA transfected cells (**c**, **d**); foci appeared 21 days after transfection (**c**); tumor produced by pZVH14 DNA induced foci at 20 days after subcutaneous inoculation of 5×10^6 cells in nude mice (**e**). Growth in agarose (**b**, **d**): 5×10^4 cells were plated per 60 mm dish in growth medium with 0.3% agarose and 10% fetal calf serum; colonies photographed at 18 days

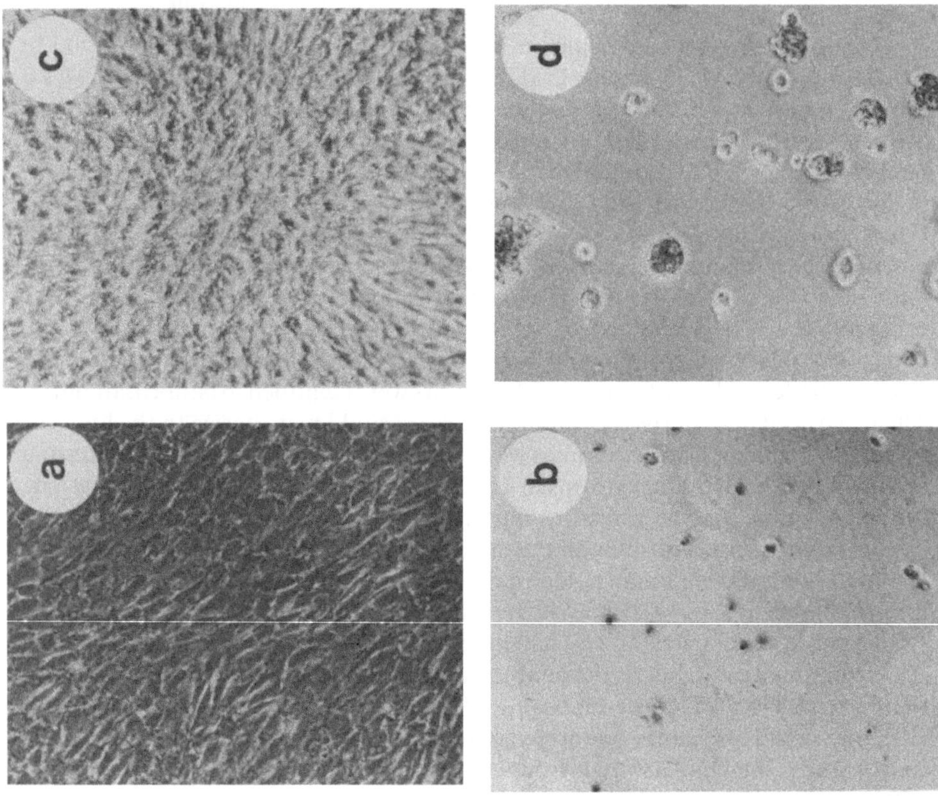

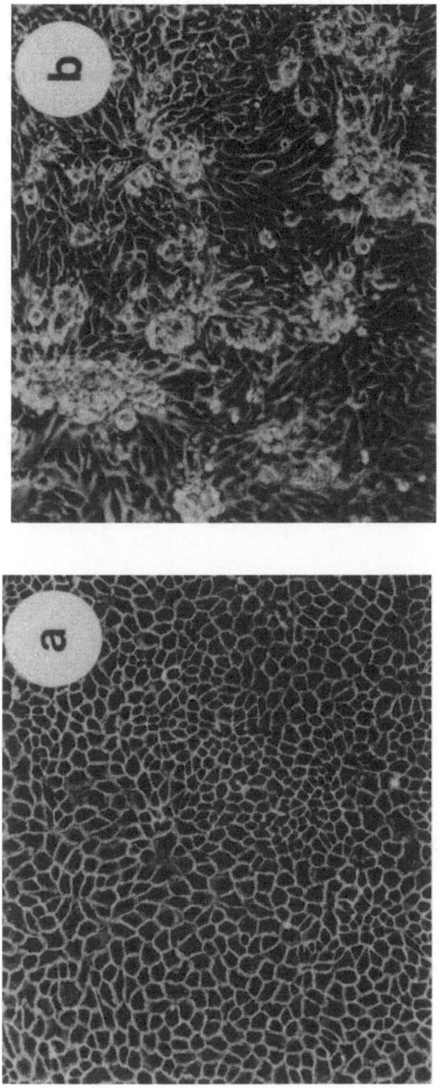

Fig. 3. a, b. Comparison of the morphology of control vector DNA transfected RHEK-1 cells (**a**) with pZVH14 DNA-induced tumor cells (**b**) in nude mice

The data on the transforming functions of HHV-6 suggest that HHV-6 is an oncogenic virus. The fact that pZVH14 DNA sequences have been detected in several human lymphoma and hepatitis cases and that this sequence can transform cells in culture suggest that HHV-6 may be involved in human malignancies. In addition, its association with human tumors and lymphoproliferative disorders may result from its tropism for lymphoid cells.

The results obtained with HHV-6 DNA resemble that found with other human herpesviruses which have been implicated in certain human neoplasia and are capable of oncogenic conversion of cells in vitro (Review, Macnab 1987). The mechanism of transformation by human herpesviruses is still unclear. No viral oncogene or protein product has yet been identified in HSV and HCMV transformed cells although some DNA sequences are shown to be retained. Similarly, we detected HHV-6 DNA sequences in the G418 selected cells but not in the focus derived transformants by Southern blot analysis. This suggests that HHV-6 DNA sequence may not be required for the maintenance of transformed phenotype but may be required for the initiation of transformation. The specific genes or sequences involved in transformation by pZVH14 or pZVB70 remain to be determined.

References

Ablashi DV, Lusso P, Huang CL, Salahuddin SZ, Josephs SF, Llana T, Kramarsky B, Biberfeld P, Markham PD, Gallo RC (1988) Utilization of human hematopoetic cell lines for the propagation and characterization of HBLV (herpesvirus-6). Int J Cancer 42: 787–791

Ablashi DV, Balachandran N, Josephs SF, Huang CL, Krueger GRF, Kramarsky B, Salahuddin SZ, Gallo RC (1991) Genomic polymophism, growth properties, and immunologic variations in human herpesvirus-6 isolates. Virology 184: 545–552

Asano Y, Suga S, Yoshikawa T, Urisu A, Yazaki T (1989) Human herpesvirus type 6 infection (exanthem subitum) without fever. J Pediatr 115: 264–265

Asano Y, Yoshikawa T, Suga S, Yazaki T, Kondo K, Yamanishi K (1990) Fatal fulminant hepatitis in an infant with human herpesvirus-6 infection. Lancet ii: 862–863

Becker WB, Engelbrecht S, Becker MLB, Piek C, Robson BA, Wood L, Jacobs P (1988) Isolation of a new human herpesvirus producing a lytic infection of helper (CD4) T lymphocytes in peripheral blood lymphocyte cultures-another cause of acquired immunodeficiency? S Afr Med J 74: 610–614

Becker WB, Engelbrecht S, Becker MLB, Piek C, Robson BA, Wood L, Jacobs P (1989) New T lymphotropic human herpesviruses. Lancet i: 41

Biberfeld P, Petrea AL, Eklund A, Lindemalm C, Barkhem T, Ekman M, Ablashi DV, Salahuddin SZ (1988) Human herpesvirus-6 (HHV-6,HBLV) in sarcoidosis and lymphoproliferative disorders. J Virol Methods 21: 49–50

Briggs M, Fox J, Tedder RS (1988) Age prevalence of antibody to human herpesvirus 6. Lancet i: 1058–1059

Brown NA, Sumaya CV, Liu CR, Ench Y, Kovacs A, Coronesi M, Kaplan MH (1988) Fall in human herpesvirus-6 seropositivity with age. Lancet ii: 396

Burd EM, Knox KK, Carrigan DR (1993) Human herpesvirus-6-associated suppression of growth factor-induced macrophage maturation in human bone marrow cultures. Blood 81: 1645–1650

Carrigan DR (1991) Human herpesvirus six and bone marrow transplantation. In: Ablashi DV, Krueger GRF (eds) Biology, epidemiology, molecular biology and clinical pathology of HHV-6. Elsevier Science, Amsterdam, pp 281–301

Carrigan DR, Knox KK, Tapper MA (1990) Suppression of human immunodeficiency type 1 replication by human herpesvirus-6. J Infect Dis 162: 844–851
Carrigan DR, Drobyski WR, Russler SK, Tapper MA, Knox KK, Ash RC (1991) Interstitial pneumonitis associated with human herpesvirus-6 infection after marrow transplantation. Lancet 338: 147–149
Downing RG, Sewankambo N, Honess RW, Crawford D, Jarrett R, Griffin BE (1987) Isolation of human lymphotropic herpesvirus from Uganda. Lancet ii: 390
Drobyski WR, Dunne WM, Burd EM, Knox KK, Ash RC, Horowitz MM, Flomenberg N, Carrigan DR (1993) Human herpesvirus-6 (HHV-6) infection in allogeneic bone marrow transplant recipients: evidence of a marrow-suppressive role for HHV-6 in vivo. J Infect Dis 167: 735–739
Dubedat S, Kappagoda N (1989) Hepatitis due to human herpesvirus-6. Lancet ii: 1463
Eizuru Y, Minematsu T, Kikuchi M, Yamanishi K, Takahashi M, Kurata T (1989) Human herpesvirus-6 in lymph nodes. Lancet i: 40
Farr TJ, Harnett GB, Pietroboni GR, Bucens MR (1990) The distribution of antibodies to HHV-6 compared with other herpesviruses in young children. Epidemiol Infect 105: 603–607
Fox JD, Briggs M, Ward PA, Tedder RS (1990) Human herpesvirus-6 in salivary glands. Lancet ii: 590–593
Huang LM, Courtier AM, Lowry JW, Buckley NH, White KT, Hoegerman SF (1990) Human herpesvirus-6 associated with fatal haemophagocytic syndrome. Lancet 336: 60–61
Jarrett RF, Gledhill S, Qureshi F, Crae SH, Madhock R, Brown I, Evans I, Krajewski A, O'Brien CJ, Cartwright RA, Venables P, Onions DE (1988) Identification of human herpesvirus-6-specific DNA sequences in two patients with non-Hodgkin's lymphoma. Leukemia 2: 496–502
Josephs SF, Buchbinder A, Streicher HZ, Ablashi DV, Salahuddin SZ, GuoH-G, Wong-Staal F, Cossman J, Raffeld M, Sundeen J, Levine P, Biggar R, Krueger GRF, Fox RI, Gallo RC (1988) Detection of human B lymphotropic virus (human herpesvirus 6) sequences in B cell lymphoma tissues of three patients. Leukemia 2: 132–135
Josephs SF, Ablashi DV, Salahuddin SZ, Jagodzinski LL, Wong-Staal F, Gallo RC (1991) Identification of the human herpesvirus 6 glycoprotein H and putative large tegument protein genes. J Virol 65: 5597–5604.
Kido S, Kondo K, Kondo T, Morishima T, Takahashi M, Yamanishi K (1990) Detection of human herpesvirus 6 DNA in throat swabs by polymerase chain reaction. J Med Virol 32: 139–142
Kondo K, Kondo T, Okuno T, Takahashi M, Yamanishi K (1991) Latent human herpesvirus 6 infection of human monocytes/macrophages. J Gen Virol 72: 1401–1408
Knox KK, Carrigan DR (1992) In vitro suppression of marrow progenitor cell differentiation by human herpesvirus six infection. J Infect Dis 165: 925–929
Krueger GRF, Koch B, Ablashi DV (1987) Persistent fatigue and depression in patients with antibody to human B lymphotropic virus. Lancet ii: 36.
Krueger GRF, Manak M, Bourgeois et al. (1989) Persistent active herpesvirus infection associated with atypical polyclonal lymphoproliferation (APL) and malignant lymphoma. Anticancer Res 9: 1457–1476.
Krueger GRF, Wasserman K, de clerk LS, Stevens WJ, Bourgeois N, Ablashi DV, Josephs SF, Balachandran N (1990) Latent herpesvirus-6 in salivary glands and bronchial glands. Lancet ii: 1255–1256.
Lawrence GL, Chee M, Craxton MA, Gompels UA, Honess RW, Barrell BG (1990) Human herpesvirus 6 is closely related to human cytomegalovirus. J Virol 64: 287–299
Levy JA, Ferro F, Greenspan D, Lennette ET (1990a) Frequent isolation of HHV-6 from saliva and high seroprevalence of the virus in the population. Lancet ii: 1047–1050.
Levy JA, Landay A, Lennette ET (1990b) Human herpesvirus-6 inhibits human immunodeficiency virus type 1 replication in cell culture. J Clin Microbiol 28: 2362–2364.
Lopez C, Pellett P, Stewart J, Goldsmith C, Sanderlin K, Black J, Warfield D, Feorino P (1988) Characteristics of human herpesvirus-6. J Infect Dis 157: 1271–1273
Luka J, Okano M, Thiele G (1990) Isolation of human herpesvirus-6 from clinical specimens using human fibroblast cell cultures. J Clin Lab Analysis 4: 483–486

Lusso P, Markham PD, Tschachler E, Veronese FDM, Salahuddin AZ, Ablashi DV, Pahwa S, Krohn K, Gallo RC (1988) In vitro cellular tropism of human B lymphotropic virus (human herpesvirus-6). J Exp Med 167: 1659–1670

Lusso P, Ensoli B, Markham PD, Ablashi DV, Salahuddin SZ, Tschachler E, Wong-Staal F, Gallo RC (1989) Productive dual infection of human CD4+ T lymphocytes by HIV-1 and HHV-6. Nature 337: 1659–1670

Macnab JCM (1987) Herpes simplex virus and human cytomegalovirus: their role in morphological transformation and genital cancers. J Gen Virol 68: 2525–2550

Neipel F, Ellinger K, Fleckenstein B (1991) The unique region of the human herpesvirus type 6 is essentially colinear to the UL segment of human cytomegalovirus. J Gen Virol 72: 2293–2297

Okuno T, Takahashi K, Balachandra K, Shiraki K, Yamanishi K, Takahashi M, Baba K (1989) Seroepidemiology of human herpesvirus-6 infection in normal children and adults. J Clin Microbiol 27: 651–653

Okuno T, Higashi K, Shiraki K, Yamanishi K, Takahashi M, Kokado Y, Ishibashi M, Takahar S, Sonoda T, Tanaka K, Baba K, Yabucchi H, Kurata T (1990) Human herpesvirus-6 infection in renal transplantation. Transplantation 49: 519–522

Okuno T, Mukai T, Baba K, Oshumi Y, Takahashi M, Yamanishi K (1991) Outbreak of exanthem subitum in an orphanage. J Pediatr 119: 759–761

Pietroboni GR, Harnett GB, Bucens MR, Honess RW (1988a) Antibody to human herpesvirus-6 in saliva. Lancet i: 1059

Pietroboni GR, Harnett GB, Farr TJ, Bucens MR (1988b) Human herpesvirus type 6 (HHV-6) and its in vitro effect on human immunodeficiency virus (HIV). J Clin Pathol 41: 1310–1312

Pruksananonda P, Hall CB, Insel RA, McIntyre K, Pellett PE, Long CE, Schnabel KC, Pincus PH, Stamey FR, Dambaugh TR, Stewart JA (1992) Primary human herpesvirus 6 infection in young children. N Engl J Med 326: 1445–1450

Razzaque A (1990) Oncogenic potential of human herpesvirus-6 DNA. Oncogene 5: 1365–1370

Razzaque A, Puri RK (1992) Human herpesvirus-6:tumorigenicity and tumor infiltrating lymphocytes. Cancer Lett 61: 111–118

Razzaque A, Williams O, Wang J, Rhim JS (1993) Neoplastic transformation of immortalized human epidermal keratinocytes by two HHV-6 DNA clones. Virology 195: 113–120

Rhim JS, Jay G, Arnstein P, Price FM, Sanford KK, Aronson SA (1985) Neoplastic transformation of human epidermal keratinocytes by AD12-SV40 and kirsten sarcoma viruses. Science 227: 1250–1252

Rhim JS, Park JB, Gilbert J (1989) Neoplastic transformation of human keratinocytes by polybrene-induced DNA-mediated transfer of an activated oncogene. Oncogene 4: 1403–1409

Salahuddin SZ, Ablashi DV, Markham PD, Josephs SF, Sturzenegger S, Kaplan M, Halligan G, Biberfeld P, Wong-Staal F, Kramarsky B, Gallo RC (1986) Isolation of a new virus, HBLV, in patients with lymphoproliferative disorders. Science 234: 596–601

Saxinger C, Polesky H, Eby N, Grufferman S, Murphy R, Tegtmeir G, Parekh V, Memon S, Huang C (1988) Antibody reactivity with HBLV (HHV-6) in US population. J Virol Methods 21: 199–208

Schirmer EC, Wyatt LS, Yamanishi K, Rodriguez WJ, Frenkel N (1991) Differentiation between two distinct classes of viruses now classified as human herpesvirus 6. Proc Natl Acad Sci USA 88: 5922–5926

Segondy M, Astruc J, Atoui N, Echenne B, Robert C, Agut H (1992) Herpesvirus 6 infection in young children. N Engl J Med 327: 1099–1100

Steeper TA, Horwitz CA, Ablashi DV, Salahuddin SZ, Saxinger C, Saltzman R, Schwartz B (1990) The spectrum of clinical and laboratory findings due to human herpesvirus-6 in patients with mononucleosis-like illness not due to EBV or CMV. Am J Clin Pathol 93: 776–783

Takahashi K, Sonoda S, Higashi K, Kondo T, Takahashi H, Takahashi M, Yamanishi K (1989) Predominant CD4 lymphocyte tropism of human herpesvirus-6 related virus. J Virol 63: 3161–3163

Tedder RS, Briggs M, Cameron CH, Honess RW, Rodertson D, Whittle H (1987) A novel lymphotropic herpesvirus. Lancet ii: 390–392

Torelli G, Marasca R, Luppi M, Selleri L, Ferrari S, Narni F, Mariano MT, Federico M, Ceccherini-Nelli L, Bendinelli M, Montagnani G, Montorsi M, Artusi T (1991) Human herpesvirus-6 in human lymphomas: Identification of specific sequences in Hodgkins's lymphomas by polymerase chain reaction. Blood 77: 2251–2258

Yamanishi K, Okuno T, Shiraki K, Takahashi M, Kondo T, Asano Y, Kurata T (1988) Identification of human herpesvirus-6 as a causal agent for exanthem subitum. Lancet i: 1065–1067

Yanagi K, Harada S, Ban F, Oya A, Okabe N, Tobina K (1990) High prevalence of antibody to human herpesvirus-6 and decrease in titer with increase in age in Japan. J infect Dis 161: 153–154

Yoshikawa T, Suga S, Asano Y, Nakashima T, Yazaki T, Sobue R, Hibrano M, Fukuda M, Kojima S, Matsuyama T (1991) Human herpesvirus-6 infection in bone marrow transplantation. Blood 78: 1381–1384 [illegible] M, [illegible] H, [illegible] JW, Robinson D, [illegible] (19[illegible]) [illegible] lymph [illegible] herpesvirus [illegible]

[illegible] M, [illegible] J, [illegible] M, [illegible] M, [illegible] (19[illegible]) [illegible] human [illegible] polymerase [illegible]

[illegible] M, [illegible] human [illegible]

[illegible]

[illegible] human [illegible]

E. Trends

Chapter 21 Herpes Simplex Virus Type 1 Genes Involved in Virus Pathogenicity: A Review

Tamir Ben-Hur

Summary

Herpes simplex virus type 1 (HSV-1) has both dermatotropic and neurotropic properties and may be neurovirulent in humans. Several viral genes that determine the biological behavior of this herpesvirus in infected animals have been characterized. The function of each virulence gene is determined by specific interactions with the host at different stages of the herpetic disease. The mapping of the pathogenicity genes in the HSV-1 genome is described, as is the mechanism behind the function of both the HSV-1 DNA sequences and genes that determine viral virulence in the eyes, peritoneum, and skin, leading to invasion of the nervous system and encephalitis.

Introduction

Herpes simplex virus type 1 (HSV-1) is a member of the neurotropic and neurovirulent human herpesvirus family. These viruses may cause acute, latent, and recurrent infections, a concept that was already suggested by Goodpasture (1929). Subsequent studies have focused on the pathogenesis of HSV-1 infection and its interaction with host defense systems. This review will discuss viral molecular factors which determine the biological behavior of HSV-1 in infected animals.

Many experimental systems were used to investigate the neural spread of HSV-1. Footpad, ear, genitalia, cornea, nose, peritoneum, and muscle are some of the loci that have been used for primary inoculation. The virus was found to use sensory, autonomic, motor, and cranial nerve routes to penetrate the nervous system. Sensory (dorsal root), autonomic and trigeminal ganglia, as well as the central nervous system, become infected both acutely and latently (Kristensson 1970; Baringer and Swoveland 1973; Cook and Stevens 1973; Baringer 1974; Knotts et al. 1974; Warren et al. 1978; Sequiera et al. 1979; Fraser et al. 1981; Martin and Dolivo 1983; Rock and Fraser 1983; Shimeld et al. 1985; Kristensson et al. 1986). It is therefore clear that HSV-1 can infect a variety of peripheral tissues

Departments of Neurology and Molecular Virology, Faculty of Medicine, Hebrew University of Jerusalem, Jerusalem Israel and is not limited to a specific type of neuron for travel by axonal transport or for producing infection. However, the outcome of HSV-1 infection is highly dependent on host factors such as mouse strain (Lopez 1975), age (Johnson 1964; Hirsch et al. 1970; Ben-Hur et al. 1983), route of inoculation (Lopez 1981; Becker and Ben-Hur 1985), and virus strain.

What, therefore, are the HSV-1 molecular mechanisms of virulence and how do they interact with host factors to produce disease? Several virus genes have been shown to play a role in the virulence phenotype. It should be noted that many virus genes are active in infection of cell lines in vitro. For example, viral glycoproteins B and D are required for virus penetration of infected cells (Cai et al. 1988; Ligas and Johnson 1988); immediate-early gene 4 (designated α22) is necessary for virus growth in rodent cells but not human or monkey cells (Sears et al. 1985); and two genetic loci in virus DNA were shown to mediate shutoff of host cell protein synthesis (Daksis and Chan 1987; Fenwick and Everett 1990). As such, these and other genes may have a general effect on HSV-1 pathogenicity in vivo. However, the term "pathogenicity genes" may be given to genes that do not have an essential role in virus replication in vitro but are important in the pathogenesis of infection in vivo at a specific stage of the disease.

We have learned that each virulence gene functions by a specific interaction of the virus with the host at different sites of infection. In order to map a virulence gene and investigate its mechanism of action, one must design an experimental system in which the specific gene has exclusive control over the phenotype of viral virulence. Accordingly, these genes may be termed intraperitoneal virulence genes, intracerebral virulence genes, neuroinvasiveness genes, and latency associated genes. They will be reviewed according to progression of the disease in acute HSV-1 infection following inoculation at a specific peripheral site.

Viral Genes Determining the Disease at the Inoculation Site

Such sequences in the viral genome were defined in two experimental systems: ocular HSV-1 keratitis and intraperitoneal HSV-1 infection. There are also several other sites for primary infection.

Ocular Keratitis

Centifanto-Fitzgerald et al. (1982) showed that the ocular disease pattern in rabbits is determined by the HSV-1 genomic region between 0.7 and 0.83 map units. This was done by producing recombinant viruses in which DNA sequences from the stromal disease-producing strain MP were transferred to the epithelial disease-producing strain F. This resulted in HSV-1 intratypic recombinants based on strain F with disease characteristics of the donor MP strain. The severity of epithelial and stromal diseases and the morphology of the dendritic lesions differed in the various recombinants, showing that several viral genetic loci affect the clinical picture of herpetic keratitis.

The mechanism by which viral genes determine the ocular herpetic disease pattern is not known. They are probably not related to neuroinvasiveness since HSV-1 MP is highly neuroinvasive and neurovirulent following ocular inoculation but the F(MP) recombinants that gained the MP corneal disease pattern were not neurovirulent. However, the HSV-1 induced corneal inflammation is highly dependent on host factors. Athymic nude mice (nu/nu) are susceptible to rapid invasion of HSV-1 from the eye to the nervous system but do not develop stromal disease in the eye, as opposed to parental Balb/C mice (Metcalf et al. 1979; Kapoor et al. 1982). Corneal edema and inflammation are induced primarily by immune T lymphocytes which infiltrate the stroma (Metcalf and Kaufman 1976; Kumano et al. 1988). The variability in stromal eye disease may be the result of the activation of different T cell subpopulations by different HSV-1 strains (Hendricks and Tumpey 1990). HSV-1 infects nerve endings in the eye and can spread to the nervous system via sensory, sympathetic, and optic pathways (Shimeld et al. 1985). The local keratitis produces degeneration of corneal innervation with anesthetic areas (Thompson et al. 1988) and correlates with depletion of the neuropeptide substance P from type C nerve endings which exclusively innervate the cornea (Tullo et al. 1983).

HSV-1 Intraperitoneal Virulence

Intraperitoneal (IP) injection is a common route of systemic inoculation of HSV-1 into laboratory animals. In our experience (Ben-Hur 1988), like that of others (Lopez 1975), the outcome of IP infection is highly dependent on mouse and virus strains. The LD_{50} varies, therefore, by more than 10^6 pfu/mouse in different

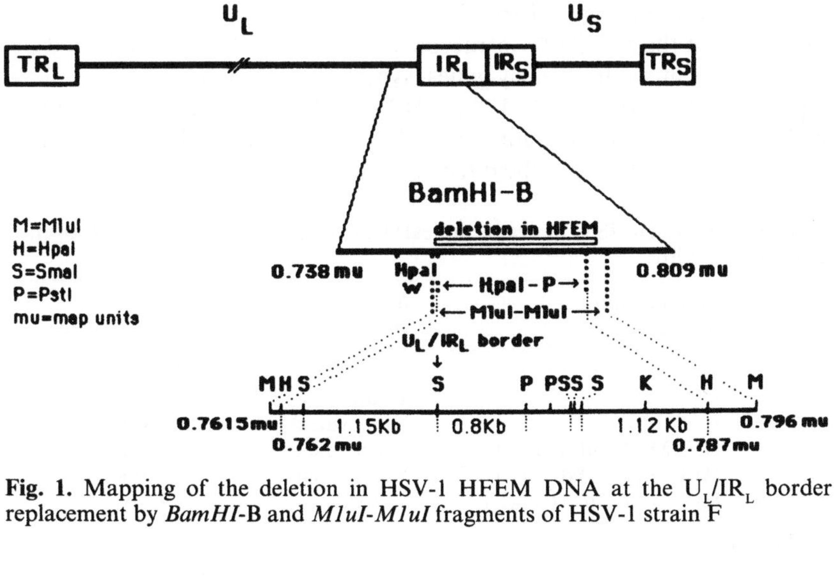

Fig. 1. Mapping of the deletion in HSV-1 HFEM DNA at the U_L/IR_L border and its replacement by *BamHI*-B and *MluI-MluI* fragments of HSV-1 strain F

experimental setups. The viral and host mechanisms involved in the pathogenesis of IP infection have been subject to investigation. While studying HSV-1 IP virulence in tree shrews Rösen and Darai (1985) found that the avirulence of strain HFEM correlated with a naturally occurring deletion in the viral genome in coordinates 0.762–0.789 map units. The exact location of the deletion was mapped by sequencing the DNA in this genomic region in strain HFEM in comparison to strain F (Koch et al. 1987). Cotransfection of unit length HFEM DNA with fragments *Bam*HI-B (0.738–0.809 map units) and *Mlu*I–*Mlu*I (0.7615–0.796 map units) cloned from the virulent strain F yielded intratypic HFEM/F recombinant viruses (Fig. 1). These recombinants were inoculated into the peritoneum of tree shrews and some of them were found to be virulent (Rösen et al. 1985).

In tree shrews HSV-1 induces a systemic disease, while in mice it is neurotropic, reaching the brain to cause encephalitis; thus mice are a somewhat more relevant experimental system in relation to humans. When these virus strains and recombinants were inoculated IP into inbred strain A mice, HSV-1 HFEM was completely avirulent and strain F induced lethal infection. One of the HFEM/F recombinants (R-MlC1) was found to have regained partial IP virulence (Becker et al. 1986; Rösen et al. 1986). HSV-1 IP virulence was, therefore, mapped in a specific region of the viral genome.

The mechanism by which the HSV-1 IP virulence gene functions was studied. Following IP inoculation, the virulent strain F persisted in the peritoneum for at least 2 days and reached the brain to cause lethal encephalitis. In contrast, avirulent strain HFEM was rapidly cleared from the peritoneum without any signs of disease except for acquired resistance to challenge with a virulent strain (Ben-Hur et al. 1988). Since the peritoneum is the site of blockage of HSV-1 HFEM, the interaction of viruses with peritoneal macrophages was studied. It was found that IP virulence correlated directly with the specific ability of the virus to induce a cytopathic effect (CPE) in macrophages that were infected in vitro (Table 1). HSV-1 HFEM can induce a CPE in many cell types, including mouse L-cells and mouse embryo fibroblasts. However, it is completely and specifically noncytopathic to macrophages (Table 2).

We showed that the HFEM/F intratypic recombinant R-MlC1 regained the ability to express its immediate-early genes, synthesize DNA, and induce a CPE in macrophages (Ben-Hur et al. 1988). HSV-1 IP virulence may function, therefore, by tissue-specific recognition of HSV-1 by macrophages, enabling expression of the viral genome in the infected macrophages. Without this specific

Table 1. Correlation between HSV-1 intraperitoneal virulence in vivo and its cytopathogenicity to macrophages in vitro

Virus strain	*Recombinant structure*	*Intraperitoneal virulence*	*CPE in macrophages*
KOS	–	Virulent	Full
F	–	Virulent	Full
HFEM	–	Nonvirulent	None
R-15	HFEM+*Bam*HI-B (F)	Nonvirulent	None
R-19	HFEM+*Bam*HI-B (F)	Nonvirulent	None
R-MlC1	HFEM+*MluI-MluI* (F)	Virulent	Full

CPE, cytopathic effect.

Table 2. HSV-1 interactions with various cell types in vitro

Cell type	*Induction of cytopathic effect by HSV-1*	
	Strain F	*Strain HFEM*
BSC-1 (monkey kidney)	Full	Full
L-929 (mouse fibroblast)	Full	Full
Mouse embryo fibroblasts	Full	Full
Resident peritoneal macrophages		
Adult strain A mice	Full	None
Adult strain C57Bl/6 mice	Partial	None
Suckling strain A mice, 1 week old	Full	None
Suckling C57Bl/6 mice, 1 week old	Partial	None
Bone marrow strain A macrophages	Full	None

DNA sequence, the virus is destroyed by the macrophage like any other invader. The defect in HSV-1 HFEM is intrinsic to the virus: it is noncytopathic to macrophages of different sources, which include peritoneum and bone marrow, macrophages from suckling mice and those from both susceptible and resistant mouse strains (Ben-Hur 1988).

The virulent HSV-1 F strain also expressed immediate-early genes in macrophages from HSV-1-resistant C57Bl/6 mice as it did in susceptible strain A macrophages, although it produced only a partial CPE in C57Bl/6 macrophages in vitro and did not kill these mice when inoculated IP (Ben-Hur 1988). This indicates that the natural resistance of C57Bl/6 mice to HSV-1 is not directed toward the IP virulence function but acts at a different stage of infection in the macrophage and in the animal. This is in agreement with Lopez and Dudas (1979), who showed that genetic resistance of mice to HSV-1 is not directly related to susceptibility of macrophages infected in vitro. Perhaps the induction of interleukin-1 genes by virus infection of peritoneal macrophages (Sprecher and Becker 1990), which further mediates the release of other cytokines (Dinarello 1988), has a role in the natural resistance of animals to HSV-1.

Although virulent HSV-1 strains and recombinants were able to efficiently express immediate-early genes and produce a complete CPE in mouse strain A macrophages, DNA synthesis and virus replication were barely detectable. Macrophages are generally resistant to HSV-1 (Stevens and Cook 1971; Mogensen 1979; Morahan 1984), and virus replication is limited in these cells even from susceptible mouse strains (Frank et al. 1978; Lopez and Dudas 1979; Sethi and Brandis 1979). However, we were able to induce viral DNA synthesis by incubation of macrophages prior to infection, with L-cell conditioned medium (Ben-Hur et al. 1988) containing colony-stimulating factors (CSFs), and especially CSF-1 (M-CSF) (Stanley and Heard 1977; Waheed and Shadduck 1979; Burgess et al. 1985). CSF-1 induces macrophage entry into S phase and cellular DNA synthesis (Wille and Willecke 1980; Tushinsky and Stanley 1985) and this may enable viral DNA replication as well.

Sethi and Brandis have reported (1979) that L-cell conditioned medium did not have any effect on the permissiveness of macrophages to HSV-1 replication. However, they used activated macrophages, which are already in a relatively permissive state (Morahan 1984). Morahan et al. (1989) showed that only a small percentage of infected resident macrophages show transcription of viral immediate-early RNA and that infection is aborted generally prior to viral DNA synthesis. Macrophages activated with thioglycolate showed greater viral DNA synthesis than resident macrophages (Sit et al. 1988), and this could not be attributed to the action of interferon or tumor necrosis factor. In addition, HSV-1 replicated better in monocytes after incubation for several days in vitro (Linnavuori and Hovi 1983). These observations may all be effects of CSF-1 on the macrophages, as was shown in our experiments. In agreement, it was suggested (Leary et al. 1985) that macrophage permissiveness to HSV-1 replication may serve as an indicator of its differentiative state. However, our studies also suggested that induction of a CPE in macrophages (and therefore pathogenesis of HSV-1 IP infection) is completely independent of viral replication in these cells.

A central question that may be raised is why the viral function of macrophage cytopathogenicity expresses itself as an IP virulence phenotype. Many studies attempted to define the specific cell type responsible for host resistance to HSV-1. While T cells were found to play a central role in HSV-1 ocular disease, they are probably not important in the peritoneum since athymic nude mice were resistant to HSV-1 IP infection, similar to the parental mouse strain (Lopez 1978; Zawatzki et al. 1979). In contrast, macrophages have an important role as a primary defense line in the peritoneum. This was shown by selective removal of these cells (Zisman et al. 1970) and adoptive transfer experiments (Hirsch et al. 1970). It may be that the high susceptibility of strain A mice to HSV-1 when inoculated by the IP route is related to the defect in their macrophages' function (Boraschi and Meltzer 1980). However, macrophages are probably not the only defense of the peritoneum, since other factors, such as NK cells (Zawatzki et al. 1979; Lopez 1981; Habu et al. 1984) and interferon (Bukowski and Welsh 1986) were also found to participate.

After HSV-1 overcomes the first line of host defense in the peritoneum, i.e., the macrophages, it spreads in order to penetrate the nervous system. Several studies have pointed to the adrenal gland as the route by which the virus penetrates the nervous system (Hill et al. 1986; Potratz et al. 1987). HSV-1 replicates mainly in the adrenal coretx (Nachtigal and Caulfield 1984; Potratz et al. 1987). This may be a result of the high local concentration of steroid hormones, suppressing immunologic and inflammatory responses. Another possibility is that HSV-1 has a special tropism to adrenal cortex cells. An additional route of neural spread of HSV-1 from the peritoneum may be replication in the myenteric plexus. While the adrenal gland serves as the entrance to the spinal cord, HSV-1 may be transported from the myenteric plexus via vagal nerves directly to the brainstem (Irie et al. 1989). Whatever route is used, it is clear from our studies that penetration of the nervous system demands another viral function in addition to the IP virulence function. We showed that the virulent HFEM/F intratypic recombinant R-MlC1 could not be isolated from brains of mice that died after IP inoculation (Ben-Hur et al. 1988). However, the R-MlC1 recombinant was easily isolated from brains of mice that died after direct intracerebral inoculation. Thus, although this recombinant virus regained the IP virulence function, it must have been defective in penetrating the nervous system. Indeed, it was shown that R-MlC1 replicated well in the adrenal gland following IP inoculation but for a shorter period of time and was unable to invade the spinal cord like other virulent strains (Peles et al. 1990). This defect may explain why R-MlC1 regained only partial IP virulence.

What is the molecular mechanism by which a specific viral DNA sequence enables virus recognition and infection in macrophages? There may be structural genes encoded by this genomic region. In fact, several novel RNA transcripts were mapped in this region, with some differences between virulent and avirulent HSV-1 strains (Rösen-Wolff et al. 1988; Ben-Hur et al. 1989). Some of these RNA transcripts have a role in HSV-1 latent infection (Spivack and Fraser 1987; Rock et al. 1987; Wagner et al. 1988) and reactivation (Steiner et al. 1989). Two models may be suggested for the participation of such genes in the infection of macrophages:

1. There may be a very early viral polypeptide that specifically induces the expression of immediate-early genes in macrophages. Since viral α genes were transcribed in macrophages in the presence of cycloheximide, which inhibits translation, this possibility is unacceptable (Ben-Hur 1988).

2. There may be a late viral polypeptide which is already present with the virus at the time of infection and specifically induces expression of viral α genes in macrophages. This peptide may be encoded by the DNA sequence referred to above or, alternatively, by another region of the viral genome which acts on this DNA sequence. We have tested this option by coinfection of macrophages with the avirulent HSV-1 strain HFEM and the UV-inactivated virulent strain KOS. Since no CPE was found in the macrophages coinfected with HFEM and inactivated KOS, we think this mechanism unlikely. However, it may still be that such a peptide exists and acts at a stage of infection prior to α gene transcription, such as virus entrance, uncoating, or protection against virus degradation by the macrophages.

Another mechanism may be the activation of specific macrophage transcription factor(s) by a cis-acting DNA sequence element. Several viral DNA sequences were shown to regulate transcription by binding of cellular (Sp1, CTF) (Jones and Tjian 1985; Jones et al. 1985a; Kristie and Roizman 1987) and viral (α TIF) (Campbell et al. 1984; Dalrymple et al. 1985; Pellet et al. 1985) factors. It may be that a tissue-specific enhancer element is missing in HSV-1 HFEM, thus preventing expression of α genes in macrophages. It should be noted that analyses of this genomic region showed repeat sequences downstream to IE-1 (αO) (Rixon et al. 1984) and others. This region is homologous to human rDNA (Jones et al. 1985b; Parks et al. 1986) and shows great variability in size between virus strains (Locker and Frenkel 1979; Post et al. 1980; Moyal et al. 1988). The relevance of these observations is not known.

HSV-1 Infection of the Skin

Several sites in the skin and mucus membranes were used for inoculation with HSV-1. A drop of virus placed on the nasal mucosa penetrates the brain, probably via olfactory routes, and rapidly produces lethal encephalitis (Ben-Hur 1988). Infection of the ear pinna was used in models for studying the reactivation of latent HSV-1 (Hill et al. 1975,1978). It was shown that HSV-1 pathogenicity in footpad and ear skin of mice depends on host factors (Langerhans cell density and mouse genetics) as well as on virus strain (Sprecher and Becker 1987). The viruses differed in the degree of necrosis, edema, and hair loss that they produced in the skin lesions. The viral genes that affect these factors have not yet been identified.

HSV-1 Genes Involved in Its Neural Spread

The neural spread of HSV-1 from the site of primary infection to the brain has been mentioned several times as the second stage in disease progression. This stage requires distinct viral functions that will be discussed below.

HSV-1 Thymidine Kinase Enzyme

HSV-1 thymidine kinase (TK) was the first viral gene that was related to virulence. It was shown that mutant viruses, deficient in TK activity, had reduced pathogenicity (Field and Wildy 1978). TK⁻ mutants replicated efficiently like their parental strains in the cornea but were defective in replication in the trigeminal ganglion or central nervous system and therefore did not kill infected animals (Tenser and Dunstan 1979; Tenser et al. 1979). Similar results were obtained in other experimental systems such as infection of the superior cervical ganglion (Price 1979; Price and Kahn 1981; Price and Schmitz 1979). Using a mutant expressing partial TK activity we showed that HSV-1 replication in neural tissues and neurovirulence correlated with the degree of TK activity (Gordon et al. 1983). We found that TK⁻ strains were nonneurovirulent by several routes of inoculation, including corneal, IP, and nasal routes, although high titers of virus were used (Becker et al. 1984a, b). It was thought that TK⁻ mutants are also unable to cause latent infection, but recently it was shown that TK is not essential for establishment of latent infection but may be necessary for virus reactivation in neural tissues (Coen et al. 1989; Efstathiou et al. 1989; Tenser et al. 1989).

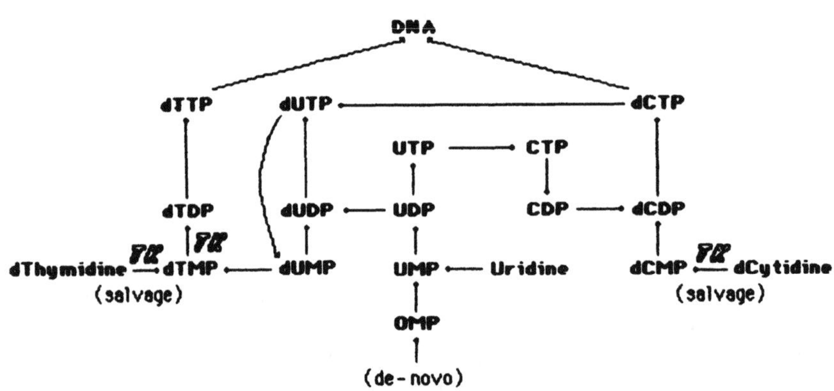

Fig 2. Stages in pyrimidine nucleotide synthesis in which the viral thymidine kinase enzyme participates

Viral TK has three phosphorylative functions: (1) as a thymidine kinase, (2) as a thymidylate kinase, and (3) as a deoxycytidine kinase (Chen et al. 1979). It therefore has major control over the synthesis of deoxypyrimidine nucleotides, which are required for incorporation into replicating DNA (Fig. 2). It is accepted that mutants lacking TK activity are dependent on cellular mechanisms for DNA synthesis. Since nerve cells do not have replicative potential and cellular TK is undetectable it is clear why TK^- mutants are defective in replication in the nervous system. However, we showed that in several experimental designs TK^- mutants were pathogenic:

1. Following direct intracerebral (IC) inoculation (Ben-Hur et al. 1983): Although 100-fold higher virus titers were needed, mice succumbed to IC inoculation of TK^- mutants. Presumably, by this route, the virus reaches glial cells which have a better capacity to support viral DNA synthesis than do neural cells.
2. Following peripheral inoculation of suckling mice during their first week of life (Ben-Hur et al. 1983): At this period of time there is high cellular TK activity in the developing brain (Sung 1971; Yamagami et al. 1972; Weichsel 1974). Afterwards, when cellular TK activity disappears, the TK^- strains rapidly lose their virulence although the immunologic system does not mature until the age of 3 weeks (Lopez 1981; Kohl and Loo 1982; Zawatzky et al. 1982).
3. Following corneal inoculation of adult mice with one of our TK^- isolates (Ben-Hur 1988): This virus may reach the brain via neural pathways, leave the neuron, and infect other cell types before it can replicate.

Although TK^- mutants are virulent, the kinetics of mouse death following either IC or ocular inoculation were slower with TK^- mutants than with parental TK^+ strains.

It may be concluded that loss of viral TK activity impaired but did not abolish completely the ability of HSV-1 to invade the nervous system and replicate there.

HSV-1 DNA Polymerase and Ribonucleotide Reductase Enzymes

HSV-2 strain 186 is nonneuroinvasive after ocular inoculation although it produces normal TK activity levels and is virulent after direct intracerebral inoculation. An HSV-1/HSV-2 intertypic recombinant, in which HSV-1 strain 17 DNA sequences from coordinates 0.31–0.44 were exchanged with HSV-2 sequences, acquired the ability to spread from the eye to the brain (Oakes et al. 1986). This function was later attributed to the viral DNA polymerase gene (Day et al. 1988). In addition, mutant viruses deficient in DNA polymerase activity showed impaired replication in the central nervous system but efficient replication in the skin, like parental strains (Field and Coen 1986). Ribonucleotide reductase is another key enzyme in the de novo synthesis of precursors for DNA synthesis. It reduces the four ribonucleotide diphosphates to 2′-deoxyribonucleotide diphosphates. A ribonucleotide reductase-deleted HSV-1 mutant exhibited reduced virulence in comparison to wild-type virus, but this may have been by a generalized severe impairment in replication in all host tissues (Cameron et al. 1988). However, it was still able to grow in peripheral tissues of adult mice and in brains of newborn mice, but not in brains of adult mice (Yamada et al. 1991). The virus mutants deleted in DNA polymerase or ribonucleotide reductase seem to depend on cellular machinary for DNA synthesis, in resemblance to the TK^- mutants. This suggests a similar mechanism of action in virulence.

Other Viral DNA Sequences Involved in Neuroinvasiveness

Several investigators used the mouse footpad as the site of primary infection with HSV-1 and observed the virus invading the spinal cord (causing paralysis) and eventually the brain (causing lethal encephalitis). The DNA sequences between coordinates 0.25 and 0.53 map units of the viral genome were shown to be responsible for this neuroinvasiveness function in $KOS/17SYN^+$ intratypic recombinants (Thompson et al. 1986). This function was distinct from the TK gene but its relation to the DNA polymerase gene (which is mapped to within this region) is unknown.

Another study showed complementation between two nonneuroinvasive strains (ANG and KOS) which resulted in invasion of and replication in the spinal cord following footpad inoculation with a mixture of these virus strains (Javier et al. 1986; Sedarati et al. 1988). This indicates at least two viral genes, in addition to the TK gene, involved in penetration of the nervous system following peripheral inoculation. One of these genes may be the glycoprotein D (gD) of HSV-1, located in the short unique segment of the genome. Further studies in this experimental system showed (Izumi and Stevens 1990) a single amino acid substitution, from alanine to glycine, at position 84 of gD between HSV-1 strains ANG and ANG path. This substitution restored neuroinvasiveness of HSV-1 ANG path following rear footpad inoculation as compared with the inability of strain ANG to invade the nervous system.

HSV-1 Genes Involved in Virulence to the Central Nervous System

The first step in characterization of viral functions involved in central nervous system disease was made by generation and characterization of HSV-1/HSV-2 intertypic recombinants. A specific recombinant, RE6, was found to be completely and specifically nonneurovirulent following intracerebral inoculation, although it replicated equally as well as the parental viruses in nonneural tissues and expressed normal TK activity (Thompson and Stevens 1983). Neurovirulence could be rescued by HSV-1 DNA sequences between 0.71 and 0.83 map units (Thompson et al. 1983).

In further experiments Thompson et al. (1985) showed that the DNA fragment *Hind*III-C, spanning coordinates 0.64–0.83, could rescue IC virulence. Cleavage of this fragment with *Eco*RI eliminated its capacity to restore virulence. When DNA fragment *Bam*HI-L (coordinates 0.71–0.74, that include a single *Eco*RI site at 0.72 map units) was added to *Eco*RI-cleaved *Hind*III-C fragment,

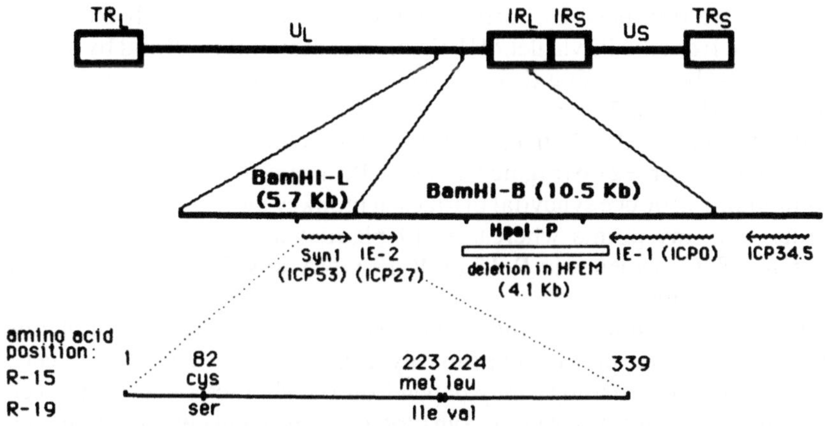

Fig 3. Location of *Bam*HI fragment L in the HSV-1 genome, in proximity to *Bam*HI-B, with the Synl (U_L53) gene at its right side. The differences in amino acid sequence of U_L53 gene between R-19 and avirulent R-15 isolates are noted. Within the *Bam*HI-B fragment are shown two viral immediate-early genes (IE-1, IE-2) and , in between, the deletion in the HFEM genome making this strain avirulent to mice by the intraperitoneal route of infection. Upstream to IE-1 ICP3.5, another viral neurovirulence gene, is located. The *black lines* in the HSV-1 DNA indicate unique sequences (U_L and U_S) and the *boxes* for terminal (TR_L and TR_S) and internal (IR_L and IR_S) repeat sequences

IC virulence was again rescued. This suggested that the specific function which induces central nervous system disease resides in the *Bam*HI-L region of the viral genome, around coordinate 0.72. This could not be proved, however, since the *Bam*HL-L fragment alone did not rescue virulence.

Further experiments showed that partial IC virulence could be rescued in RE6 by a 3.2 kb DNA fragment mapping between 0.698 and 0.721 map units (Thompson and Wagner 1989). The new recombinant still had very low pathogenicity as compared with the wild-type 17SYN$^+$ parent strain. Probably additional viral functions residing in the 0.70–0.83 map unit region were damaged during the generation of the RE6 recombinant. Indeed, the DNA sequence mapping between 0.82 and 0.832 map units cloned from strain 17SYN$^+$ could restore the neurovirulent phenotype in RE6 as well (Thompson et al. 1989). This problem emphasizes the limitations of the approach using intertypic recombinants. Indeed, it has been shown that most heterotopic combinations between HSV-1 and HSV-2 are attenuated, independent of the virulence of the parental viruses (Halliburton et al. 1987).

For direct and specific mapping of the IC virulence function we used an HSV-1 intratypic recombinant which was generated by cotransfection of unit length HFEM DNA with the *Bam*HI-B fragment cloned from strain F (Rösen et al. 1985). This recombinant, R-15, was found to be completely avirulent when inoculated directly into the brains of mice. Presumably, mutation(s) near the edges of *Bam*HI-B (0.738–0.809) damaged this virulence function. Therefore, new recombinants were produced by cotransfection of R-15 DNA with *Bam*HI-L fragments cloned from the IC virulent HFEM/F recombinant R-19 and by in vivo selection in mouse brains. Using this approach, IC virulence was mapped to the *Bam*HI-L region (0.71–0.74 map units) and further located in the right 2.0 kb *Nru*I-*Bam*HI fragment (Ben-Hur et al. 1987). The Synl gene for a cell fusion protein resides in the same 2 kb fragment (Fig. 3). Both avirulent R-15 and virulent R-19 recombinants produce syncytial plaques in monkey kidney BSC-1 cells, but R-15 plaques are tiny points whereas R-19 plaques are considerably larger. Indeed, rescue of IC virulence in R-15 by R-19 DNA was linked to transfer of R-19 plaque size phenotype.

Mutations in cell fusion genes may alter either plaque morphology or plaque size. The size of the plaque may result primarily from the degree of cell fusion gene activity. The plaque morphology may depend on the characteristics of the cell fusion proteins. Point mutations in the Syn1 locus, resulting in changes in the cell fusion protein configuration, can cause reversion of the syncytial plaque phenotype to that of clumped round cell plaques (Debroy et al. 1985; Pogue-Geile and Spear 1987). However, in the clumped round cell plaques there are still cell to cell connections by the membrane-bound cell fusion protein.

The syncytial phenotype of HSV-1 is determined by both several Syn loci and viral glycoproteins and is also dependent on the infected cell type (Ruyechan et al. 1979; Little and Schaffer 1981; Bond and Person 1984; Machuca et al. 1986). When we directly intracranially inoculated different virus strains varying in plaque morphology (from large syncytia to small round cell plaques), they were all highly virulent. However, our studies linked the Syn1 locus with the ability of the virus to cause lethal encephalitis (Ben-Hur et al. 1987). The nucleotide sequence was determined for the Syn1 gene in R-15 and R-19 and compared to reported sequences of this gene in other HSV-1 strains. Three mutations were found in the amino acid sequence of the cell fusion protein in R-15 which are responsible for its avirulence and small plaque morphology (Moyal et al. 1992). These are: the transition of serine to cysteine at amino acid position 82 and isoleucine-valine to methionine-leucine in positions 223–224 (Fig. 3). The Syn1 gene in recombinant R-19 is highly conserved when compared to sequences of other neurovirulent HSV-1 strains (Debroy et al. 1985; Pogue-Geile and Spear 1987). Cell fusion and neurovirulence may, therefore, be the function of the same protein.

Is herpes encephalitis caused by brain cell fusion leading to derangement of electrical activity and virus spread through syncytia, while protected from defense mechanisms? According to this proposed model, encephalitis is not caused by viral replication per se but rather by the activity of a cell fusion protein behaving as a neurotoxin and using the virus as a vector. By this line of thought it is not the type of cell fusion protein (determining plaque morphology) but the neurotoxic activity or amount of this protein (determining plaque size) that has a role in the progression of encephalitis. Although this mechanism has yet to be proved there are several observations with which it is consistent. Syncytial strains were most virulent (by IC inoculation) when various strains were compared (Dix et al. 1983). In addition, correlation between cell fusion activity and neurovirulence has been reported in variants of the Miyama strain of HSV-1 (Yamada et al. 1986). Hints as to the possible role of cell fusion in electrophysiological brain abnormalities as an encephalitogenic mechanism were given in several studies. A syncytial strain of HSV-1 induced spontaneous electrical activity in rat sensory neuron cultures infected in vitro (Mayer et al. 1985). It was suggested that virus-induced cell fusion of neurons induced electrical coupling between them, resulting in an electrical network that supported spontaneous activity. In animal models an association was found between the abililty of HSV-1 to induce electrophysiological abnormalities in the brain, including seizures, and its ability to spread in the brain and cause mortality (Schlitt et al. 1988; Stroop and Schaefer 1989). The concept that HSV-1 encephalitis may not be directly dependent on viral replication in the brain has been suggested by others as well. Javier et al. (1987, 1988) used HSV-1/HSV-2 intertypic recombinants to map an HSV-1 neurovirulence function in the viral genome. In one of the recombinants, which was 10^4-fold less virulent than wild-type HSV-1 or HSV-2 parents, neurovirulence could be rescued with a 9.1 kb HSV-1 DNA fragment spanning coordinates 0.079–0.143. This specific neurovirulence function was found to be dissociated from high titer virus replication in the brain since the nonneurovirulent recombinant replicated in mouse brains to peak titers equivalent to those of its virulent parental strains. In their discussion of this phenomenon Javier et al. (1988) suggested restricted replication in specific cell types or restricted spread (independent of replication) as possible explanations of why an avirulent strain can replicate well in the brain. Another possibility, which they did not raise, is the neurotoxin mechanism such as the "cell fusion neurotoxin" that has been suggested here.

One of the manifestations of virus-host interaction in the brain is the host inflammatory response to invading HSV-1. Morphological studies showed that during acute and latent HSV infection there is severe central nervous system demyelination, while peripheral myelin remains intact (Kristensson and Wisniewski 1978; Townsend and Baringer 1978; Kristensson et al. 1979). This neuropathology is a result of delayed-type hypersensitivity reactions against HSV-1 and against neural factors such as myelin basic protein (Kristensson and Wisniewski 1978; Gudmundsdotter et al. 1986; Bishop and Hill 1989). Indeed, in several experimental systems, immunosuppression reduced viral pathogenicity. Anti-thymocytic serum caused enhanced virulence in some cases (Rager-Zisman and Allison 1976; Blyth et al. 1980) and decreased virulence in others (Nahmias et al. 1969). Cyclosporin A reduced the incidence of ear paralysis following inoculation of the pinna (Bishop and Hill 1991). Immunosuppression and activation of T_S cells rather than T_H cells protected mice from demyelination and lethal encephalitis without significant change in virus titers or virus spread in the brain (Townsend and Baringer 1978; Altmann and Blyth 1985). This is, therefore, another important mechanism for the pathogenesis of herpes encephalitis independent of viral replication. Furture studies will probably connect the host delayed-type hypersensitivity demyelinative response to specific viral antigens. An example for this direction may be the induction of humoral antibody response mediated by HSV-1 DNA region in coordinates 0.7615–0.796. While strain HFEM does not induce antibody formation, the intratypic HFEM/F recombinant R-MlC1, in which this DNA fragment transferred the ability to infect macrophages (Ben-Hur et al. 1988), also regained the antibody production response in the infected animal like the parental F strain (Wollert et al. 1991).

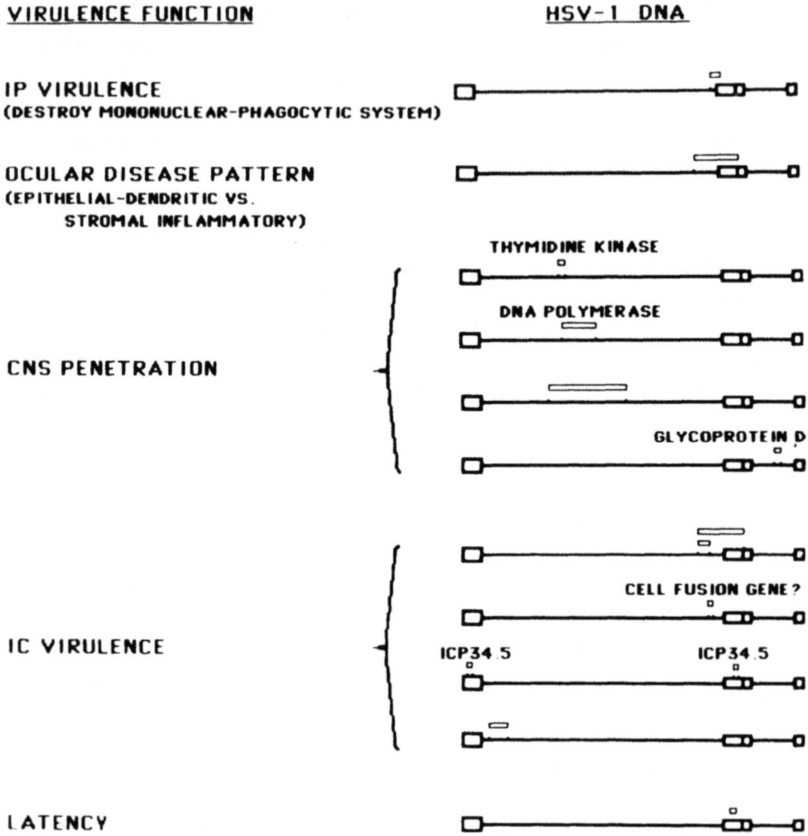

Fig 4. Summary of HSV-1 DNA sequences controlling virulence phenotypes

Recently another late (γ) viral gene, designated ICP34.5, was found to be involved in HSV-1 neurovirulence (Chou et al. 1990). This gene is located within the inverted repeat sequences surrounding the U_L region, upstream to ICP0 (immediate-early 1), and therefore is present in two copies in the viral genome (Fig. 3). Inactivation of the two copies of ICP34.5, by deleting the gene or insertion of a stop codon inside it, generated recombinant viruses which grew well in culture (somewhat less than parental strain F) but failed to replicate efficiently in the brain. They were avirulent to mice after intracerebral inoculation. Reinsertion of the genes in these recombinants restored the neurovirulent phenotype. The localization of HSV-1 neurovirulence and viral ability to replicate in neural cells to the same DNA sequences was reported by other groups as well (Thompson et al. 1989; MacLean et al. 1991). ICP34.5 represents, therefore, a typical virulence gene by its specific neurovirulent property while being non-essential for HSV-1 growth in cell cultures. The specific function of this gene is yet unknown.

Apart from the genes that we have reviewed here, additional neurovirulence properities of various HSV-1 mutants have been studied. In contrast to previous data, it was recently shown that the viral glycoprotein C is not a virulence determinant (Sunstrum et al. 1989). Viruses deleted in immediate-early gene α22 (IE-4), protein kinase, and glycoprotein G exhibited reduced virulence in comparison to wild-type virus (Sears et al. 1985; Meignier et al. 1988b). In addition, the regulatory genes U_L46 and U_L47, which modulate α-TIF activity, were found to enhance viral neuropathogenicity and efficiency of reactivation from latency.

Conclusions

Some of the viral genes affecting HSV-1 pathogenic behavior in infected animals have been reviewed here. Figure 4 summarizes the mapping of these sequences in the viral genome. Again, it is evident that the HSV-1 pathogenic phenotype is influenced by the action of many genes at various sites and stages of infection. These genes were presented according to the progression of infection from the primary site of inoculation, through viral spread and invasion of the nervous system, to the induction of encephalitis. Specifically, the molecular genetics of the HSV-1 ocular disease pattern and its virulence in the peritoneum, its neuroinvasiveness from the eye, skin, and peritoneum and, finally, viral encephalitis were described. It was not the intention of this review to discuss recent advances in the understanding of the molecular basis for latent infection.

Each virulence gene has a specific interaction with the host. In the peritoneum, the result of HSV-1 encounter with macrophages is determined by the IP virulence gene. Penetration of the nervous system is dependent on the ability of the virus to provide the machinery for DNA replication because the host cannot do so. The severity of encephalitis and ocular disease is dependent to a large extent on the host inflammatory response that is induced by specific viral genes. The virulence genes may have a role in the life cycle of infected cells. For example TK and DNA polymerase take part in DNA synthesis and Syn1 in plaque formation. However, these are not essential genes for efficient replication in vitro but rather give HSV-1 specific advantages in producing disease in vivo. The complexity of herpesvirus-infected host interactions is the result of millions of years of coevolution. This has led to the ability of the virus to produce a persistent infection throughout life in its latent form and frequent low virulence reactivations, with fatal encephalitis being the exception.

The recognition of virulence genes could eventually lead to the preparation of a genetically engineered attenuated live HSV-1 vaccine. This vaccine will be deleted of virulence genes, in place of which other herpesvirus antigens will be inserted. Meignier et al. (1988a) have recently developed two recombinant viruses based on strain F, both of which have reduced virulence or ability to establish latency in rodents but are able to induce protective immunity. These recombinants were inoculated vaginally and corneally to owl monkeys, which are highly sensitive to HSV and serve as a model for herpetic disease in immunocompromised hosts and neonates (Katzin et al. 1967). While wild-type virus was lethal after inoculation with a low dose, these recombinants produced only local lesions without dissemination even after high dose inoculation (Meignier et al. 1990). The deletions that were made in the viral genome to develop these recombinants (R7017 and R7020) included the deletion in strain HFEM, in addition to other DNA sequences. The biological behavior of these recombinants resembles that of strains F and HFEM in tree shrews (as a model for systemic herpetic disease) and mice (in which HSV-1 is neurotropic) (Rösen and Darai 1985; Rösen et al. 1985; Becker et al. 1986). These large-scale experiments with recombinant viruses in monkeys are an important step towards development of an attenuated live HSV vaccine. As has been suggested (Meignier et al. 1990), this vaccine should induce high and long-lasting immunity, should be safe in the most susceptible of human hosts, and, of course, should not cause recurrent disease. The rapidly accumulating data will bring this application closer to reality.

Acknowledgements. The assistance of Prof. Yechiel Becker in conducting the research program and preparation of this manuscript is gratefully acknowledged. I thank Prof. G. Darai, Dr. A. Rösen-Wolff, Dr. J. Hadar and Dr. M. Moyal for fruitful scientific cooperation. I also wish to thank the Foundation for Molecular Virology and Cell Biology, Phoenix, Arizona, for their continuous support.

References

Altmann DM, Blyth WA (1985) Protection from herpes simplex virus induced neuropathology in mice showing delayed hypersenitivity tolerance. J Gen Virol 66: 1297–1303
Baringer JR (1974) Recovery of herpes simplex virus from human sacral ganglions. N Engl J Med 291: 829–830
Baringer JR, Swoveland P (1973) Recovery of herpes simplex virus from human trigeminal ganglions. N Engl J Med 288: 648–650
Becker Y, Ben-Hur T (1985) Genetic resistance to herpes simplex virus-1 infections and the immune system. In: Pastoret PP, Thirty E, Saliki J (eds) Immunity to herpesvirus infections of domestic animals. Commission of the European communities, Brussels, Luxemburg, EUR 9737 EN, pp 43–50
Becker Y, Shtram Y, Barasofski A, Haber M, Scemama A, Tabor E, Ben-Hur T, Gilden D, Hadar J (1984a) Role of the promoter sequence of the thymidine kinase (tk) gene of HSV-1 in biochemical transformation of cells, gene expression and neurovirulence in mice. In: Giraldo G, Beth G (eds) The role of viruses in human cancer, vol II. Elsevier Science, Amsterdam, pp 37–58
Becker Y, Shtram Y, Hadar J, Ben-Hur T, Barasofski A, Asher Y, Tabor E, Scemama A, Gilden D (1984b) Herpes simplex virus type 1 thymidine kinase gene controls virus pathogenesis and latency in the nervous system. In: Kohn A, Fuchs P. (eds) Mechanisms of viral pathogenesis: from gene to pathogen. Developments in molecular virology, vol 3. Nijihoff, The Hague, pp 293–308
Becker Y, Hadar J, Tabor E, Ben-Hur T, Raibstein I, Rösen A, Darai G (1986) A sequence in HpaI-P fragment of herpes simplex virus-1 DNA determines intraperitoneal virulence in mice. Virology 149: 255–259
Ben-Hur T (1988) Mechanisms in the pathogenesis of herpes simplex virus infection. Ph D thesis, Hebrew University, Jerusalem
Ben-Hur T, Hadar J, Shtram Y, Gilden DH, Becker Y (1983) Neurovirulence of herpes simplex virus type 1 depends on age in mice and thymidine kinase expression. Arch Virol 78: 303–308
Ben-Hur T, Asher Y, Tabor E, Darai G, Becker Y (1987) HSV-1 virulence for mice by the intracerebral route is encoded by the BamHI-L DNA fragment containing the cell fusion gene. Arch Virol 96: 117–122

Ben-Hur T, Rösen-Wolff A, Lamade W, Darai G, Becker Y (1988) HSV-1 DNA sequence determining intraperitoneal pathogenicity in mice is required for transcription of viral immediate early genes in macrophages. Virology 163: 397–404
Ben-Hur T, Moyal M, Rösen-Wolff A, Darai G, Becker Y (1989) Characterization of DNA transcripts from herpes simplex virus-1 DNA fragment BamHI-B. Virology 169: 1–8
Bishop SA, Hill TJ (1989) Herpes simplex virus infection and damage in the central nervous system: immune response to myelin basic protein. J Neurol Sci 91: 109–118
Bishop SA, Hill TJ (1991) Herpes simplex virus infection and damage in the central nervous system: immunomodulation with adjuvant, cyclophosphamide and cyclosporin A. Arch Virol 116: 57–68
Blyth WA, Harbour DA, Hill TJ (1980) Effect of immunosuppression on recurrent herpes simplex in mice. Infect Immun 29: 902–907
Bond VC, Person S (1984) Fine structure physical map locations of alterations that affect cell fusion in herpes simplex virus type 1. Virology 132: 368–376
Boraschi D, Meltzer MS (1980) Defective tumoricidal capacity of macrophages from A/J mice. III. Genetic analysis of the macrophage defect. J Immunol 124: 1050–1053
Bukowski JF, Welsh RM (1986) The role of natural killer cells and interferon in resistance to acute infection of mice with herpes simplex virus type 1. J Immunol 136: 3481–3485
Burgess AW, Metcalf D, Kozka IJ, Simpson RJ, Vairo G, Hamilton JA, Nice EC (1985) Purification of two forms of colony-stimulating factor from mouse L-cell conditioned medium. J Biol Chem 260: 16004–16011
Cai WH, Gu B, Person S (1988) Role of glycoprotein B of herpes simplex virus type 1 in viral entry and cell fusion. J Virol 62: 2596–2604
Cameron JM, McDougall I, Marsden HS, Preston VG Ryan DM, Subak-Sharp JH (1988) Ribonucleotide reductase encoded by herpes simplex virus is a determinant of the pathogenicity of the virus in mice and a valid antiviral target. J Gen Virol 69: 2607–2612
Campbell ME, Palfreyman JW, Preston CM (1984) Identification of herpes simplex virus DNA sequences which encode a transacting polypeptide responsible for stimulation of immediate early transcription. J Mol Biol 180: 1–19
Centifanto-Fitzgerald YM, Yamaguchi T, Kaufman HE, Tognon M, Roizman B (1982) Ocular disease pattern induced by herpes simplex virus is genetically determined by a specific region of viral DNA. J Exp Med 155: 475–489
Chen MS, Summers WP, Walker J, Summers WC, Prusoff WH (1979) Characterization of pyrimidine deoxyribonucleoside kinase (thymidine kinase) and thymidylate kinase as a multifunctional enzyme in cells transformed by herpes simplex virus type 1 and in cells infected with mutant strains of herpes simplex virus. J Virol 30: 942–945
Chou J, Kern ER, Whitley RJ, Roizman B (1990) Mapping of herpes simplex virus-1 neurovirulence to gamma 34.5, a gene nonessential for growth in culture. Science 250: 1262–1266
Coen DM, Kosz-Vnenchak M, Jacobson JG, Leib DA, Bogard CL, Schaffer PA, Tyler KL, Knipe DM (1989) Thymidine kinase-negative herpes simplex virus mutants establish latency in mouse trigeminal ganglia but do not reactivate. Proc Natl Acad Sci USA 86: 4736–4740
Cook ML, Stevens JG (1973) Pathogenesis of herpetic neuritis and ganglionitis in mice: evidence for intra-axonal transport of infection. Infect Immun 7: 272–288
Daksis JI, Chan VL (1987) Physical mapping of two herpes simplex virus type 1 host shutoff loci: rescue of each its mutation with two unique cloned regions of the viral genome. J Virol 61: 143–150
Dalrymple MA, McGeoch DI, Davison AI, Preston CM (1985) DNA sequence of the herpes simplex virus type 1 gene whose product is responsible for transcriptional activation of immediate early promoters. Nucleic Acid Res 13: 7865–7879
Day SP, Lausch RN, Oakes JE (1988) Evidence that the gene for the herpes simplex virus type 1 DNA Polymerase accounts for the capacity of an intertypic recombinant to spread from eye to central nervous system. Virology 163: 16–173
Debroy C, Pederson N, Person S (1985) Nucleotide sequence of a herpes simplex virus type 1 gene that causes cell fusion. Virology 145:36-48
Dinarello CA (1988) Biology of interleukin 1. FASEB J 2: 108–115

Dix RD, McKendall RR, Baringer JR (1983) Comparative neurovirulence of herpes simplex virus type 1 strains after peripheral or intracerebral inoculation of Balb/C mice. Infect Immun 40: 103–112
Efstathiou S, Kemp S, Darby G, Minson AC (1989) The role of herpes simplex type 1 thymidine kinase in pathogenesis. J Gen Virol 70: 869–879
Fenwick ML, Everett RD (1990) Inactivation of the shutoff gene (U_L41) of herpes simplex virus type 1 and 2. J Gen Virol 71: 2961–2967
Field HJ, Wildy P (1978) The pathogenicity of thymidine kinase deficient mutants of herpes simplex virus in mice. J Hyg (Camb) 81: 267–277
Field HJ, Coen DM (1986) Pathogenicity of herpes simplex mutants containing drug resistance mutations in the viral DNA polymerase gene. J Virol 60: 286–289
Frank U, Schindling B, Lindermann J, Falke D (1978) Multiplication of HSV types 1 and 2 in macrophages of NMRI and C57B1 mice. Acta Virol 22: 193–202
Fraser NW, Lawrence WC, Wroblewska Z, Gilden DH, Koprowski H (1981) Herpes simplex virus type 1 DNA in human brain tissue. Proc Natl Acad Sci USA 78: 6461–6465
Goodpasture EW (1929) Herpetic infection with a special reference to involvement of the nervous system. Medicine (Baltimore) 8: 223–243
Gordon Y, Gilden DH, Shtram Y, Asher Y, Tabor E, Wellish M, Devlin M, Snipper D, Hadar J, Becker Y (1983) A low thymidine kinase producing mutant of herpes simplex virus type 1 causes latent trigeminal ganglia infections in mice. Arch Virol 76: 39–49
Gudmundsdotter S, Svennerholm B, Kristensson K, Lycke E (1986) Herpes simplex virus-enhanced production of autoantibodies against myelin basic protein in mice. Arch Virol 88: 37–47
Habu S, Akamatsu KI, Tamaoki N, Okumara K (1984) In vivo significance of NK cells on resistance against virus (HSV-1) infections in mice. J Immunol 133: 2743–2747
Halliburton IW, Honess RW, Killington RW (1987) Virulence is not conserved in recombinants between herpes simplex virus types 1 and 2. J Gen Virol 68: 1435–1440
Hendricks RL, Tumpey TM (1990) Contribution of the virus and immune factors to herpes simplex virus type 1 induced corneal pathology. Invest Opthalmol Vis Sci 31: 1929–1939
Hill TJ, Field HJ, Blyth WA (1975) Acute and recurrent infection with herpes simplex virus in the mouse: a model for studying latency and recurrent disease. J Gen Virol 28: 341–353
Hill TJ, Blyth WA, Harbour DA (1978) Trauma to the skin causes recurrence of herpes simplex in the mouse. J Gen Virol 39: 21–28
Hill TJ, Yirrell DL, Blyth WA (1986) Infection of the adrenal gland as a route to the central nervous system after viraemia with herpes simplex virus in the mouse. J Gen Virol 67: 309–320
Hirsch MS, Zisman B, Allison AC (1970) Macrophages and age dependent resistance to herpes simplex virus in mice. J Immunol 104: 1160–1165
Irie H, Harada Y, Yoshihashi H, Kimura T, Kojima M, Kataoka M, Saito M, Sugawara Y, Mori W (1989) Spread of herpes simplex virus type-1 (Miyama+gC strain) to the central nervous system after intraperitoneal inoculation: the role of the myenteric plexus of the gut. Arch Virol 105: 247–257
Izumi KM, Stevens JG (1990) Molecular and biological characterization of a herpes simplex virus type 1 (HSV-1) neuroinvasiveness gene. J Exp Med 172: 487–496
Javier RT, Sedarati F, Stevens JG (1986) Two avirulent herpes simplex viruses generate lethal recombinants in vivo. Science 234: 746–748
Javier RT, Thompson RL, Stevens JG (1987) Genetic and biological analyses of a herpes simplex virus intertypic recombinant reduced specifically for neurovirulence. J Virol 61: 1978–1984
Javier RT, Izumi KM, Stevens JG (1988) Localization of a herpes simplex virus neurovirulence gene dissociated from high-titer virus replication in the brain. J Virol 62: 1381–1387
Johnson RT (1964) The pathogenesis of herpesvirus encephalitis. II. A cellular basis for the development of resistance with age. J Exp Med 120: 359–374
Jones KA, Tjian R (1985) Sp1 binds to promoter sequences and activates herpes simplex virus 'immediate-early' gene transcription in vitro. Nature 317: 179–182

Jones KA, Yamamoto KR, Tjian R (1985a) Two distinct transcription factors binds to the HSV thymidine kinase promoter in vitro. Cell 42: 559–572

Jones TR, Parks CL, Spector DI, Hyman RW (1985b) Hybridization of herpes simplex virus DNA and human ribosomal DNA and RNA. Virology 144: 384–397

Kapoor AK, Nash AA, Wildy P, Phelan J, McLean CS, Field HJ (1982) Pathogenesis of herpes simplex virus in congenitally athymic mice: the relative roles of cell-mediated and humoral immunity. J Gen Virol 60: 225–233

Katzin DS, Connor JD, Wilson LA, Sexton RS (1967) Experimental herpes simplex infection in the owl monkey. Proc Soc Exp Biol Med 125: 391–398

Knotts FB, Cook ML, Stevens JG (1974) Pathogenesis of herpetic encephalitis in mice after ophthalmic inoculation. J Infect Dis 130: 16–27

Koch HG, Rösen A, Ernst F, Becker Y, Darai G (1987) Determination of the nucleotide sequence flanking the deletion (0.762–0.789 map units) in the genome of an intraperitoneally avirulent HSV-1 strain HFEM. Virus Res 7: 105–115

Kohl S, Loo LS (1982) Protection of neonatal mice against herpes simplex infection: probably in vivo antibody-dependent cellular cytotoxicity. J Immunol 129: 370–376

Kristensson K (1970) Morphological studies of the neural spread of herpes simplex virus to the central nervous system. Acta Neuropathol (Berl) 16: 54–63

Kristensson K, Wisniewski HM (1978) Arrest of myelination and demyelination in rabbit retina induced by herpes simplex virus infection. Neuropathol Appl Neurobiol 4: 25–36

Kristensson K, Svennerholm B, Persson L, Vahlne A,Lycke E (1979) Latent herpes simplex virus trigeminal ganglionic infection in mice and demyelination in the central nervous system. J Neurol Sci 43: 253–264

Kristensson K, Lycke E, Röyttä M, Svennerholm B, Vahlne A (1986) Neuritic transport of herpes simplex virus in rat sensory neurons in vitro. Effects of substances interacting with microtubular unction and axonal flow [ocodazole, taxol and erythro-9-3(2-hydroxynonyl)-adenine]. J Gen Virol 67: 2023–2028

Kristie TM, Roizman B (1987) Host cell proteins bind to the cis-acting site required for virion mediated induction of herpes simplex virus 1 α genes. Proc Natl Acad Sci USA 84:71–75

Kumano Y, Yamamoto M, Iwasaki M, Ishibashi T, Inomata H, Mori R (1988) Participation of T lymphocyte in corneal edema in the early stage of herpetic stromal keratitis. Ophthalmologica 196: 113–125

Leary K, Connor JR, Morahan PS (1985) Comparison of herpes simplex virus type 1 DNA replication and virus production in murine bone-marrow derived and resident peritoneal macrophages. J Gen Virol 66: 1123–1129

Ligas MW, Johnson DC (1988) A herpes simplex virus mutant in which glycoprotein D sequences are replaced by beta-galactosidase sequences binds to but is unable to penetrate into cells. J Virol 62: 1486–1494

Linnavuori K, Hovi T (1983) Restricted replication of herpes simplex virus in human monocyte cultures: role of interferon. Virology 130: 1–9

Little SP, Schaffer PA (1981) Expresson of the syncitial (syn) phenotype in HSV-1, strain KOS: genetic and phenotypic studies of mutants in two syn loci. Virology 112: 686–702

Locker H, Frenkel N (1979) BamHI, KpnI and SalI restriction enzyme maps of the DNAs of herpes simplex virus strains Justin and F: Occurrence of heterogeneities in defined regions of the viral genome. J Virol 32: 429–441

Lopez C (1975) Genetics of natural resistance to herpesvirus infections in mice. Nature 258: 152–153

Lopez C (1978) Genetic resistance to herpes simplex virus type 1 in the mouse is mediated by the M-cell of allogeneic resistance. Fed Proc 37: 1560

Lopez C (1981) Resistance to herpes simplex virus type 1 (HSV-1). In: Haller O (ed) Natural resistance to tumors and viruses. Springer, Berlin Heidelberg New York, pp 15–24 (Current topics in microbiology and immunology, vol 92)

Lopez C, Dudas G (1979) Replication of herpes simplex virus type 1 in macrophages from resistant and susceptible mice. Infect Immun 23: 432–437

Machuca I, Jaquemont B, Epstein A (1986) Multiple adjacent or overlapping loci affecting the level of gC and cell fusion mapped by intratypic recombinants of HSV-1. Virology 150: 117–125
MacLean AR, Ul-Fareed M, Robertson L, Harland J, Brown SM (1991) Herpes simplex virus type 1 deletion variants 1714 and 1716 pinpoint neurovirulence – related sequences in Glasgow strain 17+ between immediate early gene 1 and the 'a' sequence. J Gen Virol 72: 631–639
Martin X, Dolivo M (1983) Neuronal and transneuronal tracing in the trigeminal system of the rat using the herpes virus suis. Brain Res 273: 253–276
Mayer ML, James MH, Russel RJ, Kelly JS, Wise JCM, Pasternak CA (1985) Spontaneous electrical activity induced by herpes virus infection in rat sensory neuron cultures. Brain Res 341: 360–364
Meignier B, Longnecker R, Roizman B (1988a) In vivo behavior of genetically engineered herpes simplex viruses R7017 and R7020: construction and evaluation in rodents. J Infect Dis 158: 602–614
Meignier B, Longnecker R, Mavromara-Nazos P, Sears A, Roizman B (1988b) Virulence of and establishment of latency by genetically engineered deletion mutants of herpes simplex virus I. Virology 162: 251–254
Meignier M, Martin B, Whitley RJ, Roizman B (1990) In vivo behavior of genetically engineered herpes simplex viruses R7017 and R7020 II. Studies in immunocompetent and immunosuppressed owl monkeys (Aotus Trivirgatus). J Infect Dis 162: 313–321
Metcalf JF, Kaufman HE (1976) Herpetic stromal keratitis – evidence for cell mediated immunopathogenesis. Am J Ophthalmol 82: 827–834
Metcalf JF, Hamilton DS, Reichert RW (1979) Herpetic keratitis in athymic (nude) mice. Infect Immun 26: 1164–1171
Mogensen SC (1979) Role of macrophages in natural resistance to virus infections. Microbiol Rev 43: 1–26
Morahan PS (1984) Interactions of herpesviruses with mononuclear phagocytes. In: Rouse BT, Lopez C (eds) Immunobiology of herpes simplex infection. CRC Press, Boca-Raton, pp 71–89
Morahan PS, Connor JR, Leary KR (1985) Viruses and the versatile macrophage. Br Med Bull 41: 15–21
Morahan PS, Mama S, Anaraki F, Leary K (1989) Molecular localization of abortive infection of resident peritoneal macrophages by herpes simplex virus type 1. J Virol 63: 2300–2307
Moyal M, Raibstein I, Rösen A, Darai G, Becker Y (1988) Heterogeneity of BamHI DNA fragments B and E in several HSV-1 strains and recombinants. Virus Genes 1: 165–174
Moyal M, Berkowitz C, Rosen-Wolff A, Darai G, Becker Y (1992) Mutations in U_L53 gene of HSV-1 abolish viral neurovirulence to mice by the intracerebral route of infection. Virus Res 26: 99–112
Nachtigal M, Caulfield JB (1984) Early and late pathologic changes in the adrenal glands of mice after infection with herpes simplex virus type 1. Am J Pathol 115: 175–185
Nahmias AJ, Hirsch MS, Kramer JH, Murphy FA (1969) Effect of antithymocyte serum on herpesvirus hominis (type 1) infection in adult mice. Proc Soc Exp Biol Med 132: 696–698
Oakes JE, Gray WL, Lausch RN (1986) Herpes simplex virus type 1 DNA sequences which direct spread of virus from cornea to central nervous system. Virology 150: 513–517
Parks CL, Jones TR, Gonzalez IL, Schmickel RD, Hyman RW, Spector DJ (1986) A simple repetitive sequence common to herpes simplex virus type 1 and human ribosomal DNA's. Virology 154: 381–388
Peles E, Rosen H, Darai G, Rosen-Wolff A, Becker Y (1990) Importance of the HpaI-P sequence for herpes simplex virus-1 replication in the adrenal glands. Arch Virol 113: 151–163
Pellet PE, McKnight JLC, Jenkins FJ, Roizman B (1985) Nucleotide sequence and predicted amino acid sequence of a protein encoded in a small herpes simplex virus DNA fragment capable of trans-inducing a genes. Proc Natl Acad Sci USA 82: 5870–5874

Pogue-Geile KL, Spear PG (1987) The single base pair substitution responsible for the syn phenotype of herpes simplex virus type 1, strain MP. Virology 157: 67–74
Post LE, Conley AI, Mocarsky ES, Roizman B (1980) Cloning of reiterated and nonreiterated herpes simplex virus 1 sequences as BamHI fragments. Proc Natl Acad Sci USA 77: 4201–4205
Potratz D, Brake B, Dienes HP, Schulz TF, Hosp M, Dierich MP, Falke D (1987) Herpes simplex virus type 1 and 2 in the adrenal glands: replication and histopathology. Arch Virol 90: 207–222
Price RW (1979) 6-Hydroxydopamine potentiates acute herpes simplex virus infection of the superior cervical ganglion in mice. Science 205: 518–520
Price RW, Kahn A (1981) Resistance of peripheral autonomic neurons to in-vivo productive infection by herpes simplex virus mutants deficient in thymidine kinase activity. Infect Immun 34: 571–580
Price RW, Schmitz J (1979) Route of infection, systemic host resistance, and integrity of ganglionic axons influence acute and latent herpes simplex virus. Infect Immune 23: 373–383
Rager-Zisman B, Alllison AC (1976) Mechanism of immunologic resistance to herpes simplex virus 1 (HSV-1). J Immunol 116: 35–40
Rixon FJ, Campbell ME, Clements JB (1984) A tandemly reiterated DNA sequence in the long repeat region of herpes simplex virus type 1 found in close proximity to immediate-early mRNA. J Virol 52: 715–718
Rock DL, Fraser NW (1983) Detection of HSV-1 genome in central nervous system of latently infected mice. Nature 302: 523–525
Rock DL, Nesburn AB, Ghiasi H, Ong J, Lewis TL, Lokensgard JR, Wechsler SL (1987) Detection of latency-related viral RNAs in trigeminal ganglia of rabbits latently infected with herpes simplex virus type 1. J Virol 61: 3820–3826
Rösen A, Darai G (1985) Mapping of the deletion in the genome of HSV-1 strain HFEM responsible for its avirulent phenotype. Med Microbiol Immunol 173: 329–343
Rösen A, Gelderblom H, Darai G (1985) Transduction of virulence in herpes simplex virus type 1 from a pathogenic to an apathogenic strain by a cloned viral DNA fragment. Med Microbiol Immunol 173: 257–278
Rösen A, Ernst F, Koch HG, Gelderblom H, Darai G, Hadar J, Tabor E, Ben-Hur T, Becker Y (1986) Replacement of the deletion in the genome (0.762–0.789 mu) of avirulent HSV-1 HFEM using cloned Mlul DNA fragment (0.7615–0.796 mu) of virulent HSV-1 F leads to generation of virulent intratypic recombinant. Virus Res 5: 157–175
Rösen-Wolff A, Ben-Hur T, Becker Y, Darai G (1988) Comparative analysis of the transcripts mapped in the BamHI DNA fragment B of the avirulent HSV-1 HFEM, virulent HSV-1 F, and their intratypic recombinant viruses. Virus Res 10: 315–329
Ruyechan WT, Morse LS, Knipe DM, Roizman B (1979) Molecular genetics of herpes simplex virus. II. Mapping of the major viral glycoproteins and of the genetic loci specifying the social behavior of infected cells. J Virol 29: 677–697
Schlitt M, Bucher AP, Stroop WG, Pindak F, Bastian FO, Jennings RA, Lakeman AD, Whitley RJ (1988) Neurovirulence in an experimental focal herpes encephalitis: relationship to observed seizures. Brain Res 440: 293–298
Sears AE, Halliburton IW, Meignier B, Silver S, Roizman B (1985) Herpes simplex virus 1 mutant deleted in the α22 gene: growth and gene expression in permissive and restrictive cells and establishment of latency in mice. J Virol 55: 338–346
Sedarati F, Javier RT, Stevens JG (1988) Pathogenesis of a lethal mixed infection in mice with two nonneuroinvasive herpes simplex virus strains. J Virol 62: 3037–3039
Sequiera LW, Jennings LC, Carrasso LH, Lord MA, Curry A, Sutton RN (1979) Detection of herpes simplex virus genome in brain tissue. Lancet 2: 609
Sethi KK, Brandis H (1979) In vitro acquisition of resistance against herpes simplex virus by permissive murine macrophages. Arch Virol 59: 157–172
Shimeld C, Tullo AB, Hill TJ, Blyth WA, Easty DL (1985) Spread of herpes simplex virus and distribution of latent infection after intraocular infection of the mouse. Arch Virol 85: 175–187
Sit MF, Tenney DJ, Rothstein JL, Morahan PS (1988) Effect of macrophage activation on

resistance of mouse peritoneal macrophages to infection with herpes simplex virus types 1 and 2. J Gen Virol 69: 1999–2010
Spivack JG, Fraser NW (1987) Detection of herpes simplex virus type 1 transcripts during latent infection in mice. J Virol 61: 3841–3847
Sprecher E, Becker Y (1987) Herpes simplex virus type 1 pathogenicity in footpad and ear skin of mice depend on langerhans cells density, mouse genetics and virus strain. J Virol 61: 2515–2522
Sprecher E, Becker Y (1990) Interleukin-1 alpha and beta gene expression in mouse peritoneal exudate cells after intraperitoneal infection with herpes simplex virus-1. Arch Virol 110: 259–269
Stanley ER, Heard PM (1977) Factors regulating macrophage production and growth. Purification and some properties of the colony stimulating factor from medium conditioned by mouse L-cells. J Biol Chem 252: 4305–4312
Steiner I, Spivack JG, Lirette RP, Brown S, MacLean AR, Subak-Sharp JH, Fraser NW (1989) Herpes simplex virus type 1 latency – associated transcripts are evidently not essential for latent infection. EMBO J 8: 505–511
Stevens JG, Cook ML (1971) Restriction of herpes simplex virus by macrophages. J Exp Med 133: 19–38
Stroop WG, Schaefer DC (1989) Neurovirulence of two clonally related herpes simplex virus type 1 strains in a rabbit seizure model. J Neuropathol Exp Neurol 48: 171–183
Sung SC (1971) Thymidine kinase in the developing rat brain. Brain Res 35: 268–271
Sunstrum JC, Chrisp CE, Levine M, Glorioso JC (1989) Pathogenicity of glycoprotein C negative mutants of herpes simplex virus type 1 for the mouse central nervous system. Virus Res 11: 17–32
Tenser RB, Dunstan ME (1979) Herpes simplex virus thymidine kinase expression in infection of the trigeminal ganglion. Virology 99: 417–422
Tenser RB, Miller RL, Rapp F (1979) Trigeminal ganglion infection by thymidine kinase negative mutants of herpes simplex virus. Science 205: 915–917
Tenser RB, Hay KA, Edris WA (1989) Latency associated transcript but not reactivable virus is present in sensory ganglion neurons after inoculation of thymidine kinase – negative mutants of herpes simplex virus type 1. J Virol 63: 2861–2865
Thompson HW, Dupuy B, Beuerman RW, Hill JM, Rootman DS, Haruta Y (1988) Neural activity from the acutely infected HSV-1 rabbit cornea. Curr Eye Res 7: 147–155
Thompson RL, Stevens JG (1983) Biological characterization of a herpes simplex virus intertypic recombinant which is completely and specifically non-neurovirulent. Virology 131: 171–179
Thompson RL, Wagner EK (1989) Partial rescue of herpes simplex virus neurovirulence with a 3.2 kb cloned DNA fragment. Virus Genes 1: 261–273
Thompson RL, Wagner EK, Stevens JG (1983) Physical location of a herpes simplex virus type 1 gene function (s) specifically associated with a 10 million fold increase in HSV neurovirulence. Virology 131: 180–192
Thompson RL, Devi-Rao GV, Stevens JG, Wagner EK (1985) Rescue of a herpes simplex virus type 1 neurovirulence function with a cloned DNA fragment. J Virol 55: 504–508
Thompson RL, Cook ML, Devi-Rao GB, Wagner EK, Stevens JG (1986) Functional and molecular analyses of the avirulent wild type herpes simplex virus type 1 strain KOS. J Virol 58: 203–211
Thompson RL, Rogers SK, Zerhusen MA (1989) Herpes simplex virus neurovirulence and productive infection of neural cells is associated with a function which maps between 0.82 and 0.832 map units on the HSV genome. Virology 172: 435–450
Townsend JJ, Baringer JR (1978) Alteration of herpes simplex induced CNS lesions by immunosuppression. J Neuropathol Exp Neurol 37: 255–262
Tullo AB, Keen P, Blyth WA, Hill TJ, Easty DL (1983) Corneal sensitivity and substance P in experimental herpes simplex keratitis in mice. Invest Opthalmol Vis Sci 24: 596–598
Tushinsky RJ, Stanley ER (1985) The regulation of mononuclear phagocyte entry into S phase by the colony-stimulating factor CSF-1. J Cell Physiol 122: 221–228
Wagner EK, Devi-Rao G, Feldman LT, Dobson AT, Zhang Y, Flanagan WM, Stevens JG

(1988) Physical characterization of the herpes simplex virus latency – associated transcript in neurons. J Virol 62: 1194–1202
Waheed A, Shadduck RK (1979) Purification and properties of L-cell-derived colony stimulating factor. J Lab Clin Med 94: 180–194
Warren KG, Brown SM, Wroblewska Z, Gilden D, Koprowski H, Subak-Sharp J (1978) Isolation of latent herpes simplex virus from the superior cervical and vagus ganglions of human beings. N Eng J Med 298: 1068–1070
Weichsel ME (1974) Effect of thyroxine on DNA synthesis and thymidine kinase activity during cerebellar development. Brain Res 78: 455–465
Wille W, Willecke K (1980) Stimulation of DNA synthesis in cultured mouse peritoneal macrophages after fusion injection of macrophage growth factor. Exp Cell Res 130: 95–99
Wollert K, El-Ansari T, Rösen-Wolff A, Falke D (1991) The region 0.7615–0.796 m.u. of the HSV-1 genome determines suppression of humoral antibody formation against herpes simplex virus. Arch Virol 116: 175–184
Yamada M, Uno F, Nii S (1986) In vitro cytopathology and pathogenicity to inbred mice shown by five variants of a laboratory strain of type 1 herpes simplex virus. Arch Virol 90: 183–196
Yamada Y, Kimura H, Morishima T, Daikoku T, Maeno K, Nishiyama Y (1991) The pathogenicity of ribonucleotide reductase – null mutants of herpes simplex virus type 1 in mice. J Infect Dis 164: 1091–1097
Yamagami S, Mori K, Kawakita Y (1972) Changes of thymidine kinase in the developing rat brain. J Neurochem 19: 369–376
Zawatzky R, Hilfenhaus J, Kirchner H (1979) Resistance of nude mice to herpes simplex virus and correlation with in vitro production of interferon. Cell Immunol 47: 424–428
Zawatzky R, Engler H, Kirchner H (1982) Experimental infection of inbred mice with herpes simplex virus III. Comparison between newborn and adult C57Bl/6 mice. J Gen Virol 60: 25–29
Zisman B, Hirsch MS, Allison AC (1970) Selective effects of anti-macrophage serum, silica, and anti-lymphocyte serum on pathogenesis of herpes virus infection of young adult mice. J Immunol 104: 1155–1159

Chapter 22 Live Herpesvirus Vaccines: Serendipity or Engineering of the Virus Genome

Yechiel Becker

Summary

The concept that a live attenuated virus can be used for immunization of humans to protect against a virus disease caused by a pathogenic wild type was set forward by Edward Jenner about 200 years ago. In this century, the search for apathogenic virus isolates led to the development of live virus vaccines such as the yellow fever 17D vaccine virus and Sabin's polioviruses for vaccination of humans against poliomyelitis. Apathogenic herpesvirus of turkeys and an apathogenic Marek's disease virus-1 (Rispens) were found to immunize chickens against Marek's disease of poultry. Recently, a naturally occurring varicella zoster virus (VZV) isolate (OKa strain) came under study for its usefulness as a human vaccine. Developments in molecular virology and genetic engineering and studies on virus pathogenicity genes of the poxvirus vaccinia and herpesviruses has led to the development of a new generation of virus recombinants for human vaccination. These virus recombinants may carry foreign virus genes to protect against a number of human virus pathogens. The usefulness as potential vaccines of herpesvirus recombinants from which the pathogenicity genes were deleted is discussed.

Introduction

The concept for the search for apathogenic human virus vaccines was set forward by Edward Jenner about 200 years ago, when he demonstrated that a virus from cowpox could protect humans against smallpox by variolation. During the classical period of virology, the first 50–60 years of the twentieth century, live attenuated virus vaccines were developed, mainly Theiler's yellow fever 17D and Sabin's poliovirus vaccine. In the last 30 years live vaccine viruses were developed against Marek's disease (MD) in poultry, caused by Marek's disease virus (MDV), using naturally occurring herpesvirus of turkeys (HVT) (Churchill et al. 1969) and a naturally occurring isolate of apathogenic MDV-1 (Rispens et al. 1972) which immunizes chickens against MD. Recently, the OKa strain of varicella zoster virus

Department of Molecular Virology, Faculty of Medicine, The Hebrew University of Jerusalem, Jerusalem, Israel

(VZV) (Takahashi et al. 1985) was found to be highly attenuated and capable of protecting humans from VZV-caused diseases (see below). In contrast to the search for naturally occurring apathogenic viruses there is the concept that genetic engineering technology can be used to modify the herpesvirus genome by deletion of genes which are responsible for virus pathogenicity (designated pathogenicity genes) and are not essential for virus replication. At the sites of the deleted virus genes in the genome, foreign virus genes coding for structural proteins can be inserted. Virus recombinants of vaccinia virus, e.g., recombinant vaccinia virus carrying the VZV gI gene (Ludvikova et al. 1991), express an array of viral structural proteins and may produce humoral and cellular responses in the immunized individual against a number of pathogenic viruses. Such virus recombinants were developed on the basis of the vaccinia virus genome. It was assumed that since vaccinia virus is a human virus vaccine there is no need to remove any pathogenicity genes. Yet, Buller et al. (1988) reported that deletion of a growth factor gene from vaccinia virus resulted in further attenuation of the pathogenicity of vaccinia virus to mice. The need to characterize pathogenicity genes in vaccinia virus has already been discussed elsewhere (Becker 1987). In animals, recombinant vaccines of the herpesvirus Suid I (pseudorabies virus) were developed (Kit 1990; van Oireschot et al. 1990; Post et al. 1990; Sedegah et al.1992).

The aim of the present analysis is to evaluate the approach of using an apathogenic human virus-like VZV OKa isolate that was found by serendipity and which is being used as a live virus vaccine. There has been minimal analysis of its properties, in contrast to the genetic engineering approach of developing an apathogenic herpes simplex-1 virus (HSV-1), which involves indentifying its pathogenicity genes as candidates for deletion and insertion sites of foreign viral glycoprotein genes.

Varicella Zoster Virus Live Virus Vaccine (OKa Strain)

Genomic Organization

Hayakawa et al. (1986) studied the restriction enzyme pattern of cleaved viral DNA profiles of OKa vaccine virus after 10 and 85 passages in human embryo cells. The OKa virus differed in the *Hpa*I-F fragment of the viral DNA when compared to wild-type VZV isolates. The authors concluded that three variable fragments (*Hpa*I-K, -F, and -G) are present in the OKa virus DNA as compared to wild-type virus isolates. On in vitro passaging of the OKa isolate and after human transmission, the *Hpa*I-K DNA fragment was stable but was different from the pattern among all wild-type viruses isolated from unrelated virus outbreaks. The *Hpa*I-F DNA fragment was unstable on passage in vitro and in human infections by either vaccine or wild-type strains. The HpaI-G fragment was not influenced by in vitro passage of the virus. The genomic changes in VZV DNA were localized to coordinate 0.16 (*Hpa*I-K), 0.35 (*Hpa*I-F) and near the 3′end of the DNA genome (*Hpa*I-G). Comparing the changes in OKa virus DNA to the genomic map determined by Davison and Scott (1986) it was possible to note that the change at 0.16 map units corresponds to a DNA fragment, *Hpa*I-K, that contains gene 13 (thymidine kinase). The change at 0.35 (*Hpa*I-F) contains genes 23, 24, 25. The right-hand terminal contains gene 71, a homologue of the immediate-early gene of HSV-1 (1E-175). The study by Hayakawa et al. (1986) clearly showed that the VZV DNA genome in OKa vaccine virus contains changes that reduce its pathogenicity. However, the sequence of OKa virus DNA at the changed sites is not yet known.

Varicella Virus Vaccines

Takahashi et al. (1985) and Tsolia et al. (1990) reported on the use of the OKa strain of VZV in clinical trials in normal and high risk children in Japan, the USA and Europe. The vaccine gave good results, leading to the licensing of the virus as a human virus vaccine. Bergen et al. (1990) reported on the use of the OKa/Merck varicella vaccine. Children who were given 950 pfu of the live vaccine virus had an initial seroconversion rate of 87% but VZV cell-mediated immunity was diminished at 8 weeks. The current vaccine preparations elicit cellular and humoral immune responses to VZV antigen, and memory T lymphocytes specific for major viral proteins were detectable 1 year after immunization. Dennehy et al. (1991) compared the immunogenicity and safety of varicella vaccine by subcutaneous or intramuscular injection. The results of the study revealed that vaccination by both routes resulted in seroconversion. In both groups of vaccinees 21% complained of reactions at the injection site and 7% had a varicella-like rash. Kuter et al. (1991) reported on the results of a large, double-blind, randomized placebo-controlled trial with live attenuated OKa/Merck varicella vaccine in healthy children. The efficiency of the vaccine was 98%. After the breaking of the code, the vaccine group was followed for the development of varicella. Some 95% of the vaccinees were varicella-free for the 7 years of follow-up. The persistence of antibody was 100% after 6 years. Clinical studies with the varicella vaccine (VARIVAX) was reported by White et al. (1991). The vaccination resulted in 96% efficacy and 99% of the vaccinees maintained antibodies at 1 year. The incidence of varicella following household exposure in vaccinees was approximately 12% compared to an 87% infection incidence in nonvaccinated household contacts. Nearly all the vaccinees who developed varicella had a clinically modified disease. Hardy et al. (1991) reported on a study of leukemic children immunized with the OKa strain of varicella and found that the subsequent zoster incidence was lower in vaccinees than in nonvaccinated leukemic children. Transmission of the live attenuated varicella virus from vaccinated children with leukemia was studied. It was found that the vaccine virus was released from skin lesions of vaccinees and infected varicella-susceptible close contacts by the respiratory tract route. Yet, the mildness of the illness in contacts suggested to the authors that the vaccine virus is attenuated and reversion of the virus to virulence was not found. Diaz et al. (1991) studied the safety of administering the live attenuated OKa/Merck varicella vaccine to siblings of children with malignancies. No evidence of vaccine virus transmission was found in the immunocompromised children. Hardy and Gershon (1990) and Gershon and Steinberg (1990) suggested that the population of

VZV-negative adults, which is 5%–10% of the adult population, is at risk of being infected with VZV and their primary disease is more severe than that in children. The authors concluded that vaccination of this subpopulation with the live attenuated varicella vaccine might be most beneficial.

OKa Varicella Vaccine Expressing Hepatitis B Virus Surface Antigen

Shiraki et al. (1991) inserted part of the hepatitis B surface antigen gene into the varicella virus OKa virus thymidine kinase gene by cotransfection of the virus genome together with a bacterial plasmid carrying the hepatitis B virus surface antigen (HBS) gene. A recombinant OKa virus carrying the portion of the HBS gene was isolated. The authors suggested that OKa varicella vaccine which expresses hepatitis B may be a good candidate for a combined hepatitis B virus and VZV vaccine.

Properties of Varicella Zoster Virus Attenuated Vaccine

The studies on the attenuated VZV OKa isolate suggested that this virus is a mutant of the pathogenic varicella virus, with mutations that were identified by restriction enzyme analysis and not by nucleotide sequencing. However, the virus was found to have properties of attenuated VZV when injected into vaccinees intradermally or intramuscularly. In some vaccinees the virus produces a mild disease which can then spread from the skin lesion of the vaccinee to contacts. Once airborne, the virus can infect through entry into the respiratory tract, similar to the infection with wild-type VZV. However, infection in household contacts results in mild varicella infections. It was clearly shown that the vaccine virus may maintain its pathogenicity genes. Since the immune status of the vaccinee may lead to better or worse protection against the vaccine virus, it is necessary to define which pathogenicity genes are still operative in the live VZV vaccine virus. The degree of seroconversion in adults and leukemic children is very high. In these studies the ability of the vaccine virus to cause latent infections was not studied for obvious reasons. However, the ability of VZV to cause latent infections and the conditions for reactivation of the attenuated vaccine virus are not known. Since the human population is now composed of healthy individuals, a growing number of HIV-1-infected and AIDS patients, and individuals with conditions that decrease the cellular immune response, special care to eliminate the residual pathogenicity of the VZV live vaccine virus is needed.

Cytomegalovirus Live Vaccine

Towne Strain of CMV as a Vaccine in Solid Organ Transplant Patients

Cytomegalovirus is a human pathogen which can be reactivated in immunosuppressed transplant patients, causing a life-threatening infection (see Becker et al. 1993, devoted to CMV and CMV diseases). Starr et al. (1991) and Plotkin et al.

(1991) evaluated the properties of the live attenuated CMV Towne strain in healthy volunteers and renal transplant patients. Both groups responded with humoral and cell-mediated immune responses. In immunized renal transplant patients the incidence of moderately severe and severe CMV diseases due to reactivation of the virus was significantly reduced compared to unimmunized individuals.

Live CMV Vaccine in the Prevention of Congenital Disease

Porath et al. (1990) noted that, per 100 000 pregnant women per year, there are 24 cases of symptomatic congenital CMV infections at birth and a similar number of cases with late sequelae (mainly deafness). If seronegative women were selectively immunized, newborn infants exposed to CMV would be protected. Such immunization could lead to savings that exceed the expenses of administering the live VZV vaccine to pregnant women.

Vaccination with CMV Live Virus

Research on the development of live attenuated CMV vaccines is lagging behind studies on vaccination with live attenuated VZV OKa virus vaccine. However, the promising results from the varicella vaccine studies provide direction for the development of a live attenuated CMV vaccine. The latency properties of the CMV Towne strain live vaccine are not known and information is lacking on its ability to reactivate in immunosuppressed individuals, e.g., those who develop AIDS after infection with HIV-1 or who have undergone treatment with immune suppressive drugs for solid organ transplantation (Merigan and Resta 1990). The parameters of CMV attenuation need to be investigated to allow development of a safe CMV vaccine.

Can Attenuated Live Herpes Simplex Virus be Used as a Human Vaccine?

The Approach to the Study of HSV-1 Pathogenicity Genes

In contrast to VZV and CMV, which infect humans through the upper respiratory tract, HSVs infect through the mucus membranes. Furthermore, serendipity was not on the side of HSV researchers and naturally occurring apathogenic HSV-1 isolates were not discovered. HSV-1 is a human pathogen that causes infections in early childhood and the virus becomes latent in the trigeminal ganglia and in the brain; however, live attenuated vaccine viruses have not been developed. Similarly, an HSV-2 virus (which causes genital infections) with attenuated properties was not found. Mutant viruses, e.g., thymidine kinase (TK)-negative

Table 1. Genomic localization of HSV-1 pathogenicity genes

Name	ORF	Position
Thymidine-kinase (TK)	UL23	47802–46674
Glycoprotein-B (gB)	UL27	50809–52714
Ribonucleotide reductase (α)	UL39	86444–89855
Ribonucleotide reductase (β)	UL40	89926–90946
Virus host shutoff	UL41	92637–91170
Glycoprotein-C (gC)	UL44	96311–97844
Cell fusion protein	UL53	112179–113193
Intraperitoneal pathogenicity	UL56	116923–116264
LAT (latency-associated)		119461– ----
Glycoprotein-D (gD)	US6	138419–139601

ORF, open reading frame.

HSV-1 mutants were generated in laboratories and it was shown that such mutants were less (or not) pathogenic in mice. A reduction in HSV-1 pathogenicity was noted in mice when TK-negative HSV-1 mutants were used for infections (Ben-Hur et al. 1983). However, it was demonstrated that newborn mice, which express cellular TK in their developing brains, were highly sensitive to the HSV-1 TK-negative mutants (Ben Hur et al. 1983). Thus, HSV-1 TK-negative mutants still remained a threat to newborn and young children.

Together with G. Darai an approach to creating an attenuated HSV-1 was developed from the idea that it is necessary to identify those virus genes coding for tissue-specific viral proteins which damage the infected organs in HSV-1-mediated diseases. Studies from many laboratories (reviewed by T. Ben-Hur, this volume) indicated that several genes in HSV-1 are involved in virus pathogenicity (Fig. 1, Table 1). Removal of such pathogenicity genes from the virus genome may lead to the development of an apathogenic virus which will lack pathogenicity genes and which will not be able to become latent in vaccinees. Yet, if the HSV-1 TK gene is mutated to 5%–10% level of expression, the attenuated property of the virus may not be changed and a low level of the TK enzyme will be produced in infected cells. These infected cells will retain the virus' susceptibility to treatment with anti-herpetic drugs like acyclovir. This form of HSV-1, attenuated by attenuation of one pathogenicity gene (TK) and removal of all other pathogenicity genes, may be used as a vaccine in humans. As a safeguard against possible uncontrolled replication of the HSV-1 live vaccine virus in immunosuppressed individuals, the residual cell activity will allow treatment with acyclovir. In this respect an engineered HSV-1 live apathogenic vaccine virus may have marked advantages over other recombinant vaccine viruses based on the vaccinia virus.

HSV-1 Pathogenicity Genes

The position of the genes which are regarded as HSV-1 pathogenicity genes is presented in Fig. 1. Ten genes are thought to be involved in virus pathogenicity to mice:

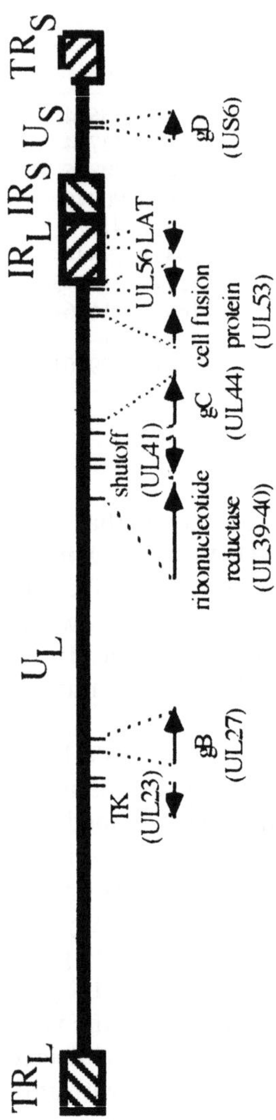

Fig. 1. HSV-1 genome and the location of genes for pathogenicity. The coordinates of the genes in HSV-1 genome are indicated in Table 1

1. The ability of the virions to interact with and enter into the infected cells is partly due to the structural virus glycoprotein genes *gB*, *gC*, and *gD* (Table 1). These structural genes are involved in the initiation of infection and determination of the virus' ability to enter into specific tissues (e.g., nerve endings).

2. The *UL53* gene (Table 1) codes for a protein which is involved in cell fusion: The UL53 protein is needed for the intracerebral pathogenicity of the virus when inoculated into mouse brain. Mutants that express mutated UL53 products can replicate in the infected mouse brain but mice do not die as a result of virus infection (Ben Hur et al. 1987; Moyal et al. 1992).

3. The *UL56* gene (Rösen-Wolff and Darai, and Grifman and Becker, this volume) is responsible for the ability of HSV-1 to cause intraperitoneal disease in tree shrews and mice after intraperitoneal inoculation. The infection of the mouse adrenal glands by the virus is dependent on the activity of *UL56* (Peles et al. 1990).

4. The HSV-1 genes coding for the ribonucleotide reductase (RR) subunits (*UL39* and *UL40*) and for the thymidine kinase (*UL23*) were found to be critical for HSV-1 replication in neural tissues.

5. A virus host shutoff gene, *UL41,* codes for a protein which is present in released virions and which is capable of inhibiting cellular mRNA utilization in infected cells.

6. The *latency-associated gene* (LAT) is expressed in latent ganglia but its role in HSV-1 latency is still undecided.

Is an Engineered HSV-1 DNA Genome Lacking the Pathogenicity Gene a Possible Candidate for a Live Virus Vaccine?

Characterization of pathogenicity genes of HSV-1 provides us with information on the connection between viral genes and the disease-causing mechanisms displayed by virus infection of different tissues and organs in the mouse. Genetic engineering techniques allow removal of critical pathogenicity genes from the HSV-1 genome and their replacement with genes coding for coat proteins from VZV,CMV or other pathogenic viruses. Such virus recombinants may induce humoral and cellular immunity when inoculated into the skin of a vaccinee, directed not only against HSV-1/HSV-2 but also against additional pathogenic viruses. The residual TK activity due to a TK gene with a low expression level will act as a safeguard against virus replication, since the latter will be inhibited by available anti-HSV drugs.

A Herpesvirus Cytokine Gene with a Potential for Virus Pathogenicity

A recent study by Rode et al. (1993) showed that the equine herpesvirus-2 DNA genome contains an interleukin-10 (IL-10) gene homologous to the cellular IL-10 gene in Epstein-Barr virus (EBV). Fiorentino et al. (1991) showed that the IL-10 cytokine can shut off the expression of cytokines in macrophages, thus preventing macrophages from defending the organism against pathogenic viruses. The presence of genes in herpesviruses which produce viral proteins capable of interfering with the cellular immune responses suggests that additional virus genes may be involved in disease processes i.e., those which counteract the immune system. Discovery and elimination of such genes may lead to further improvement in the development of live herpesvirus vaccines to protect humans and livestock.

Conclusions

Using the live VZV OKa vaccine, immunization against VZV herpesvirus-mediated diseases is now possible, not only in healthy individuals but also in immunosuppressed leukemic children. However, more information on the VZV genes which were changed in the OKa virus genome and their role in the attenuation of the virus is needed to understand the reasons for VZV attenuation. A similar approach should be taken to develop CMV live attenuated virus vaccine.

The strategy that we developed together with Prof. G. Darai was to map and characterize the pathogenicity genes of HSV-1. Removal of the pathogenicity genes from HSV-1 genome serves as a rationale for the development of a fully attenuated recombinant HSV-1. It seems that with the available information on HSV-1 genes it may be possible to genetically engineer an apathogenic HSV-1 vaccine virus, which may be useful in vaccination of children and adults to prevent primary HSV-1 infection and with the hope of eliminating HSV-1 latency and recurrences. Using the engineered HSV-1 genome for inclusion of foreign virus genes may add to the attractiveness of this approach to the development of a multivirus protective HSV-1 vaccine for humans use.

Acknowledgements. The author is grateful to members of the Department of Molecular Virology, Yael Asher, Eynat Tabor, Michal Moyal, Mirta Grifman, Cheryl Berkowitz, Eli Sprecher and Tamir Ben-Hur, for their excellent collaboration in the research on HSV-1 pathogenicity genes. This research was done in collaboration with Prof. G. Darai and Dr. Angela Rösen-Wolff of Heidelberg University and their collaborators. The constant support from the Foundation for the Study of Molecular Virology and Cell Biology, President Mrs. Ronnie Bendheim, Phoenix, Arizona, USA, is greatly appreciated.

References

Becker Y (1987) The need for characterization of pathogenicity genes of recombinant DNA viruses used as human vaccines. Virus Genes 1: 117–120

Becker Y, Darai G, Huang E-S, (eds) (1993) Molecular aspects of human cytomegalovirus diseases (Frontiers of Virology vol 2). Springer, Berlin Heidelberg New York

Ben Hur T, Hadar J, Shtram Y, Gilden DH Becker Y. (1983) Neurovirulence of Herpes simplex type 1 depends on age in mice and thymidine kinase expression. Arch Virol 78: 303–308

Ben Hur T, Asher Y, Tabor E, Darai G, Becker Y (1987) HSV-1 virulence for mice by the intracerebral route is encoded by the BamHI-L DNA fragment containing the cell fusion gene Arch Virol 96: 117–122.

Bergen RE, Diaz PS, Arvin AM (1990) The immunogenicity of the OKa/Merck varicella vaccine in relation to infectious varicella-zoster virus and relative viral antigen content. J Infect Dis 162: 1049–1054
Buller RML, Chakrabarti S, Cooper JA, Twardzik DR, Moss B (1988) Deletion of the vaccinia virus growth factor gene reduces virus virulence. J Virol 62: 866–874
Churchill AE, Payne LN, Chubb RC (1969) Immunization against Marek's disease using a live attenuated virus. Nature 221: 744–747
Davison AJ, Scott JE (1986) The complete DNA sequence of varicella-zoster virus. J Gen Virol 67: 1759–1816
Dennehy PH, Reisinger KS, Blatter MM, Veloudis BA (1991) Immunogenicity of subcutaneous versus intramuscular OKa/Merck varicella vaccination in healthy children. Pediatrics 88: 604–607
Diaz PS, Au D, Smith S, Amylon M, Link M, Smith S, Arvin AM (1991) Lack of transmission of the live attenuated varicella vaccine virus to immunocompromised children after immunization of their siblings. Pediatrics 87: 166–70
Fiorentino DF, Zlotnik A, Nosmann TR, Howard M, O'Garrra A (1991) IL-10 inhibits cytokine production by activated macrophages. J Immunol 147: 3815–3822.
Gershon AA, Steinberg SP (1990) Live attenuated varicella vaccine: protection in healthy adults compared with leukemic children National Institute of Allergy and Infectious Diseases Varicella Vaccine Collaborative Study Group. J Infect Dis 161: 661–666.
Hardy I, Gershon AA (1990) Prospects for use of a varicella vaccine in adults. Infects Dis Clin North Am 4: 159–173
Hardy I, Gershon AA, Steinberg SP, LaRussa P (1991) The incidence of zoster after immunization with live attenuated varicella vaccine. A study in children with leukemia. Varicella Vaccine Collaborative Study Group. N Engl J Med 325: 1545–1550
Hayakawa Y, Yamamoto T, Yamanishi K, Takahashi M (1986) Analysis of Varicella-Zoster virus DNAs of clinical isolates by endonuclease HpaI. J Gen Virol 67: 1817–1829.
Kit S (1990) Genetically engineered vaccines for control of Aujeszky's disease Vaccine 8: 420-424
Kuter BJ, Weibel RE, Guess HA, Matthews H, Morton DH, Neff BJ, Provost PJ, Watson BA, Starr SE, Plotkin SA (1991) OKa/Merck varicella vaccine in healthy children: final report of a 2-year efficacy study and 7-year follow-up studies. Vaccine 9: 643–647.
Ludvilova V, Kunke D, Hamsikova E, Kutinova L, Vonka V (1991) Immunogenicity in mice of varicella-zoster glycoprotein I expressed by a vaccinia virus-varicella-zoster virus recombinant. J Gen Virol 72: 1445–1449
Merigan TC, Resta S (1990) Cytomegalvirus: where have we been and where are we going? Rev Infect Dis 12: 5693–5700
Moyal M, Berkowitz C, Rösen-Wolff A, Darai G, Becker Y (1992) Mutations in the UL53 gene of HSV-1 abolish virus neurovirulence to mice by the intracerebral route of infection. Virus Res 26: 99–112
Peles E, Rosen H, Darai G, Rösen-Wolff A, Becker Y, (1990) Importance of the HpaI-P sequence of herpes simplex virus-1 replication in the adrenal glands. Arch Virol 113: 151–163
Plotkin SA, Starr SE, Friedman HM, Brayman K, Harris S, Jackson S, Tustin NB, Grossman R, Dafoe D, Barker C (1991) Effect of Towne live virus vaccine on cytomegalovirus disease after renal transplant. A controlled trial. Ann Intern Med 114: 525–531
Porath A, McNutt RA, Smiley LM, Weigle KA (1990) Effectiveness and cost benefit of a proposed live cytomegalovirus vaccine in the prevention of congenital disease. Rev Infect Dis 12: 31–40
Post LE, Thomsen DR, Petrovskis EA, Meyer AL, Berlinski PJ, Wardley RC (1990) Genetic engineering of the pseudorabies virus genome to construct live vaccines. J Reprod Fertil Suppl 41: 97–104
Rispens BH, Van Vloten M, Maas HJ, Hendrik JL (1972) Control of Marek's disease in the Netherlands. I. Isolation of an avirulent Marek's disease virus strain (CVI 988) and its use in laboratory vaccination trials. Avian Dis 16: 108–125

Rode H-J, Janssen W, Rösen-Wolff A, Bugert JJ, Thein P, Becker Y (1993) The genome of equine herpesvirus type 2 harbors an interleukin-10 (IL-10) gene. Virus Genes (in press)

Sedegah M, Chiang, CH, Weiss WR, Mellouk, S, Cochran MD, Houghton RA, Beaudoin RL, Smith D, Hoffman SL (1992) Recombinant pseudorabies virus carrying a plasmodium gene: herpesvirus as a new live viral vector for inducing T-and B-cell immunity. Vaccine 10: 578–584

Shiraki K, Hayakawa Y, Mori H, Namazue J, Takamizawa A, Yoshida I, Yamanishi K, Takahashi M (1991) Development of immunogenic recombinant OKa varicella vaccine expressing hepatitis B virus surface antigen. J Gen Virol 72: 1393–1399

Starr SE, Friedman HM, Plotkin SA (1991) The status of cytomegalovirus vaccine. Rev Infect Dis 13: 5964-5965

Takahashi M, Asano Y, Kamiya H, Baba K (1985) Varicella vaccine: case studies. Microbiol Sci 2: 249–254

Tsolia M, Gershon AA, Steinberg SP, Gelb L (1990) Live attenuated varicella vaccine: evidence that the virus is attenuated and the importance of skin lesions in transmission of varicella-zoster virus. National Institute of Allergy and Infections Diseases Varicella Vaccine Collaborative Study Group. J Pediatr 16: 184–189

van Oirschot JT, Terpstra C, Moormann RJ, Berns AJ, Gielkens AL (1990) Safety of an Aujeszky's disease vaccine based on deletion mutant strain 783 which does not express thymidine kinase and glycoprotein I Vet Rec 127: 443–446

White CJ, Kuter BJ, Hildebrand CS, Isganitis KL, Matthrew H, Miller WJ, Provost PJ, Ellis RW, Gerety RJ, Calandra GB (1991) Varicella vaccine (VARIVAX) in healthy children and adolescents: results from clinical trials, 1987 to 1989. Pediatrics 87: 604–610

Subject Index

MIX
Papier aus verantwortungsvollen Quellen
Paper from responsible sources
FSC® C105338

If you have any concerns about our products,
you can contact us on
ProductSafety@springernature.com

In case Publisher is established outside the EU,
the EU authorized representative is:
Springer Nature Customer Service Center GmbH
Europaplatz 3, 69115 Heidelberg, Germany

Printed by Libri Plureos GmbH
in Hamburg, Germany